THE EARLY DEVELOPMENT OF MAMMALS

THE SECOND SYMPOSIUM OF
THE BRITISH SOCIETY
FOR DEVELOPMENTAL BIOLOGY

THE EARLY DEVELOPMENT OF MAMMALS

EDITED BY

M. BALLS
Senior Lecturer in
Human Morphology
University of Nottingham

&

A. E. WILD
Lecturer in Zoology
University of Southampton

CAMBRIDGE UNIVERSITY PRESS
CAMBRIDGE
LONDON · NEW YORK · MELBOURNE

Published by the Syndics of the Cambridge University Press
The Pitt Building, Trumpington Street, Cambridge CB2 1RP
Bentley House, 200 Euston Road, London NW1 2DB
32 East 57th Street, New York, N.Y. 10022, USA
296 Beaconsfield Parade, Middle Park, Melbourne 3206, Australia

Library of Congress Catalogue Card Number: 75-2728

ISBN: 0 521 20771 1

First published 1975

Printed in Great Britain
by R. & R. Clark Ltd., Edinburgh

CONTENTS

PREFACE

BY M. BALLS AND A. E. WILD

The second symposium meeting of the British Society for Developmental Biology, on *The Early Development of Mammals*, was held at the University of East Anglia, Norwich, on 9–12 September 1974. During the last fifteen years or so, rapid advances have taken place in our understanding of early mammalian development. In the case of pre-implantation stages, this has been largely brought about by the development of in-vitro culture techniques, which have enabled the early embryo, normally hidden within the female reproductive tract, to be viewed directly and, perhaps more importantly, to be manipulated. Experimentally, the mammalian embryo can now, for a time, enjoy a free-living existence, paralleling that normally experienced by anamniotes. The concepts derived from the classical studies on lower vertebrate and invertebrate development can thus more readily be applied to mammals. However, apart from the natural desire by embryologists to observe in the living state what could previously only be inferred from fixed, sectioned material, a major driving force has been the need to understand, control and, in some situations, aid human development. Before this aim can be achieved, we must refine and exploit the new methods that are now available to us, and it is our hope and belief that this conference, and the volume which results from it, will be a significant stimulus to discussion and experiment.

The opening chapters deal with problems related to the fertilisation, culture and storage of mammalian ova (Whittingham), with the latest developments in experimental parthenogenesis in the mouse (Kaufman), and with methods for maintaining embryos *in vitro* during the time when implantation would normally occur *in vivo* (McLaren & Hensleigh) and after implantation has taken place *in vivo* (Steele).

The totipotency of the blastomeres, up to and including the 8-cell stage, has clearly been shown in experiments reviewed and described by Wilson & Stern and also by Kelly, pointing to labile epigenetic control of the early differentiation of two distinct cell populations (trophoblast and inner cell mass). The further development of these two cell types is then considered by Gardner & Papaioannou, by Ansell, and by Sherman. Much is now known about the synthesis of various macromolecules in the pre-implantation embryo, and Pikó describes the nuclear and mitochondrial gene activity underlying these events.

The effects on development of mutations of the important and complex

T-locus of the mouse are discussed by Hillman, who proposes that metabolic errors are of primary importance, whereas Bennett suggests that this locus controls a specific and highly significant cell surface antigen. Billington & Jenkinson review the evidence supporting the view that histocompatibility antigens are absent from trophoblast cells, and Jacob discusses antigenic expression on teratocarcinoma cells. Methods for producing benign teratomas and re-transplantable teratocarcinomas, and the factors regulating them, are described by Solter, Damjanov & Koprowski, and the characteristics of established cell lines from teratocarcinomas are outlined by Evans. The central theme of this group of papers is that studies on the abnormal development of mutants and of teratomas/teratocarcinomas will help in furthering our understanding of normal developmental processes. In the same vein, Ford discusses the topic of genetic imbalance and its effects, and outlines a model system for its study; Cattanach reviews sex reversal in mammals.

The next group of papers are concerned with the importance of cell–cell interactions in the morphogenesis of organs such as the kidney (Saxén & Karkinen-Jääskeläinen), in pre-natal erythropoiesis (Cole), and in pre-natal lymphopoiesis (Ritter). Pantelouris argues, from his studies on the effects of genetic athymia, that the thymus might have an important endocrine function in female reproductive physiology. In the concluding paper, McKenzie presents evidence that early cell death might be the underlying cause of the human morphological abnormalities previously thought to be due to 'amniotic bands'.

The idea that the second symposium meeting of the Society should be on *The Early Development of Mammals* was conceived during the first symposium meeting at Bristol in 1972. The outline programme was planned a few weeks later, in collaboration with Dr C. F. Graham, and we would especially like to thank him for his interest and enthusiastic support throughout this venture. The conference was by far the most international of the Society's meetings up to that time; the 150 participants included thirty visitors from Denmark, Finland, France, Hungary, Israel, Italy, The Netherlands, Sweden, Switzerland, the USA and West Germany. We are grateful to the authors for the high quality of their contributions, to the staff of Cambridge University Press for another enjoyable collaboration, and to the staff of the University of East Anglia for their hospitality.

FERTILISATION, EARLY DEVELOPMENT AND STORAGE OF MAMMALIAN OVA *IN VITRO*

BY D. G. WHITTINGHAM*

Physiological Laboratory, Downing Street, Cambridge

The mammalian oocyte and early embryo remain 'free' or unattached within the lumen of the female genital tract from ovulation to implantation. In most mammals the completion of oocyte maturation and sperm penetration are concomitant processes occurring within the ampular region of the oviduct. The resulting zygote continues to develop as it passes along the oviduct and enters the uterus, generally between three and four days after conception at stages varying between 4-cell and early blastocyst depending upon the species (Blandau, 1961; Biggers & Stern, 1973). The time of implantation varies greatly between species, occurring from one day to several weeks after the embryo enters the uterus.

The culture of the mammalian embryo during its 'free-living' phase has posed a challenge to biologists since the advent of tissue culture techniques over sixty years ago, but it is only in the past fifteen years that reliable methods have been developed for the fertilisation of oocytes *in vitro* (see review by Brackett, 1971) and the culture of the pre-implantation stages (see reviews by Biggers, Whitten & Whittingham, 1971; Whittingham, 1971*a*; Brinster, 1972). The main reason for the recent progress results from intensified research into the field of reproductive biology stimulated by the concern for controlling the rapidly increasing human population and the need to increase food supplies by improving the reproductive capacity of our domestic animals. In addition, techniques for super-ovulation have provided larger numbers of oocytes and embryos for in-vitro studies and improved microbiochemical methods have allowed detailed analyses to be made of the synthetic and metabolic processes occurring during pre-implantation (Biggers & Stern, 1973). As a result the mammalian embryo has taken its place in the field of developmental biology.

In the following account, the present status of work on fertilisation *in vitro*, embryo culture, and low temperature storage, is examined in various mammalian species. Particular consideration is given to defining

* Present address: MRC Mammalian Development Unit, Wolfson House, 4 Stephenson Way, London NW1 2HE.

the conditions necessary for obtaining development *in vitro* and the viability of cultured embryos after transfer to recipient foster mothers.

GENERAL CONSIDERATIONS

Unlike the environment *in vivo*, the closed culture system used for fertilisation and embryo culture (for details of methods see Brinster, 1963, and Whitten, 1971) permits little exchange of materials between the medium and external environment except for gases. Therefore inhibitory effects may be produced in embryonic development by:

(*a*) the products of metabolism accumulating in the medium;
(*b*) the depletion of exogenous substrates below optimal levels caused either by rapid embryo utilisation or instability of chemicals in the medium;
(*c*) the leaching of essential nutrients or inorganic ions from the embryo when these substances are below optimal concentrations or absent from the medium.

The importance of chemically-defined media for analysing the nutritional requirements and metabolic activity of cells *in vitro* is well-documented (Biggers, Rinaldini & Webb, 1957; Waymouth, 1972). Until recently such studies were confined to established cell lines which have few distinguishing characteristics between them. The present interest in the culture of specific cell types has already shown that the individual requirements are varied and complex (Paul, 1972), emphasising the importance of careful selection and construction of the proper medium in each case (Evans, Bryant, Kerr & Schilling, 1964). The mammalian oocyte and early embryo exemplify such specific requirements for development *in vitro*. Since Whitten (1956) discovered that 8-cell mouse embryos develop to the blastocyst stage *in vitro* in a simple chemically-defined medium based on Krebs–Ringer bicarbonate, almost all subsequent studies have used this type of medium. The most comprehensive analysis of the nutrient requirements and metabolic activity have been made on mouse embryos (Biggers *et al.*, 1971; Whitten, 1971; Whittingham, 1971*a*; Brinster, 1972, 1973; Biggers & Stern, 1973). At present the specific requirements for other species except the rabbit (Brinster, 1970; Kane & Foote, 1971; Kane, 1972) are ill-defined, limited to some extent by their refractoriness to culture, illustrating the necessity to discover the specific requirements peculiar to each species in order to obtain successful pre-implantation development *in vitro*.

FERTILISATION *IN VITRO*

Since Chang (1959) first reported the successful fertilisation of rabbit eggs *in vitro*, the fertilisation of mammalian eggs *in vitro* has been extended to include the hamster (Yanagimachi & Chang, 1964), mouse (Whittingham, 1968), rat (Toyoda & Chang, 1968), human (Edwards, Bavister & Steptoe, 1969), cat (Hamner, Jennings & Sojka, 1970), pig (Harms & Smidt, 1970), guinea-pig (Yanagimachi, 1972), gerbil (Noske, 1972), squirrel monkey (Gould, Cline & Williams, 1973) and ferret (Whittingham, 1974*a*). So far it is only in the rabbit (Chang, 1959), mouse (Whittingham, 1968) and rat (Toyoda & Chang, 1974*a*) that fertilisation *in vitro* has been unequivocally attained with the birth of live young after transfer to recipient foster mothers (Table 1). The most important factors which have led to these recent achievements have arisen from a greater understanding of sperm capacitation with the realisation that epididymal and ejaculated sperm could be used in the in-vitro system, the improved culture techniques for manipulation and culture of the gametes, and the increased knowledge of the nutritional requirements for culture of the oocyte and embryo.

Table 1. *Fertilisation* in vitro

	Subsequent development	
Species	*In vitro*	*In vivo*
Rabbit	Blastocyst	Birth
Hamster	2-cell	None
Mouse	Blastocyst	Birth
Rat	2-cell + few 4-cell	Birth
Human	Blastocyst	Not tested
Cat	16-cell	Not tested
Pig	2-cell	Not tested
Guinea-pig	2-cell	Not tested
Gerbil	2-cell	Not tested
Squirrel monkey	2-cell	Not tested
Ferret	Blastocyst	Not tested

CULTURE MEDIA FOR FERTILISATION *IN VITRO*

These fall into two main categories: (*a*) physiological salt solutions plus biological fluids, e.g. serum, follicular fluid, oviducal and uterine fluid; (*b*) physiological salt solutions plus albumin (Table 2).

Table 2. *Some typical chemically defined media used for fertilisation* in vitro

Compound	Mouse/ Rat	Rabbit	Hamster	Human
NaCl	94.59	112.05	112.40	128.31
KCl	4.78	4.02	2.20	5.23
KH_2PO_4	1.19	–	–	–
NaH_2PO_4	–	0.83	0.29	0.37
$CaCl_2.2H_2O$	1.71	2.25	1.48	1.80
$MgSO_4.7H_2O$	1.19	–	–	–
$MgCl_2.6H_2O$	–	0.52	0.41	0.60
$NaHCO_3$	25.07	37.00	36.08	39.22
Na pyruvate	0.50	0.01	0.10	0.10
Na lactate	21.58	–	–	–
Glucose	5.56	13.90	4.56	5.56
Bovine serum albumin	32 mg ml^{-1}	3 mg ml^{-1}	9 mg ml^{-1}	3.6 mg ml^{-1}
Penicillin	100 U ml^{-1}	31 μg ml^{-1}	100 U ml^{-1}	100 U ml^{-1}
Streptomycin	50 μg ml^{-1}	–	–	–
Phenol red	10 mg	–	20 mg ml^{-1}	17 mg ml^{-1}

Sources: mouse, Kaufman & Whittingham (1972), Kaufman (1973); rat, Toyoda & Chang (1974*a*, *b*); rabbit, Mills, Jeitles & Brackett (1973); hamster, Whittingham & Bavister (1974); human, Fowler & Edwards (1973).

So far the evidence shows that media in the first category have been used chiefly to obtain the necessary preparatory changes in the sperm for fertilisation to take place, i.e. capacitation and acrosome reaction. Capacitation of sperm is a necessary prerequisite of the fertilisation process in mammals requiring up to six hours in the rabbit *in vivo* but shorter times in hamsters, rats and mice (see reviews by Chang, 1968; Austin, 1969; Bedford, 1970; Brackett, 1971). Initially, fertilisation *in vitro* was obtained with sperm previously capacitated in the uterus (Chang, 1959; Whittingham, 1968) but subsequently it was found that epididymal sperm undergo capacitation and the acrosome reaction in the presence of various body fluids e.g. follicular and oviducal fluids (for hamster, Barros & Austin, 1967*a*, b; Yanagimachi, 1969: for mouse, Iwamatsu & Chang, 1970; Mukherjee, 1972: for guinea-pig, Yanagimachi, 1972). With these complex media, an analysis of the conditions for fertilisation *in vitro* is virtually impossible, but now that the preparatory changes have been recognised and can be separated from the actual processes of sperm penetration and syngamy, the physical and chemical conditions controlling these processes can be more accurately determined by means of chemically defined media.

The chemically-defined media outlined in Table 2 are based upon the two physiological salt solutions used most extensively for fertilisation *in vitro*. For mouse and rat fertilisation, modifications of Krebs–Ringer bicarbonate solution have been used (Kaufman & Whittingham, 1972; Kaufman, 1973; Hoppe & Pitts, 1973; Toyoda & Chang, 1974*a*, *b*) and for rabbit, hamster and human, modifications of the salts of Tyrode's solution have been used (Mills, Jeitles & Brackett, 1973; Fowler & Edwards 1973; Whittingham & Bavister, 1974). Capacitation *in vitro* has been achieved in these media without the addition of various complex body fluids where the only macromolecular component of the medium has been bovine serum albumin – BSA (for mouse, Toyoda, Yokoyama & Hosi, 1971*a*, *b*; for hamster, Bavister, 1969; or for rat, Toyoda & Chang, 1974*a*, *b*; for human, Edwards *et al.*, 1969).

Although the cumulus cells were present in these studies and their role in the capacitation of epididymal sperm *in vitro* has been postulated (Pavlok & McLaren, 1972), the high rates of fertilisation *in vitro* in their absence clearly demonstrates that they are not an essential part of the in-vitro system (Hoppe & Pitts, 1973; Hoppe & Whitten, 1974*a*; Kaufman, 1973). Even in the absence of cumulus cells, some contributing factor to sperm capacitation from the diluted epididymal secretions may play an important role, although this seems unlikely as rabbit epididymal fluid contains a decapacitation factor (Weinman & Williams, 1964) and epididymal extracts have been shown to have similar activity in the mouse (Iwamatzu & Chang, 1971). However, in the latter report, the 'minced up' epididymal extracts may have contained many factors detrimental to fertilisation *in vitro*, e.g. degradative enzymes.

Little attempt has been made to distinguish between the requirements for capacitation, acrosome reaction and sperm penetration. The time interval between the addition of sperm to eggs and cumulus, and penetration, has been shortened by pre-incubation of epididymal sperm in chemically-defined media plus BSA (for mouse, Toyoda *et al.*, 1971*b*; Hoppe & Whitten, 1974*b*) and by replacing BSA by polyvinylpyrrolidone – PVP (for hamster, Austin, Bavister & Edwards, 1973). In the latter report, the acrosome reaction did not occur until sperm were incubated with eggs and cumulus cells. Hoppe & Whitten (1974*b*) were unable to substitute synthetic polymers (e.g. PVP) for BSA, fatty acid free BSA, fraction V BSA or crystalline egg albumin in chemically-defined media for fertilisation of mouse oocytes, although Kaufman (1973) had earlier obtained some fertilisation when BSA was replaced by PVP. Since capacitation of hamster sperm *in vitro* occurs with PVP as the only macromolecular component, the role of proteins may be only in the acrosome reaction and the maintenance of viability of the oocyte by, for example, providing support to membrane stability. The presence of proteins in the oviduct fluid is well documented (Hamner & Fox, 1969) and it is known that the mouse embryo can only develop *in vitro* in the absence of protein from the 2-cell stage (Cholewa & Whitten, 1970). In the rat, pre-incubation of epididymal sperm in a chemically-defined medium (plus BSA) with a high potassium/sodium ratio (0.32), similar to the levels found in rat uterine fluid at pro-oestrus (Howard & DeFeo, 1959), significantly reduced the time for penetration (Toyoda & Chang, 1974*b*). The authors produced similar effects with αcAMP alone or in combination with high potassium levels and they concluded that both conditions accelerated the capacitation process. Oliphant & Brackett (1973) found that brief exposure of mouse epididymal sperm to chemically-defined media (+BSA) of elevated ionic strength increased the number of oocytes cleaving to the 2-cell stage twenty-two hours after semination with the pretreated sperm. Clearly more investigations on the role of inorganic ions in the capacitation process are warranted.

In other work, where the separate processes involved in fertilisation have not been separated, fertilisation up to the stage of pronuclear formation was obtained in media with wide ranges of osmolarity (for mouse and hamster, Miyamoto & Chang, 1973; for mouse, Edwards, 1973) and hydrogen ion concentrations (for hamster, Bavister, 1969; for mouse, hamster and rat, Miyamoto, Toyoda & Chang, 1974; for mouse, Edwards, 1973); see Table 3. Although optimal ranges of osmolarity and pH are shown in the data (Table 3), the design of the experiments makes it impossible to calculate a true optimal value. In some of the experiments on hamster fertilisation (Miyamoto & Chang, 1973; Miyamoto *et al.*, 1974) hamster serum was incorporated in the medium (1 : 2 and 1 : 3 respectively). Apart fom the work mentioned above on the effect of potassium/sodium ration on capacitation, little attention has been paid to the relative importance of the inorganic ions during fertilisation *in vitro*. Iwamatzu & Chang (1971) indicated that a calcium concentration of 1.80 mM was most favourable for penetration of mouse eggs *in vitro*.

There is no detailed analysis of the energy requirements for fertilisation *in vitro*. The inclusion of pyruvate was found to be beneficial for human fertilisation *in vitro* (Edwards, 1973) and lactate as well as pyruvate and glucose in the rat (Toyoda & Chang, 1974*a*). Any analysis is confounded by the separate requirements for sperm and oocytes and possibly the accummulation of metabolites of sperm, since sperm are required in such high concentrations to obtain satisfactory rates of fertilisation *in vitro* (up to 10^6 sperm ml^{-1}). The energy requirements for the matura-

Table 3. *The effect of osmolarity and hydrogen ion concentration on fertilisation* in vitro

Species	Range over which fertilisation occurs	Optimal range	Reference
Osmolarity (mosmol)			
Mouse	256–418	308–372	Miyamoto & Chang, 1973
Mouse	272–400*	~316	Edwards, 1973
Hamster	232–452	292–392	Miyamoto & Chang, 1973
Hydrogen ion			
Mouse	6.6–8.0	7.0–7.8	Edwards, 1973
Mouse	6.0–8.5	7.3–7.7	Miyamoto, Toyoda & Chang, 1974
Hamster	7.2–7.8*	~7.6	Bavister, 1969
Hamster	6.7–8.7	6.8–8.2	Miyamoto, Toyoda & Chang, 1974
Rat	6.75–8.3	~7.8	Miyamoto, Toyoda & Chang, 1974

* Data only available for these ranges.

tion of the mouse oocyte and first cleavage division of the zygote are known, and only take place in chemically-defined media containing pyruvate or oxaloacetate if the cumulus cells are removed (Biggers, Whittingham & Donahue, 1967). However, glucose will support these processes when cumulus cells are present and Donahue & Stern (1968) have shown that they produce pyruvate from glucose *in vitro*. The effect of sperm metabolites on oocyte maturation and first cleavage warrants investigation.

Kaufman (1973) has shown that low oxygen concentration (5 %) is not required for fertilisation of mouse oocytes *in vitro* but he did not compare subsequent development to the blastocyst at different oxygen tensions. Whitten (1971) found high concentration of oxygen (20 %) prevented development of 1-cell oocytes (fertilised *in vivo*) to the blastocyst stage *in vitro*. The optimal concentration of oxygen for development was 5 %.

All effective media for fertilisation *in vitro* are buffered with sodium bicarbonate. Low rates of fertilisation were obtained with low bicarbonate concentrations in mouse and hamster (Miyamoto *et al.*, 1974) and none in its absence, but these results are confounded by variations in pH. So far no study has examined its importance although it is an essential requirement for continued pre-implantation development *in vitro* (Quinn & Wales, 1973).

FURTHER DEVELOPMENT OF OOCYTES FERTILISED *IN VITRO*

In culture

In the rabbit (Ogawa, Satoh, Hamada & Hashimoto, 1972), mouse (Mukherjee & Cohen, 1970; Miyamoto & Chang, 1972; Hoppe & Pitts, 1973), human (Steptoe, Edwards & Purdy, 1971) and ferret (Whittingham, 1974*a*) oocytes fertilised *in vitro* have been cultured to the blastocyst stage. With the exception of the rat (Toyoda & Chang, 1974*a*) and cat (Hamner *et al.*, 1970) development in other species has ceased at the 2-cell stage (Table 1). This is due to the technical problems encountered in culturing the embryos of the latter species. Development in culture cannot be taken as conclusive evidence of fertilisation *in vitro* as parthenogenetic activation can occur in this system (see Kaufman, this volume). Proof may be obtained by chromosomal analysis of the embryo with the demonstration of the male Y chromosome with the correct complement of autosomes plus a single female X chromosome.

After transfer to foster mothers

The birth of genetically marked live young of both sexes from oocytes fertilised *in vitro* and subsequently transferred to recipient foster mothers is the ultimate proof of fertilization *in vitro*. This has only been demonstrated in two species – the rabbit (Chang, 1959) and mouse (Whittingham, 1968). Live young of both sexes, not genetically marked, were born after the transfer of rat oocytes fertilised *in vitro* (Toyoda & Chang, 1974*a*); the recipient females were made pseudopregnant by electrical stimulation and the oocytes transferred at the 2-cell stage – which is sufficient evidence for fertilisation *in vitro*. Birth of live young or further development other than that obtained *in vitro* has yet to be reported for the species where fertilisation *in vitro* has been accomplished. Recently Whittingham & Bavister (1974) attempted to establish the viability of hamster eggs fertilised *in vitro* by transfer to the oviduct. No further development of pronuclear or 2-cell embryos took place; in fact the 2-cell ova produced by fertilisation *in vitro* were all degenerate when recovered from the oviduct forty-eight hours later. However, pronucleate and 2-cell ova fertilised *in vivo* continued development after transfer when the medium for collection and manipulation of the ova was phosphate buffered at a stable pH of approximately 7.2. The authors concluded that the high pH necessary for optimal fertilisation *in vitro* in the hamster produced irreparable damage in the resulting zygote.

The survival of mouse oocytes fertilised *in vitro* after transfer at the 2-cell and blastocyst stage is extremely variable (Table 4). The highest

Table 4. *Comparison of development of mouse oocytes fertilised* in vitro *after transfer to pseudopregnant recipients*

Stage at transfer	No. embryos transferred	No. foetuses and liveborn (%)	Estimated % of original oocytes surviving to term	Reference
2-cell	54	9 (17)	7	Whittingham, 1968
2-cell	55	13 (24)	22	Cross & Brinster, 1970
2-cell	80	10 (13)	10	Miyamoto & Chang, 1972
2-cell	78	42 (54)	44	Kaufman & Whittingham, cited by Kaufman, 1973
Blastocyst	23	11 (48)	18	Mukherjee & Cohen, 1970
Blastocyst	11	5 (45)*	10	Mukherjee, 1972
Blastocyst	299	111 (37)	34	Hoppe & Pitts, 1973
Blastocyst	95	12 (20)	16	Kaufman & Whittingham, cited by Kaufman, 1973
Blastocyst	81	49 (60)	49	Whittingham, unpublished observations

* Oocytes matured and fertilised *in vitro*.

survival after transfer at the 2-cell stage was 54 % or 42/78 (Kaufman & Whittingham, cited by Kaufman, 1973), and at the blastocyst stage 60 % or 49/81 (Whittingham, unpublished observations). When survival to term is estimated from the original population of oocytes with which fertilisation *in vitro* was attempted the highest rates of survival from transfers at the 2-cell and blastocyst stages are similar (44 % versus 49 %). The increased survival of transfers at the blastocyst stage may reflect the improved culture conditions where low oxygen (5 %) was used to culture the zygote after fertilisation (Hoppe & Pitts, 1973; Whittingham, unpublished observations). However, the best survival rates are still below the values obtained after the transfer of embryos obtained after fertilisation and development *in vivo* (Mullen & Carter, 1973).

So far, there is only one report of the transfer of rat oocytes fertilised *in vitro* (Toyoda & Chang, 1974*a*); 21 % (43/103) of the 2-cell embryos transferred developed to 19-day foetuses and birth. In the rabbit, oocytes fertilised *in vitro* have been transferred at the 2–4-cell stage, and approximately 20 % survive to term (Frazer & Dandekar, 1973; Brackett, Mills & Jeitles, 1972; Mills, *et al.*, 1973). Estimated survival to term from the original population of oocytes with which fertilisation was attempted is low in all cases – approximately 10 % – when compared to the transfer of embryos fertilised *in vivo*; e.g. 54 % or 293/538 (Maurer, Hunt, Van Vleck & Foote, 1968). These results indicate some inadequacy in the fertilisation and/or culture technique which should be further investigated.

EARLY DEVELOPMENT OF MAMMALIAN EMBRYOS *IN VITRO*

As mentioned earlier, the nutritional requirements for in-vitro development of the pre-implantation mouse embryo are well established. The following account is restricted to the consideration of some current problems involved in culturing mammalian embryos; interspecies comparisons are made where information is available on nutrient requirements and metabolic activity.

The pioneering work of Whitten (1956, 1957) and Brinster (1963, 1965) using chemically-defined media has formed the basis for all subsequent studies with the mouse and with most of the other species investigated. A brief summary of embryonic development *in vitro* with chemically-defined media for the species studied to date is given in Table 5. Blastocysts can develop from the continuous culture of fertilised 1-cell ova in the mouse (Whitten & Biggers, 1968), rabbit (Kane & Foote, 1971; Ogawa, Satoh & Hashimoto, 1971), ferret (Whittingham, 1974*a*), and sheep (Tervit, Whittingham & Rowson, 1972). Embryonic development in other species is much more restricted, and in the sheep, development of 1-cell sheep ova as far as the early blastocyst stage was very limited (Tervit *et al.*, 1972; Tervit & Rowson, 1974). The addition of foetal calf serum to media has a beneficial effect on the development of blastocysts from 1-cell human ova, previously fertilised *in vitro* (Steptoe *et al.*, 1971) and 8-cell rat ova (Brinster, 1968*a*; Mayer & Fritz, 1974). Although the serum components exerting this effect were not determined it is known that the addition of various essential amino acids can enhance or replace serum for the initiation of hatching and trophoblast differentiation in the mouse (Gwatkin, 1966; Spindle & Pederson, 1973). Perhaps the culture of cattle and sheep ova can be similarly improved by the addition of serum components or amino acids.

Table 5. *Development of pre-implantation embryos* in vitro *in chemically-defined media*

Species	Development achieved *in vitro*	Reference
Mouse	1-cell to blastocyst	Whitten & Biggers, 1968
Rabbit	1-cell to blastocyst	Kane & Foote, 1971; Ogawa, Satoh & Hashimoto, 1971
Human	1-cell to 8-cell	Edwards, Steptoe & Purdy, 1970
Sheep	1 cell-to early blastocyst	Tervit, Whittingham & Rowson, 1972; Tervit & Rowson, 1974
Cow	1-cell to morula 8-cell to blastocyst	Tervit, Whittingham & Rowson, 1972
Ferret	1-cell to blastocyst	Whittingham, 1974*a*
Pig	1-cell to 4-cell 8–16-cell to blastocyst	Polge, unpublished observations
Rat	1-cell to 4-cell	Brinster, 1968*a*
	8-cell to blastocyst	Folstad, Bennett & Dorfman, 1969
Hamster	1-cell to 2-cell	Whittingham & Bavister, 1974
	8-cell to blastocyst	Whittingham, unpublished observations
Gerbil		Fisher & Fischer, 1973

The culture of one-cell mouse embryos

Fertilised 1-cell ova from certain F1 hybrid and several inbred strains of mice develop routinely to the blastocyst stage *in vitro* (Whitten & Biggers, 1968; Whitten, 1971). Maximum responses (100 %) were obtained with an oxygen concentration of 5 % (Whitten, 1971) but only 50–60 % developed to blastocysts in an atmosphere of 20 % oxygen (Biggers, 1971; Kaufman, 1973). Much more variable and limited success was obtained with random bred mice from various sources (Biggers, 1971; Cross & Brinster, 1973; Whittingham, unpublished observations). Cross & Brinster (1973) obtained 47 % blastocysts when random bred 1-cell ova were cultured in media containing 0.25 mM pyruvate during the first cleavage division and 30 mM lactate plus 0.25 mM pyruvate for the remaining culture period. Contact of the 1-cell ova with cumulus cells may provide these conditions *in vivo* since pyruvate is produced by the cumulus cells (Donahue & Stern, 1968). A recent attempt to solve this problem (Whittingham & Whitten, unpublished observations) is reported here. Initially a straightforward comparison was made between the development of fertilised 1-cell ova from F1 hybrid (C57BL × A2G) and random bred (CFLP produced by Carworth Europe) mice. Fertilised 1-cell ova from naturally mated mice were cultured either in test tubes (Whitten, 1971) or in drops of medium with a paraffin oil overlay (Brinster, 1963) in both cases with a gas phase of 4 % O_2, 5 % CO_2 and 91 % N_2. The results are summarised in Table 6. Development was similar for

Table 6. *Comparison of the development of 1-cell mouse embryos of F1 hybrid and random bred strains to the blastocyst stage* in vitro

Strain	Culture system	No. 1-cell cultured (No. replicates)*	No. blastocysts	% (range)
F1(C57BL × A2G)	Tube	113 (13)	106	94 (78–100)
	Dish	87 (10)	83	95 (74–100)
Random bred CFLP	Tube	162 (12)	27	17 (0–55)
	Dish	105 (8)	22	21 (0–36)

Gas phase: 4 % O_2, 5 % CO_2, 91 % N_2.
* Each replicate consisted of one clutch of ova from naturally mated females varying from 7–11 and 10–16 for F1 and CFLP respectively.

both methods with each strain but the percentage of random bred developing to blastocysts was significantly less than the F1 hybrid, thus confirming earlier observations. Also, lowering the oxygen concentration did not improve the previous results (Biggers, 1971; Cross & Brinster, 1973). Comparisons were made between the composition of media formulated by Whitten (1971) and Whittingham (1971*a*) and distilled water derived from two different sources (Bar Harbor, Maine and Cambridge, England), but no significant differences in the development of F1 and random bred ova were observed. In addition, development of random bred ova was not improved by: (1) controlling the pH during recovery and manipulation (but when this was done with a phosphate buffered medium development was completely inhibited after the 2-cell stage with the random bred and reduced by approximately 50 % with F1 hybrid embryos); (2) shortening of the time interval between recovery and placing in culture to three minutes or less; or (3) recovering the 1-cell ova closer to the first cleavage division.

The effect of increasing the potassium concentration and potassium/sodium ratio of the medium was examined without success; see Table 7. No development of random bred 1-cell ova occurred above 6 mM potassium; with F1 hybrid 1-cell ova development was reduced in 12 mM potassium and completely inhibited in 24 mM potassium. This finding contrasts with the findings of Wales (1970) who showed only a small linear reduction in the development of 2-cell mouse embryos to blastocysts in concentrations of potassium ranging between 6 and 48 mM.

Table 7. *Effect of increasing the potassium concentration and potassium/sodium ratio on the development of 1-cell mouse embryos to the blastocyst stage*

Potassium concn (mM)	K/Na ratio	No. blastocysts/No. 1-cell cultured	
		F1(C57BL × A2G)	Random bred CFLP
6	0.04	20/21	5/28
12	0.09	12/19	0/27
24	0.19	1/19	0/27
48	0.48	0/20	0/29

Each treatment was replicated twice.

In conclusion, the problem of 1-cell development in some strains of mice is still unsolved, but even if this is an artifact of the present culture system this example serves to demonstrate the sensitivity of the 1-cell mouse embryo to culture conditions and its complete dependency upon the correct environment for further development. Similar specialised conditions of culture have to be found for the successful development of embryos of other species *in vitro* where at present development is limited to one or two cleavage divisions.

Culture of rat embryos

The 8-cell rat embryo can develop to the blastocyst stage in chemically-defined media (Folstad, Bennett & Dorfman, 1969) but hatching of the blastocyst was only observed in media supplemented with foetal calf serum (Brinster, 1968*a*; Mayer & Fritz, 1974). In addition, development from the 8-cell to blastocyst can take place in media devoid of energy sources and BSA (BSA replaced by Ficoll) but the blastocysts formed after 24 hours in culture degenerate soon afterwards. Blastocysts formed from 8-cell embryos in media containing only BSA, glucose, pyruvate or lactate continued to expand in culture between 24 and 48 hours but subsequently degenerated and no hatching was observed – similar to results obtained with the complete medium (Whittingham, unpublished observations).

Attempts to culture the 1-cell ovum have been limited to one or two divisions (Brinster, 1968*a*; Mayer & Fritz, 1974; Toyoda & Chang, 1974*a*). Brinster (1968*a*) noted that the 1-cell 'appeared to have a requirement for pyruvate or lactate'. Culture of 1-cell ova in chemically-defined media with lowered oxygen concentrations (1 %, 2 % or 4 % O_2; 5 % CO_2 and the balance with nitrogen) was no more successful than previous attempts with 20 % oxygen (Whittingham & Whitten, unpublished observations).

Toyoda & Chang (1974*a*) showed that 2-cell rat embryos developing in culture after fertilisation *in vitro* are capable of continuing development to birth after transfer to uterine foster mothers, thus establishing that fertilisation and the first cleavage division could take place normally *in vitro*.

A contemporaneous comparison was made of the compounds supporting the first cleavage division of the rat and mouse, in an effort to gain some indication of the culture requirements for early development in the rat. A modified Krebs–Ringer bicarbonate solution (Whittingham, 1971*a*) was used for these studies and the data are summarised in Table 8. In marked contrast to the mouse, a wide range of energy sources appears to support the first cleavage division in the rat in the absence of BSA, but the 2-cell embryos degenerated after 48 hours of culture in α-ketoglutarate,

Table 8. *Comparison of compounds supporting the first cleavage division in the rat and mouse*

Compound	Mouse	Rat[a]
Pyruvate	+	+[c]
Lactate	−[b]	+[c]
Oxaloacetate	+	+[c]
Phosphoenol-pyruvate	−[b]	+[c]
α-ketoglutarate	−	+
Malate	−	±
Citrate	−	±
Acetate	−	±
Glucose	−	+
BSA	−	+[c]
Ficoll	−	±

[a] All compounds tested in a medium where BSA was replaced by Ficoll; [b] ova in these media did not degenerate in the first 24 hours; [c] ova in these media did not degenerate in the first 48 hours.

+, More than 50 % ova cleaved; ±, less than 50 % ova cleaved.

malate, citrate, acetate and glucose. The first cleavage division also took place in media containing BSA only and degeneration of the 2-cell ova was prevented by increasing the albumin concentration from 1 mg ml^{-1} to 4 mg ml^{-1}. Moreover, up to 50 % cleaved in the basic salt solution devoid of energy or BSA when the BSA was replaced by Ficoll, but these degenerated rapidly after 24 hours in culture. Development to the 4-cell stage occurred in media containing pyruvate, lactate, oxaloacetate or glucose plus BSA. Do these results indicate less restricted patterns of energy metabolism in the rat and the possible use of endogenous materials for the support of the first cleavage division? This is difficult to predict as conditions for continued development *in vitro* are unknown. Several differences are apparent in the early metabolic activity of the rat and mouse embryo. There is greater oxygen uptake in the presence of glucose and pyruvate in the rat and unlike the mouse the pentose shunt appears to be very active during early cleavage (Sugawara & Takeuchi, 1973) similar to the rabbit embryo (Brinster, 1968*b*). In the rat, carbon dioxide production from labelled pyruvate was greater than that produced from glucose (Sugawara & Takeuchi, 1973). This is similar to the situation in the mouse and rabbit (Brinster, 1968*a*), although the increase in production between the 1-cell and blastocyst stage was much less than in the rabbit and mouse. Lactate dehydrogenase activity in the early rat embryo is less than half that found in the mouse (Brinster, 1967). Some of these differences may account for the different conditions required for the development of rat embryos in culture, but such conditions have yet to be determined.

Culture of rabbit embryos

Rabbit embryos will develop from the 1-cell stage in chemically-defined media (Kane & Foote, 1971; Ogawa *et al.*, 1971; Kane, 1972). Initially the addition of pyruvate was shown to improve development (Kane & Foote, 1971), but more recently Kane (1972) found that a high percentage (81 %) of 1-cell ova developed into expanded blastocysts in media containing BSA and no energy sources. Previously Brinster (1970) found that 2-cell rabbit ova develop to the morula stage in media containing only BSA, oxidised glutathione or certain amino acids and he also indicated a beneficial effect from the addition of pyruvate and lactate to the media.

Culture of ferret embryos

One-, 2- and 8-cell ferret embryos develop to the blastocyst stage in chemically-defined media (Whittingham, 1974*a*); see Table 9. Development of all stages into expanded blastocysts occurred in media containing BSA and no other sources of energy but development was retarded compared with similar embryos developing *in vivo*. When BSA concentrations were increased maximum unretarded development occurred with 8 mg ml^{-1} and above and no beneficial effect was found by the addition of pyruvate, lactate and glucose.

Table 9. *Development of ferret embryos* in vitro *in a chemically-defined medium*

Developmental stage when recovered (hours *post coitum*)		Duration of culture (h)	Blastocysts	(%)
1-cell	(48)	120	8/15	(53)
2–8-cell	(72)	96	20/28	(71)
8-cell	(96)	72	127/162	(78)

After Whittingham (1974a).

Culture of hamster embryos

One-cell hamster embryos only develop to the 2-cell stage *in vitro* and the viability of these embryos has yet to be established (Whittingham & Bavister, 1974). Eight-cell embryos will develop to blastocysts in a modified Tyrode's medium (Whittingham, 1971*a*) with much more limited development occurring in the normal modifications of Krebs-Ringer bicarbonate medium used so successfully in the mouse (Table 10). However, nothing is known regarding the specific requirements for development of the hamster embryo *in vitro* except that lowered oxygen tensions, increased albumin concentration or variations in the K/Na ratio do not improve development (Whittingham, unpublished observations).

Table 10. *Development of 8-cell hamster embryos* in vitro *in chemically-defined media*

Medium	No. 8-cell cultured	No. blastocysts	(%)
Tyrode (T6) (Whittingham, 1971*a*)	167	116	(69)
Krebs–Ringer (Whittingham, 1971*a*)	54	3	(6)
Krebs–Ringer (Whitten, 1971)	41	4	(10)

Development of cultured embryos after transfer

The results of transferring embryos fertilised *in vivo* and grown in culture for the greater part of the pre-implantation period are similar to those obtained for embryos fertilised *in vitro* and cultured for similar periods (see earlier discussions). The validity of rabbit blastocysts cultured from the 1-cell stage has not yet been established. Reduced viability of cultured 2-cell rabbit embryos has been shown after 48 hours *in vitro* (Maurer, Onuma & Foote, 1970) and in general, the developmental potential of cultured rabbit embryos reported so far is low (Mills *et al.*, 1973). Similar results were obtained in the sheep (Tervit & Rowson, 1974). Development to term of 8-cell embryos after 3 days, 2–4 cell after 5 days, 2–4 cell after 6 days and 1-cell after 6 days in culture was 56 %, 37 %, 25 % and 0 % respectively. In the mouse, development to term after culture much more closely approximates the results obtained with the transfer of embryos obtained after development *in vivo* (Gates, 1965; Whitten, unpublished observations). Reports of the transfer of other mammalian embryos cultured *in vitro* are few. Forty-four per cent (8/18) of the ferret blastocysts cultured from 8-cell embryos developed into foetuses on examination 13 days after transfer (Whittingham, 1974*a*). Less than 10 % (3/32) of the blastocysts cultured from the 8-cell hamster embryos developed into 12-day foetuses after transfer. Two of four cattle embryos cultured from the 8-cell stage for 4 days continued development at least to day 35 of pregnancy (Tervit *et al.*, 1972). Pig blastocysts cultured from 8- to 16-cell embryos in chemically-defined media developed to blastocysts after 48 hours in culture. Fifty-one out of 229 blastocysts transferred (22 %) developed into normal foetuses when examined at day 18 of pregnancy (C. Polge, personal communication).

CONCLUSIONS

The mammalian embryo is particularly exacting with regard to the conditions required for development *in vitro* however simple they may be. It appears that the embryo, unlike many other cell types, is unable to adapt easily to the conditions imposed upon it *in vitro*. If it is assumed that the developmental sequence to the blastocyst stage is programmed in the oocyte, these events can only take place when the necessary key factors are provided by the environment of the oviducal lumen or the culture medium. Differences are apparent in the requirements for development *in vitro* between the species studied so far and in order to interpret these findings more information is required on the biochemistry of the embryos and the oviducal secretions.

Inadequacies in the present culture media are also reflected in the developmental potential of embryos grown successfully over the greater part of the pre-implantation period; except for the mouse, the development to term of cultured embryos after transfer (where it is known) is extremely low.

STORAGE OF MAMMALIAN EMBRYOS AT LOW TEMPERATURE

The first mammalian embryo reported to survive freezing and thawing was the mouse (Whittingham, 1971*b*), but in this initial study embryos failed to survive periods of more than 30 minutes at −79 °C. When the cryobiological factors which influence survival (suspending medium, cryoprotective agents, cooling rate, final storage temperature and warming rate) were examined in more detail an effective technique for freezing mouse embryos to −196 °C and −269 °C was developed, resulting in high survival rates (50–70 %) after storage at −196 °C for up to eight days (Whittingham, Leibo & Mazur, 1972). These findings were independently confirmed by Wilmut (1972). More recently, survival rates of up to 100 % were obtained with embryos stored at −196 °C for up to eight months (Whittingham & Whitten, 1974), which demonstrates that mouse embryos, similar to other tissue cells, show no deterioration in viability when stored for prolonged periods at −196 °C (Meryman, 1966). The formation of mouse embryo banks similar to semen and other cell banks provides a unique opportunity for the conservation of genetic material which might otherwise be lost. For the first time in mammals, genetic pedigree standards can be established enabling a check to be made for genetic drift in subsequent generations. Other practical applications, discussed more fully elsewhere (Whittingham, 1974*b*), include the banking of: (*a*) inbred strains as a protection against their loss by fire, disease, or other hazard, (*b*) mutations and recombinant inbred strains which are very often uneconomic to maintain; and (*c*) embryos for use in early developmental studies – e.g. biochemical analyses – and to obtain sufficient numbers of various strain combinations for the production of chimaeras.

Details of the techniques for freezing, thawing and subsequent transfer have appeared elsewhere (Whittingham *et al.*, 1972; Whittingham, 1972, 1974*b*).

The finding that mouse embryos were extremely sensitive to rapid freezing and thawing was a major discovery in the embryo freezing technique. Subsequently it has led to the successful freezing of several other mammalian species (Table 11). Rabbit 8-cell embryos and morulae have been frozen and thawed and have successfully developed to term after transfer (Whittingham & Adams, 1974). Two calves have been born from

Table 11. *Low temperature storage of mammalian embryos at −196 °C*

Species	Stages successfully frozen and thawed
Mouse	All stages (1-cell to blastocyst)
Rat	2-cell and 8-cell
Rabbit	8-cell and morula
Cow	Blastocyst
Sheep	Morula and blastocyst

embryos frozen and thawed at the blastocyst stage (Wilmut & Rowson, 1973). More recently the birth of live lambs resulting from the transfer of frozen/thawed sheep blastocysts was reported (Willadsen, Polge, Rowson & Moor, 1974). Two- and 8-cell rat embryos also survive freezing and thawing; 19-day foetuses were obtained after the transfer of frozen/thawed 8-cell embryos (Whittingham, 1974*c*). In contrast to these successful reports, the 8-cell embryo of the pig does not survive cooling below approximately 15 °C (Polge, Wilmut & Rowson, 1974). This is another indication, as mentioned earlier in the discussion of culture requirements of embryos, that there are major differences in the responses of various mammalian embryos to culture and other manipulative procedures.

In conclusion, the field of embryo storage is a rapidly advancing area and it is predicted that this technique will make as great a contribution to biological research as the banking of tissue culture cell lines and storage of semen have done over the past twenty years.

REFERENCES

Austin, C. R. (1969). Variations and anomalies of fertilization. In *Fertilization*, ed. C. B. Metz & A. Monroy, vol. 2, pp. 437–65. New York & London: Academic Press.

Austin, C. R., Bavister, B. D. & Edwards, R. G. (1973). Components of capacitation. In *The Regulation of Mammalian Reproduction*, ed. S. J. Segal, R. Crozier, P. A. Corfman & P. G. Condliffe, pp. 247–54. Springfield, Illinois: Charles C. Thomas.

Barros, C. & Austin, C. R. (1967*a*). *In vitro* fertilization of golden hamster ova. *Anatomical Record*, **157**, 209–10.

BARROS, C. & AUSTIN, C. R. (1967*b*). *In vitro* fertilization and the sperm acrosome reaction in the hamster. *Journal of Experimental Zoology*, **166**, 317–23.

BAVISTER, B. D. (1969). Environmental factors important for *in vitro* fertilization in the hamster. *Journal of Reproduction and Fertility*, **18**, 544–5.

BEDFORD, J. M. (1970). Sperm capacitation and fertilization in mammals. *Biology of Reproduction*, **2**, Supplement 2, 128–58.

BIGGERS, J. D. (1971). New observations on the nutrition of the mammalian oocyte and the preimplantation embryo. In *The Biology of the Blastocyst*, ed. R. J. Blandau, pp. 319–25. Chicago & London: University of Chicago Press.

BIGGERS, J. D. & STERN, S. (1973). Metabolism of the preimplantation embryo. *Advances in Reproductive Physiology*, **6**, 1–59.

BIGGERS, J. D., RINALDINI, L. R. & WEBB, M. (1957). The studies of growth factors in tissue culture. *Symposium of the Society for Experimental Biology* **11**, 264–97.

BIGGERS, J. D., WHITTEN, W. K. & WHITTINGHAM, D. G. (1971). The culture of mouse embryos *in vitro*. In *Methods of Mammalian Embryology*, ed. J. C. Daniel Jr, pp. 86–116. San Francisco: W. H. Freeman.

BIGGERS, J. D., WHITTINGHAM, D. G. & DONAHUE, R. P. (1967). The pattern of energy metabolism in the mouse oocyte and zygote. *Proceedings of the National Academy of Sciences, USA*, **58**, 560–7.

BLANDAU, R. J. (1961). Biology of eggs and implantation. In *Sex and Internal Secretions*, ed. W. C. Young, pp. 797–882. Baltimore: William & Wilkins.

BRACKETT, B. G. (1971). Recent progress in investigations of *in vitro* fertilization. In *The Biology of the Blastocyst*, ed. R. J. Blandau, pp. 329–48. Chicago & London: University of Chicago Press.

BRACKETT, B. G., MILLS, J. A. & JEITLES, G. G. (1972). *In vitro* fertilization of rabbit ova recovered from ovarian follicles. *Fertility and Sterility*, **23**, 898–909.

BRINSTER, R. L. (1963). A method for *in vitro* cultivation of mouse ova from two-cell to blastocyst. *Experimental Cell Research*, **32**, 205–8.

BRINSTER, R. L. (1965). Studies on the development of mouse embryos *in vitro*. IV. Interactions of energy sources. *Journal of Reproduction and Fertility*, **10**, 227–40.

BRINSTER, R. L. (1967). Lactate dehydrogenase activity in preimplantation rat embryo. *Nature, London*, **214**, 1246–7.

BRINSTER, R. L. (1968*a*). *In vitro* culture of mammalian embryos. *Journal of Animal Science*, **27**, Supplement 1, 1–14.

BRINSTER, R. L. (1968*b*). Carbon dioxide production from glucose by the preimplantation rabbit embryo. *Experimental Cell Research*, **51**, 330–44.

BRINSTER, R. L. (1970). Culture of two-cell rabbit embryos to morulae. *Journal of Reproduction and Fertility*, **21**, 17–22.

BRINSTER, R. L. (1972). Cultivation of the mammalian embryo. In *Growth, Nutrition and Metabolism of Cells in Culture*, ed. G. H. Rothblat & V. J. Cristofalo, vol. 2, pp. 251–86. New York & London: Academic Press.

BRINSTER, R. L. (1973). Protein synthesis and enzyme constitution of the preimplantation mammalian embryo. In *The Regulation of Mammalian Reproduction*, ed. S. J. Segal, R. Crozier, P. A. Corfman & P. G. Condliffe, pp. 302–16. Springfield, Illinois: Charles C. Thomas.

CHANG, M. C. (1959). Fertilization of rabbit ova, *in vitro*. *Nature, London*, **184**, 466–7.

CHANG, M. C. (1968). *In vitro* fertilization of mammalian eggs. *Journal of Animal Science*, **27**, Supplement, 1, 15–22.

CHOLEWA, J. A. & WHITTEN, W. K. (1970). Development of two-cell mouse embryos in the absence of a fixed-nitrogen source. *Journal of Reproduction and Fertility*, **22**, 553–5.

CROSS, P. C. & BRINSTER, R. L. (1970). *In vitro* development of mouse oocytes. *Biology of Reproduction*, **3**, 298–307.

CROSS, P. C. & BRINSTER, R. L. (1973). The sensitivity of one-cell mouse embryos to pyruvate and lactate. *Experimental Cell Research*, **77**, 57–62.

DONAHUE, R. P. & STERN, S. (1968). Follicular cell support of oocyte maturation: production of pyruvate *in vitro*. *Journal of Reproduction and Fertility*, **17**, 395–8.

EDWARDS, R. G. (1973). Physiological aspects of human ovulation, fertilization and cleavage. *Journal of Reproduction and Fertility*, Supplement 18, 87–101.

EDWARDS, R. G., BAVISTER, B. D. & STEPTOE, P. C. (1969). Early stages of fertilization *in vitro* of human oocytes matured *in vitro*. *Nature, London*, **221**, 632–5.

EDWARDS, R. G., STEPTOE, P. C. & PURDY, J. M. (1970). Fertilization and cleavage *in vitro* of pre-ovulator human oocytes. *Nature, London*, **227**, 1307–9.

EVANS, V. J., BRYANT, J. C., KERR, H. A. & SCHILLING, E. L. (1964). Chemically defined media for cultivation of long-term cell strains from four mammalian species. *Experimental Cell Research*, **36**, 439–74.

FISHER, D. L. & FISCHER, T. V. (1973). Recovery and *in vitro* culture of pre-implantation Mongolian gerbil embryos. *Teratology*, **7**, A14–15.

FOLSTAD, L., BENNETT, J. P. & DORFMAN, R. I. (1969). The *in vitro* culture of rat ova. *Journal of Reproduction and Fertility*, **18**, 145–6.

FOWLER, R. E. & EDWARDS, R. G. (1973). The genetics of early human development. *Progress in Medical Genetics*, **9**, 49–112.

FRAZER, L. R. & DANDEKAR, P. V. (1973). Fertilization of rabbit eggs *in vitro* without supplemental CO_2 in the atmosphere. *Journal of Reproduction and Fertility*, **33**, 159–61.

GATES, A. H. (1965). Rate of ovular development as a factor in embryonic survival. In *Preimplantation Stages of Pregnancy*, ed. G. E. Wolstenholme & M. O'Connor, pp. 270–88. London: Churchill.

GOULD, K. G., CLINE, M. & WILLIAMS, W. L. (1973). Observations on the induction of ovulation and fertilization *in vitro* in the squirrel monkey (*Simiri sciureus*). *Fertility and Sterility*, **24**, 260–8.

GWATKIN, R. B. L. (1966). Amino acid requirements for attachment and outgrowth of the mouse blastocyst *in vitro*. *Journal of Cellular and Comparative Physiology*, **68**, 335–44.

HAMNER, C. E. & FOX, S. B. (1969). Biochemistry of oviductal secretions. In *The Mammalian Oviduct*, ed. E. S. E. Hafez & R. J. Blandau, pp. 333–55. Chicago & London: Chicago University Press.

HAMNER, C. E., JENNINGS, L. L. & SOJKA, N. J. (1970). Cat (*Felus cattus*) spermatozoa require capacitation. *Journal of Reproduction and Fertility*, **23**, 477–80.

HARMS, E. & SMIDT, D. (1970). *In vitro* fertilization of follicular and tubal ova of pigs. *Berliner und Münchener Tierärztliche Wochenschrift*, **83**, 269–75.

HOPPE, P. C. & PITTS, S. (1973). Fertilization *in vitro* and development of mouse ova. *Biology of Reproduction*, **8**, 420–6.

HOPPE, P. C. & WHITTEN, W. K. (1974*a*). Maturation of mouse sperm *in vitro*. *Journal of Experimental Zoology*, **188**, 133–6.

HOPPE, P. C. & WHITTEN, W. K. (1974*b*). An albumen requirement for fertilization of mouse eggs *in vitro*. *Journal of Reproduction and Fertility*, **39**, 433–6.

HOWARD, E. & DEFEO, V. J. (1959). Potassium and sodium content of uterine and seminal vesicle secretions. *American Journal of Physiology*, **196**, 65–8.

IWAMATSU, T. & CHANG, M. C. (1970). Further investigation of capacitation of sperm and fertilization of mouse eggs *in vitro*. *Journal of Experimental Zoology*, **175**, 271–82.

IWAMATSU, B. & CHANG, M. C. (1971). Factors involved in the fertilization of mouse eggs *in vitro*. *Journal of Reproduction and Fertility*, **26**, 197–208.

KANE, M. T. (1972). Energy substrate and culture of single cell rabbit ova to blastocyst. *Nature, London*, **238**, 468–9.

KANE, M. T. & FOOTE, R. H. (1971). Factors affecting blastocyst expansion of rabbit zygotes and young embryos in defined media. *Biology of Reproduction*, **4**, 41–7.

KAUFMAN, M. H. (1973). Cytogenetic analyses of the first cleavage mitosis in parthenogenetic and fertilised eggs of the mouse. PhD Thesis, University of Cambridge.

KAUFMAN, M. H. & WHITTINGHAM, D. G. (1972). Viability of mouse oocytes ovulated within 14 hours of an injection of pregnant mare's serum gonadotrophin. *Journal of Reproduction and Fertility*, **28**, 465–8.

MAURER, R. R., HUNT, W. L., VAN VLECK, L. D. & FOOTE, R. H. (1968). Developmental potential of superovulated rabbit ova. *Journal of Reproduction and Fertility*, **15**, 171–5.

MAURER, R. R., ONUMA, H. & FOOTE, R. H. (1970). Viability of cultured and transferred rabbit embryos. *Journal of Reproduction and Fertility*, **21**, 417–22.

MAYER, J. F. & FRITZ, H. I. (1974). The culture of preimplantation rat embryos and the production of allophenic rats. *Journal of Reproduction and Fertility*, **39**, 1–9.

MERYMAN, H. T. (1966). Review of biological freezing. In *Cryobiology*, ed. H. T. Meryman, pp. 1–114. New York & London: Academic Press.

MILLS, J. A., JEITLES, G. & BRACKETT, B. G. (1973). Embryo transfer following *in vitro* and *in vivo* fertilization of rabbit ova. *Fertility and Sterility*, **24**, 602–608.

MIYAMOTO, H. & CHANG, M. C. (1972). Development of mouse eggs fertilized *in vitro* by epididymal spermatozoa. *Journal of Reproduction and Fertility*, **30**, 135–7.

MIYAMOTO, H. & CHANG, M. C. (1973). Effect of osmolality on fertilization of mouse and golden hamster eggs *in vitro*. *Journal of Reproduction and Fertility*, **33**, 481–7.

MIYAMOTO, H., TOYODA, Y. & CHANG, M. C. (1974). Effect of hydrogen-ion concentration on *in vitro* fertilization of mouse, golden hamster and rat eggs. *Biology of Reproduction*, **10**, 487–93.

MUKHERJEE, A. B. (1972). Normal progeny from fertilization *in vitro* of mouse oocytes matured in culture and spermatozoa capacitated *in vitro*. *Nature, London*, **237**, 397–8.

MUKHERJEE, A. B. & COHEN, M. M. (1970). Development of normal mice by *in vitro* fertilization. *Nature, London*, **228**, 472–3.

MULLEN, R. J. & CARTER, S. C. (1973). Efficiency of transplanting normal, zona-free and chimaeric embryos to one and both uterine horns of inbred and hybrid mice. *Biology of Reproduction*, **9**, 111–15.

NOSKE, I. G. (1972). *In vitro* fertilization of the Mongolian gerbil egg. *Experientia*, **28**, 1348–50.

OGAWA, S., SATOH, K., HAMADA, M. & HASHIMOTO, H. (1972). *In vitro* culture of rabbit ova fertilized by epididymal sperms in chemically defined media. *Nature, London*, **238**, 270–1.

OGAWA, S., SATOH, K. & HASHIMOTO, H. (1971). *In vitro* culture of rabbit ova from the single cell to the blastocyst stage. *Nature, London*, **233**, 422–4.

OLIPHANT, G. & BRACKETT, B. G. (1973). Capacitation of mouse spermatozoa in media with elevated ionic strength and reversible decapacitation with epididymal extracts. *Fertility and Sterility*, **24**, 948–55.

PAUL, J. (1972). General introduction. In *Growth, Nutrition and Metabolism of Cells in Culture*, ed. G. H. Rothblat & V. J. Cristofalo, vol. 1, pp. 1–9. New York & London: Academic Press.

PAVLOK, A. & MCLAREN, A. (1972). Role of cumulus cells and the zona pellucida in fertilization of mouse eggs *in vitro*. *Journal of Reproduction and Fertility*, **29**, 91–7.

POLGE, C., WILMUT, I. & ROWSON, L. E. A. (1974). The low temperature preservation of cow, sheep and pig embryos. Paper given at the Eleventh Annual Meeting of the Society for Cryobiology, London, 1974.

QUINN, P. & WALES, R. G. (1973). Growth and metabolism of preimplantation mouse embryos cultured in phosphate-buffered medium. *Journal of Reproduction and Fertility*, **35**, 289–300.

SPINDLE, A. I. & PEDERSEN, R. A. (1973). Hatching, attachment and outgrowth of mouse blastocysts *in vitro*: fixed nitrogen requirements. *Journal of Experimental Zoology*, **186**, 305–18.

STEPTOE, P. C., EDWARDS, R. G. & PURDY, J. M. (1971). Human blastocysts grown in culture. *Nature, London*, **229**, 132–3.

SUGAWARA, S. & TAKEUCHI, S. (1973). On glycolysis in rat eggs during preimplantation stages. *The Tohoku Journal of Agricultural Research*, **24**, 76–85.

TERVIT, H. R. & ROWSON, L. E. A. (1974). Birth of lambs after culture of sheep ova *in vitro* for up to six days. *Journal of Reproduction and Fertility*, **38**, 177–179.

TERVIT, H. R., WHITTINGHAM, D. G. & ROWSON, L. E. A. (1972). Successful culture *in vitro* of sheep and cattle ova. *Journal of Reproduction and Fertility*, **30**, 493–7.

TOYODA, Y. & CHANG, M. C. (1968). Sperm penetration of rat embryos *in vitro* after dissolution of zona pellucida with chymotrypsin. *Nature, London*, **220**, 589–90.

TOYODA, Y. & CHANG, M. C. (1974*a*). Fertilization of rat eggs *in vitro* by epididymal spermatozoa and the development of eggs following transfer. *Journal of Reproduction and Fertility*, **36**, 9–22.

TOYODA, Y. & CHANG, M. C. (1974*b*). Capacitation of epididymal spermatozoa in a medium with high K/Na ratio and cyclic AMP for the fertilization of rat eggs *in vitro*. *Journal of Reproduction and Fertility*, **36**, 125–34.

TOYODA, Y., YOKOYAMA, M. & HOSI, T. (1971*a*). Studies on the fertilization of mouse eggs *in vitro*. I. *In vitro* fertilization of eggs by post-epididymal sperms. *Japanese Journal of Animal Reproduction*, **16**, 147–51.

TOYODA, Y., YOKOYAMA, M. & HOSI, T. (1971*b*). Studies on the fertilization of mouse eggs *in vitro*. II. Effect of *in vitro* pre-incubation of spermatozoa on time of sperm penetration of mouse eggs *in vitro*. *Japanese Journal of Animal Reproduction*, **16**, 152–7.

WALES, R. G. (1970). Effects of ions on the development of the pre-implantation mouse embryo *in vitro*. *Australian Journal of Biological Sciences*, **23**, 421–9.

WAYMOUTH, C. (1972). Construction of tissue culture media. In *Growth, Nutrition and Metabolism of Cells in Culture*, ed. G. H. Rothblat & B. J. Cristofalo, vol. 1, pp. 11–47. New York & London: Academic Press.

WEINMAN, D. E. & WILLIAMS, W. L. (1964). Mechanism of capacitation of rabbit spermatozoa. *Nature, London*, **203**, 423–4.

WHITTEN, W. K. (1956). Culture of tubal mouse ova. *Nature, London*, **177**, 96.

WHITTEN, W. K. (1957). Culture of tubal ova. *Nature, London*, **179**, 1081–2.

WHITTEN, W. K. (1971). Nutrient requirements for the culture of preimplantation embryos. *Advances in the Biosciences*, **6**, 129–39.

WHITTEN, W. K. & BIGGERS, J. D. (1968). Complete development *in vitro* of the preimplantation stages of the mouse embryo in a simple chemically defined medium. *Journal of Reproduction and Fertility*, **17**, 399–401.

WHITTINGHAM, D. G. (1968). Fertilization of mouse eggs *in vitro*. *Nature, London*, **220**, 592–3.

WHITTINGHAM, D. G. (1971*a*). Culture of mouse ova. *Journal of Reproduction and Fertility*, Supplement 14, 7–21.

WHITTINGHAM, D. G. (1971*b*). Survival of mouse embryos after freezing and thawing. *Nature, London*, **233**, 125–6.

WHITTINGHAM, D. G. (1972). Low temperature preservation of mouse embryos. Fellowship thesis, Royal College of Veterinary Surgeons, London.

WHITTINGHAM, D. G. (1974*a*). Fertilization and culture of ferret eggs *in vitro*. *Journal of Reproduction and Fertility*, in press.

WHITTINGHAM, D. G. (1974*b*). Embryo banks in the future of developmental genetics. *Genetics*, supplement, in press.

WHITTINGHAM, D. G. (1974*c*). Freeze preservation of mammalian embryos. In *Basic Aspects of Freeze Preservation of Mouse Strains*. Proceedings of UNESCO-ICLA-ICRO-Workshop. Stuttgart: Gustav Fischer Verlag.

WHITTINGHAM, D. G. & ADAMS, C. E. (1974). Low temperature preservation of rabbit embryos. *Eleventh Annual Meeting of the Society for Cryobiology, London*, 1974, Abstract 75.

WHITTINGHAM, D. G. & BAVISTER, B. D. (1974). Development of hamster eggs fertilized *in vitro* or *in vivo*. *Journal of Reproduction and Fertility*, **38**, 489–92.

WHITTINGHAM, D. G., LEIBO, S. P. & MAZUR, P. (1972). Survival of mouse embryos frozen to −196 °C and −269 °C. *Science*, **178**, 411–14.

WHITTINGHAM, D. G. & WHITTEN, W. K. (1974). Long-term storage and aerial transport of frozen mouse embryos. *Journal of Reproduction and Fertility*, **36**, 433–5.

WILLADSEN, S. M., POLGE, C., ROWSON, L. E. A. & MOOR, R. M. (1974). Preservation of sheep embryos in liquid nitrogen. *Eleventh Annual Meeting of the Society for Cryobiology, London*, 1974, Abstract 73.

WILMUT, I (1972). The effect of cooling rate, warming rate, cryoprotective agent and stage of development on survival of mouse embryos during freezing and thawing. *Life Sciences*, **11**, 1071–9.

WILMUT, I. & ROWSON, L. E. A. (1973). Experiments on the low-temperature preservation of cow embryos. *Veterinary Record*, **92**, 686–90.

YANAGIMACHI, R. (1969). *In vitro* acrosome reaction and capacitation of golden hamster spermatozoa with bovine follicular fluid and its fractions. *Journal of Experimental Zoology*, **170**, 269–80.

YANAGIMACHI, R. (1972). Fertilization of guinea-pig eggs *in vitro*. *Anatomical Record*, **174**, 9–20.

YANAGIMACHI, R. & CHANG, M. C. (1964). *In vitro* fertilization of golden hamster ova. *Journal of Experimental Zoology*, **156**, 361–76.

THE EXPERIMENTAL INDUCTION OF PARTHENOGENESIS IN THE MOUSE

BY M. H. KAUFMAN

Department of Genetics, Weizmann Institute of Science, Rehovot, Israel

The aim of this article is to review the latest developments in the field of mouse parthenogenesis. Some of the recent technical advances will be considered which now make it possible to increase the incidence of selected classes of haploid and diploid parthenogenones. The various stimuli which are capable of initiating mouse parthenogenetic development are discussed, as well as the relevant experimental data which might suggest the possible underlying mechanisms in each case. Current ideas on the factors governing the fate of mouse parthenogenones, and a range of problems which may be investigated using activated oocytes, will be discussed.

Most of the reports which have appeared since Tarkowski's (1971) recent review are concerned with advances in methodology in areas already known to induce mouse parthenogenetic development. Thus further reports on in-vitro activation employing hyaluronidase may be found in Graham (1972), Kaufman (1973*a*, *c*, *d*), Biczysko, Solter, Graham & Koprowski (1974), Graham & Deussen (1974), Kaufman & Gardner (1974), Kaufman & Surani (1974), Phillips & Kaufman (1974) and Solter *et al.* (1974). Studies on the effect of heat shock *in vitro* have been reported by Komar (1973). Detailed observations relating to earlier work by Tarkowski, Witkowska & Nowicka (1970) on in-vivo activation by electrical shock stimulation of the oviduct have recently been published by Witkowska (1973*a*, *b*). Mintz & Gearhart (1973) have compared the properties of the zonae pellucidae of fertilised and parthenogenetic embryos also using electrical shock treatment. To this list must be added ether anaesthesia (Braden & Austin, 1954*a*), and heat shock to the oviducts (Braden & Austin, 1954*b*), which are both capable of initiating a limited degree of parthenogenetic development *in vivo*.

Avertin anaesthesia (Kaufman, unpublished data) is also capable of inducing parthenogenetic activation. The embryos produced are capable of achieving a limited degree of post-implantation development similar to those resulting from electrical stimulation of the oviduct (Tarkowski *et al.*, 1970; Witkowska, 1973*b*) following the transfer of eggs activated *in vitro* to pseudopregnant recipients (Kaufman & Gardner, 1974) and the spontaneous activation of ovulated ova in LT/ChReSv mice (Stevens & Varnum, 1974). All these reports confirm that a wide range of stimuli are capable of initiating mouse parthenogenetic development.

The recent observation that spontaneous parthenogenetic development may commonly be observed in the ovaries or, more rarely, in the oviducts or uteri of LT/ChReSv and related strains of mice (Stevens & Varnum, 1974) is especially important in relation to the very high incidence of ovarian teratomas reported in these mice.

The main advances in this field over the past few years have come in defining some of the factors which can increase the proportion of eggs activated, and control the pathways of development taken by parthenogenones following activation *in vitro*. Thus by altering the postovulatory age of oocytes at the time of activation (Kaufman, 1973*a*), and the osmolarity of the medium during the first 2–3 hour period of culture when activated eggs would be completing meiosis II (Graham, 1972; Kaufman & Surani, 1974), selected types of parthenogenones can be preferentially obtained.

Very little information is available on the underlying mechanisms involved when mammalian eggs are activated to develop parthenogenetically either *in vivo* or *in vitro*, and it is not yet possible to relate the diverse range of stimuli which are capable of activating eggs to the early normal events associated with fertilisation.

Activation may be carried out either *in vivo* or *in vitro*. The advantages and disadvantages of these two approaches have been discussed by Tarkowski (1971). In order to simplify the discussion on the possible underlying mechanisms involved in these two approaches, each will be dealt with separately.

IN-VITRO ACTIVATION

Hyaluronidase treatment

Kaufman (1973*a*) noted that sperm-free filtrates can activate aged mouse eggs. However, fertilisation *in vitro* occurred only when freshly ovulated eggs were used (Kaufman, 1973*d*). This suggested that certain changes were occurring in the zona within 6–8 hours of ovulation which prevented spermatozoa from penetrating aged eggs *in vitro*. The activating factor in sperm suspensions was probably released from the acrosome region. This factor was capable of inducing a type of 'zona reaction' which blocked sperm entry. *In vivo*, in contrast to the situation *in vitro*, approximately 70 % of eggs are penetrated up to 27 hours after HCG (human chorionic gonadotrophin), though 80 hours later less than 30 % of these ova appeared to be cleaving normally (Marston & Chang, 1964).

Graham (1970, 1971, 1972) demonstrated that a proportion of mouse eggs treated with hyaluronidase 24–9 hours after HCG became activated. More detailed observations by Kaufman (1973*a*) demonstrated that an increasing proportion of activated eggs were obtained at 16, 18 and 20–1 hours after HCG. The activation frequency in populations of eggs stimulated approximately 25 hours after HCG is similar to that observed at 20–1 hours, although different classes of parthenogenetic embryos are obtained. This is especially marked between 16–20 and 25 hours after HCG, while an intermediate response is observed at 21 hours.

It is not clear whether the zona of a freshly ovulated egg is permeable to hyaluronidase, or whether a change occurs in response to ageing. In addition, very little is known about the properties of the egg's vitelline membrane. Measurements of the electrical changes in the surface of mouse eggs during maturation, or following fertilisation, or during pre-implantation development (Cross, Cross & Brinster, 1973), have demonstrated a pattern of events which is similar to that observed in other species. This suggests that similar events are initiated at fertilisation in all species (Epel, 1972). One interesting observation was that there was no significant change in the membrane potential of unfertilised eggs when this was determined at 13–15 hours and 20–2 hours after HCG. This is of importance since membrane potential changes are one of the first events which can be demonstrated after fertilisation, followed slightly later by the cortical reaction (Epel, 1972).

The effect of pronase

A further demonstration of the effect of ageing was observed in a recent series of experiments on the effect of removing the zona pellucida with pronase (Kaufman, 1973*d*). Freshly ovulated (HCG + 14.5–15.0 hours) and more aged eggs (HCG + 21.5–22.0 hours) were treated for 10–15 minutes with 0.25% pronase in phosphate buffered saline. Approximately half of the eggs had been pretreated for 10–15 minutes with medium containing hyaluronidase, and eggs were examined 6–8 hours later. The results of this series of experiments are presented in Table 1. This demonstrates that almost all of the eggs in the two aged groups treated with pronase became activated, whereas no activation was observed in the two treatment groups where freshly ovulated eggs were involved. It seems likely that pronase causes certain changes in the oocyte plasma membrane, in addition to its proteolytic effect on the zona pellucida.

The unexpectedly high rate of activation observed in the aged eggs (see Table 1) suggests that the cell surface change induced by pronase is related to the induction of activation in these eggs. As freshly ovulated eggs treated with pronase do not respond in this way, it seems likely that the surface change which takes place acts as a trigger mechanism. The membrane changes induced by the activating spermatozoon are presumably dissimilar, in that the triggering of the specific changes associated with fertilisation can occur in both freshly ovulated and relatively aged oocytes.

Table 1. *The effect of pronase on the activation of mouse oocytes*

		Classes of parthenogenones			
HCG + hours	Treatment	1 pronucleus +2nd polar body	2 pronuclei	Immediate cleavage	Overall activation frequency* (%)
14.5–15.0	Pronase		0	0	0/70 (0)
	Pronase + hyaluronidase	0	0	0	0/47 (0)
21.5–22.0	Pronase	66	0	6	72/73 (98.6)
	Pronase + hyaluronidase	41	2	4	47/48 (97.9)

* Observations made 4–5 hours after treatment.

Some recent information has been provided by Pienkowski & Koprowski (1974). They have shown that a similar amount of concanavalin A (Con A) binds to the surface of fertilised and unfertilised eggs. They found that fertilised eggs in culture agglutinate in a concentration of 10 μg/ml of Con A, while unfertilised eggs require 2000 μg/ml before they will agglutinate. Exposure of unfertilised eggs to pronase for 10 minutes changed their surface properties in such a way that they then agglutinated at the lower concentration of Con A.

The effect of heat shock

Komar (1973) has recently reported on the effect of heating mouse oviducts containing ova to temperatures between 43.0 and 45.5 °C for periods of time varying from 5–10 minutes. Females were either superovulated and their oviducts excised between 14.5 and 17.5 hours after HCG, or ovulated spontaneously, and mated with vasectomised males. In the spontaneously ovulating group, dissections were carried out between 08.00 and 11.00 hours on the day a vaginal plug was detected. The response observed depended on both the temperature and the duration of treatment. At all temperatures tested, haploid eggs predominated over diploids. Of the heat-treated eggs 15.4 % developed to the morula or blastocyst stage after 4 days in organ culture, compared with 57.4% of fertilised eggs.

As neither the lighting schedule nor the normal time of spontaneous ovulation in the strains examined was reported, it is impossible to be certain whether heat shock applied *in vitro* is effective in activating recently ovulated eggs (within 2–3 hours of ovulation). While the number of eggs undergoing different classes of reaction have been tabulated, there is no mention of whether these eggs originated from induced or spontaneously ovulating mice. There is also no information as to whether they were isolated from 'A' strain or Swiss Albino females. The absence of this information considerably reduces the value of the data. Had these various groups been tabulated separately, it might have been possible to establish whether heat is the primary stimulating agent in electric shock treatment *in vivo*.

FACTORS WHICH MODIFY THE INITIAL PATHWAYS OF DEVELOPMENT OF PARTHENOGENONES

The spontaneous central migration of the second meiotic spindle in response to postovulatory ageing of the oocyte

One of the earliest histological changes which may be observed in ageing oocytes is the rotation and migration of the second meiotic spindle from the periphery to the centre of the egg (Szollosi, 1971).

As the developmental pathway taken by activated eggs depends on the location of the spindle at the time of activation, eggs in which the spindle is peripheral will normally extrude a second polar body. Those in which

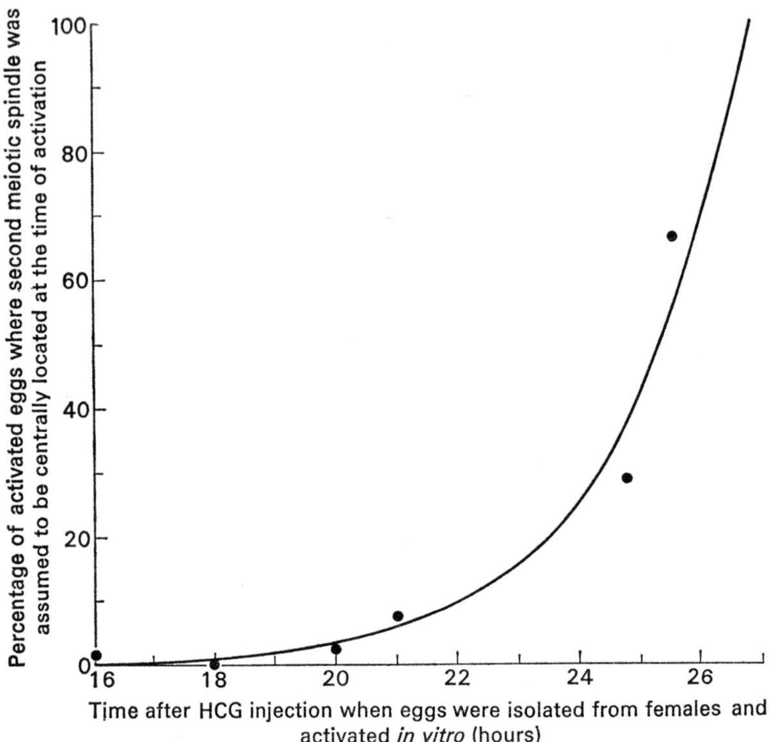

Fig. 1. Regression line demonstrating the spontaneous central migration of the second meiotic spindle in response to postovulatory ageing. The points represent the percentage of activated eggs in each time group undergoing immediate cleavage, and with two pronuclei (without a second polar body).

the spindle is centrally located will either undergo immediate cleavage or, if cytokinesis does not occur, develop two haploid, or a single, diploid pronucleus. Indirect evidence of spindle migration was first detected about 17 hours after HCG (approximately 5 hours after ovulation). In Fig. 1 the proportion of eggs in which the spindle was assumed to be centrally located at the time of activation has been plotted against the time after HCG when eggs were isolated and activated *in vitro*. These data indicate that the spindle would be centrally located in almost all ova by approximately 27 hours after HCG. The very low incidence of eggs undergoing immediate cleavage or with two pronuclei observed up to 21 hours after HCG, suggests that hyaluronidase treatment does not accelerate spindle migration.

The effect of culture in hypotonic medium

When eggs were cultured in hypotonic medium during the first few hours after hyaluronidase treatment, retention of the second polar body and the formation of two pronuclei occurred in a high proportion of the activated eggs (Kaufman & Surani, 1974). No increase in activation frequency was observed following culture in hypotonic medium compared with untreated controls. These eggs are potentially diploid, as the two pronuclear chromosome sets unite on the first cleavage spindle (Kaufman, 1973*c*). While second polar body retention had been observed following culture in hypotonic medium, Graham (1972) recorded a high incidence of activated eggs with a single (presumed diploid) pronucleus, and a much smaller proportion with two pronuclei (see also Graham & Deussen, 1974). Similarly, Graham & Deussen (1974) observed that culture in hypotonic medium following hyaluronidase treatment increased the number of activated eggs in some mouse strains. Whether these differences are due to strain variation in the mice employed, or differences in experimental procedure, is not clear.

The time of entry into the first cleavage mitosis of 50 % of activated eggs cultured in hypotonic medium was 15 hours after hyaluronidase treatment (Kaufman & Surani, 1974). Pronucleus formation occurred at about the normal time, that is, within 3–4 hours of activation, though the entry of these eggs into the first cleavage mitosis was delayed by approximately the duration of culture in hypotonic medium. Amino acid incorporation is significantly reduced during the culture period in hypotonic medium, and this may be related to a reduction in carrying capacity, due to the low Na^+ in the medium (Schultz & Curran, 1970).

Retention of the second polar body is probably due to a direct inhibitory effect on the oocyte plasma membrane and microfilament system. Thus culture in hypotonic medium superficially resembles the effect of cytochalasin B on cells (Defendi & Stoker, 1973) and early cleavage embryos (Snow, 1973). As a result of failure of extrusion of the second polar body, both products of anaphase II are retained within the egg vitellus. However, the spindle apparatus remains functionally intact. From their earliest appearance at prometaphase until the onset of metaphase of the first cleavage mitosis, the two pronuclear chromosome sets always appear to be synchronous in their degree of condensation (Kaufman, 1973*c*). This contrasts with the situation in fertilised eggs, where chromosome condensation in the male and female pronuclei is asynchronous until the two groups unite on the spindle equator (McGaughey & Chang, 1971; Donahue, 1972*a*; Kaufman, 1973*b*). Graham (1971) originally stated that eggs with two pronuclei underwent a process of delayed immediate cleavage, which invariably produced haploid embryos, and that diploid embryos were exclusively formed from haploid eggs by doubling of the haploid set of chromosomes. The occurrence of haploid–diploid mosaics is evidence that regulation to diploidy certainly does occur in some blastomeres (Tarkowski *et al.*, 1970; Witkowska, 1973*a*). The majority of diploid embryos probably result from suppression of second polar body extrusion, rather than suppression of the first cleavage division. Delayed immediate cleavage was not observed by Kaufman (1973*c*) following serial observations on large numbers of eggs with two pronuclei, or by Witowska (1973*a*).

IN-VIVO ACTIVATION

Electrical stimulation of the oviduct

The underlying mechanisms which induce eggs to undergo parthenogenetic activation *in vivo* are more difficult to examine. At least two distinct classes of stimuli may be differentiated – those which are capable of activating recently ovulated eggs, and those which are only capable of activating aged eggs. Electrical stimulation of the oviduct (Tarkowski *et al.*, 1970) belongs to the first class of stimuli. This type of stimulation may also induce aged eggs to develop parthenogenetically, but research on this topic has not been reported to date. Most of the other stimuli which activate eggs *in vivo* belong to the second class, and these include ether anaesthesia (Braden & Austin, 1954*a*), heat shock to the oviduct (Braden & Austin, 1954*b*) and avertin anaesthesia (Kaufman, unpublished data).

All of these stimuli must alter, either directly or indirectly, the biochemical environment within the oviduct. Electrical stimulation may induce activation by a local heating effect or by causing a change in the ionic composition of the oviduct fluid, or by a direct effect on the oocyte's membrane potential.

Gwatkin, Williams, Hartmann & Kniazuk (1973) demonstrated that electrical stimulation of hamster oocytes *in vitro* caused the cortical granules to rupture and release a trypsin-like protease into the perivitelline space. Presumably activation was not induced, as these authors did not comment on polar body extrusion or nucleus formation in these eggs. Cortical granule breakdown may occur when freshly ovulated mouse eggs are activated in this way, when the cortical granules are still peripherally located. The demonstration of a partial zona reaction by Mintz & Gearhart (1973) suggests that either their stimulation was suboptimal, or that a proportion of the cortical granules had already spontaneously migrated from the periphery of the eggs in response to postovulatory ageing. The normal mouse zona reaction may be a quantitative response which is only induced when the products of all the cortical granules are released.

One of the first changes which takes place following sperm penetration in normal fertilisation is the zona reaction (Braden, Austin & David, 1954). This is probably caused by the release of a trypsin-like protease from the cortical granules into the perivitelline space. This substance interacts with the zona pellucida to cause a change in its physical properties, rendering it impermeable to spermatozoa (Austin & Braden, 1956; Barros & Yanagimachi, 1971; Gwatkin *et al.*, 1973). It has long been recognised that a change occurs in the zona pellucida following fertilisation, which renders it resistant to proteolytic enzymes (Smithberg, 1953; Chang & Hunt, 1956; Mintz, 1962). Mintz & Gearhart (1973) demonstrated that parthenogenetic embryos induced by electrical stimulation of the oviduct showed a response in the time taken to dissolve the zona with pronase intermediate between that seen in unfertilised and fertilised eggs. It remains to be seen whether this 'incomplete' zona reaction plays any role in the death of these embryos. Migration of the cortical granules from the periphery of the unfertilised egg occurs spontaneously in response to ageing in the oviduct following ovulation (Szollosi, 1971), so that delayed mating would, in any case, induce an abnormal response. This does not, however, completely preclude embryo viability (see Marston & Chang, 1964), and would be analogous to the situation in parthenogenones induced *in vitro*, where postovulatory ageing is an important factor in the induction of high rates of activation.

The effect of avertin anaesthesia

Of more general interest has been the recent demonstration (Kaufman, unpublished data) that avertin anaesthesia is capable of inducing a relatively high rate of parthenogenetic activation and development. An activation frequency of 46 % was observed in a spontaneously ovulating population of (C57Bl $\times A_2G$)F1 females anaesthetised 13 hours after the 'mid dark point' of their photoperiod. Ovulation was assumed to have occurred at about this time (see Braden, 1957). Females were killed 20–4 hours after anaesthesia, and the number of cleavage embryos recorded. A high proportion of the embryos were at the 3- or 4-cell stage at the time of examination, suggesting that they underwent immediate cleavage. This is consistent with previous observations on the activation of eggs isolated approximately 13 hours after ovulation (Kaufman, 1973*a*, HCG+25-hour group). When 18 spontaneously ovulating mice anaesthetised 13 hours after ovulation were killed on days 5, 6 or 7 (day of anaesthesia referred to as day 0), the overall mean number of implants per horn was 0.50 (18/36), or 0.90 (18/20) if only the females with implants are considered. Approximately 12.5 % of all the eggs ovulated, or 27.3 % of all the eggs activated, survived at least until the moment of implantation.

The pattern of activation following avertin anaesthesia was similar to that observed after hyaluronidase treatment of oocytes *in vitro*, in that an increasing incidence of activation was observed when females were anaesthetised at 4, 6.5, 9 or 13 hours after the mid dark point. Neither the injection of saline nor the stimulus of mating to vasectomised males induced activation.

A similar implantation rate was observed in superovulated females anaesthetised 20 hours after HCG, where the initial activation frequency was 29.6 %. However, nearly all of the activated eggs developed a single pronucleus and extruded a second polar body. When examined on day 7, out of 5 females 2 had no implants, and a total of 6 implants was observed in the remaining 3 females.

It is interesting that the implantation rates observed in this series are similar to those observed when eggs activated *in vitro* were transferred at the pronuclear stage to pseudopregnant recipients (Kaufman & Gardner, 1974). Twenty-nine per cent of all immediate cleavage embryos transferred evoked decidual reactions, or 37.5 % if only the females with implants are considered. Corresponding figures for haploid embryos with a single pronucleus and second polar body were 19.8 % and 35.1 % respectively.

Braden & Austin (1954*a*) noted that 17 out of 98 mouse eggs became activated following ether anaesthesia carried out 10–14 hours after ovulation, though development was only observed to the 2- or 4-cell stage. It is not clear whether all anaesthetics given at the appropriate period after ovulation would induce parthenogenetic development. No activation was observed in eggs isolated from control oviducts of mice anaesthetised with nembutal (Tarkowski *et al.*, 1970; Witkowska, 1973*a*), though anaesthesia may have been performed during the period when these eggs would have been refractory to stimulation by any anaesthetic agent.

THE ULTRASTRUCTURE OF PARTHENOGENONES

Recently several ultrastructural analyses of parthenogenones induced by hyaluronidase treatment of oocytes *in vitro* have been carried out (Biczysko *et al.*, 1974; Solter *et al.*, 1974), and it is now possible to compare the morphological changes occurring in these embryos with those occurring during meiotic maturation (Calarco, Donahue & Szollosi, 1972), in unfertilised (Zamboni, 1970) and fertilised 1-cell eggs (Hillman & Tasca, 1969; Zamboni, 1971, 1972; Zamboni, Chakraborty & Smith, 1972), and early cleavage embryos.

In their ultrastructural study Solter *et al.* (1974) examined 1-cell and cleaving embryos which had been activated *in vitro*, and demonstrated numerous differences between the parthenogenones and fertilised eggs at similar stages of development. No cortical reaction was observed in these parthenogenones, and numerous cortical granules and vacuolated mitochondria were usually found near the cell membrane. Primary nucleoli persisted during mitosis, cytokinesis was generally irregular, and fragments of cytoplasm were commonly seen between dividing cells. This analysis suggested that numerous disturbances affecting various cell systems were occurring during early cleavage of the parthenogenones. Apart from the abnormal cortical reaction and the persistence of primary nucleoli, these authors suggested that RNA and protein synthesis were probably also disturbed.

Considering how morphologically grossly abnormal all these parthenogenones appeared to be, it is difficult to see how such embryos could survive beyond implantation. One explanation might be that the parthenogenones which were examined would not, in any case, have survived beyond a few cleavage divisions. Thus, most of the observations may have been carried out on potentially inviable embryos, as only approximately 5–10 % of eggs activated *in vitro* are capable of development to the blastocyst stage in culture (Graham, 1971). A more favourable approach might be the examination of parthenogenetic embryos following oviduct transfer (Kaufman & Gardner, 1974), where 37–60 % of embryos, depending on their ploidy, survive beyond implantation.

THE DEVELOPMENT OF PARTHENOGENONES

Measurement of the duration of the first cleavage mitosis in haploid and diploid parthenogenones and fertilised eggs (Kaufman, 1973*b*, *c*, *d*) demonstrated that the overall duration was related to the ploidy, rather than to whether eggs were of parthenogenetic origin or not. Thus the duration of the first cleavage mitosis in fertilised eggs and diploid parthenogenones was about 120 minutes, compared with about 160 minutes in haploid embryos. Later, Kaufman & Surani (1974) demonstrated that both classes of parthenogenones entered the first cleavage mitosis at approximately the same time after activation. Further observations will be required to determine whether subsequent mitoses and intermitotic intervals in haploid and diploid parthenogenones follow a similar pattern. If this were the case, by the time the blastocyst stage is reached immediate cleavage embryos would be expected to be one cleavage division ahead of haploids derived from eggs in which the second polar body was extruded.

Immediate cleavage embryos may reach the blastocyst stage earlier if both classes cleave at the same rate, and cell number is the only factor involved in cavitation. This may not prove to be the case, and early-cavitating blastocysts of immediate cleavage origin may prove to have twice as many blastomeres as haploids originating from eggs in which extrusion of the second polar body occurs. This would be consistent with previous findings (Tarkowski & Wroblewska, 1967), which demonstrated that blastocoel formation was initiated when blastomeres of the appropriate age were present. Cavitation occurred despite the fact that some embryos contained only a third or a quarter the normal number of cells.

These observations on fertilised eggs (Tarkowski & Wroblewska, 1967) are not strictly correlated with parthenogenetic development. Witkowska (1973*a*) found that many parthenogenones did not cavitate despite the accumulation of very large numbers of cells. This analysis also suggested that the transformation of morulae into blastocysts was not connected with ploidy, as haploid, haploid–diploid mosaic and diploid blastocysts contained on the average more cells than morulae. The mitotic activity of most of these parthenogenones was only slightly less than in fertilised embryos, when measured in numbers of cell cycles. The number of nuclei present in parthenogenones recovered on the 5th day was correlated with the degree of ploidy; haploids had the greatest number of cells, diploids the smallest, and mosaics were intermediate. The relationship between ploidy and cell number had previously been predicted by Beatty & Fischberg (1951) and confirmed by Edwards (1958), though with less convincing data.

Detailed observations were also made on the progress of these embryos (Witkowska, 1973*a*), which showed that the development of eggs from spontaneously ovulating pseudopregnant females was superior to that of eggs from superovulated females. No comparable analyses have been performed on the pre-implantation stages of development of parthenogenones induced by any other technique, apart from the studies on the time of entry into the first cleavage division and the overall duration of the first cleavage mitosis referred to above. In a subsequent paper on the post-implantation development of parthenogenones induced by electrical stimulation (Witkowska, 1973*b*) over 150 implantation sites were examined from the 5th to the 10th day of pseudopregnancy. The number of living embryos recovered decreased steadily with every day, while the most advanced embryos recovered were two living embryos found on the 9th and 10th day. This provides strong evidence that development may occasionally proceed beyond this stage.

SPONTANEOUS PARTHENOGENETIC DEVELOPMENT

Stevens & Varnum (1974) have recently reported that LT/ChReSv (referred to as LT) mice show a very high incidence of ovarian teratomas, and they present evidence which suggests that these probably originate from oocytes which develop parthenogenetically within the ovary. A smaller incidence of implantation sites evoked by oocytes which activated spontaneously following ovulation was also noted. The spontaneous parthenogenones which were implanted died between days 5 and 7 at approximately the same developmental stage as most experimentally induced parthenogenones.

The early stages of development of ovarian embryos were observed in all mice over 18 days of age, while in older mice, some parthenogenones reached stages comparable to normal 6–7 day embryos, after which stage they became disorganised. As teratomas were only observed in mice old enough to have developed corpora lutea, these authors hypothesised that a hormonal factor might be a possible aetiological agent. Impressive as the circumstantial evidence appears to be, that these teratomas arise from the spontaneous activation of ovarian oocytes, incontravertible proof of their parthenogenetic origin would only be provided by the demonstration of haploid metaphases in cell lines derived from these tumours. At the present time, it is impossible to exclude the possibility that normal ovarian tissue in the region of degenerating parthenogenetic embryos may be induced to develop into teratomas, because teratomas in these mice always develop near to parthenogenetic islands of cells.

The fact that the parthenogenones which develop from ovulated eggs implant in virgin females, suggests that their hormonal status is in some way disturbed. Thus, in addition to spontaneous activation, these females also seem to be capable of becoming spontaneously pseudopregnant. This may be the mechanism by which spontaneously activated blastocysts are allowed to implant.

THE USE OF CHROMOSOME ANALYSIS OF THE FIRST CLEAVAGE MITOSIS OF PARTHENOGENONES

The normal morphology of the first cleavage mitosis in haploid parthenogenones has recently been described (Kaufman, 1973*c*). This has allowed a comparison to be made between the appearance of the chromosomes at this stage, and at metaphase II, in the parallel events in fertilised eggs (McGaughey & Chang, 1971; Donahue, 1972*a*, *b*; Kaufman, 1973*b*), and in the subsequent pre-implantation cleavage mitoses (Tarkowski, 1966). This study (Kaufman, 1973*c*) clearly demonstrated the greater morphological simplicity of the chromosomes at the first cleavage metaphase compared to metaphase II, and to the chromosomes of cleavage embryos. It is frequently very difficult to analyse metaphase II chromosome groups, even to the extent of determining the exact number of dyads present.

The chromosomes observed at the first cleavage metaphase of the haploid parthenogenone represent half of those participating in the second meiotic cleavage. Both products of anaphase II may be examined if oocytes are induced to develop as immediate cleavage embryos, or as potentially diploid eggs with two haploid pronuclei. In both these types of parthenogenones the two pronuclei enter prometaphase approximately synchronously, so that air-dried preparations of these eggs shortly after their pronuclear outlines have disappeared will allow the two chromosome groups to be examined. This type of analysis would be particularly useful for detecting gross morphological damage to chromosomes induced during oogenesis by chemical agents or X-irradiation, as chromosomal aberrations would be easier to assess at the first cleavage metaphase than at metaphase II (Rohrborn & Hansmann, 1971). When the second cleavage division in fertilised eggs is used for this type of analysis (Rohrborn, Kuhn, Hansmann & Thon, 1971) other difficulties arise, due to the increase in asynchrony between blastomere nuclei and the additional presence of a male chromosome set.

Oocytes from 'Bpa' mice (Phillips, Hawker & Moseley, 1973) and related stocks were examined (Phillips & Kaufman, 1974) to determine the mechanism underlying the high production of XO female offspring. Oocytes were induced to develop parthenogenetically, and both products of anaphase II examined as outlined above. X-chromosome loss was found to be due to non-disjunction, which occurred in about one-third of the oocytes from 'Bpa' females, mainly at meiosis I, though evidence of non-disjunction at meiosis II was also observed. This type of analysis has allowed meiotic events occurring within the ovary to be examined in detail for the first time.

A further important use of this approach could be the examination of oocytes, matured *in vitro*, to determine whether these show any evidence of chromosome imbalance which could explain why so few are capable of fertilisation *in vitro* (Cross & Brinster, 1970). Similarly, the chromosome constitution of oocytes ovulated by aged female mice should be re-examined to determine whether the reduced chiasma frequency observed with increasing maternal age (Henderson & Edwards, 1968) manifests itself by this stage. This would confirm or refute the observations of Jagiello & Polani (reported in Fowler & Edwards, 1973, p. 69, as personal communication) who stated that all mouse oocytes examined at metaphase II were diploid (see also Fowler & Edwards, 1973, for discussion).

GENERAL DISCUSSION

Very little information is available which could help to establish why mouse parthenogenones do not survive beyond the early post-implantation period. Various theories have been proposed to account for their premature death (Graham, 1971, 1974; Tarkowski, 1971; Mintz & Gearhart, 1973). In these parthenogenones gene expression might be disordered (possibly because of an altered nuclear–cytoplasmic ratio), or their extensive genetic homozygosity may expose recessive lethal genes. Alternatively a compound from the male gamete may be necessary for normal development to take place. This component may be cytoplasmic or nuclear, and may function in normal X-inactivation (Brown & Chandra, 1973).

The observation that their slower rate of development compared to normal embryos would result in asynchrony between the embryo and the uterus, does not seem to be a major factor in causing their premature death. Embryos isolated at the morula and blastocyst stage which were transferred to recipients at an earlier stage of pseudopregnancy implanted, but an increased survival rate was not obtained (Witkowska, 1973*b*; Kaufman & Gardner, 1974).

The decidual response induced by these embryos appears to be normal, which suggests that their trophoblastic component at least is functioning normally at the time of implantation. Thus one explanation which could account for the early death of these embryos, is the failure of normal development and functional capacity of their inner cell mass or embryonic component. A balance between these two entities is essential for normal embryonic growth to take place (Gardner, 1971; Gardner & Johnson, 1972; Gardner, Papaioannou & Barton, 1973), while direct contact with the uterine mucosa is probably necessary for inner cell mass development (Tarkowski, 1962).

The microsurgical transfer of parthenogenetic inner cell masses to trophoblast vesicles derived from normal blastocysts, and the reciprocal transfer of inner cell masses from normal blastocysts to trophoblast vesicles from parthenogenetic embryos, would give invaluable information on the developmental potential of both these components. The transfer of these reconstituted chimaeric blastocysts to suitable recipients may be the only means of obtaining advanced parthenogenetic embryonic growth. Chromosome and enzyme markers would serve to differentiate between the two components of the chimaera. The transfer of chimaeric embryos formed by the fusion of parthenogenetic to fertilised morulae to pseudopregnant recipients (Graham, 1970) gave inconclusive results, as no evidence of colour marker genes known to be present in the parthenogenones were expressed in the live-born young.

In addition to the ultrastructural anomalies found by Solter *et al.* (1974), evidence of abnormal development may be observed in some embryos during early cleavage. A small proportion of embryos develop binucleate blastomeres following electrical stimulation of the oviduct (Tarkowski, Witkowska & Nowicka, 1970). This presumably accounts for the haploid–diploid mosaic blastocysts observed after this treatment. Similar mosaic embryos may also occur following other types of activation, but comparable detailed analyses have not been reported. It is also unclear whether the various classes of haploid embryos are equally susceptible to this form of anomalous development. Transfer of selected types of embryos at the pronuclear stage, followed by detailed serial cytological examination during early cleavage, and chromosome analysis at the blastocyst stage, would provide this information. However, the possibility remains that the cytoplasmic damage which results in failure of cytokinesis in some blastomeres may be due to the type of stimulation employed, rather than to a particular susceptibility of one class of embryo compared to another. To date, no examples of diploid–tetraploid mosaics have been reported. It is also equally unclear whether the 'incomplete' zona reaction observed by Mintz & Gearhart (1973) plays any significant role in the post-implantation death of these embryos.

It is curious that post-implantation embryos obtained by electrical stimulation of the oviduct, avertin anaesthesia, in-vitro activation and embryo transfer at the pronuclear stage, or those resulting from spontaneous activation, all appear to die shortly after implantation. The egg-cylinder stage is quite commonly seen, but only very rarely are more advanced embryos observed.

A great deal more information will be required on the normal pre- and post-implantation stages of development of parthenogenones before any meaningful comparisons can be made with the events occurring in normal embryos.

The different types of stimuli which have been employed to activate mouse eggs *in vivo* and *in vitro*, and the possible underlying mechanisms involved in each case have been considered. A correspondingly much smaller proportion of this article has been concerned with the application of these techniques to the understanding of early development. It is hoped that within the next few years further improvements in technique will overcome the present limitations of growing post-implantation parthenogenetic embryos.

I would like to thank Dr C. F. Graham for drawing my attention to certain papers 'in press', and Dr E. Huberman for his helpful criticism of the manuscript. The author is a recipient of a Royal Society–Israel Academy of Sciences Programme Fellowship.

REFERENCES

AUSTIN, C. R. & BRADEN, A. W. H. (1956). Early reactions of the rodent egg to spermatozoon penetration. *Journal of Experimental Biology*, **33**, 358–65.

BARROS, C. & YANAGIMACHI, R. (1971). Induction of zona reaction in golden hamster eggs by cortical granule material. *Nature, London*, **233**, 268–9.

BEATTY, R. A. & FISCHBERG, M. (1951). Cell number in haploid, diploid and polyploid mouse embryos. *Journal of Experimental Biology*, **28**, 541–52.

BICZYSKO, W., SOLTER, D., GRAHAM, C. & KOPROWSKI, H. (1974). Synthesis of endogenous Type-A virus particles in parthenogenetically stimulated mouse eggs. *Journal of the National Cancer Institute*, **52**, 483–9.

BRADEN, A. W. H. (1957). The relationship between the diurnal light cycle and the time of ovulation in mice. *Journal of Experimental Biology*, **34**, 177–88.

BRADEN, A. W. H. & AUSTIN, C. R. (1954*a*). Reactions of unfertilized mouse eggs to some experimental stimuli. *Experimental Cell Research*, **7**, 277–80.

BRADEN, A. W. H. & AUSTIN, C. R. (1954*b*). Fertilization of the mouse egg and the effect of delayed coitus and of hot-shock treatment. *Australian Journal of Biological Sciences*, **7**, 552–65.

BRADEN, A. W. H., AUSTIN, C. R. & DAVID, H. A. (1954). The reaction of the zona pellucida to sperm penetration. *Australian Journal of Biological Sciences*, **7**, 391–409.

BROWN, S. W. & CHANDRA, H. S. (1973). Inactivation system of the mammalian X chromosome. *Proceedings of the National Academy of Sciences, USA*, **70**, 195–9.

Calarco, P. G., Donahue, R. P. & Szollosi, D. (1972). Germinal vesicle breakdown in the mouse oocyte. *Journal of Cell Science*, **10**, 369–85.

Chang, M. C. & Hunt, D. M. (1956). Effects of proteolytic enzymes on the zona pellucida of fertilized and unfertilized mammalian eggs. *Experimental Cell Research*, **11**, 497–9.

Cross, P. C. & Brinster, R. L. (1970). *In vitro* development of mouse oocytes. *Biology of Reproduction*, **3**, 298–307.

Cross, M. H., Cross, P. C. & Brinster, R. L. (1973). Changes in membrane potential during mouse egg development. *Developmental Biology*, **33**, 412–416.

Defendi, V. & Stoker, M. G. P. (1973). General polyploid produced by Cytochalasin B. *Nature New Biology*, **242**, 24–6.

Donahue, R. P. (1972*a*). Cytogenetic analysis of the first cleavage division in mouse embryos. *Proceedings of the National Academy of Sciences, USA*, **69**, 74–7.

Donahue, R. P. (1972*b*). Fertilization of the mouse oocyte: sequence and timing of nuclear progression to the two-cell stage. *Journal of Experimental Zoology*, **180**, 305–18.

Edwards, R. G. (1958). The number of cells and cleavages in haploid, diploid, polyploid and other heteroploid mouse embryos at 3½ days gestation. *Journal of Experimental Zoology*, **138**, 189–207.

Epel, D (1972). Activation of a Na^{+} dependent amino acid transport system upon fertilization of sea urchin eggs. *Experimental Cell Research*, **72**, 74–89.

Fowler, R. E. & Edwards, R. G. (1973). The genetics of early human development. In *Progress in Medical Genetics*, ed. A. G. Steinberg & A. G. Bearn, vol. 9, pp. 49–112. New York: Grune & Stratton, Inc.

Gardner, R. L. (1971). Manipulations on the blastocyst. *Advances in the Biosciences*, **6**, 279–96.

Gardner, R. L. & Johnson, M. H. (1972). An investigation of inner cell mass and trophoblast tissues following their isolation from the mouse blastocyst. *Journal of Embryology and Experimental Morphology*, **28**, 279–312.

Gardner, R. L., Papaioannou, V. E. & Barton, S. C. (1973). Origin of the ectoplacental cone and secondary giant cells in mouse blastocysts reconstituted from isolated trophoblast and inner cell mass. *Journal of Embryology and Experimental Morphology*, **30**, 561–72.

Graham, C. F. (1970). Parthenogenetic mouse blastocysts. *Nature, London*, **226**, 165–7.

Graham, C. F. (1971). Experimental early parthenogenesis in mammals. *Advances in the Biosciences*, **6**, 87–97.

Graham, C. F. (1972). Genetic manipulation of mouse embryos. *Advances in the Biosciences*, **8**, 263–77.

Graham, C. F. (1974). The production of parthenogenetic mammalian embryos and their use in biological research. *Biological Reviews*, **49**, 399–422.

Graham, C. F. & Deussen, Z. A. (1974). In-vitro activation of mouse eggs. *Journal of Embryology and Experimental Morphology*, **31**, 497–512.

Gwatkin, R. B. L., Williams, D. T., Hartmann, J. F. & Kniazuk, M. (1973). The zona reaction of hamster and mouse eggs: production *in vitro* by a trypsin-like protease from cortical granules. *Journal of Reproduction and Fertility*, **32**, 259–65.

Henderson, S. A. & Edwards, R. G. (1968). Chiasma frequency and maternal age in mammals. *Nature, London*, **218**, 22–8.

Hillman, N. & Tasca, R. J. (1969). Ultrastructural and autoradiographic studies of mouse cleavage stages. *American Journal of Anatomy*, **126**, 151–74.

KAUFMAN, M. H. (1973*a*). Parthenogenesis in the mouse. *Nature, London*, **242**, 475–6.

KAUFMAN, M. H. (1973*b*). Timing of the first cleavage division of the mouse and the duration of its component stages: a study of living and fixed eggs. *Journal of Cell Science*, **12**, 799–808.

KAUFMAN, M. H. (1973*c*). Timing of the first cleavage division of haploid mouse eggs, and the duration of its component stages. *Journal of Cell Science*, **13**, 553–66.

KAUFMAN, M. H. (1973*d*). Cytogenetic analysis of the first cleavage mitosis in parthenogenetic and fertilized eggs of the mouse. PhD Thesis. University of Cambridge.

KAUFMAN, M. H. & GARDNER, R. L. (1974). Diploid and haploid mouse parthenogenetic development following *in vitro* activation and embryo transfer. *Journal of Embryology and Experimental Morphology*, **31**, 635–42.

KAUFMAN, M. H. & SURANI, M. A. H. (1974). The effect of osmolarity on mouse parthenogenesis. *Journal of Embryology and Experimental Morphology*, **31**, 513–26.

KOMAR, A. (1973). Parthenogenetic development of mouse eggs activated by heat-shock. *Journal of Reproduction and Fertility*, **35**, 433–43.

MCGAUGHEY, R. W. & CHANG, M. C. (1971). Chromosomes at prometaphase and metaphase of the first cleavage in mouse and hamster eggs. *Journal of Experimental Zoology*, **177**, 31–40.

MARSTON, J. H. & CHANG, M. C. (1964). The fertilizable life of ova and their morphology following delayed insemination in mature and immature mice. *Journal of Experimental Zoology*, **155**, 237–52.

MINTZ, B. (1962). Experimental study of the developing mammalian egg: removal of the zona pellucida. *Science*, **138**, 594–5.

MINTZ, B. & GEARHART, J. D. (1973). Subnormal zona pellucida changes in parthenogenetic mouse embryos. *Developmental Biology*, **31**, 178–84.

PHILLIPS, R. J. S., HAWKER, S. G. & MOSELEY, H. J. (1973). Bare-patches, a new sex-linked gene in the mouse, associated with a high production of XO females. I. A preliminary report of breeding experiments. *Genetical Research*, **22**, 91–9.

PHILLIPS, R. J. S. & KAUFMAN, M. H. (1974). Bare-patches, a new sex-linked gene in the mouse, associated with a high production of XO females. II. Investigations into the nature and mechanism of the XO production. *Genetical Research*, in press.

PIENKOWSKI, M. & KOPROWSKI, H. (1974). Study of the growth regulation of pre-implantation mouse embryos using Concanavalin A. *Journal of the National Cancer Institute*, in press.

ROHRBORN, G. & HANSMANN, I. (1971). Induced chromosome aberrations in unfertilized oocytes of mice. *Humangenetik*, **13**, 184–98.

ROHRBORN, G., KUHN, O., HANSMANN, I. & THON, K. (1971). Induced chromosome aberrations in early embryogenesis of mice. *Humangenetik*, **11**, 316–22.

SCHULTZ, S. G. & CURRAN, P. F. (1970). Coupled transport of sodium and organic solutes. *Physiological Reviews*, **50**, 637–718.

SMITHBERG, M. (1953). The effect of different proteolytic enzymes on the zona pellucida of mouse ova. *Anatomical Record*, **117**, 554.

SNOW, M. H. L. (1973). Tetraploid mouse embryos produced by cytochalasin B during cleavage. *Nature, London*, **244**, 513–15.

SOLTER, D., BICZYSKO, W., GRAHAM, C., PIENKOWSKI, M. & KOPROWSKI, H. (1974). Ultrastructure of early development of mouse parthenogenones. *Journal of Experimental Zoology*, **188**, 1–23.

STEVENS, L. C. & VARNUM, D. S. (1974). The development of teratomas from parthenogenetically activated ovarian mouse eggs. *Developmental Biology*, **37**, 369–80.

SZOLLOSI, D. (1971). Morphological changes in mouse eggs due to aging in the fallopian tube. *American Journal of Anatomy*, **130**, 209–26.

TARKOWSKI, A. K. (1962). Inter-specific transfers of eggs between rat and mouse. *Journal of Embryology and Experimental Morphology*, **10**, 476–95.

TARKOWSKI, A. K. (1966). An air-drying method for chromosome preparations from mouse eggs. *Cytogenetics*, **5**, 394–400.

TARKOWSKI, A. K. (1971). Recent studies on parthenogenesis in the mouse. *Journal of Reproduction and Fertility*, Supplement, **14**, 31–9.

TARKOWSKI, A. K., WITKOWSKA, A. & NOWICKA, J. (1970). Experimental parthenogenesis in the mouse. *Nature, London*, **266**, 162–5.

TARKOWSKI, A. K. & WROBLEWSKA, J. (1967). Development of blastomeres of mouse eggs isolated at the 4- and 8-cell stage. *Journal of Embryology and Experimental Morphology*, **18**, 155–80.

WITKOWSKA, A. (1973*a*). Parthenogenetic development of mouse embryos *in vivo*. I. Preimplantation development. *Journal of Embryology and Experimental Morphology*, **30**, 519–45.

WITKOWSKA, A. (1973*b*). Parthenogenetic development of mouse embryos *in vivo*. II. Postimplantation development. *Journal of Embryology and Experimental Morphology*, **30**, 547–60.

ZAMBONI, L. (1970). Ultrastructure of mammalian oocytes and ova. *Biology of Reproduction*, Supplement, **2**, 44–63.

ZAMBONI, L. (1971). *Fine Morphology of Mammalian Fertilization*. New York: Harper & Row.

ZAMBONI, L. (1972). Fertilization in the mouse. In *Biology of Mammalian Fertilization and Implantation*, ed. K. S. Moghissi & E. S. E. Hafez, pp. 213–62. Springfield, Illinois: C. C. Thomas.

ZAMBONI, L., CHAKRABORTY, J. & SMITH, D. M. (1972). First cleavage division of the mouse zygote. An ultrastructural study. *Biology of Reproduction*, **7**, 170–93.

CULTURE OF MAMMALIAN EMBRYOS OVER THE IMPLANTATION PERIOD

BY A. McLAREN* AND H. C. HENSLEIGH*

ARC Unit of Animal Genetics & Institute of Animal Genetics
West Mains Road,
Edinburgh, EH9 3JN

IN-VITRO DEVELOPMENT BEFORE AND AFTER IMPLANTATION

The last ten years have seen major advances in the culture of mammalian embryos during the pre-implantation period. Mouse oocytes have undergone maturation *in vitro*, followed by fertilisation *in vitro* and cleavage *in vitro*; after transfer to the uteri of foster-mothers at the blastocyst stage, successful development and birth have ensued (Mukherjee & Cohen, 1970; Hoppe & Pitts, 1973). Use of chemically defined media based on Krebs–Ringer bicarbonate solutions has allowed sophisticated biochemical analyses to be carried out, at least in the mouse and rabbit (see Whitten, 1971; Biggers & Stern, 1973). Similar culture techniques are now being extended to other species, including man.

Once the embryo has implanted in the uterus and has entered on the period of organogenesis, it can be removed and once more maintained in culture for a day or two with good growth and differentiation. Thus in the rat, late egg-cylinder stages (8 days *post coitum*) can be grown *in vitro* through to the establishment of a yolk sac circulation ($10\frac{1}{2}$ days *p.c.*), 10-day embryos can be grown to the hind limb bud stage ($11\frac{1}{2}$ days *p.c.*), and so on up to $13\frac{1}{2}$-day foetuses, which develop in culture for a further 24 hours. For the later stages, circulating medium has to be used to allow adequate oxygenation of the tissues. Comparable results have been obtained with the mouse, rabbit, opossum and other species. The types of analysis familiar in amphibian and chick material are beginning now to be extended to mammalian in-vitro systems, for example in the fields of experimental embryology (e.g. Deuchar, 1971) and teratology (e.g. Morriss & Steele, 1974). A comprehensive account of studies on mammalian embryos *in vitro* during the period of organogenesis has recently been published by New (1973), who himself pioneered many of the relevant culture systems.

* Present address: MRC Mammalian Development Unit, University College, Wolfson House, 4 Stephenson Way, London NW1 2HE.

IN-VITRO DEVELOPMENT OVER THE IMPLANTATION PERIOD

There remains a crucial period, lasting two to three days in mouse and rat, during which time *in vivo* the embryo is establishing its connections with the maternal organism. Attempts to achieve *in vitro* anything approaching normal embryonic development during this period are still in their infancy. The claims of Petrucci, to have obtained prolonged embryonic development of human eggs fertilised *in vitro*, created a sensation when they appeared in the popular scientific press in the late 1950s, but although they earned excommunication for their author they gained little scientific credence.

In defined media, on a glass or plastic surface, development of the pre-implantation mouse embryo ceases at the blastocyst stage. The zona pellucida ruptures, the blastocyst hatches, cell division stops, and no further development occurs, though the dormant blastocysts may survive for at least ten days in culture (Gwatkin, 1966*a*). An early indication that the critical factor might lie in the culture medium came from studies using complex media. Bryson (1964) grew cleavage stage embryos in diffusion chambers in the peritoneal cavity of recipient mice: at the blastocyst stage they hatched from the zona and attached to the membrane, showing extensive trophoblast outgrowth and some inner cell mass development, including an indication of differentiation into endoderm and ectoderm. Mintz (1964) mentioned that some cellular outgrowth occurred when mouse blastocysts were placed in a culture medium containing 50 % calf serum.

Cole & Paul (1965) reported that mouse blastocysts maintained in Waymouth's medium supplemented with calf and human serum were capable of extensive trophoblast outgrowth on a glass or plastic surface. A small proportion of blastocysts (5–10 % if isolated directly from the uterus, 2 % if cleaved *in vitro*) formed 'structures superficially resembling "egg-cylinder" stages'. The presence of a feeder layer of HeLa cells in no way improved post-blastocyst development. More recently, Salomon & Sherman (1974) have examined mouse embryo development on a cellular monolayer, either of mouse uterine cells, or of one of several established cell types from different species. The outgrowing trophoblast displaced the underlying cells, perhaps by cytolytic as well as mechanical activity. Attachment took somewhat longer than on a plastic surface; inner cell mass development was less; trophoblast outgrowth was similar.

Rabbit embryos show much more differentiation *in vitro* than do mouse embryos, perhaps because *in vivo* they achieve a more advanced stage of development before implantation. If the zona pellucida was removed by enzyme treatment, the oxygen tension kept high by maintaining the embryonic disc near the gas phase interface, and trophoblast deliberately prevented from attaching and outgrowing, embryonic development proceeded to closure of the amniotic folds and formation of a beating heart (Cole & Paul, 1965). Several primary cell strains were established, using isolated embryonic discs grown on a collagen substrate (Cole, Edwards & Paul, 1965).

Guinea-pig blastocysts cultured in the presence of serum and chick embryo extract showed extensive trophoblast outgrowth, and occasional expansion of the yolk sac, but no differentiation of the inner cell mass (Blandau, 1971).

The requirements for outgrowth of trophoblast from mouse blastocysts *in vitro* were studied systematically by Gwatkin & Meckley (1965, 1966) and Gwatkin (1966*a*, *b*). The addition of 10 % foetal calf serum to a standard tissue culture medium permitted outgrowth, and differentiation of a population of trophoblast giant cells with prominent nucleoli. Cellular proliferation declined after about four days, but the cultures survived for two weeks or more. The calf serum evidently satisfied both a macromolecular requirement and a requirement for amino acids. The macromolecular requirement could be met also by various fractions (but not all) of dialysed calf serum, including the α-globulin fetuin, but neither bovine plasma albumen nor the synthetic polyvinylpyrrolidone was capable of supporting outgrowth. Certain amino acids proved to be essential (arginine, cystine, histidine, leucine, threonine), but neither glutamine nor vitamins were required. The contrast with pre-implantation development, which can take place from the 2-cell to the blastocyst stage in the complete absence of exogenous amino acids (Cholewa & Whitten, 1970), is striking.

The dependence of mouse trophoblast outgrowth on the presence of an inner cell mass, established for the situation *in vivo* by Gardner & Johnson (1972), has been confirmed *in vitro* by Ansell & Snow (1974). Normal blastocysts outgrown in medium containing foetal calf serum show an increase in the total number of trophoblast cells, while retaining a population of small, diploid cells. In blastocysts deprived of their inner cell mass by prior culture in [^{3}H]thymidine, all the cells undergo giant cell transformation, but the total number never exceeds that in the pre-hatching blastocyst.

Recently, Jenkinson & Wilson (1973) have suggested that calf serum may facilitate trophoblast outgrowth by providing a protein coat on the surface of the culture vessel rather than by playing any nutritional role. They found that outgrowth from mouse blastocysts would occur without any serum in the medium on a surface of reconstituted collagen, but on an agar surface no outgrowth occurred even when foetal calf serum was present.

TOWARDS A THIRD DIMENSION

In vivo, trophoblast invades the uterine stroma during implantation, forming a three-dimensional interlocking network of maternal and embryonic tissues. On a glass or plastic surface, however, trophoblast grows out in a two-dimensional monolayer.

In an attempt to provide a more normal situation for in-vitro growth over the implantation period, Glenister (1961*a*,*b*) maintained strips of rabbit endometrium in organ culture, and placed 6½-day rabbit blastocysts on top of them, after removal of the zona pellucida. The trophoblast cells penetrated through the uterine epithelium and into the stroma, forming a syncitium which by electron microscopy appeared similar to the analogous structure *in vivo* (Glenister, 1964). Provided the oxygen concentration was high enough, more than 30 % of blastocysts produced well differentiated embryonic structures, such as pulsating heart, somites and neural tube, but the organisation of the embryos was 'profoundly disturbed' (Glenister, 1962). Attachment and trophoblast invasion occurred equally readily whether the endometrium was taken from a pregnant or non-pregnant doe, whether it was oriented with the epithelial surface upwards or downwards, or even when endometrium was replaced by strips of bladder wall; when chemically defined culture medium was used, the behaviour of the trophoblast was unaffected by the addition of oestrogen and progesterone; and the orientation of the blastocyst, with embryonic or abembryonic pole in contact with the endometrium, also proved to be irrelevant. The organ culture system was, therefore, only partially successful as a model for implantation. Attempts to repeat the experiments using other species, including the mouse, proved unsuccessful (Glenister, 1967).

A more successful in-vitro model for implantation in the mouse was devised by Grant (1973). Entire uterine horns from sexually immature mice were maintained in organ culture, in chemically defined medium supplemented with foetal calf serum, and 3½-day blastocysts were inserted into their lumina. Attachment occurred, followed by trophoblast invasion through the epithelium and into the uterine stroma in a small percentage of blastocysts. Cultures were terminated after two to three days. Embryonic development included the differentiation and downgrowth of endoderm, but soon became abnormal owing, perhaps, to the collapse of the yolk sac. No obvious Reichert's membrane was seen between endoderm and trophoblast. The addition of a high concentration of progesterone to the culture medium increased the proportion of explants showing trophoblast invasion of the epithelial and stromal tissues; no in-vitro effect of oestrogen was detected. The optimal oxygen concentration for both trophoblast invasion and embryonic development was 26–40 %. The uterine stroma did not undergo a decidual cell reaction.

Using an alternative approach, Jenkinson & Wilson (1970) attempted to provide a three-dimensional support system by injecting mouse blastocysts into a network of bovine lens fibres, and obtained rather consistent development up to the early egg-cylinder stage, with differentiation of proximal and distal endoderm, and even formation of a cavity thought to correspond to the pro-amniotic cavity. Development became disorganised before primitive streak formation, and mesoderm was never seen.

HSU'S NEWS

In 1971 came the dramatic first report from Hsu of mouse embryo development from the early blastocyst to the 'beating heart' stage, including differentiation of red blood cells and contractile elements, *in vitro*. Later reports (Hsu, 1972, 1973; Hsu, Baskar, Stevens & Rash, 1974) extended the period of culture, described more precisely the structures that developed, and gave further details of the technique. Development *in vitro* could begin at the 2-cell stage; at the blastocyst stage the embryos were transferred to collagen-coated plastic Petri dishes and the culture medium (Eagle's minimal essential) was supplemented with 10 % foetal calf serum.

Under these conditions (Hsu, 1971) attachment and trophoblast outgrowth took place, with the inner cell mass developing into an early egg-cylinder covered by endoderm. In 80–95 % of outgrowths development ceased at this stage, but in the remaining few, the embryo rapidly expanded to form a fluid-filled vesicle 2–3 mm in diameter, with a protruding ectoplacental cone attached to the underlying collagen. Extra-embryonic ectoderm could be distinguished from embryonic, the relations of amniotic cavity, exocoel and ectoplacental cavity appeared normal, and blood islands were formed on the vesicle after 10–14 days' culture, with primitive blood vessels leading to a rhythmically contracting area. Although initially described as trophectoderm and distal endoderm, the walls of the vesicle were later identified as extra-embryonic ectoderm and proximal endoderm (Hsu *et al.*, 1974), making the formation of blood islands in the intervening mesoderm more comparable to the situation *in vivo*. Traces of distal endoderm and Reichert's membrane material were found in association with the ectoplacental cone.

The variation in the ability of different batches of foetal calf serum to support the differentiation of endoderm and ectoderm was stressed (Hsu, 1972). In Hsu's later work foetal calf serum was replaced by calf serum supplemented, after loss of the zona pellucida, with 20 % human cord serum, and the medium changed daily. A small minority of embryos (1–3 %, or 3–5 %, if cultured from the 2-cell or blastocyst stage respectively) would then show fairly normal development up to about the 10-somite stage, the transition from the egg-cylinder stage to an embryo with neural tube and somites being apparently dependent on the presence of cord serum (Hsu, 1973). Unlike foetal calf serum, different batches of human cord serum gave consistent results.

In Hsu's system, the ectoplacental cone develops at the point of attachment of the blastocyst to the substrate, and the egg-cylinder grows upwards. The trophoblast covering the upper surface of the egg-cylinder migrates away, and development of giant cells, distal endoderm and Reichert's membrane, all associated *in vivo* with mural trophoblast, is limited and abnormal. The egg-cylinder is thus left free to expand and differentiate, unencumbered by any covering, in contrast to the situation in the classical 'trophoblast outgrowth' system of Gwatkin, in which the inner cell mass is flattened by the surrounding trophoblast and undergoes little further development.

ANALYSIS

Hsu's approach has been innovatory rather than analytical. His technique, or some modification of it, has now been adopted by other laboratories. The low proportion of embryos that develop to the somite stage, even in the most successful experiments, suggests that factors as yet unknown are exerting a critical influence on development.

Spindle & Pedersen (1973) investigated the amino acid requirements for trophoblast outgrowth and embryonic development in the mouse, using Eagle's basal medium and a collagen-coated surface in a system essentially similar to that of Hsu. They found that hatching, attachment and outgrowth required the presence of all the 'essential' amino acids except for isoleucine, but was unaffected by 'non-essential' amino acids except for glutamine. In comparison with the earlier studies by Gwatkin (1966*b*), the requirements for lysine, methionine, tyrosine and valine proved more stringent, and for arginine, leucine and histidine, less stringent. Even if all amino acids were present in the medium at optimal concentrations, serum was still required for extensive trophoblast outgrowth and for further differentiation of the inner cell mass. In medium supplemented with 1 % dialysed foetal calf serum, the blastocyst cavity collapsed and no differentiation of the inner cell mass was seen, but when the serum concentration was raised to 10 %, differentiation of ectoderm and endoderm occurred. Under optimal culture conditions the collagen substrate proved unnecessary. In the same culture system that gave good post-blastocyst development of trophoblast and inner cell mass from genetically normal embryos, presumed A^y/A^y embryos (homozygous lethal yellow) showed only limited trophoblast outgrowth and complete disintegration of the inner cell mass (Pedersen, 1974).

If mouse blastocysts were treated with pronase to remove the zona pellucida, and subsequently incubated for 30 minutes in 0.05 % trypsin, more than 50 % proved capable after five days of giving rise to egg-cylinders with proximal endoderm, and ectoderm differentiated into embryonic and extra-embryonic parts (Pienkowski, Solter & Koprowski, 1974). Culture was in Eagle's minimal essential medium, pH 7.4, supplemented with 10 % foetal calf serum, on a plastic surface. The rate of egg-cylinder formation fell to 24 % without trypsin incubation, and to 14 % without removal of the zona pellucida. Culture beyond five days was not examined. In contrast to the findings of Hsu, the blastocysts were reported to attach to the substrate by the abembryonic pole, with the egg-cylinders growing downwards rather than upwards. When the pH was lowered to 7.0, trophoblast outgrowth occurred, but no egg-cylinders developed.

EXPERIMENTS IN PROGRESS

We have recently made an attempt to optimise and standardise the culture conditions required for the first stage of post-implantation embryonic development in the mouse, namely growth of the egg-cylinder and differentiation of endoderm and ectoderm. Blastocysts are taken from randomly bred Q-strain females $3\frac{1}{2}$ days *post coitum*, and grown in 30 × 10 mm Falcon plastic Petri dishes in 2 ml of medium.

Attachment of the blastocysts is predominantly at the abembryonic pole. Trophoblast cells undergo giant cell transformation, rapidly attaching to the plastic surface and stretching into a monolayer. The inner cell mass, usually located near the centre of the trophoblast outgrowth, differentiates into endoderm and ectoderm, as in the egg-cylinder stage *in vivo*. Further development is variable: a few develop vesicles resembling a yolk sac enveloping an exocoelom; these are often attached by a base resembling an ectoplacental cone consisting of giant trophoblast cells, parietal endoderm and/or ectoderm, and pockets of Reichert's membrane. Structures resembling an embryo proper are very seldom seen.

Variables that we have so far examined include the type of culture medium, its amino acid content, the surface for attachment, the source of serum and its concentration. Development was assessed using a scoring system similar to that of Spindle & Pedersen (1973). Scores were allotted separately for trophoblast and for inner cell mass development, and subsequently combined to give a single figure. Results in terms of these scores are given in Fig. 1.

In combination with 20 % foetal calf serum, Eagle's minimal essential medium (MEM) was compared with NCTC-135 and with Eagle's basal medium (BME), alone or supplemented with amino acids (AA) to achieve the concentrations found to be optimal by Spindle & Pedersen (1973). MEM proved significantly better than either of the other media tested, and the amino acid supplements did not enhance development in BME (Fig. 1*a*). Coating the plastic surface of the Petri dish with reconstituted collagen, either dry or as a moist layer, had no effect on development (Fig. 1*b*). Varying the concentration of foetal calf serum from 0 to 50 % gave a maximum score with 20 % serum, though the improvement over the scores at 10 % and 50 % was not statistically significant at the 5 % level (Fig. 1*c*). Different batches of foetal calf serum gave significantly different scores (Fig. 1*d*). In further trials, the control embryos were cultured on a plastic surface, using MEM supplemented with 20 % foetal calf serum from a good batch.

The marked variation in the ability of different batches of foetal calf serum to support development should be amenable to biochemical analysis, but so far the critical factor or factors have not been identified. The sera may contain varying proportions of steroid hormones, known to affect implantation *in vivo*; it seems unlikely, however, that these stimulate or inhibit embryonic development *in vitro*, since Salomon & Sherman (1974) demonstrated that removal of all detectable steroids by treatment with dextran-coated norit did not affect the ability of foetal calf serum to support attachment and outgrowth of mouse blastocysts. We have tested 'good' and 'bad' batches of foetal calf serum for protein patterns on polyacrylamide gel electrophoresis and for amino acid concentrations, but have so far failed to find any consistent differences. Total protein content varies from batch to batch, but the relative ability of different batches to support

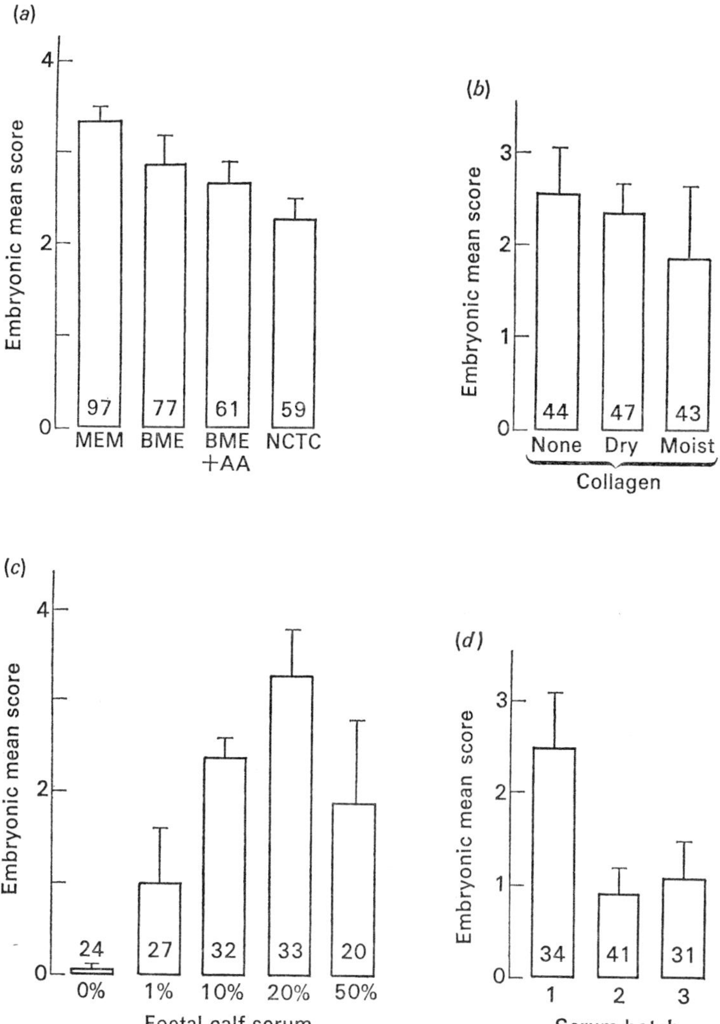

Fig. 1. Relative outgrowth of mouse embryos after 72 h in different culture conditions. Number of embryos in each group is given within the appropriate rectangle. Lines indicate one standard error. (*a*) Comparison of different culture media (20% foetal calf serum used with each medium). (*b*) Use of reconstituted rat tail collagen as a surface for outgrowth (culture medium was MEM and 20% foetal calf serum). (*c*) Effect of varying the percentage of foetal calf serum (culture medium was MEM). (*d*) Comparison of development in three batches of foetal calf serum (all tested at 20% in MEM).

outgrowth remains the same when they are adjusted to a standard protein content.

The overall score gives a useful measure for comparing different treatments at a gross level, but does not reveal whether, for example, development of trophoblast has been favoured at the expense of inner cell mass. We therefore calculated also the percentage of blastocysts that, during three days of culture, (1) hatched from the zona pellucida, (2) showed trophoblast outgrowth, (3) showed some development of the inner cell mass (Plate 1*a*, *b*) or (4) showed extensive development of the inner cell mass (larger than about 4000 μm^2). Embryos with extensive development of the inner cell mass usually showed differentiation of ectoderm and endoderm (Plate 1*c*, *d*), and sometimes also fluid-filled vesicles bounded by a yolk sac membrane.

Figure 2 illustrates the application of this percentage scale to cultures using foetal calf serum (controls), adult bovine serum, foetal and adult mouse serum, and bovine plasma albumen. About 80 % of the control embryos hatched, almost all of these showed trophoblast outgrowth, and about three-quarters of them also gave inner cell mass (ICM) development, of which about a fifth were judged to constitute extensive development.

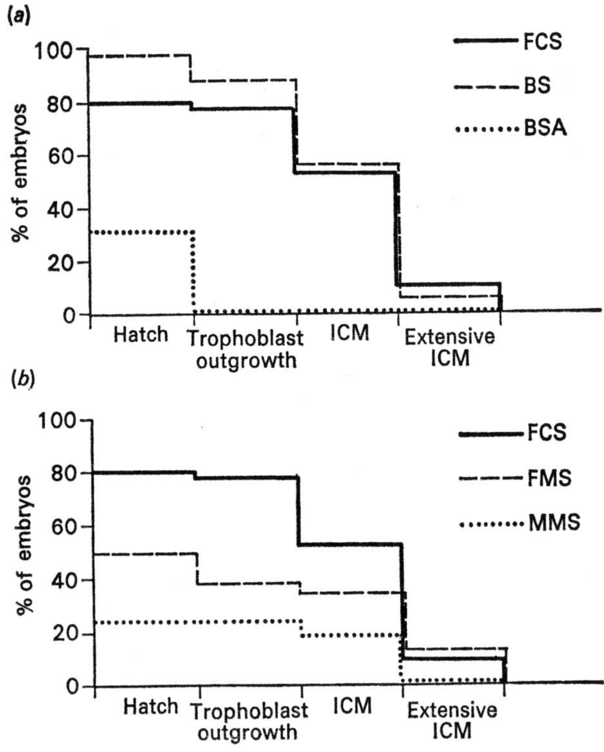

Fig. 2. (*a*) The percentage of mouse embryos reaching progressive stages of development in bovine serum (BS) and bovine serum albumin (BSA) in relation to a control group in foetal calf serum (FCS). (*b*) The percentage of mouse embryos reaching progressive stages of development in foetal mouse serum (FMS) and maternal mouse serum (MMS) in relation to a control group in foetal calf serum (FCS).

With bovine serum albumen about a third hatched, but almost no further development was seen, the embryos remaining as expanded blastocysts. Adult bovine serum gave a better hatching rate than the controls, and a somewhat lower percentage of extensive inner cell mass development. Foetal mouse serum gave a much lower hatching rate than the controls, and adult mouse serum a lower rate still. Most of those that hatched showed both trophoblast and some inner cell mass development. The lower hatching rate with mouse than with calf serum is statistically significant ($P < 0.001$). Mouse serum has been reported to be toxic to blastocysts in culture (Sherman & Chew, 1972); on the other hand, Dunn (1974) found that serum of adult mice, regardless of sex and strain, would support blastocyst attachment and outgrowth unless the serum donors had been carrying the ascites form of a teratoma.

Figure 3 gives the same results expressed in terms of the percentage of hatched embryos. Neither the differences in trophoblast outgrowth nor

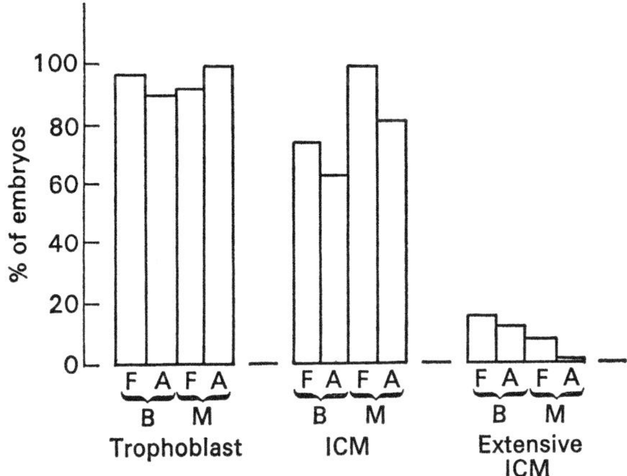

Fig. 3. Comparison of the percentage of hatched embryos developing trophoblast outgrowth, some inner cell mass, and extensive inner cell mass (see text), in medium supplemented with foetal (F) or adult (A) bovine (B) or mouse (M) serum.

those in extensive inner cell mass development were statistically significant. Mouse serum appeared to support inner cell mass development better than did bovine serum ($P < 0.05$), perhaps because a higher proportion of the less vigorous embryos had failed to hatch in mouse serum. Both for mouse and for bovine serum, more inner cell mass development occurred in the presence of foetal than adult serum; the percentages (Fig. 3) were not significantly different, but development continued further with foetal serum, as judged by the appearance of the outgrowths (Plate 1*a*, *b*).

GROWTH RATE *IN VITRO*

To determine the growth rate of explanted mouse embryos in our culture conditions, the area of trophoblast outgrowth and the extent of the inner cell mass region were measured from the 2nd day to the 6th day of culture, using a micrometer eyepiece grid. In a second experiment, the measurements were continued to the 8th day. At the end of the culture period the embryos were freeze-dried, removed from the plastic culture dish and weighed on a quartz-fibre balance (Hensleigh & Weitlauf, 1974).

Four groups of Q-strain embryos were examined: (1) blastocysts taken directly from the uterus on the 4th day of pregnancy, (2) blastocysts cultured for 48 hours from the 8-cell stage after removal of the zona pellucida, (3) blastocysts cultured from the 8-cell stage after zona removal and aggregation in pairs, and (4) blastocysts cultured from the 8-cell stage after zona removal and aggregation in fours. Culture from the 8-cell stage, removal of the zona pellucida and aggregation were as described by Bowman & McLaren (1970).

The two experiments gave closely similar results, as shown in Fig. 4.

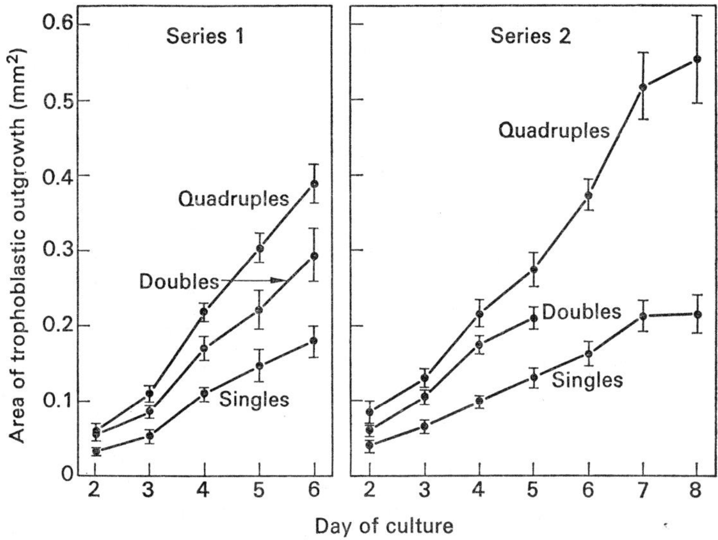

Fig. 4. Change in area of trophoblast outgrowths of single blastocysts cultured from the 8-cell stage after removal of the zona pellucida (singles), blastocysts cultured from the 8-cell stage after zona removal and aggregation in pairs (doubles), and blastocysts cultured from the 8-cell stage after zona removal and aggregation in fours (quadruples). Day 0 of culture refers to the day on which the blastocysts were transferred to the outgrowth medium. In Series 2, the doubles were terminated on day 5 because the cultures became infected. Lines indicate one standard error above and below the mean.

The area covered by the outgrowing trophoblast increased steadily throughout the period of measurement. Unlike the trophoblast outgrowth, which is a monolayer of cells, the inner cell mass develops in three dimensions. Not surprisingly, little change was seen in the mean area of the inner cell mass during the period of measurement. However, dry weights of the outgrowths were correlated with the area of the inner cell mass ($P < 0.01$) while showing no correlation with the area of the trophoblast.

Blastocysts taken directly from the uterus started to outgrow earlier than those which had been cultured for 48 hours. This may reflect the lower metabolic rate of embryos cultured up to the blastocyst stage in our system (Menke & McLaren, 1970). Once outgrowth had begun, the rate was similar in both groups. Comparisons of the single, double and quadruple embryos (Fig. 4) showed that growth was proportional to the number of cells present initially. When the areas were expressed on a logarithmic scale, the three regression lines were parallel to one another. At the end of the culture period, the weights were significantly related to the initial number of embryos in the aggregate ($P < 0.01$). The mean weight of the double embryos (counting only those in which both trophoblast and inner cell mass had developed) was 1.7 times that of the singles, and the mean of the quadruple embryos was 1.9 times that of the doubles. There was thus little sign of the growth regulation observed by Buehr & McLaren (1974) in the egg-cylinders of double embryos *in vivo*. In comparison with the weight of control blastocysts, the embryos increased approximately eight-fold in culture, during a period of six days, indicating a doubling time of about 48 hours.

CONCLUSIONS

Results from a number of laboratories indicate that both mouse and rabbit embryos can undergo extensive growth and differentiation *in vitro* over the period when implantation would be occurring *in vivo*. It is therefore unlikely that there is any special feature of the uterine environment that cannot be reproduced *in vitro*. On the other hand, embryonic development *in vitro* is always abnormal and to a greater or lesser extent disorganised, and the variable results obtained, even in the same laboratory and with a single batch of embryos, bear witness to the inadequacy of our understanding of the factors regulating development during this critical period.

With the techniques available at present, it should be possible to extend to the post-implantation period some of the biochemical studies that have yielded so much information on pre-implantation embryos. The activation of the genome, and the appearance of paternal isozyme markers, might be rewarding to investigate. It is also during this period that X-chromosome activation or inactivation is thought to occur. If the culture system could be improved so that embryonic development occurred regularly and repeatably to the same extent as is at present seen only on rare occasions, many problems concerned with the emergence of the basic architecture of the foetus, including the establishment of bilateral symmetry, would become accessible to experimental analysis.

We are grateful to the Ford Foundation for financial support.

REFERENCES

ANSELL, J. D. & SNOW, M. H. L. (1974). The development of trophoblast in the absence of the inner cell mass. *Journal of Embryology and Experimental Morphology*, in press.

BIGGERS, J. D. & STERN, S. (1973). Metabolism of the preimplantation mammalian embryo. *Advances in Reproductive Physiology*, **6**, 1–59.

BLANDAU, R. J. (1971). Culture of guinea pig blastocyst. In *The Biology of the Blastocyst*, ed. R. J. Blandau, pp. 59–69. Chicago: University of Chicago Press.

BOWMAN, P. & McLAREN, A. (1970). Viability and growth of mouse embryos after *in vitro* culture and fusion. *Journal of Embryology and Experimental Morphology*, **23**, 693–704.

BRYSON, D. L. (1964). Development of mouse eggs in diffusion chambers. *Science*, **144**, 1351–3.

BUEHR, M. & McLAREN, A. (1974). Size regulation in chimaeric mouse embryos. *Journal of Embryology and Experimental Morphology*, **31**, 229–34.

CHOLEWA, J. A. & WHITTEN, W. K. (1970). Development of two-cell mouse embryos in the absence of a fixed-nitrogen source. *Journal of Reproduction and Fertility*, **22**, 553–5.

COLE, R. J., EDWARDS, R. G. & PAUL, J. (1965). Cytodifferentiation in cell colonies and cell strains derived from cleaving ova and blastocysts of the rabbit. *Experimental Cell Research*, **37**, 501–4.

COLE, R. J. & PAUL, J. (1965). Properties of cultured preimplantation mouse and rabbit embryos, and cell strains derived from them. In *Ciba Foundation Symposium* on *Preimplantation Stages of Pregnancy*, ed. G. E. W. Wolstenholme & M. O'Connor, pp. 82–112. London: J. & A. Churchill.

DEUCHAR, E. M. (1971). The mechanism of axial rotation in the rat embryo: an experimental study *in vitro*. *Journal of Embryology and Experimental Morphology*, **25**, 189–201.

DUNN, G. R. (1974). Inhibition of blastocyst outgrowth *in vitro* by serum from mice with ascites teratoma. *Journal of Reproduction and Fertility*, **39**, 93–5.

GARDNER, R. L. & JOHNSON, M. H. (1972). An investigation of inner cell mass and trophoblast tissues following their isolation from the mouse blastocyst. *Journal of Embryology and Experimental Morphology*, **28**, 279–312.

GLENISTER, T. W. (1961*a*). Organ culture as a new method for studying the implantation of mammalian blastocysts. *Proceedings of the Royal Society, London*, Ser. B, **154**, 428–31.

GLENISTER, T. W. (1961*b*). Observations on the behaviour in organ culture of rabbit trophoblast from implanting blastocysts and early placentae. *Journal of Anatomy*, **95**, 474–84.

GLENISTER, T. W. (1962). Embryo–endometrial relationships during nidation in organ culture. *Journal of Obstetrics and Gynaecology of the British Commonwealth*, **69**, 809–14.

GLENISTER, T. W. (1964). Ultrastructure of earliest embryo–endometrial contacts and attachments *in vivo* and *in vitro*. *Journal of Anatomy*, **98**, 470.

GLENISTER, T. W. (1967). Organ culture and its combination with electron microscopy in the study of nidation processes. Proceedings of the 5th World Congress on Fertility and Sterility. *Excerpta Medica International Congress Series*, **133**, 385–94.

GRANT, P. S. (1973). The effect of progesterone and oestradiol on blastocysts cultured within the lumina of immature mouse uteri. *Journal of Embryology and Experimental Morphology*, **29**, 617–38.

GWATKIN, R. B. L. (1966*a*). Defined media and development of mammalian eggs *in vitro*. *Annals of the New York Academy of Science*, **139**, 79–90.

GWATKIN, R. B. L. (1966*b*). Amino acid requirements for attachment and outgrowth of the mouse blastocyst *in vitro*. *Journal of Cellular Physiology*, **68**, 335–44.

GWATKIN, R. B. L. & MECKLEY, P. E. (1965). *In vitro* culture of mouse ova beyond the blastocyst stage: effect of substrate and medium components. *Journal of Cell Biology*, **27**, 136A–137A.

GWATKIN, R. B. L. & MECKLEY, P. E. (1966). Chromosomes of the mouse blastocyst following its attachment and outgrowth *in vitro*. *Annales Medicinae Experimentalis et Biologicae Fennicae*, **44**, 125–7.

HENSLEIGH, H. C. & WEITLAUF, H. M. (1974). Effect of delayed implantation on dry weight and lipid content of mouse blastocysts. *Biology of Reproduction*, **10**, 315–20.

HOPPE, P. C. & PITTS, S. (1973). Fertilization *in vitro* and development of mouse ova. *Biology of Reproduction*, **8**, 420–6.

HSU, Y.-C. (1971). Post-blastocyst differentiation *in vitro*. *Nature, London*, **231**, 100–2.

HSU, Y.-C. (1972). Differentiation *in vitro* of mouse embryos beyond the implantation stage. *Nature, London*, **239**, 200–2.

HSU, Y.-C. (1973). Differentiation *in vitro* of mouse embryos to the stage of early somite. *Developmental Biology*, **33**, 403–11.

HSU, Y.-C., BASKAR, J., STEVENS, L. C. & RASH, J. E. (1974). Development *in vitro* of mouse embryos from the two-cell egg stage to the early somite stage. *Journal of Embryology and Experimental Morphology*, **31**, 235–45.

JENKINSON, E. J. & WILSON, I. B. (1970). *In vitro* support system for the study of blastocyst differentiation in the mouse. *Nature, London*, **228**, 776–8.

JENKINSON, E. J. & WILSON, I. B. (1973). *In vitro* studies on the control of trophoblast outgrowth in the mouse. *Journal of Embryology and Experimental Morphology*, **30**, 21–30.

MENKE, T. M. & MCLAREN, A. (1970). Mouse blastocysts grown *in vivo* and *in vitro*: CO_2 production and trophoblast outgrowth. *Journal of Reproduction and Fertility*, **23**, 117–27.

MINTZ, B. (1964). Formation of genetically mosaic mouse embryos, and early development of 'lethal (t^{12}/t^{12})–normal' mosaics. *Journal of Experimental Zoology*, **157**, 273–92.

MORRISS, G. M. & STEELE, C. E. (1974). The effect of excess vitamin A on the development of rat embryos in culture. *Journal of Embryology and Experimental Morphology*, in press.

MUKHERJEE, A. B. & COHEN, M. M. (1970). Development of normal mice by *in vitro* fertilization. *Nature, London*, **228**, 472–3.

NEW, D. A. T. (1973). Studies on mammalian fetuses *in vitro* during the period of organogenesis. In *The Mammalian Fetus* in vitro, ed. C. R. Austin, pp. 15–65. London: Chapman & Hall.

PEDERSEN, R. A. (1974). Development of lethal yellow (A^y/A^y) mouse embryos *in vitro*. *Journal of Experimental Zoology*, **188**, 307–20.

PIENKOWSKI, M., SOLTER, D. & KOPROWSKI, H. (1974). Early mouse embryos: growth and differentiation *in vitro*. *Experimental Cell Research*, **85**, 424–8.

SALOMON, D. S. & SHERMAN, M. I. (1974). Implantation and invasiveness of mouse blastocysts on uterine monolayers. *Experimental Cell Research*, in press.

SHERMAN, M. I. & CHEW, N. J. (1972). Detection of maternal esterase in mouse embryonic tissues. *Proceedings of the National Academy of Sciences, USA*, **69**, 2551–5.

SPINDLE, A. I. & PEDERSEN, R. A. (1973). Hatching, attachment, and outgrowth of mouse blastocysts *in vitro*: fixed nitrogen requirements. *Journal of Experimental Zoology*, **186**, 305–18.

WHITTEN, W. K. (1971). Nutrient requirements for the culture of preimplantation embryos *in vitro*. In *Advances in the Biosciences*, ed. G. Raspe, vol. 6, pp. 129–141. Oxford: Pergamon Press.

EXPLANATION OF PLATE

PLATE 1

(*a*) and (*b*) Embryonic development in (*a*) maternal mouse serum and (*b*) foetal mouse serum (both day 3 of culture in MEM).

(*c*) Development of layers of tissue in an inner cell mass (day 8 of culture in MEM with 50 % foetal calf serum).

(*d*) Section of embryo showing development of embryonic cavities and layers of tissue (day 7 of culture in MEM with 20 % foetal calf serum). We provisionally identify the large central cavity as the proamniotic cavity, the two small adjacent cavities as exocoelom, the mass of tissue in the lower part of the picture as ecto-placental cone, and the three layers above as proximal endoderm (outer), mesoderm (loose mesenchyme) and ectoderm (bounding the 'proamniotic cavity').

PLATE I

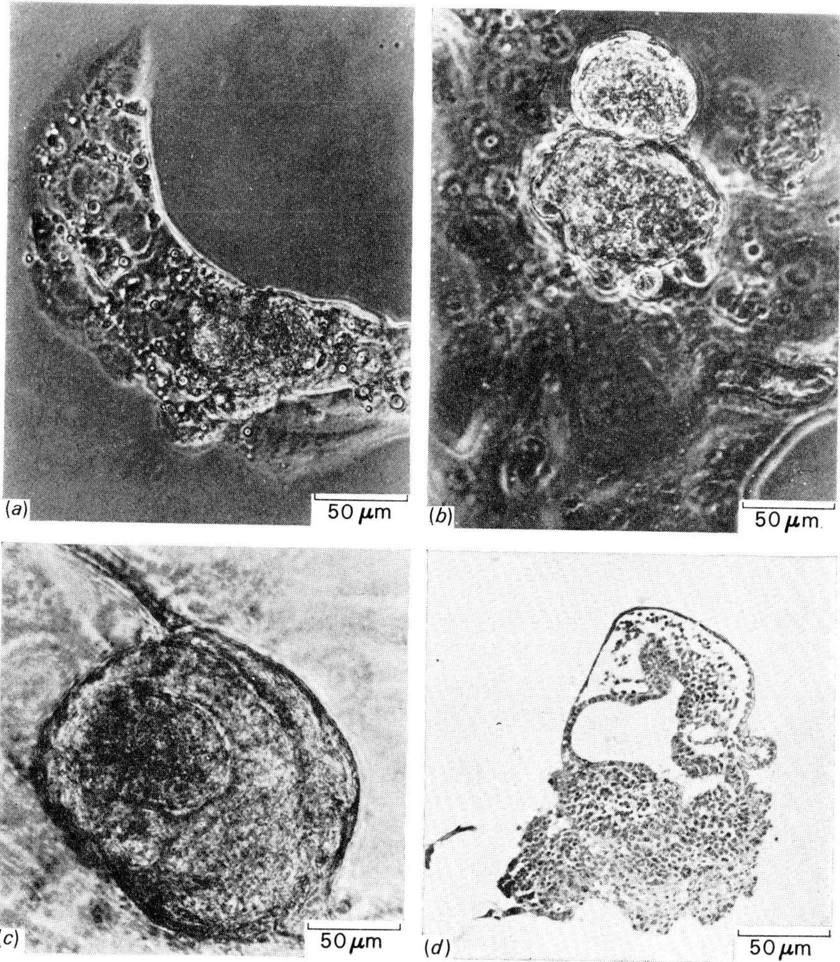

(*Facing p.* 60)

THE CULTURE OF POST-IMPLANTATION MAMMALIAN EMBRYOS

BY C. E. STEELE*

Biologie de la Reproduction, Faculté des Sciences, Jussieu, Paris VI, France

Culture techniques for post-implantation mammalian embryos are less advanced than those for pre-implantation stages. The reasons for this probably include the greater availability of pre-implantation ova (which can be obtained in large numbers by superovulation), their relative independence from the mother and the disappointing results of early attempts to culture post-implantation embryos. The resulting inaccessibility of post-implantation stages has meant that in-vitro studies of the causal analysis of morphogenesis have largely been carried out on chick and amphibian embryos, and to a lesser extent on those of fish and reptiles. Proficient culture methods for these groups in which a close relationship between mother and foetus does not exist have long been available (New, 1966*a*).

In the absence of satisfactory culture methods, other techniques restricted to the later stages of gestation have been used to study mammalian development (Austin, 1973). The in-utero operation (Hooker & Nicholas, 1930; Nicholas, 1934; Jost, 1947; Wells, 1947; Brambell, 1958; Leissring & Anderson, 1961) is relatively difficult to perform; however the foetus can withstand a considerable amount of surgery providing the placental blood supply is left intact. Alternatively, the foetus can be exposed for several hours of direct study leaving the placenta attached to the uterus (Nemeth, 1971). With larger species, e.g. sheep, man, the foetus (or the foeto-placental unit) can be perfused *in vitro* (Zapol *et al.*, 1969; Lerner, Saxena & Diczfalusy, 1971). More recently techniques such as amniocentesis, amnioscopy, ultrasonography and X-radiography have been developed for monitoring the later stages of gestation (Austin, 1973). With younger embryos the transfer of embryonic tissues to an ectopic site has greatly increased our knowledge of development (reviewed by Kirby, 1970).

The unique contribution of culture techniques is to give direct access to the embryo during the critical stages of organogenesis by the elimination of the maternal system.

The earliest attempts to achieve this objective were made in the 1930s

* Present address: Department of Anatomy, Downing Street, Cambridge.

by three groups using the rat (Nicholas & Rudnick, 1934, 1938; Jolly & Lieure, 1938), guinea-pig (Jolly & Lieure, 1938) and rabbit (Waddington & Waterman, 1933) explanted at various stages after implantation as far as the early somite stage. These attempts met with limited success; the most advanced stages reached in culture were 6 somites for the rabbit and 16 somites for the rat. Similar results were obtained with the guinea-pig where the majority of embryos ceased development within 36 hours of explantation. Nicholas (1938) later claimed improved development of rat embryos in a circulating medium but did not give details.

Although this limited development *in vitro* was restricted to a narrow range of embryonic ages (and inferior to that in the chick embryo) these techniques were used by experimental embryologists to study heart development (Goss, 1935; Dwinnel, 1939) and embryonic induction (Törö, 1938).

In the last decade, largely as a result of the work of New and his colleagues, post-implantation culture has advanced rapidly such that a wide range of embryonic ages from several species can now be cultured *in vitro* for periods of up to four days. In this article I will describe the present culture methods and their uses, and also provide new experimental data concerning the optimum culture conditions for embryos explanted soon after implantation. Embryos will be considered to be in the 1st day of gestation on the day sperm are found in the vaginal smear.

CULTURE EQUIPMENT

The choice of apparatus for the culture is determined by the stage of development of the embryo and the duration of the culture period. Embryos can be grown in static or flowing medium.

Static medium

Embryos explanted on or before the 11th day (rat) grow well in a watchglass containing liquid medium (New, 1971). Each watchglass is contained within a Petri dish which, in turn, is stored in a simple gas chamber consisting of an inverted 800 ml beaker on a dish containing liquid paraffin. The atmosphere is humidified by placing moist gauze in each Petri dish. This is essentially the standard technique for organ culture (Paul, 1970).

'Circulators'

A simple apparatus – a 'circulator' (Plate 1) – for the continuous circulation of the medium without a mechanical pump has been devised by New

(1967, see New, 1971 for review). A 'circulator' permits constant observation of the explants. Rat embryos can be explanted into this apparatus at any stage between the early egg-cylinder ($7\frac{1}{2}$ d) and 50–5 somites ($13\frac{1}{2}$ d). A 'circulator' gives optimum development of all stages and the final stages are more advanced than in watchglass cultures.

Daniel (1970) modified the embryo chamber in order to make it suitable for rabbit embryos.

Other, more complex, circulating systems, which require mechanical pumps, have been devised by Tamarin & Jones (1968) and Robkin, Shepard & Tanimura (1972). Although these systems have certain advantages, such as avoidance of pulsatile flow and frothing of the medium, development is not superior to that in the simpler 'circulator'.

Ogawa, Satoh, Hamada & Hashimoto (1972) have cultured rabbit blastocysts in a medium which was recirculated by a convection current induced by point heating.

Roller tubes

The continuous rotation – at higher speeds than those used for cell culture (Paul, 1970) – of a specimen tube or small glass bottle provides a simple method of growing embryos in a flowing medium (New, Coppola & Terry, 1973). However, the disadvantage of this method is that it is inconvenient for continuous observation and it should only be used when a final assessment of development is all that is required. Providing the culture vessel is an adequate size, development of rat embryos explanted at $10\frac{1}{2}$ and $11\frac{1}{2}$ days gestation (approximately 15 and 25 somites) is equal to that in a 'circulator' (New *et al.*, 1973).

EXPERIMENTAL ANIMAL

The culture technique makes use of the simple yolk sac placenta found in rodents, lagomorphs and marsupials. Other factors dictating the choice of experimental animal include the duration of gestation (short time preferred), size of litter (large size preferred), the ability of the experimenter to accurately time the pregnancy and the availability of the animal.

The majority of culture experiments have used the rat, and the stages at explantation and the stage attained in culture for this species are shown in Fig. 1. The mouse has been used less extensively (New & Stein, 1964; Clarkson, Doering & Runner, 1969; Hernandez-Verdun & Legrand, 1971). There has been no recent work on the guinea-pig embryo and the only published results are those of Jolly & Lieure (1938).

Givelber & DiPaolo (1968) obtained the equivalent of 24–30 hours growth *in vivo* with hamster embryos explanted at $7\frac{1}{2}$ days of gestation

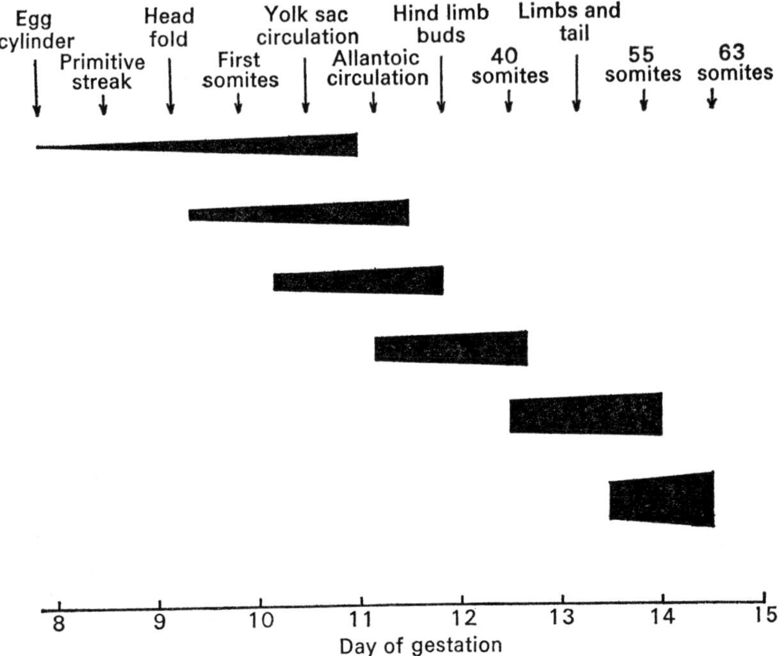

Fig. 1. Periods of development (represented by black areas) of rat foetuses in culture. From New (1973).

(head-fold–early somite). Attempts to culture embryos of $6\frac{1}{2}$ and $8\frac{1}{2}$ days gestation resulted in failure. This is disappointing since the hamster is in some instances more useful as an experimental animal than the mouse or rat, e.g. for the study of viral infections of the foetus.

The rabbit embryo is more advanced at the time of implantation than the other species mentioned here. Daniel (1970) has modified the 'circulator' method to culture pre-implantation $6\frac{3}{4}$ day rabbit embryos (early or pre-primitive streak stages); 95 % reached the 5 somite stage and 40 % grew as far as the 20 somite stage. There was no blood circulation even when the heart was beating. Daniel (1968; 1971) has also used static medium and a culture vessel consisting of a flask of medium through which gas is bubbled for 7 day embryos, but the stages achieved in culture were less advanced than those in circulating medium.

Most metatherian mammals (marsupials) do not have a complex allantoic placenta and rely almost entirely on the yolk sac which becomes much more expanded than in rodents. The neonate marsupial is only slightly more advanced, with the exception of a few precociously developed organs, than the 14 day rat foetus. New & Mizell (1972) have made a preliminary study of the growth of late opossum embryos in culture. Poor results were obtained with embryos explanted at 8–9 days gestation (8–10 somites) but with those explanted at 10–11 days (28 somites) development continued at about the same rate as *in vivo* for periods of up to 20 hours such that the most advanced stage in culture was within 24 hours of the end of normal gestation (12¾ d). The authors suggested that, with further improvements in technique, the opossum will be useful for assessing the post-natal effects of experiments on the foetus.

GAS PHASE

Regulation of the gas phase in contact with nutrient medium has been an important factor in the improvement of culture techniques, New & Stein (1964) found that the addition of 4–5 % CO_2 was beneficial to mouse (7 d) and rat (9–10½ d) embryos cultured on plasma clots and it is now standard practice to use a gas mixture containing 5 % CO_2 (reduced for hyperbaric cultures to maintain a similar partial pressure) for embryos of all ages in order to regulate the pH of the medium.

The proportion of oxygen in the gas phase has turned out to be critical and varies with type of culture (static or flowing medium) and the age of the embryo. In watchglass cultures, where a gradient in oxygen tension exists between the surface of the medium and the bottom of the watchglass, the only means of increasing the oxygen available to the embryo is to increase the proportion in the gas phase and/or the pressure in the culture vessel. Thus, 7½ and 8½ day rat embryos grow well in medium equilibrated with 20–40 % O_2 (Steele, 1972), 9½ day embryos with 60–95 % O_2, 10½ day with 95 % O_2 (New & Coppola, 1970*a*) but 11½ day embryos require hyperbaric oxygen (New & Coppola, 1970*b*). Embryos explanted after 11½ days do not grow in static medium.

Embryos at all stages develop further (i.e. synthesise more protein and somites) when cultured in circulating medium, where an additional factor – the rate of circulation – can also be used to alter the oxygen supply to the conceptus. Thus the optimum amount of oxygen (or its partial pressure) in the gas phase in contact with circulating medium is less than that for static medium for embryos of a given age. Embryos of 12½ and 13½ days are capable of development. As in static medium there is a gradual increase in the proportion of O_2 resulting in optimum growth (Table 1). These data could be explained on the assumption that ‘embryonic tissues at all stages require an O_2 pressure similar to that provided by blood in equilibrium with air' (New & Coppola, 1970*a*) and can be explained in terms of two factors.

Firstly, the change in metabolic pathways utilised by embryonic tissues. Shepard, Tanimura & Robkin (1969, 1970) have used the culture technique to show that in the 11 somite ($10\frac{1}{3}$ d) rat embryo there is a high rate of glucose utilisation yielding large amounts of lactate, a lack of oxygen dependence and low activity of the terminal electron transport system. This suggests that glycolysis is taking place by means of the Embden–Meyerhof pathway (anaerobic). Embryos explanted at the 25 and 35 somite stages showed a decreased glucose utilisation and lactate production, appearance of oxygen dependence and increased activity in the terminal electron transport system.

The second factor is the failure of the chorioallantoic placenta to grow in culture, thereby forcing an increased oxygen carrying load onto the yolk sac placenta. The respiratory and nutritional functions of this placenta become less important *in vivo* after the onset of the chorioallantoic circulation (11th day of gestation).

Table 1. *Growth (protein synthesis) of rat foetuses explanted, with visceral yolk sac intact, at $9\frac{1}{2}$–$13\frac{1}{2}$ d of gestation and cultured for 2 d in serum equilibrated with different oxygen concentrations**

Age	Oxygen concentration				
(d)	5 %	20 %	95 %	2 atm	3 atm
$9\frac{1}{2}$	28	37	<20		
$10\frac{1}{2}$		26	17		
$11\frac{1}{2}$		8	22	21	
$12\frac{1}{2}$			7	14	10
$13\frac{1}{2}$				<5	

From New (1973).
* Results are expressed as number of hours *in vivo* required for the same amount of protein synthesis.

Since at all stages an excess of oxygen is harmful to the yolk sac and therefore detrimental to embryonic development, the beneficial effects of merely increasing the oxygen supply reaches a limit, occurring with embryos explanted at $12\frac{1}{2}$ days gestation. These will develop to the 50–5 somite stage but after this 2 atmospheres is inadequate for the oxygen needs of the embryo and higher pressures rapidly destroy the yolk sac. However, Cockroft (1973) has improved the growth of $12\frac{1}{2}$ day embryos and successfully cultured 13½ day embryos to the 60–3 somite stage, both at atmospheric pressure, as a result of opening the yolk sac. Presumably this improvement is brought about by the exposure of the foetal surface to the medium thus supplementing the oxygenation by the yolk sac. Hyperbaric oxygen (1.5 or 2 atm) proves to be toxic to the 13½ day embryo with an opened yolk sac.

OXYGEN REQUIREMENTS OF EGG-CYLINDERS IN CULTURE

The experiments to be described here assess the development of 8½ day rat egg-cylinders in circulating and static medium equilibrated with gas mixtures of various proportions of oxygen in order to determine the optimum conditions *in vitro*. A comparison with development *in vivo* will also be made.

In view of the correlation between developmental stage and O_2 requirements *in vitro* it seemed probable that the development of embryos in circulating medium, like that of 9½ day embryos, would not be improved by increasing the proportion of O_2 in the gas phase above 20 %. However, in static medium, because of the gradient in oxygen tension between the surface of the medium and the bottom of the watchglass, gas mixtures containing more than 20 % O_2 were tested.

The methods were as previously described (Steele, 1972), the serum being prepared from blood which was allowed to clot and stand overnight.

RESULTS

Circulating medium for 24 hours

During the first 24 hours in culture maximum protein synthesis occurred in serum equilibrated with 20 % O_2 (Table 2). However, the difference between this and 5 % O_2 was not significant. The cultured embryos were considerably smaller than those of a similar stage *in vivo* (Plate 2) and the protein content was approximately one-half that *in vivo* (25–30 μg). All the embryos were at or near the headfold stage but in most the amniotic cavity and the ectoplacental cavity were still continuous so that the exocoelom was not fully formed.

Table 2. *Development of 8½ day rat embryos grown* in vitro *for 24 h in circulating medium with 5 or 20% O_2*

% O_2	Rate of flow of medium (ml/min)	No. of embryos	Total protein (μg) mean ± S.E.	Size of conceptus (mm) mean ± S.E. Length	Width
5	<0.5	15	11.5 ± 1.2	0.92 ± 0.03	0.52 ± 0.04
5	1.0	14	12.6 ± 1.4	0.98 ± 0.03	0.55 ± 0.03
20	1.0–2.5	15	14.6 ± 1.6	0.99 ± 0.02	0.55 ± 0.03

Comparing total protein:
20 % O_2→5 % O_2, <0.5 ml/min $P = 0.05–0.10$
20 % O_2→5 % O_2, 1.0 ml/min $P = 0.15–0.20$

Circulating medium for 72 hours

The amount of protein synthesised was greater in explants grown in medium equilibrated with 20% O_2 than with 5% O_2 (Table 3) but the difference was not significant either for embryonic protein or membrane

Table 3. *Development of 8½ day rat embryos grown* in vitro *for 72 h in circulating medium: 5 or 20% O_2 for first 24 h, then all media were equilibrated with 20% O_2*

% O_2 for first 24 h	Rate of flow of medium (ml/min)	No. of embryos	Protein (μg) mean ± S.E. Embryo	Membranes	Total	Yolk sac diameter (mm) mean ± S.E.
5	1.0	15	19 ± 3	42 ± 4	61 ± 6	1.44 ± 0.09
20	2.5	15	23 ± 3	46 ± 5	69 ± 7	1.70 ± 0.11

Comparing protein content: 20 % O_2→5 % O_2
Embryo $P = 0.20–0.25$
Membranes $P = 0.30–0.35$
Total $P = 0.20–0.25$

The figures used in this table are a combination of those in Tables 4 and 5.

protein. A reduction in the proportion of O_2 to 2% resulted in a smaller average protein content (Table 4) but again this was not significant compared with explants grown in 20% O_2 or 5% O_2. Explants grown in serum equilibrated with 5% CO_2 in nitrogen were very retarded. There was a significant difference in protein content (Table 5) between these explants and those grown in serum equilibrated with 20% or 5% O_2.

Embryos cultured for 72 hours developed a double heart, confirming the result of New & Daniel (1969). In 20% and 5% O_2 the great majority

Table 4. *Development of $8\frac{1}{2}$ day rat embryos grown* in vitro *for 72 h in circulating medium: 2, 5 or 20 % O_2 for the first 24 h of culture then all media were equilibrated with 20% O_2*

% O_2 for first 24 h	Rate of flow of medium (ml/min)	No. of embryos	Total protein (μg) mean ± S.E.	Yolk sac diameter (mm) mean ± S.E.
2	1.0	8	52 ± 7	1.47 ± 0.12
5	1.0	8	62 ± 10	1.50 ± 0.16
20	2.5	8	63 ± 9	1.64 ± 0.15

Comparing total protein:
20 % O_2→2 % O_2 $P = 0.15–0.20$
5 % O_2→2 % O_2 $P = 0.20–0.25$

Table 5. *Development of $8\frac{1}{2}$ day rat embryos grown* in vitro *for 72 h in circulating medium: 0 (5 % CO_2 in N_2), 5 % or 20 % O_2 for the first 24 h of culture then all media were equilibrated with 20 % O_2*

% O_2 for first 24 h	Rate of flow of medium (ml/min)	No. of embryos	Total protein (μg) mean ± S.E.	Yolk sac diameter (mm) mean ± S.E.
0	1.0	7	39 ± 3	0.94 ± 0.02
5	1.0	7	61 ± 7	1.38 ± 0.10
20	2.5	7	75 ± 10	1.76 ± 0.17

Comparing total protein:
20 % O_2→0 % O_2 $P = 0.0025–0.005$
5 % O_2→0 % O_2 $P = 0.01–0.0125$

of embryos developed a heartbeat (14/15 and 13/15 respectively) whereas in 2 % and 0 % O_2 it occurred less frequently (3/8 and 2/7 respectively).

Static medium

In serum in contact with a gas phase containing 20 or 40 % O_2 embryos developed well for the first 48 hours reaching the early somite stages, similar to development in circulating medium. They usually died during the next 24 hours. An increase in the proportion of O_2 to 95 % retarded development and the embryos survived less than 36 hours.

In general the somites do not develop well in embryos explanted at the egg-cylinder stage and cannot be reliably used to assess development.

The results, showing development in oxygen concentrations from 2–20 % in circulating medium and 20–40 % in static medium, indicate that the rat egg-cylinder is similar to some pre-implantation embryos in its oxygen requirements *in vitro*. Both are harmed by 95 % oxygen and can maintain growth and differentiation at low levels of oxygen but fail to develop without oxygen.

The greatly reduced protein content of cultured rat egg-cylinders, which is apparent after less than 24 hours, is unlikely to be due to insufficient oxygen reaching the embryo. There must be a need for modification of the nutrient medium and/or the physical forces acting upon the embryo; procedures allowing this have proved effective with some cultured pre-implantation ova (Krishnan & Daniel, 1967; Jenkinson & Wilson, 1970).

The variation in embryonic protein content of cultured egg-cylinders is considerable, sometimes varying by a factor of three (compare results in this paper with Steele, 1972). This is a characteristic of embryo culture, although with older embryos the variation is not quite as great (see Steele & New, 1974, for growth of 9½ day embryos in delayed-centrifuged serum).

NUTRIENT MEDIUM

Development is more normal in a liquid medium (New, 1966*b*) than on the surface of a plasma clot (New & Stein, 1964) or a medium stiffened with agar (Smith, 1964) which mechanically disrupt the conceptus. The liquid most commonly used is homologous serum since heterologous sera are usually inferior and often lethal (New, 1966*b*). As in pre-implantation culture (Chang, 1949) heat inactivation can reduce or abolish the lethality of heterologous sera. Steele & New (1974) have recently shown that the development of rat egg-cylinders in heat inactivated homologous medium is superior to that in medium from the same batch which has not been heat treated. However, Cockroft (1974) found no advantage to be gained by heat inactivation of serum used for the culture of 12½ day rat embryos with opened yolk sacs.

An exception to the rule concerning heterologous sera is the development of rat embryos of the 14, 22 and 35 somite stages in human serum (Shepard, Tanimura & Robkin, 1969) and the development of mouse embryos in rat serum (Clarkson *et al.*, 1969).

The source of blood used to prepare the homologous serum is not critical, since it is immaterial whether the blood is obtained from male rats, pregnant or non-pregnant females, from the same individual or another, or the same variety or another (New, 1966*b*, 1967).

Complex biological media such as serum or plasma vary from batch to batch and even within the same batch before and after freezing. Furthermore, their preparation – particularly from small mammals such as mice and hamsters – is time-consuming and they are expensive to purchase. For these and other reasons several workers have examined the growth of post-implantation embryos in media consisting only partially of serum.

Givelber & DiPaolo (1968) found that 100 % hamster serum and Eagle's minimal essential medium containing 20–60 % serum were unsatisfactory for the growth of explanted hamster embryos. McCoys 5A containing 30 % hamster serum gave better results. Twenty per cent rat serum in Waymouth's medium resulted in better development of mouse embryos than media containing 0, 10 or 100 % rat serum (Clarkson, Doering & Runner, 1969). Fifty per cent serum in Waymouth's medium has also been shown to be unsuccessful but better development occurred with the same proportions of serum and '199' medium (New, 1967).

The effect of diluting serum with simple physiological saline has also been studied. Cockroft (1973) found that serum diluted with 50 or 75 % Tyrode saline resulted in as good or better growth of 12½ and 13½ day rat foetuses with opened yolk sacs compared with 100 % serum. Foetal survival was improved by diluting the serum with 75 % Tyrode. However, contrasting results were obtained by New (1973) who showed that serum/Tyrode dilutions caused poorer development of 10½ and 11½ day rat embryos with closed yolk sacs.

Cockroft (1973) suggested that dilution of serum might improve development because of a dilution of inhibitory factors. This probably applies to substances present in the serum before culture since the likelihood of whole serum accumulating toxic quantities of excretory products (or being depleted of essential nutrients) is diminished by the observations that the re-use of serum does not affect embryonic development (New, 1967) and the failure of an increase in the volume of medium, or a transfer of the embryo to fresh medium, improves development significantly (New, 1966*b*, 1967).

No clear account of the nutritional requirements of the post-implantation embryo can be made from the few studies devoted to this problem. However, in several instances (Givelber & DiPaolo, 1968; Clarkson *et al.*, 1969; Cockroft, 1973) artificial media (or physiological saline) containing a small proportion of serum resulted in better embryonic development than whole serum or artificial medium alone. These results could be explained in terms of a balance between the dilution of inhibitory factors and essential nutrients present in the biological medium. The effects of any changes in osmolarity of the medium could also be important in this respect.

Recently, Steele (1972) found that the method of preparation of the serum was critical for the development of egg-cylinders in culture. If blood was allowed to clot and stand overnight before centrifugation and separation, then the heart primordia of the embryos grown in this medium usually failed to fuse and development was subsequently retarded. This had previously been found by New & Daniel (1969). However, if the blood was centrifuged immediately, then the embryos grown in this serum (or if grown in plasma) had a single heart and development was greatly improved. Steele & New (1974) have since shown that headfold stage embryos also developed better in immediately-centrifuged (IC) serum than in delayed-centrifuged (DC) serum, though different samples varied in this respect. They also investigated the factor(s) responsible for different heart development in IC and DC sera. It is now known that it is not the result of any difference in the sodium, potassium (Steele, 1972) or calcium (Steele & New, 1974) ion concentration of the media and complement concentration may also be unimportant, despite the fact that heat inactivation reduces the frequency of double heart formation in DC serum (Steele & New, 1974). It was found that the harmful properties of DC serum appeared rapidly (within 30 min) in contact with a normal blood clot in which the blood cells were trapped in the fibrin coagulum, but did not develop after 18 hours contact with separated blood cells and fibrin clot.

The addition of erythrocytes to serum produces poor embryonic development, possibly a result of extensive haemolysis and sedimentation (New, 1967).

GROWTH IN CULTURE

Berry (1968) has shown that growth (protein content) of 10½ day rat embryos in culture proceeds more slowly than *in vivo* whereas differentiation (somite number) is normal. This is a feature common to all studies using post-implantation culture. New (1973) has extended this study and shown that development in culture is progressively less with increasing age at explantation, protein increase being more affected than somite formation.

USES OF CULTURE TECHNIQUES

In culture there is direct access to the conceptus mechanically, chemically and visually.

Mechanical

By means of operations on the conceptus several workers have provided information on axis formation (Smith, 1964), axial rotation (Deuchar, 1971) and the functions of the embryonic membranes (Payne & Deuchar, 1972). It has been demonstrated by Payne & Deuchar that the visceral yolk sac is important in the regulation of the extra-embryonic fluids as well as for the nutrition of the embryo. They also suggested that its nutrient function is incomplete without Reichert's membrane. The latter is removed during explantation since it reduces expansion and rotation of the embryo *in vitro*.

Chemical

If a drug is found to affect development when injected *in vivo* it is not possible to determine whether it was a direct effect on the conceptus (embryo and/or extra-embryonic membranes) or whether the effect was mediated via the mother. In this instance the use of culture techniques can, by eliminating the maternal system, be of use. Of the limited number of substances so far tested in culture all have been found to have a direct effect on the conceptus.

Turbow (1966) demonstrated a direct effect of trypan blue on the embryo and found that the yolk sac has a protective effect at low doses by the accumulation of the dye. Other workers argue that the teratogenic action of trypan blue results from disrupted yolk sac nutrition (Beck & Lloyd, 1966). This is presumably the explanation for an embryopathic effect of antibody to yolk sac shown by New & Brent (1972) to have its primary effect on the yolk sac endoderm (the surface in contact with the nutrient medium). In these studies the teratogen was injected into the yolk sac cavity; in this case trypan blue was found to be active but antibody to yolk sac inactive.

Berry (1971) has used the culture technique to study the action of an antiserum to the contractile proteins of the heart and its effects on development. The poorer development of experimental embryos was attributed to disturbed nutrition resulting from reduced blood flow since the myocardium had been damaged. The effect of erythropoeitin on haem synthesis has also been studied on the 8th–9th day mouse embryo *in vitro* (Cole & Paul, 1966).

Before closure of the yolk sac the embryo is in direct contact with the nutrient medium. With embryos of the egg-cylinder stage, Morriss & Steele (1974) have shown a direct, teratogenic effect of added vitamin A upon the embryo at doses very much lower than those found to be teratogenic *in vivo*. This has raised the possibility that vitamin A may be responsible for human malformations resulting from only minor variations in plasma vitamin A levels.

Another vitamin recently studied *in vitro* is vitamin E, commonly referred to as the anti-sterility vitamin since rats lacking this vitamin are sterile. In the female this effect takes the form of gestation–resorption. There has been extended controversy (see Green & Bunyan, 1969) as to whether this vitamin acts directly on the conceptus to support gestation, probably as an antioxidant, or whether its presence in the maternal system is necessary for the synthesis of some compound such as a steroid hormone that influences gestation. By means of the culture technique, Steele, Jeffery & Diplock (1974) have shown that vitamin E has a direct effect on the conceptus in maintaining embryonic growth. As *in vivo*, the action of the antioxidant N,N′-diphenyl-p-phenylenediamine (DPPD) was similar to vitamin E but ethoxyquin was inactive. It therefore seems likely that vitamin E acts *in vivo*, with respect to gestation–resorption, solely as an antioxidant, although the possibility that DPPD, but not ethoxyquin, possesses some property of vitamin E resulting in the prevention of gestation–resorption cannot yet be completely excluded. Further studies of antioxidants in culture will help solve this problem.

Visual

The visualisation of developmental processes is an important advance. The filming of normal (LeGoascogne & Brun, 1969) and abnormal development will provide an added means of analysis.

Cockroft & New have made a film of the explantation and culture of 11½ day and 13½ day (opened yolk sacs) rat embryos.

DEVELOPMENT ACROSS THE IMPLANTATION PERIOD IN CULTURE

The most successful attempts to culture embryos during and beyond implantation have been with rabbit embryos (see previous sections) presumably because they are more advanced than the rat and mouse at the time of implantation. Apart from simple culture with rabbit blastocysts, Glenister (1971) has developed techniques to study the implantation of rabbit blastocysts *in vitro* on strips of uterine endometrium and embryos can be grown to advanced stages in composite explants of mesometrial endometrium bearing the developing placenta and embryonic disc. It has not yet proved possible to achieve ovo-implantation in organ culture using other species.

With explants of the conceptus only, Jenkinson & Wilson (1970) and Wilson & Jenkinson (1974) have successfully cultured late mouse blastocysts through the implantation stages using the soft, transparent lens fibre from the bovine eye as a support medium. Hsu (1973) seems to have achieved the best results starting with mouse blastocysts by using collagen as the substrate for attachment and by changing the nutrient medium to suit the requirements at successive stages of development. Most of the blastocysts 'implanted' on the collagen and approximately 50% formed egg-cylinders. However, only 10% reached the early somite stage after as long as 10 days of gestation, by which time there was a 5 day retardation in development. With ova explanted at the 2-cell stage 95% developed into blastocysts in culture but only 10% became egg-cylinders (Hsu, Baskar, Stevens & Rash, 1974).

FUTURE OF THE CULTURE TECHNIQUE

The development of a completely artificial medium would help in the analysis of the nutrient requirements of the embryo. The subsequent elimination of variability in the nutrient medium would probably also lead to greater consistency of development in culture.

So far it has been impossible to obtain growth of the allantoic placenta in culture. This would be a major advance resulting in greatly improved growth of older embryos whose oxygen requirement cannot be met in the present culture set-up.

REFERENCES

AUSTIN, C. R. (1973). *The mammalian fetus* in vitro, ed. C. R. Austin. London: Chapman & Hall.

BECK, F. & LLOYD, J. B. (1966). The teratogenic effects of azo dyes. *Advances in Teratology*, **1**, 131–93.

BERRY, C. L. (1968). Comparison of *in vivo* and *in vitro* growth of the rat foetus. *Nature, London*, **219**, 92–3.

BERRY, C. L. (1971). The effects of an antiserum to the contractile proteins of the heart on the developing rat embryo. *Journal of Embryology and Experimental Morphology*, **25**, 203–12.

BRAMBELL, F. W. R. (1958). The passive immunity of the young mammal. *Biological Reviews*, **33**, 488–531.

CHANG, M. C. (1949). Effects of heterologous sera on fertilized rabbit ova. *Journal of General Physiology*, **32**, 291–300.

CLARKSON, S. G., DOERING, J. V. & RUNNER, M. N. (1969). Growth of post-implantation mouse embryos cultured in serum-supplemented, chemically defined medium. *Teratology*, **2**, 181–6.

COCKROFT, D. L. (1973). Development in culture of rat foetuses explanted at 12.5 and 13.5 days of gestation. *Journal of Embryology and Experimental Morphology*, **29**, 473–83.

COCKROFT, D. L. (1974). Growth of fat foetuses in culture. PhD Thesis. University of Cambridge.

COLE, R. J. & PAUL, J. (1966). The effects of erythropoeitin on haem synthesis in mouse yolk sac and cultured foetal liver cells. *Journal of Embryology and Experimental Morphology*, **15**, 245–60.

DANIEL, J. C. (1968). Oxygen concentrations for the culture of rabbit blastocysts. *Journal of Reproduction and Fertility*, **17**, 187–90.

DANIEL, J. C. (1970). Culture of the rabbit embryo in circulating medium. *Nature, London*, **225**, 193–4.

DANIEL, J. C. (1971). Culture of the rabbit blastocyst across the implantation period. In *Methods in Mammalian Embryology*, ed. J. C. Daniel, pp. 284–9. San Francisco: Freeman.

DEUCHAR, E. M. (1971). The mechanism of axial rotation: an experimental study *in vitro*. *Journal of Embryology and Experimental Morphology*, **25**, 189–201.

DWINNEL, L. A. (1939). Physiological contraction of double hearts in rabbit embryos. *Proceedings of the Society for Experimental Biology and Medicine*, **42**, 264–7.

GIVELBER, H. M. & DIPAOLO, J. A. (1968). Growth of explanted eighth day hamster embryos in circulating medium. *Nature, London*, **220**, 1131–2.

GLENISTER, T. W. (1971). Methods for studying ovoimplantation and early embryo-placental development *in vitro*. In *Methods in Mammalian Embryology*, ed. J. C. Daniel, pp. 320–33. San Francisco: Freeman.

GOSS, C. M. (1935). Double hearts produced experimentally in rat embryos. *Journal of Experimental Zoology*, **72**, 33–46.

GREEN, J. & BUNYAN, J. (1969). Vitamin E and the biological antioxidant theory. *Nutrition Abstracts and Reviews*, **39**, 321–45.

HERNANDEZ-VERDUN, D. & LEGRAND, C. (1971). Différentiation du trophoblaste au cours du développement *in vitro* d'embryon de Souris. *Journal of Embryology and Experimental Morphology*, **26**, 175–87.

HOOKER, D. & NICHOLAS, J. S. (1930). Spinal cord section in rat fetuses. *Journal of Comparative Neurology*, **38**, 315–47.

HSU, Y.-C. (1973). Differentiation *in vitro* of mouse embryos to the stage of early somite. *Developmental Biology*, **33**, 403–11.

HSU, Y.-C., BASKAR, J., STEVENS, L. C. & RASH, J. E. (1974). Development *in vitro* of mouse embryos from the two-cell stage to the early somite stage. *Journal of Embryology and Experimental Morphology*, **31**, 235–45.

JENKINSON, E. J. & WILSON, I. B. (1970). *In vitro* support system for the study of blastocyst differentiation in the mouse. *Nature, London*, **228**, 776–8.

JOLLY, J. & LIEURE, C. (1938). Recherches sur la culture des oeufs des mammifères. *Archives d'Anatomie microscopique*, **34**, 307–74.

JOST, A. (1947). Expériences de décapitation de l'embryon de lapin. *Comptes rendus, Académie des Sciences, Paris*, **225**, 322–4.

KIRBY, D. R. S. (1970). The extra-uterine mouse egg as an experimental model. In *Schering Symposium on the Mechanisms involved in Conception. Advances in the Biosciences*, vol. 4, ed. G. Raspé, pp. 255–73. Oxford: Pergamon–Vieweg.

KRISHNAN, R. S. & DANIEL, J. C. (1967). 'Blastokinin': inducer and regulator of blastocyst development in the rabbit uterus. *Science*, **158**, 490–2.

LeGoascogne, C. & Brun, J. L. (1969). Développement *in vitro* de l'embryon de rat. *Comptes rendus hebdomadaires des séances de l'Académie des Sciences, Paris*, **268**, 3195–8.

Leissring, J. C. & Anderson, J. W. (1961). The transfer of serum proteins from mother to young in the guinea pig. I. Prenatal rates and routes. *American Journal of Anatomy*, **109**, 149–55.

Lerner, U., Saxena, B. N. & Diczfalusy, E. (1971). Extracorporeal perfusion of the human foetus, placenta and foeto-placental unit. In *Perfusion Techniques. Fourth Karolinska Symposium on Research Methods in Reproductive Endocrinology*, pp. 310–30. Copenhagen: Bogtrykkeriet Forum.

Morriss, G. M. & Steele, C. E. (1974). The effect of excess vitamin A on the development of rat embryos in culture. *Journal of Embryology and Experimental Morphology*, in press.

Nemeth, A. M. (1971). Study of the exposed guinea pig fetus. In *Methods in Mammalian Embryology*, ed. J. C. Daniel, pp. 347–54. San Francisco: Freeman.

New, D. A. T. (1966*a*). *The culture of vertebrate embryos*. London: Logos Press.

New, D. A. T. (1966*b*). Development of rat embryos cultured in blood sera. *Journal of Reproduction and Fertility*, **12**, 509–24.

New, D. A. T. (1967). Development of explanted rat embryos in circulating medium. *Journal of Embryology and Experimental Morphology*, **17**, 513–25.

New, D. A. T. (1971). Methods for the culture of post-implantation embryos of rodents. In *Methods in Mammalian Embryology*, ed. J. C. Daniel, pp. 305–19. San Francisco: Freeman.

New, D. A. T. (1973). In *The Mammalian Fetus* in vitro, ed. C. R. Austin, chapter 2. London: Chapman & Hall.

New, D. A. T. & Brent, R. L. (1972). Effect of yolk-sac antibody on rat embryos grown in culture. *Journal of Embryology and Experimental Morphology*, **27**, 543–53.

New, D. A. T. & Coppola, P. T. (1970*a*). Effects of different oxygen concentrations on the development of rat embryos in culture. *Journal of Reproduction and Fertility*, **21**, 109–18.

New, D. A. T. & Coppola, P. T. (1970*b*). Development of explanted rat fetuses in hyperbaric oxygen. *Teratology*, **3**, 153–62.

New, D. A. T., Coppola, P. T. & Terry, S. (1973). Culture of explanted rat embryos in rotating tubes. *Journal of Reproduction and Fertility*, **35**, 135–8.

New, D. A. T. & Daniel, J. C. (1969). Cultivation of rat embryos explanted at 7.5 and 8.5 days of gestation. *Nature, London*, **223**, 515–16.

New, D. A. T. & Mizell, M. (1972). Opossum fetuses grown in culture. *Science*, **175**, 533–6.

New, D. A. T. & Stein, K. F. (1964). Cultivation of post-implantation mouse and rat embryos on plasma clots. *Journal of Embryology and Experimental Morphology*, **12**, 101–11.

Nicholas, J. S. (1934). Experiments on developing rats. I. Limits of foetal regeneration; behaviour of embryonic material in abnormal environments. *Anatomical Record*, **58**, 387–413.

Nicholas, J. S. (1938). The development of rat embryos in a circulating medium. *Anatomical Record*, **70**, 199–210.

Nicholas, J. S. & Rudnick, D. (1934). The development of rat embryos in tissue culture. *Proceedings of the National Academy of Sciences, USA*, **20**, 656–8.

Nicholas, J. S. & Rudnick, D. (1938). Development of rat embryos of egg cylinder to head-fold stages in plasma cultures. *Journal of Experimental Zoology*, **78**, 205–32.

OGAWA, S., SATOH, K., HAMADA, M. & HASHIMOTO, H. (1972). Culture of rabbit blastocyst in a medium with a convection current. *Nature, London*, **238**, 402–3.

PAUL, J. (1970). *Cell and tissue culture*, p. 185 & p. 282. Edinburgh & London: E. & S. Livingstone.

PAYNE, G. S. & DEUCHAR, E. M. (1972). An *in vitro* study of functions of embryonic membranes in the rat. *Journal of Embryology and Experimental Morphology*, **27**, 533–42.

ROBKIN, M. A., SHEPARD, T. H. & TANIMURA, T. (1972). A new *in vitro* culture technique for rat embryos. *Teratology*, **5**, 367–76.

SHEPARD, T. H., TANIMURA, T. & ROBKIN, M. (1969). *In vitro* study of rat embryos. I. Effects of decreased oxygen on embryonic heart rate. *Teratology*, **2**, 107–10.

SHEPARD, T. H., TANIMURA, T. & ROBKIN, M. A. (1970). Energy metabolism in early mammalian embryos. *Developmental Biology*, Supplement, **4**, 42–58.

SMITH, L. J. (1964). The effects of transection and extirpation on axis formation and elongation in the young mouse embryo. *Journal of Embryology and Experimental Morphology*, **12**, 787–803.

STEELE, C. E. (1972). Improved development of 'rat egg-cylinders' *in vitro* as a result of fusion of the heart primordia. *Nature, New Biology*, **237**, 150–1.

STEELE, C. E., JEFFERY, E. H. & DIPLOCK, A. J. (1974). The effect of vitamin E and synthetic antioxidants on the growth *in vitro* of explanted rat embryos. *Journal of Reproduction and Fertility*, **38**, 115–23.

STEELE, C. E. & NEW, D. A. T. (1974). Serum variants causing the formation of double hearts and other abnormalities in explanted rat embryos. *Journal of Embryology and Experimental Morphology*, **31**, 707–19.

TAMARIN, A. & JONES, K. W. (1968). A circulating medium system permitting manipulation during culture of post-implantation embryos. *Acta Embryologiae et Morphologiae Experimentalis*, **10**, 288–301.

TÖRÖ, E. (1938). The homeogenetic induction of neural folds in rat embryos. *Journal of Experimental Zoology*, **79**, 213–36.

TURBOW, M. M. (1966). Trypan blue induced teratogenesis of rat embryos cultivated *in vitro*. *Journal of Embryology and Experimental Morphology*, **15**, 387–95.

WADDINGTON, C. H. & WATERMAN, A. J. (1933). The development *in vitro* of young rabbit embryos. *Journal of Anatomy*, **67**, 355–70.

WELLS, L. J. (1947). Progress of studies designed to determine whether the fetal hypophysis produces hormones that influence development. *Anatomical Record*, **97**, 409.

WILSON, I. B. & JENKINSON, E. J. (1974). Blastocyst differentiation *in vitro*. *Journal of Reproduction and Fertility*, **39**, 243–9.

ZAPOL, W. M., KOLOBOW, T., PIERCE, J. E., VUREK, G. G. & BOWMAN, R. L. (1969). Artificial placenta: two days of total extrauterine support of the isolated premature lamb fetus. *Science*, **166**, 617–18.

EXPLANATION OF PLATES

PLATE 1

A 'Circulator'. The gas entering via the vertical arm on the right carries some of the medium with it along the sloping arm. The medium is oxygenated in the chamber (A) where the bubbles collapse. The rate of flow of the medium can be finely regulated by means of the stopcock. The embryos are housed in the detachable embryo chamber (B).

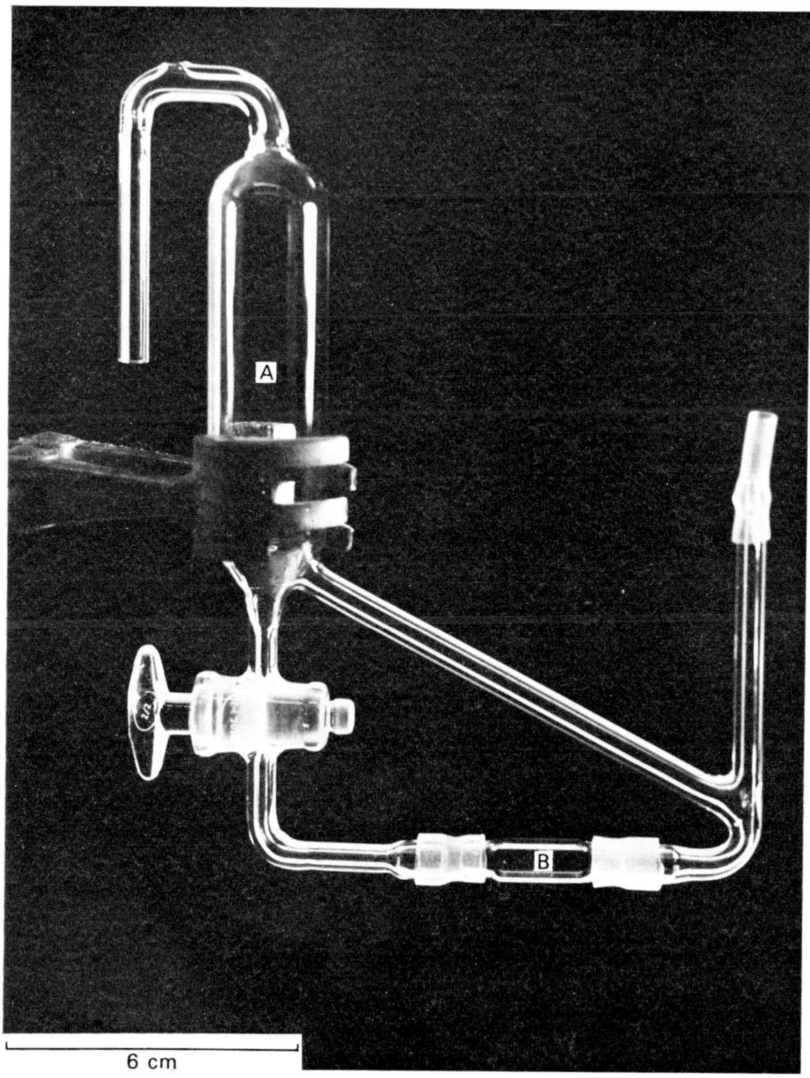

(*Facing p. 78*)

PLATE 2

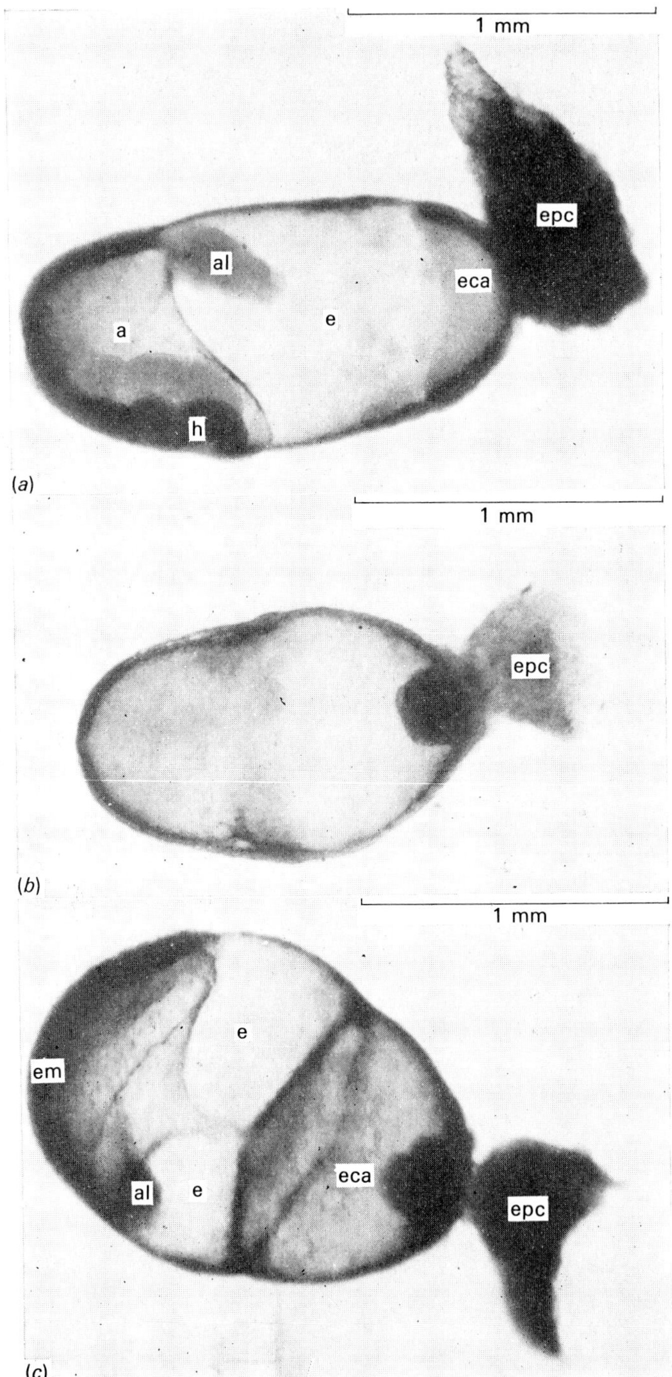

PLATE 2

(*a*) A $9\frac{1}{2}$ day (head-fold) rat embryo *ex utero*.

(*b*) and (*c*) Rat embryos explanted at $8\frac{1}{2}$ days and cultured for 24 h. These two embryos illustrate the range of development during the first 24 h in culture, (*c*) being the optimum.

epc, ectoplacental cone; eca, ectoplacental cavity; e, exocoelom; al, allantois; em, embryo; h, head-fold.

ORGANISATION IN THE PRE-IMPLANTATION EMBRYO

BY I. B. WILSON AND M. S. STERN

Department of Zoology, The Brambell Laboratories, University College of North Wales, Bangor, Gwynedd LL57 2UW

Investigations into the patterns and nature of the development of the eggs of invertebrates and lower vertebrates have a long and fruitful history (for reviews see Morgan, 1927; Huxley & De Beer, 1934; Reverberi, 1971; Bellairs, 1971). Mammals became the subject of similar studies only comparatively recently but, over the last fifteen years, interest in this field has shown a remarkable exponential growth.

To an experimental embryologist mammals are of particular interest because their eggs have become secondarily holoblastic though the embryo proper develops from only a small part of the egg which gives rise to some of the inner cell mass (ICM). The remainder of the egg yields the trophoblast and that part of the ICM which contributes to the placenta and extra-embryonic membranes. The differentiation of two cell types, trophoblast and ICM, is the first gross morphological manifestation of differentiation in the egg and it leads to the formation of the blastocyst. These early stages of development form the subject of this presentation.

HISTOLOGICAL AND HISTOCHEMICAL OBSERVATIONS

Amongst the first studies on the early cleavage stages of mammalian eggs were those of Assheton, whose work has received little comment from more recent workers although it contains some pertinent ideas on early differentiation. Assheton (1898*a*) postulated, on the basis of his observations on the sheep, that the future 'epiblast' (ICM) was actively surrounded by the 'hypoblast' (trophoblast) and that both cell types were fully differentiated by the 8-cell stage. His idea seemed to be supported by the earlier observations of Hubrecht (1889) on *Tupaia* and Van Beneden (1875, 1880, cited by Assheton, 1898*a*) on the bat and rabbit, though Heape (1886) disputed that this was the case in rabbit and mole. He could find no qualitative differences between the blastomeres of the 2-cell rabbit embryo despite Van Beneden's claim that one was larger, with hyaline contents, and gave rise to the 'epiblast' while the other was smaller, contained a dense vitelline material, and gave rise to the 'hypoblast'. Similarly, although Heape observed differences in the size of the blastomeres of the mole embryo he was unable to distinguish differences in their contents or density during cleavage and could observe no definite cleavage pattern. Dissimilarity in the size of the blastomeres was also observed in the pig (Assheton, 1898*b*) but it was not until 1929 that Heuser & Streeter extensively investigated this animal. They concluded that since the blastomeres of the 2-cell stage differed in size in the proportion 3 : 4 they were probably not identical and that some segregation of cytoplasmic substances had occurred. Progressive segregation of cytoplasmic substances from the onset of the second cleavage was thought to be reflected by a relative change in the volume of the individual cells when expressed as a percentage of the whole, e.g. at the 8-cell stage: 31, 18, 16, 11, 11, 5, 4 and 4%. Heuser & Streeter believed that the more slowly dividing ICM cells were surrounded by the more rapidly dividing trophoblast cells. However, there was no histological evidence given in support of this and, as in the studies mentioned above, arguments were based on what was known of lower forms, as scientific opinion of the time favoured 'preformation' concepts. Boyd & Hamilton (1952) gave a list of animals in which such size discrepancy between the blastomeres had been observed (e.g. mouse, rabbit, *Tarsius*, rat, dog, cat, goat, cow, pig and ferret). It was thought at that time that inequality of blastomeres in the 8-cell stage was typical of the majority of mammalian eggs. However, we have never noted this to be the case for 2-, 4- or 8-cell stages of living rat or mouse eggs and we know of no recent work confirming or denying this claim.

Histological and histochemical studies, mainly in rat, by Dalcq and his collaborators (see Dalcq, 1955, 1957 for review) and later Mulnard (1961 for review) provided the basis for an interpretation of differentiation dependent upon a progressive segregation of the egg cytoplasm which had, *ab initio*, polarity and bilaterality. These workers believed that two types of cytoplasm could be distinguished: a dense area of 'dorsal' cytoplasm (at the animal pole) rich in RNA and alkaline phosphatase, and 'ventral' cytoplasm which was both less dense and basophilic, but rich in mucopolysaccharides and acetalphosphatides, and with a vacuolated appearance. It was suggested (in accordance with Assheton's observations on the sheep) that by the 8-cell stage the blastomeres were irrevocably determined. Those blastomeres which received the 'dorsal' cytoplasm were said to divide more rapidly and to be predisposed to differentiating into ICM, while those which received 'ventral' cytoplasm formed trophoblast by sliding over and around the dorsal blastomeres. By the 16-cell stage the 'dorsal' blastomeres were thought to be completely enveloped. Similar observations and interpretations have been extended to the mouse by Dalcq (cited by Mulnard, 1965*a*, *b*).

No unequivocal evidence of polarity or bilaterality has been obtained from examination of living eggs (see Austin, 1961 for review) or by examining eggs fixed for the electron microscope (Izquierdo & Vial, 1962; Hillman, Tasca & Wileman, 1967; Krauskopf, 1968*a*, 1968*b*; Calarco & Brown, 1969). Sotelo & Porter (1959) observed particles rich in RNA distributed evenly throughout the cytoplasm of fertilised and unfertilised rat eggs while Cerisola & Izquierdo (1969) found no evidence of either polarity or bilateral symmetry when examining the distribution of RNA and other basophilic material in the mouse egg, although by the blastocyst stage the ICM was more basophilic than the trophoblast. Furthermore, although Dalcq (1957) and Mulnard (1965*a*, *b*) demonstrated the presence of acid phosphatase in the ICM of rat and mouse blastocysts and its absence in the trophoblast, the more recent work of Rodé, Damjanov & Skreb (1968), on the rat blastocyst, suggests that acid phosphatase is omnipresent. They suggest that the discrepancy between these results may be a consequence of fixation artefacts, diffusion of the enzyme during incubation or simply an optical effect resulting from the superposition of cells in the area of the ICM. To this date histological and histochemical studies have not yielded an unequivocal demonstration of differentiation-promoting constituents in the mammalian egg.

THE EXPERIMENTAL APPROACH

It is only relatively recently that extensive studies on metabolism and culture requirements of the pre-implantation embryo have facilitated its experimental manipulation *in vitro* (see Biggers & Stern, 1973 for review). However, the mouse is still the only mammal extensively used. Rat, sheep, and pig are more difficult to handle and culture (see Brinster, 1968 for review). The extent to which experimental manipulation has helped to elucidate differentiation in the pre-implantation rodent embryo will now be considered.

Early experimental studies followed the lines of previous work, on lower animals, designed to demonstrate the regulative potential of the eggs and the extent to which predetermined as opposed to epigenetic factors controlled development.

Techniques were designed for isolating blastomeres of early cleavage stages and the developmental potential of these was assessed. At first the results were interpreted to support the idea that development followed a predetermined pattern (see Mulnard, 1966; Wilson, Bolton & Cuttler, 1972, for reviews). Seidel (1956, 1960), working with rabbits, proposed that there was an 'organising centre' in the uncleaved egg and that its presence, at least in part, was necessary in blastomeres isolated from 2- and 4-cell eggs if viable blastocysts were to develop. Tarkowski (1959*a*, *b*) argued that the variability of structures developing from blastomeres isolated from similar stage mouse eggs could only be accounted for if the cleavage planes bore no constant relationship to any cytoplasmic organising substances contained within the egg. This interpretation has, however, not been substantiated by more recent work. In 1967, on the basis of further experiments on blastomeres isolated from 2-, 4- and 8-cell mouse eggs, Tarkowski & Wroblewska postulated a theory of development which seriously challenged the authenticity of the generally held view that mammalian development resulted from progressive segregation of a predetermined egg cytoplasm. The observations of these two workers indicated that differentiation of the blastomeres into trophoblast or ICM was dependent on the structure of the morula and the position of the blastomeres in it. ICM could only develop from cells confined in an intercellular environment by a surrounding envelope of cells which, because they have an 'exposed' surface, differentiate as trophoblast. This hypothesis was supported by indirect evidence gained from the studies of Mintz (1965) who aggregated varying numbers of mid-cleavage stage eggs. Pairs of synchronous eggs denuded of their zonae pellucidae and differing in cytoplasmic granularity, or with one member radioactively labelled, were made to aggregate into a single mass. The distribution of marked cells in the composite morulae and blastocysts which developed showed that selective sorting out of blastomeres did not occur and that cell movements were limited and random during aggregation (Mintz, 1962*a*, *b*, 1965). However, the only attempts to test the hypothesis directly have been by Hillman, Sherman & Graham (1972) and Wilson *et al.* (1972). Hillman *et al.* followed the fate of radioactively labelled blastomeres or whole embryos combined in aggregates with unlabelled embryos in a variety of spatial configurations. Over 90 % of the progeny of labelled disaggregated blastomeres, from 4- or 8-cell embryos, which were deliberately placed on the outside of 4- to 16-cell eggs formed part of the trophoblast at the blastocyst stage, while cells of whole embryos surrounded by unlabelled embryos were, in some cases, found exclusively in the ICM. Such aggregate blastocysts demonstrate that the properties of potential trophoblast or ICM cells can be modified by manipulation of the cellular arrangement. The ICM can be dissected out of these aggregate blastocysts and cultured separately but, although it contains most, if not all, of the potential trophoblast cells, it is not able to show the characteristic trophoblast activity of fluid accumulation. Stern (1973*a*) compressed cleaving eggs in a special culture chamber, causing them to develop as a monolayer of cells between two glass surfaces. Typical fluid accumulation started but this occurred throughout the cell plate, not just in the peripheral cells. Wilson *et al.* (1972) and Stern & Wilson (1972) used a micro-injection technique to place vital markers in cleaving eggs which were then aggregated in various pairwise combinations. They showed conclusively that potential trophoblast cells, derived from 4- or 8-cell stages, morulae or early blastocysts, are capable of contributing to the ICM of composite blastocysts. Although this demonstrates that primary regulation is possible, it remains to be proved that such cells can subsequently contribute to the formation of the embryo proper.

Tarkowski & Wroblewska's proposal, that epigenetic factors control the early stages of differentiation, is clearly supported by experimental evidence; under appropriate environmental conditions the labile blastomeres can be induced to develop into either trophoblast or ICM. Trophoblast will form if the cells are, in any way, exposed to the external environment. For ICM differentiation to occur total enclosure by other cells seems to be necessary. However, Wilson *et al.* (1972) showed that in the *intact* embryo the peripheral cytoplasm of the egg becomes the trophoblast while the central region becomes ICM. These workers suggest that this is because cleavage results merely in cellularisation and does not involve spatial disturbance of the cytoplasmic pattern. Thus, a given area of totipotent cytoplasm is always exposed to the same environmental/epigenetic influences so development normally follows an essentially predetermined pattern, though this carries the absolute minimum of instructions. The experimental studies of Snow (1973), on the effect of [^{3}H] thymidine on early stage mouse embryos, suggest that the ICM cells become metabolically distinct at the 16–32-cell stage. It is at this stage also, that apical tight junctions are apparent between peripheral cells of the embryo and that presumptive trophoblast cells become morphologically distinct (Calarco & Epstein, 1973).

THE REGULATIVE CAPACITY OF THE PRE-IMPLANTATION EMBRYO

The earliest experimental studies testing the regulative capacity of the pre-implantation mammalian embryo were those of Nicholas & Hall (1942). They found that single blastomeres from 2-cell rat embryos ($\frac{1}{2}$

blastomeres) were capable of developing at least to day 7 'egg cylinder' stages.

Subsequent studies have shown that $\frac{1}{2}$ and $\frac{1}{4}$ blastomeres of mouse (Tarkowski, 1959*a*, *b*) and rabbit (Seidel, 1956, 1960) form normal embryos after transfer to an appropriate foster mother. Although normal viable young have been produced from $\frac{1}{8}$ rabbit blastomeres (Moore, Adams & Rowson, 1968) technical difficulties have prevented, so far, an assessment of the apparently complete blastocyst developing from $\frac{1}{8}$ mouse blastomeres (Tarkowski & Wroblewska, 1967). Similarly, $\frac{1}{4}$ and $\frac{1}{6}$ pig blastomeres can form apparently normal blastocysts (Moore, Polge & Rowson, 1969).

Chimaeric (allophenic) animals have been produced from aggregated pairs of rat (Nicholas & Hall, 1942; Mayer & Fritz, 1974), sheep (Pighills, Hancock & Hall, 1968) and mouse eggs (Tarkowski, 1961, 1963, 1965; Mintz, 1962*a*, *b*, 1965). At birth these chimaeras are normal in size. In the mouse such size regulation occurs after the egg cylinder stage but prior to pro-amniotic cavity formation (Buehr & McLaren, 1974).

In the mouse, egg aggregation can be regularly achieved using embryos up to the early blastocyst stage (Stern & Wilson, 1972). Later blastocysts will not form such aggregates unless pretreated with versene (Stern, 1972*a*). This is, perhaps, because exposure to versene breaks down cell contacts in the presumptive trophoblast, thus facilitating the establishment of new contacts with the closely-adjacent egg. Thus, the importance of purely physical relationships between cells must not be underestimated when differentiative changes and regulative capacity are considered. Similarly, experiments on disaggregated late cleavage stages indicate that normal physical relationships between cells are essential if the fluid necessary for cavitation is to be discharged correctly from the trophoblast cells (Stern, 1972*a*).

In an elegant series of manipulations, Gardner (1968, 1971, 1972) demonstrated that trophoblast and ICM derived from the late blastocyst stage exhibit important differences in their properties. Pure trophoblast is capable of fluid accumulation and also of evoking a decidual response from sensitised uterine tissues. These activities are not exhibited by pure ICM which resembles the early cleavage stage embryo in its ability to fuse; in this case with other ICMs. Neither of these two cell types is capable of regulating its development to form complete blastocysts when cultured separately *in vitro*. However, despite its comparatively differentiated state the late blastocyst is still a relatively labile system. This has been shown by its capacity to regulate for the removal of substantial quantities of ICM (Lin, 1969) or trophoblast (Gardner, 1971, 1972) and by the ability of partially dissociated late stage blastocysts to form single, apparently normal, blastocysts when joined in pairs (Stern, 1972*a*). It would be interesting to know if the trophoblast and ICM of late blastocysts separated by Gardner's technique could be persuaded to regulate their development by dissociation and reaggregation, as we still do not know when the two cell types become irrevocably committed to their differentiated state.

The ease with which reaggregation and regulation occur in disaggregates in which spatial relationships are completely upset, implies that any polarity derived from the oocyte cannot be functionally essential to normal development. Indeed, it is not apparent what functional significance such 'preformation' might have as it seems likely that orientation of the ICM within the trophoblast is not determined until the time of implantation (Kirby, Potts & Wilson, 1967). At this time the embryonic axis takes up a characteristic orientation with respect to the uterine axis by relative movement of the ICM within the trophoblast shell (Jenkinson & Wilson, 1970). The early embryo can regulate its development following disturbance of the spatial arrangement of its cells by disaggregation, by addition or removal of some cells or by aggregation with a similar stage embryo. More remarkably it can regulate for a temporal discrepancy between its cells. Mulnard (1971) described the aggregation of asynchronous pairs of 4-cell and 8-cell mouse eggs which yielded blastocysts. Stern & Wilson (1972) aggregated pairs of 8-cell eggs with morulae or early blastocysts to produce blastocysts capable of developing at least to egg cylinder stages. The ICM is also capable of such 'chronological regulation' since an isolated $4\frac{1}{2}$ day ICM transferred to a $3\frac{1}{2}$ day blastocyst will be incorporated to produce an apparently 'normal' chimaera (Gardner, 1971, 1972). The most dramatic demonstration of regulative abilities of the early mammalian embryo followed these demonstrations of 'chronological' regulation. The development of mouse and rat eggs is similar in many ways but they follow different time schedules. However, aggregation of 8-cell rat eggs paired with 8-cell mouse eggs has been achieved with successful development of chimaeric blastocysts (Mulnard, 1973; Stern, 1972*b*; Zeilmaker, 1973) and 8-cell rat eggs paired with mouse 4-cell or early morula aggregated to form blastocysts (Stern, 1973*b*). It is not yet known whether such chimaeras are capable of further development. However, Gardner & Johnson (1973) have produced chimaeras by inserting ICM, isolated from 4.5 day rat blastocysts into the cavity of 3.5 day mouse blastocysts; the transferred ICM contributed to embryonic development and rat and mouse cells survived side by side in apparently 'normal' tissues at least until the 30-somite stage of embryogenesis.

FURTHER STUDIES

The observations presented above support an interpretation of early development based upon epigenetic factors controlling the differentiation of trophoblast and ICM; but how do we assay epigenetic factors? It should be apparent that most of the experiments reviewed here contribute mainly to a knowledge of the versatility of the embryo and its parts under artificially contrived conditions and they may be no more than pointers to the available pathways through which differentiation might be effected. Hence, the extent to which the properties exhibited under these conditions are related to the development of the intact embryo *per se* requires further study directed at analysis of the physiological activities within and between the cells of the differentiating morula/blastocyst.

This is not to deny that manipulations of early embryos have helped to elucidate many aspects of later embryonic and foetal development, but techniques which produce immediately analysable results in pre-implantation stages are rare. Our knowledge of changes in protein synthesis (Brinster, 1973) and nucleic acid metabolism (Graham, 1973), in particular the very early activation of the genome, is quite extensive. However, there are very few data on physiological properties of, and interactions between, cells of the cleavage and blastocyst stages.

In one of her early papers on the production of genetic chimaeras, Mintz (1964) conceived an interesting combination of (mutant + normal) eggs for aggregation into blastocysts. She paired eggs carrying the lethal mutant t^{12}/t^{12}, which naturally arrested development at the morula stage, with normal eggs and she found that the mutant cells could survive and develop in this combination, at least to the blastocyst stage. She reasoned that mosaics of this kind, containing cells which were non-viable independently, would be useful for analyses of intercellular relationships and morphogenetic interactions. Eicher & Hoppe (1973) have used chimaeras in this way in a study of X-linked genes which are lethal in the male. We have tried to follow up Mintz's idea using drug-treated eggs, rather than genetic mutants, as a means of gaining information on interactions between cells of the cleaving egg.

Most of our work has been with actinomycin D (AMD), though puromycin and colchicine have been used in a few experiments. (Some of the experiments mentioned here were performed by Honours and MSc students, as part of their training, in the past five years; in particular by Margaret Acutt, Pamela Hardman and Peter Candiah.)

The early activation of the genome in the cleaving mouse egg is made evident by, for example, the appearance of nucleoli at the 4-cell stage

(Mintz, 1964; Graham, 1973 for review). Continued development of the mammalian embryo through cleavage to the blastocyst appears to depend upon contemporaneous transcription of RNA and this is reflected in the response of the embryo to drugs which block RNA activity at transcription or translation. AMD blocks RNA transcription, and cleaving eggs exposed to a dose of 1×10^{-1} μg ml^{-1} in culture are almost immediately arrested in development (Skalko & Morse, 1969; Monesi & Molinaro, 1971).

In our experiments we have used 8-cell mouse eggs, routinely cultured in Whittingham's medium (1971). Eggs were denuded of their zonae with pronase and then cultured for a 2 h recovery period before being exposed to AMD (Sigma Chemical Company) at 1×10^{-1} μg ml^{-1} in Whittingham's medium with 10% serum added but excluding albumin and antibiotics. Higher levels of AMD appeared to have toxic side effects. The eggs were exposed to AMD for only 2 h, then they were washed three times and set up in various combinations for culture. Prior removal of the zona was a very necessary step with only a 2 h AMD treatment period. (Skalko & Morse (1969) had left zonae round eggs and treated with AMD for 48 h.) Out of 35 8-cell eggs with intact zonae exposed to AMD for 2 h, 28 (80%) showed further development (7 up to morulae, 21 up to blastocysts); 68 eggs denuded of zonae were similarly exposed for 2 h and none developed further (an abortive cleavage occurred in 4 of these). Chimaeric aggregations of 8-cell eggs were made in the following pairings: normal + normal; treated + treated; treated + normal. The pooled results of all the experiments in which at least some of the normal + normal aggregations succeeded are summarised in Table 1.

Table 1. *Aggregation of AMD treated (tr) 8-cell eggs with normal (n) partners*

Combination	No. of examples	Development achieved		
		None	Morula	Blastocyst
n + n	37		1	36
tr + tr	30	25	1	4
tr + n	105	15	4	86

Thirty-seven normal pair aggregates were made and all developed into rounded morulae, 36 of which continued development to large blastocysts. Only 5 of 30 treated pairs developed beyond the 8 + 8-cell stage, 4 of them forming blastocysts, but with several cells excluded. Most interestingly, of the 105 treated + normal pairs, 90 rounded up more or less completely and cleavage occurred in treated and untreated cells to form morulae, of which 86 went on to form large, apparently healthy, blastocysts. It is a common observation when producing these chimaeras that some of the cells of one, or both, partners may not be incorporated into the morula. This was observed in our experiments, significantly so with the treated eggs where varying numbers of the 8-cell blastomeres were not incorporated but remained loosely attached to the differentiating aggregated part; usually these cells did not cleave further and started degenerating towards the end of a 48 h culture period. Only pairs in which more than half of the treated partner was incorporated are included in the numbers of successful blastocysts shown in the table.

A preliminary attempt has been made to assess the viability of the AMD treated + untreated chimaeric blastocysts. Eggs from a pigmented strain (C3H or CBA) were treated with AMD as described and paired with untreated eggs from an albino strain (LACA or A2G). Blastocysts developing successfully from these aggregates were then transferred to albino foster mothers. Three young have been born; one single-born was, unfortunately, almost immediately eaten but another litter of two survived. At birth these two young, both apparently normal males, were larger than the usual birth size, but this could simply be due to lack of uterine 'competition'. However, their initial growth rate was substantially below normal so that at three weeks of age they weighed only about half (6.9 g and 7.9 g) the weight of males of either parental strain (approx. 13.0 g). The smaller male was sacrificed at three weeks and examined for possible chimaerism. Both had complete phenotypically albino coats and examination of the sacrificed male has yielded no evidence of chimaerism in any tissue. Gonads and reproductive tract were typically male and there was no evidence to account for the apparent 'runting'. The surviving male showed a remarkable burst of growth in its fourth and fifth week and quickly reached average weight. (It is possible that the runting was due to some undetected maternal inadequacy during suckling.) It has since been mated with an albino female and all its offspring appear to be normal albinos.

Although we have not yet shown that AMD-treated 8-cell embryos can contribute to a chimaeric foetus, the fact that they can be supported and induced (?) by normal cells to undergo cell divisions, and take part in blastocyst formation, fulfils the aim of the experiments. It should be possible to analyse the nature of the support provided and to use other drugs or metabolic analogues in this system to demonstrate the character of cellular interactions at this stage. Some rather preliminary experiments with puromycin using the same procedures have given results similar to those with AMD.

Puromycin dihydrochloride (Sigma Chemical Company), at a dose of

50 μg ml^{-1} in the culture medium, will block the development of denuded 8-cell embryos; this blockage can be overcome by aggregating treated with normal embryos and the treated component then contributes to formation of a blastocyst. Puromycin has not proved as useful as we had hoped because, under our conditions, it causes blebbing of the cell surfaces and this seems to impede aggregation.

Exposing 8-cell embryos (with or without zonae) to media containing 1×10^{-5}M colchicine for 15 min is sufficient to block cleavage. We have some evidence that, if such arrested embryos are aggregated with normal partners, the blockage can be bypassed so that the treated cells divide and form part of a blastocyst (though it is possible that the colchicine effect may be reversed simply by elution of the drug when the eggs are placed in fresh medium).

We intend to try asynchronous combinations of eggs with, for example, an 8-cell stage treated with AMD then aggregated with a normal late morula. Does the support offered carry with it an element of dominance so that the younger cells are induced to behave like the older supporting cells and to start accumulating blastocoelic fluid precociously?

Hopefully, analysis of the cellular interactions in the example mentioned and in similar material treated with a variety of specific antimetabolites or analogues could provide information on the cellular and intercellular activities which promote and control differentiation in the pre-implantation embryo.

We are grateful to the ARC, SRC and The Wellcome Trust (IBW) and to the MRC (MSS) for financial support. Miss Valmai Griffiths has provided skilled and much appreciated technical assistance.

REFERENCES

ASSHETON, R. (1898*a*). The segmentation of the ovum of the sheep with observation on the hypothesis of a hypoblastic origin for the trophoblast. *Quarterly Journal of Microscopical Science*, **41**, 205–61.

ASSHETON, R. (1898*b*). The development of the pig during the first ten days. *Quarterly Journal of Microscopical Science*, **41**, 329–60.

AUSTIN, C. R. (1961). *The Mammalian Egg*. Oxford: Blackwell Scientific.

BELLAIRS, R. (1971). *Developmental Processes in Higher Vertebrates*. London: Logos Press.

BIGGERS, J. D. & STERN, S. (1973). Metabolism of the preimplantation mammalian embryo. *Advances in Reproductive Physiology*, **6**, 1–59.

BOYD, J. D. & HAMILTON, W. J. (1952). Cleavage, early development and implantation of the egg. In *Marshall's Physiology of Reproduction*, ed. A. S. Parkes, vol. 2, pp. 1–126. London: Longman.

Brinster, R. L. (1968). *In vitro* culture of mammalian eggs. VIIIth Biennial Symposium on Animal Reproduction, *Journal of Animal Science*, **27**, Supplement **1**, 1–222.

Brinster, R. L. (1973). Protein synthesis and enzyme constitution of the preimplantation mammalian embryo. In *Regulation of Mammalian Reproduction*, ed. S. J. Segal, R. Crosier, P. A. Costman & P. G. Condliffe, pp. 302–16. Springfield, Illinois: Charles C. Thomas.

Buehr, M. & McLaren, A. (1974). Size regulation in chimaeric mouse embryos. *Journal of Embryology and Experimental Morphology*, **31**, 229–34.

Calarco, P. G. & Brown, E. H. (1969). An ultrastructural and cytological study of preimplantation development of the mouse. *Journal of Experimental Zoology*, **171**, 253–84.

Calarco, P. G. & Epstein, C. J. (1973). Cell surface changes during preimplantation development in the mouse. *Developmental Biology*, **32**, 208–13.

Cerisola, H. & Izquierdo, L. (1969). Mouse embryogenesis and RNA distribution. *Archivos de biología y medicina experimentales*, **6**, 10–16. (In Spanish.)

Dalcq, A. M. (1955). Processes of synthesis during early development of rodents' eggs and embryos. *Studies on Fertility*, **17**, 113–22.

Dalcq, A. M. (1957). *Introduction to General Embryology*. London: Oxford University Press.

Eicher, E. M. & Hoppe, P. C. (1973). Use of chimaeras to transmit lethal genes in the mouse and to demonstrate allelism of the two X-linked mouse lethal genes *jp* and *msd*. *Journal of Experimental Zoology*, **183**, 181–4.

Gardner, R. L. (1968). Mouse chimaeras obtained by the injection of cells into the blastocyst. *Nature, London*, **220**, 596–7.

Gardner, R. L. (1971). Manipulations on the blastocyst. Shering Symposium on Intrinsic and Extrinsic Factors in Early Mammalian Development, ed. G. Raspé. *Advances in Biosciences*, **6**, 279–96.

Gardner, R. L. (1972). An investigation of inner cell mass and trophoblast tissues following their isolation from the mouse blastocyst. *Journal of Embryology and Experimental Biology*, **28**, 279–312.

Gardner, R. L. & Johnson, M. H. (1973). Investigation of early mammalian development using interspecific chimaeras between rat and mouse. *Nature New Biology*, **246**, 86–9.

Graham, C. F. (1973). Nucleic acid metabolism during early mammalian development. In *Regulation of Mammalian Reproduction*, ed. S. J. Segal, R. Crosier, P. A. Costman & P. G. Condliffe, pp. 286–98. Springfield, Illinois: Charles C. Thomas.

Heape, W. (1886). The development of the mole (*Talpa europaea*), the ovarian ovum, and the segmentation of the ovum. *Quarterly Journal of Microscopical Science*, **26**, 157–74.

Heuser, C. H. & Streeter, G. L. (1929). Early stages in the development of pig embryos, from the period of initial cleavage to the time of the appearance of limb-buds. *Contributions to Embryology*, No. 109, **20**, 3–30. Carnegie Institution of Washington.

Hillman, N., Sherman, M. I. & Graham, C. (1972). The effect of spatial arrangement on cell determination during mouse development. *Journal of Embryology and Experimental Morphology*, **28**, 263–78.

Hillman, N. W., Tasca, R. J. & Wileman, G. (1967). Ultrastructural studies of preimplantation mouse embryos. *Journal of Cell Biology*, **35**, 56A.

Hubrecht, A. A. W. (1889). Studies in mammalian embryology. The placentation of *Erinaceus europeus*, with remarks on the phylogeny of the placenta. *Quarterly Journal of Microscopical Science*, **30**, 283–404.

HUXLEY, J. S. & DE BEER, G. R. (1934). *The Elements of Experimental Embryology*. London: Cambridge University Press.

IZQUIERDO, L. & VIAL, J. D. (1962). Electron microscope observations on the early development of the rat. *Zeitschrift für Zellforschung und Mikroskopische Anatomie*, **56**, 157–79.

JENKINSON, E. J. & WILSON, I. B. (1970). *In vitro* support system for the study of blastocyst differentiation in the mouse. *Nature, London*, **228**, 776–8.

KIRBY, D. R. S., POTTS, D. M. & WILSON, I. B. (1967). On the orientation of the implanting blastocyst. *Journal of Embryology and Experimental Morphology*, **17**, 527–32.

KRAUSKOPF, C. (1968*a*). Electronmikroskopische Untersuchungen über die Struktur der Oozyle und des 2-Zellenstadiums beim Kaninchen. I. Oozyle. *Zeitschrift für Zellforschung*, **92**, 275–95.

KRAUSKOPF, C. (1968*b*). Electronmikroskopische Untersuchungen über die Struktur der Oozyle und des 2-Zellenstadiums beim Kaninchen. II. Blastomeren. *Zeitschrift für Zellforschung*, **92**, 296–312.

LIN, T. P. (1969). Microsurgery of inner cell mass of mouse blastocysts. *Nature, London*, **222**, 480–1.

MAYER, J. F. & FRITZ, H. I. (1974). The culture of preimplantation rat embryos and the production of allophenic rats. *Journal of Reproduction and Fertility*, **39**, 1–9.

MINTZ, B. (1962*a*). Formation of genotypically mosaic mouse embryos. *American Zoologist*, **2**, 432.

MINTZ, B. (1962*b*). Experimental recombination of cells in the developing mouse egg: normal and lethal mutant genotypes. *American Zoologist*, **2**, 541.

MINTZ, B. (1964). Synthetic processes and early development in the mammalian egg. *Journal of Experimental Zoology*, **157**, 85–100.

MINTZ, B. (1965). Experimental genetic mosaicism in the mouse. In *Preimplantation Stages of Pregnancy*, *Ciba Foundation Symposium*, ed. G. E. W. Wolstenholme & M. O'Connor, pp. 194–207. London: Churchill.

MONESI, V. & MOLINARO, M. (1970). Macromolecular synthesis and effect of metabolic inhibitors during preimplantation development in the mouse. Schering Symposium on Intrinsic and Extrinsic Factors in early Mammalian Development, ed. G. Raspe. *Advances in Biosciences*, **6**, 101–20.

MOORE, N. W., ADAMS, C. E. & ROWSON, L. E. A. (1968). Developmental potential of single blastomeres of the rabbit egg. *Journal of Reproduction and Fertility*, **17**, 527–31.

MOORE, N. W., POLGE, C. & ROWSON, L. E. A. (1969). The survival of single blastomeres of pig eggs transferred to recipient gilts. *Australian Journal of Biological Sciences*, **22**, 979–82.

MORGAN, T. H. (1927). *Experimental Embryology*. New York: Columbia University Press.

MULNARD, J. G. (1961). Problèmes de structure et d'organisation morphogénétique de l'œuf mammifère. In *Symposium on the Germ Cells and Earliest Stages of Development*, pp. 639–88. Milan: Fondazione A. Baselli, Istituto Lombardo.

MULNARD, J. G. (1965*a*). Studies of regulation of mouse ova *in vitro*. In *Preimplantation Stages of Pregnancy*, *Ciba Foundation Symposium*, ed. G. E. W. Wolstenholme & M. O'Connor, pp. 123–38. London: Churchill.

MULNARD, J. G. (1965*b*). Aspects cytochimiques de la régulation *in vitro* de l'œuf de souris après destruction d'un des blastomères du stade II. I. La phosphomonoestérase acide. *Mémoires de L'Académie Royale de Belgique*, ser. 2, **5**, 35–67.

MULNARD, J. G. (1966). Les mécanismes de la régulation aux premiers stades du développement des mammifères. *Bulletin de la Société Zoologique de France*, **91**, 253–77.

MULNARD, J. G. (1971). Manipulation of cleaving mammalian embryo with special reference to a time-lapse cinematographic analysis of centrifuged and fused mouse eggs. Schering Symposium on Intrinsic and Extrinsic Factors in Early Mammalian Development, ed. G. Raspé. *Advances in Biosciences*, **6**, 255–274.

MULNARD, J. G. (1973). Formation de blastocystes chimériques par fusion d'embryos de rat et de souris au stade VIII. *Comptes rendus, Académie des sciences, Paris*, Ser. D, **276**, 379–81.

NICHOLAS, J. S. & HALL, B. V. (1942). Experiments on developing rats. II. The development of isolated blastomeres and fused eggs. *Journal of Experimental Zoology*, **90**, 441–60.

PIGHILLS, E., HANCOCK, J. L. & HALL, J. G. (1968). Attempted induction of chimaeras in sheep. *Journal of Reproduction and Fertility*, **17**, 543–7.

REVERBERI, G. (ed.) (1971). Experimental embryology of marine and fresh-water invertebrates. Amsterdam: North Holland Publishing Co.

RODÉ, B., DAMJANOV, I. & SKREB, N. (1968). Distribution of acid and alkaline phosphatase activity in early stages of rat embryos. *Bulletin Scientifique, Conseil des Académies de la RPF, Yugoslavia*, A, **13**, 304.

SEIDEL, F. (1956). Nachweis eines Zentrums zur Bildung der Keimscheibe im Säugetiere. *Naturwissenschaften*, **43**, 306–7.

SEIDEL, F. (1960). Die Entwicklungsfähigkeiten isolierter Furchungszellen aus dem Ei des Kaninchens *Oryctolagus cuniculus*. *Roux Archiv für Entwicklungsmechanik*, **152**, 43–130.

SKALKO, R. G. & MORSE, J. M. D. (1969). The differential response of the early mouse embryo to Actinomycin D treatment *in vitro*. *Teratology*, **2**, 47–54.

SNOW, M. H. L. (1973). Differential effect of [^{3}H]-thymidine upon 2 populations of cells in preimplantation mouse embryos. In *The Cell Cycle in Development and Differentiation*, ed. M. Balls & F. S. Billett, pp. 311–23. London: Cambridge University Press.

SOTELO, J. R. & PORTER, K. R. (1959). An electron microscope study of the rat ovum. *Journal of Biophysical and Biochemical Cytology*, **5**, 327–42.

STERN, M. S. (1972*a*). Experimental studies on the organisation of the early rodent embryo. II. Reaggregation of disaggregated embryos. *Journal of Embryology and Experimental Morphology*, **28**, 255–61.

STERN, M. S. (1972*b*). Experimental studies on the organisation of the early rodent embryo. PhD Thesis, University of Wales.

STERN, M. S. (1973*a*). Development of cleaving mouse embryos under pressure. *Differentiation*, **1**, 407–12.

STERN, M. S. (1973*b*). Chimaeras obtained by aggregation of mouse eggs with rat eggs. *Nature, London*, **243**, 472–3.

STERN, M. S. & WILSON, I. B. (1972). Experimental studies on the organisation of the preimplantation mouse embryo. I. Fusion of asynchronously cleaving eggs. *Journal of Embryology and Experimental Morphology*, **28**, 247–54.

TARKOWSKI, A. K. (1959*a*). Experiments on the development of isolated blastomeres of mouse eggs. *Nature, London*, **184**, 1286–7.

TARKOWSKI, A. K. (1959*b*). Experimental studies on regulation in the development of isolated blastomeres of mouse eggs. *Acta Therologica*, **3**, 191–267.

TARKOWSKI, A. K. (1961). Mouse chimaeras developed from fused eggs. *Nature, London*, **190**, 857–60.

TARKOWSKI, A. K. (1963). Studies on mouse chimaeras developed from eggs fused *in vitro*. *National Cancer Institute Monograph*, **11**, 51–71.

TARKOWSKI, A. K. (1965). Embryonic and postnatal development of mouse chimaeras. In *Preimplantation Stages of Pregnancy, Ciba Foundation Symposium*, ed. G. E. W. Wolstenholme & M. O'Connor, pp. 183–93. London: Churchill.

TARKOWSKI, A. K. & WROBLEWSKA, J. (1967). Development of blastomeres of mouse eggs isolated at the 4- and 8-cell stage. *Journal of Embryology and Experimental Morphology*, **18**, 155–80.

WILSON, I. B., BOLTON, E. & CUTTLER, R. H. (1972). Preimplantation differentiation in the mouse egg as revealed by microinjection of vital markers. *Journal of Embryology and Experimental Morphology*, **27**, 467–79.

WHITTINGHAM, D. G. (1971). Culture of mouse ova. *Journal of Reproduction and Fertility*, Supplement, **14**, 7–21.

ZEILMAKER, G. H. (1973). Fusion of rat and mouse morulae and formation of chimaeric blastocysts. *Nature, London*, **242**, 115–16.

STUDIES OF THE POTENCY OF THE EARLY CLEAVAGE BLASTOMERES OF THE MOUSE

BY S. J. KELLY

Department of Zoology, South Parks Road, Oxford OX1 3PS

There is still a controversy about the mechanism of development of the early mammalian embryo, and how it comes, at the blastocyst stage, to contain two distinct cell populations (see review by Wilson & Stern, this volume). Broadly, there are two opposing views about the mechanism of this differentiation: either that it occurs because different cells inherit different cytoplasmic substances, or that it occurs because the blastomeres become different as a result of their differing positions in the embryo and therefore differing micro-environments. The evidence to date for and against both these views has been presented by Drs Wilson & Stern. However, much of the evidence cited is indirect or equivocal: the histological and histochemical data on the mammalian egg and early embryo (see Wilson & Stern for references); the isolation experiments of Tarkowski & Wroblewska (1967); the aggregation experiments of Mintz (1965); the cell micro-injection experiments of Wilson, Bolton & Cuttler (1972); and the experiments of Hillman, Sherman & Graham (1972) extending as far as the blastocyst stage. Many of these experiments point to a lability of the early cleavage blastomeres which is incompatible with the first theory. However, to exclude this theory it is necessary to show a complete lack of restriction in developmental potential of all the early cleavage stage blastomeres. Direct evidence that the blastomeres are not restricted in their developmental potential would come from studies which showed that whole embryos may be obtained from isolated blastomeres. The present state of such evidence is that in the mouse and rat, one of the blastomeres at the 2-cell stage may develop into a normal embryo (Tarkowski, 1959; Nicholas & Hall, 1942), and that in the rabbit one of the blastomeres at the 2-, 4-, or 8-cell stages may form a normal rabbit (Siedel, 1960; Moore, Adams & Rowson, 1968). However, these experiments, all employing destruction of all but one of the blastomeres, do not tell us that such totipotency is a property common to all the blastomeres at each stage in question.

The experiments described here are attempts to analyse the full development of *all* the blastomeres of 4-cell stage and 8-cell stage embryos of the mouse.

EXPERIMENTAL PROCEDURE

These experiments were designed to test the potency of *all* the individual blastomeres of a particular donor embryo. The criterion used to establish the potency of any individual blastomere was its ability to contribute to the embryo, yolk sac and trophoblast on the 10th day of pregnancy, and extensively to an adult mouse, including the germline. Isolation experiments have shown that it is difficult, if not impossible, to obtain post-implantation development from mouse blastomeres isolated at stages after the 2-cell stage. Thus, in order to follow such advanced development of each blastomere after isolation from the whole embryo, they were combined individually with genetically distinct blastomeres, termed 'carrier' blastomeres. These chimaeric combinations were cultured to the blastocyst stage and transferred to pseudopregnant hosts (techniques of dissociation and culture were based on those described in Hillman *et al.*, 1972). The resulting embryos were recovered at the 10th day of pregnancy or at term.

Donor embryos were obtained from mice homozygous for the *a* allele at the glucosephosphate isomerase locus (*Gpi*-1) (DeLorenzo & Ruddle, 1969), and at the albino locus (*c*/*c*) (Green, 1966). These were 'A' type embryos. 'Carrier' embryos were obtained from mice homozygous for the *b* allele of *Gpi*-1 and for pigmentation at the albino locus (*C*/*C*). These were 'B' type embryos.

Blastomere potency at the 4-cell stage: 'Quartet' experiments

To test the potency of each blastomere at the 4-cell stage the experimental procedure was as shown in Fig. 1. Donor embryos were dissociated at the 4-cell stage, the isolated blastomeres were allowed to divide once, to show that they had suffered no immediate damage from the dissociation procedure, and then each was combined with six 8-cell stage blastomeres of the 'carrier' type. The sets of combinations with donor blastomeres derived from a single embryo were termed 'Quartets' (Plate 1*a*). Each 'Quartet' was cultured and transferred as a group. Embryos recovered on the 10th day of pregnancy were divided into three fractions – embryo, yolk sac and trophoblast of Hillman *et al.* (1972) – and analysed for their GPI-1 isozymes by the technique of Chapman, Whitten & Ruddle (1971). At term, embryos recovered were examined for eye and coat colour, and those that survived to adulthood were examined for donor contribution to the germline and GPI-1 of various tissues.

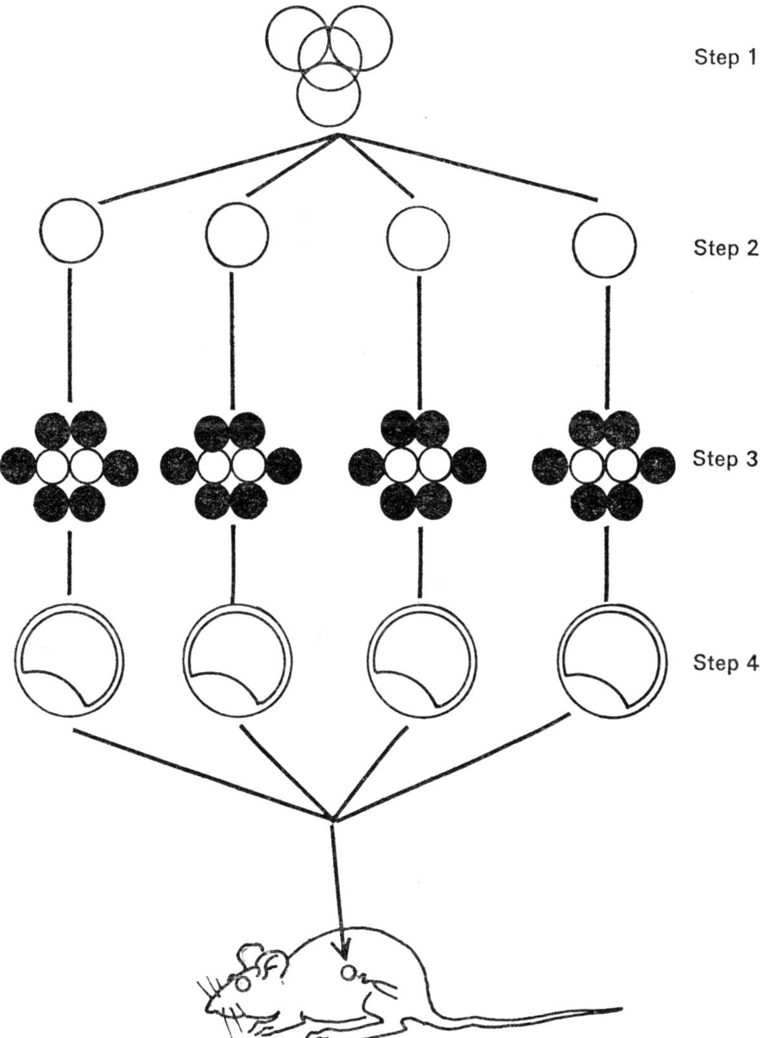

Fig. 1. Design for the 'Quartet' experiments to test the potency of all individual blastomeres of a 4-cell embryo. Step 1: Donor embryo obtained at the 4-cell stage. Step 2: The blastomeres are dissociated. Step 3: The blastomeres are combined, at the 8-cell stage, with 'carrier' type blastomeres. Step 4: The composites are cultured to the blastocyst stage and transferred as a group to a single uterine horn of a pseudopregnant recipient. ○ GPI-1A, albino embryo. ● GPI-1B, black embryo.

Blastomere potency at the 8-cell stage: 'Octet Pair' experiments

To test the potency of all the blastomeres of an 8-cell stage embryo by the direct method used above seemed an untenable proposition. Thus the procedure shown in Fig. 2 was adopted. The donor blastomeres were

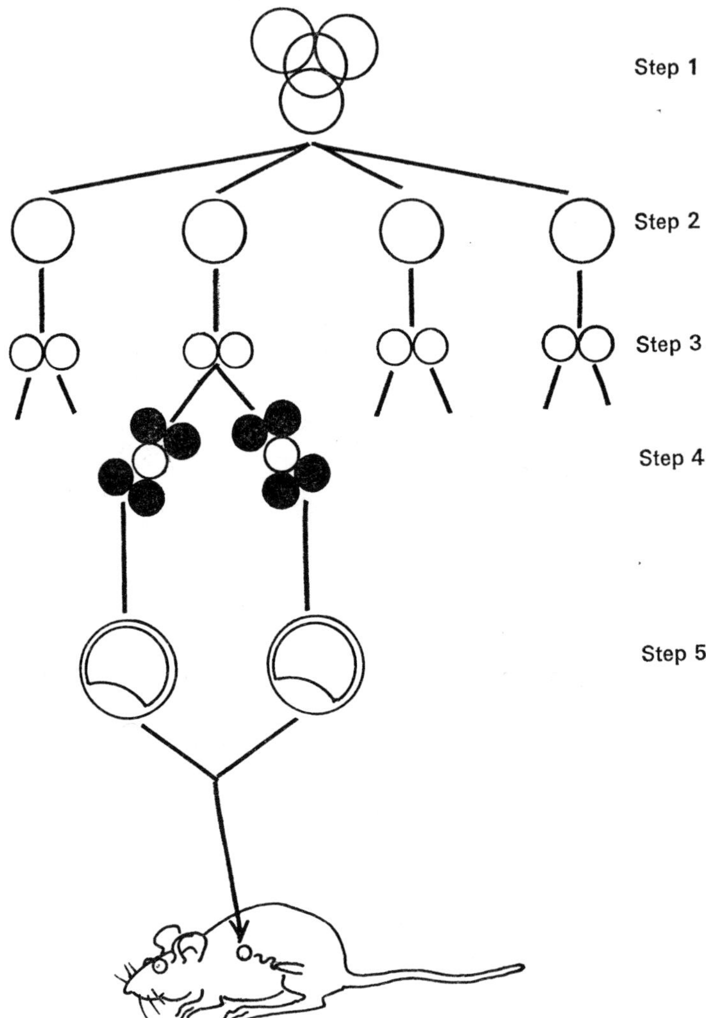

Fig. 2. Design for the 'Octet Pairs' experiments to test the potency of all individual blastomeres of an 8-cell embryo. Step 1: donor embryo obtained at the 4-cell stage. Step 2: the blastomeres are dissociated. Step 3: they divide to the 8-cell stage giving rise to four 'Octet Pairs' of blastomeres. Step 4: the pairs are further dissociated and each individual blastomere is combined with four 8-cell stage 'carrier' type blastomeres. Step 5: the pairs of composites are cultured to the blastocyst stage and transferred as groups to single uterine horns of pseudopregnant recipients. ○ GPI-1A, albino embryo. ● GPI-1B, black embryo.

separated at the 4-cell stage, each was allowed to divide once and then further separated into a pair of isolated 8-cell stage blastomeres. These were 'Octet Pairs'. The individual 8-cell stage blastomeres were each surrounded with four 8-cell stage 'carrier' type blastomeres, see Plate 1(*b*).

Each 'Octet Pair' was cultured and transferred as a group. Embryos were recovered at the 10th day of pregnancy or at term and analysed as described in the previous section.

RESULTS AND DISCUSSION

'Quartet' experiments

Recovery of the composites on the 10th day of pregnancy

One complete 'Quartet' of four embryos was among those recovered on the 10th day of pregnancy. The GPI-1 isozymes of the three fractions of each embryo are shown in Table 1. Each of the 4-cell donor blastomeres has contributed to the embryo proper and yolk sac of the embryo into which it was incorporated. Only two of the four have also contributed to the trophoblast of the recovered embryos. However, this apparent limitation of the potential of the other two blastomeres is probably artefactual since Hillman *et al.* (1972) have shown previously that at the 4-cell stage each blastomere may contribute to this tissue. These results lead to the conclusion that at the 4-cell stage there is no restriction of the developmental potential of the blastomeres in terms of their ability to contribute to the embryo, yolk sac and trophoblast on the 10th day of pregnancy.

Table 1. *Distribution of GPI-1 isozymes in embryos recovered on the 10th day of pregnancy in the 'Quartet' experiments*

Donor embryo	Embryo number	GPI-1 Analysis		
		Embryo	Yolk sac	Trophoblast
1	1	A+B	A+B	A+B
	2	A+B	A+B	A+B
	3	A+B	A+B	B
	4	A	A+B	B

Recovery of the composites at term

Forty-seven mice were recovered in these experiments. Of these, thirty-eight showed donor type characters. These included three 'Quartets' in which three out of the four embryos transferred came to term. The distribution of the donor type characters in these mice is shown in Table 2. Each mouse was scored at birth as having donor type or partially donor type eye colour. The coat colours of these mice, when they developed, were also assessed for donor contribution (albino). Mouse 6 was eaten by its foster

mother, but the complete lack of pigmentation in its eyes at birth suggested that it would also have been entirely albino as mice 4 and 5 are. The three mice from donor 1 are shown in Plate 1*c*. All the mice from each donor embryo are like sexed, as would be expected if the individual blastomeres each contributed extensively to the resulting mice. The blood of these mice has been analysed for GPI-1 isozymes. The proportion of donor type found follows coat colour closely. Indeed coat colour seems in general to be a very good indication of the overall degree of donor contribution to the mouse. The eight mice in Table 2 that reached adulthood were all found to have germlines that were exclusively of the donor type. (Of the other thirty-two mice showing donor contributions that survived to adulthood, all except one had donor type germlines. This one was an overtly chimaeric male.) Of the mice that died and were analysed for donor type GPI-1, all that had donor type eye colour also had donor GPI-1 in the tissues analysed: brain, liver, gut, spleen, kidneys, lungs, heart, skeletal muscle. These results indicate that in addition to being able to contribute to the three fractions analysed at the 10th day of pregnancy the blastomeres of the 4-cell embryo are all capable of contributing to all parts of an adult mouse, including the germline. This suggests that there is no segregation of

Table 2. *Distribution of donor type characters in mice recovered at term in the 'Quartet' experiments*

Donor embryo	Mouse number	Eye colour	Coat colour	Blood GPI-1	Sex	Germline
1	1	Donor	Donor	Donor	♂	Donor
	2	Donor	Donor	Donor	♂	Donor
	3	Donor/Carrier	Donor/Carrier	Donor/Carrier	♂	Donor
2	4	Donor	Donor	Donor	♀	Donor
	5	Donor	Donor	Donor	♀	Donor
	6	Donor	(Donor)	—	?♀	—
3	7	Donor/Carrier	Donor/Carrier	Donor/Carrier	♂	Donor
	8	Donor/Carrier	Donor/Carrier	Donor/Carrier	♂	Donor
	9	Donor/Carrier	Donor/Carrier	Donor/Carrier	♂	Donor

morphogenetic factors at the 4-cell stage in the mouse, not even for the germline, where one might expect that such factors would be readily detectable.

'Octet Pair' experiments

Recovery of the composites on the 10th day of pregnancy

A total of seven pairs of embryos were recovered on the 10th day of pregnancy. The distribution of the isozymes of GPI-1 in the embryos, yolk sacs and trophoblasts of these pairs of embryos is shown in Table 3. In all seven pairs the donor type was found in each embryo fraction. In five of the seven pairs there were donor contributions to each of the yolk sac fractions; in the other two pairs only one yolk sac in each had donor type, while the second yolk sac was carrier type only in one pair and was not scored in the other. In another five of the seven pairs there were donor contributions to both of the trophoblast fractions; in one of the remaining pairs the donor appeared in only one of the trophoblasts and in the final pair there was no donor contribution to either of the trophoblasts. These results, combined with the analysis of the unpaired embryos recovered, suggest that each of the 8-cell stage blastomeres of an embryo is capable of contributing to the embryo, yolk sac and trophoblast of embryos recovered on the 10th day of pregnancy.

Some of the embryos recovered lack a donor contribution to one of the three fractions analysed. However, this is not surprising since the donor cell made up only one fifth of the number of cells used in the original combinations, and the occasional exclusion of its progeny from one or other of the fractions might be expected.

Table 3. *Occurrence of the donor GPI-1 in seven paired embryos recovered on the 10th day of pregnancy in the 'Octet Pairs' experiments*

Occurrence of donor GPI-1 in:	Embryo	Yolk sac	Trophoblast
Both embryos of a pair	7	5	5
Only one embryo of a pair	0	2*	1
Neither embryo of a pair	0	0	1

* In one of these pairs the second yolk sac was not scored.

Recovery of the composites at term

Six mice were recovered at term (Table 4). Two of these mice came from the same donor embryo, but they were not a pair (Plate 1*d*). Five of the six mice showed donor contributions to the eye colour at birth and one did not. This latter died, and GPI-1 analysis of various tissues revealed no trace of the donor. Three of these five survived to at least one month of age, and all developed donor type coat colour. Two of these survived further, and one is breeding and has a donor type germline. The remaining two died at two days of age and GPI-1 analysis showed that one was a chimaera of donor and carrier type and the other was entirely of the donor type.

These results, although scanty, suggest that the 8-cell stage blastomere is not in any way restricted in potential and may in favourable circumstances develop into a fully fertile adult mouse.

Table 4. *Distribution of donor type characters in mice recovered at term in the 'Octet Pairs' experiments*

Donor embryo	Mouse number	Eye colour	Coat colour	Blood GPI-1	Sex	Germline
1	1	Donor	Donor	Donor	♂	Donor
2*	2	Donor	Donor	Donor	♂	?
	3	Donor	Donor	–	♂	–
3	4	Donor	–	Donor	?	–
4	5	Donor/Carrier	–	Donor/Carrier	?	–
5	6	Carrier	–	Carrier	?	–

* The two embryos derived from this donor were not from the same pair.

These experiments do not eliminate the possibility that a segregation of morphogenetic factors might be manifested at the 16-cell stage or later. However, in a simple-minded view of the differentiation of the inner cell mass and trophoblast cells in the mouse blastocyst, one must think in terms of generating two distinct populations of cells in the proportion of approximately 1 : 3 within a total number of about sixty-four cells (Graham, 1971). In a segregationist theory this would suggest that at the 16-cell stage there should be five or, at the most, six cells destined to give rise to the inner cell mass. If one then takes the simple view of segregation that particular morphogenetic factors in a cell may be inherited by only one *or* both of the cell's progeny at the next division, then, from eight *totipotent* cells, the only populations of *distinct* cells that could be generated would be in equal proportions, i.e. eight + eight. This kind of argument shows that totipotency of the blastomeres at the 8-cell stage of the mouse gives rise to difficulties for the straightforward view of the segregationist. Furthermore, totipotency at this stage of mouse development is in complete contradiction with the rigid formulation of Dalcq (1957), which states that there are two clearly distinct groups of blastomeres at the 8-cell stage, which have inherited his 'dorsal' and 'ventral' cytoplasm respectively. On the other hand, the proof of totipotency of the blastomeres of the mouse at the 8-cell stage does provide permissive conditions for, but does not prove, the positional theory of the formation of inner cell mass and trophoblast cell types in the blastocyst, since this theory requires absolutely that the blastomeres remain totipotent up until a stage at which their positions relative to each other in the whole embryo may be distinguishable.

PLATE I

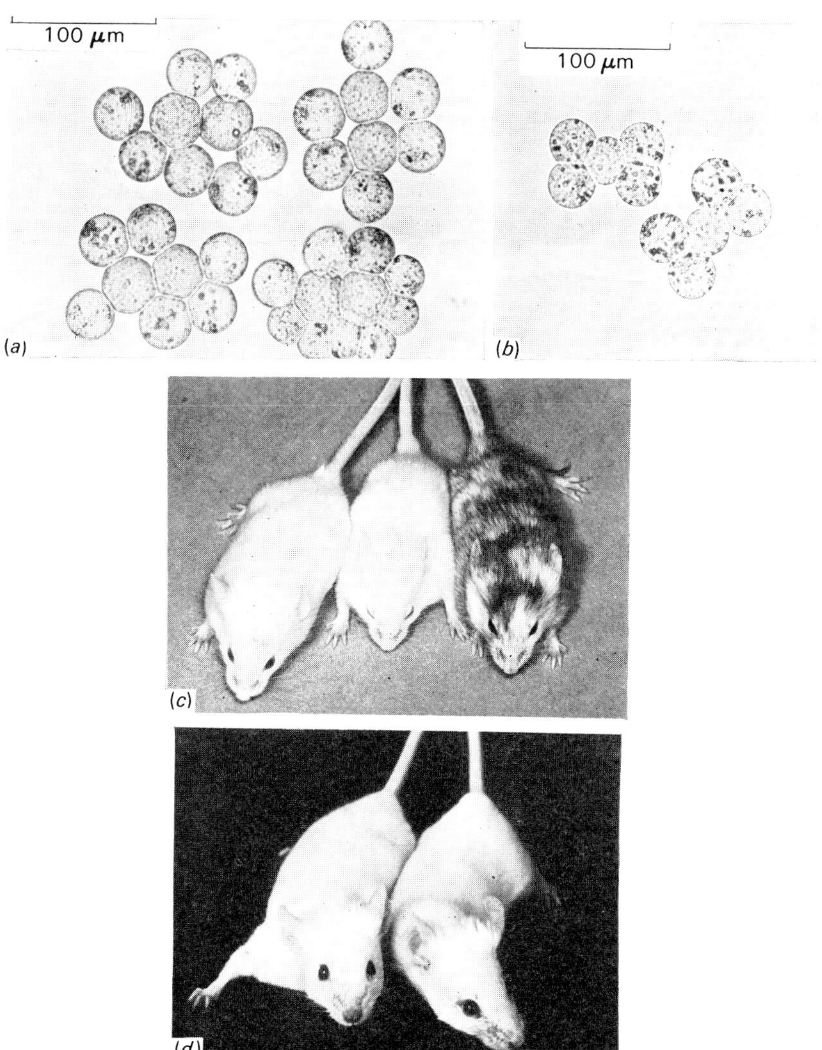

(Facing p. 104)

REFERENCES

CHAPMAN, V. M., WHITTEN, W. K. & RUDDLE, F. H. (1971). Expression of paternal glucosephosphate isomerase (*Gpi-1*) in preimplantation stages of the mouse. *Developmental Biology*, **26**, 153–8.

DALQC, A. M. (1957). *Introduction to General Embryology*. London: Oxford University Press.

DELORENZO, R. J. & RUDDLE, F. H. (1969). Genetic control of two electrophoretic variants of glucosephosphate isomerase in the mouse (*Mus musculus*). *Biochemical Genetics*, **3**, 151–62.

GRAHAM, C. F. (1971). The design of the mouse blastocyst. In *Control Mechanisms of Growth and Differentiation, Symposia of the Society for Experimental Biology*, **25**, 371–9.

GREEN, M. C. (1966). Mutant genes and linkages. In *Biology of the Laboratory Mouse*, 2nd edition, ed. E. L. Green, pp. 87–151. New York: McGraw-Hill.

HILLMAN, N., SHERMAN, M. I. & GRAHAM, C. F. (1972). The effect of spatial arrangement on cell determination during mouse development. *Journal of Embryology and Experimental Morphology*, **28**, 263–78.

MINTZ, B. (1965). Experimental genetic mosaicism in the mouse. In *Preimplantation Stages of Pregnancy, A Ciba Foundation Symposium*, ed. C. E. W. Wolstenholme & M. O'Connor, pp. 194–207. London: J. & A. Churchill.

MOORE, N. W., ADAMS, C. E. & ROWSON, L. E. A. (1968). Developmental potential of single blastomeres of the rabbit egg. *Journal of Reproduction and Fertility*, **17**, 527–31.

NICHOLAS, J. S. & HALL, B. V. (1942). Experiments on developing rats. II. The development of isolated blastomeres and fused eggs. *Journal of Experimental Zoology*, **90**, 441–61.

TARKOWSKI, A. K. (1959). Experiments on the development of isolated blastomeres of mouse eggs. *Nature, London*, **184**, 1286–7.

TARKOWSKI, A. K. & WROBLEWSKA, J. (1967). Development of blastomeres of mouse eggs isolated at the 4- and 8-cell stage. *Journal of Embryology and Experimental Morphology*, **18**, 155–80.

SEIDEL, F. (1960). Die Entwicklungsfähigkeiten isolienter Furchungszellen aus dem Ei des Kaninchens *Oryctolagus cuniculus*. *Wilhelm Roux Archiv für Entwicklungsmechanik der Organismen*, **152**, 43–130.

WILSON, I. B., BOLTON, E. & CUTTLER, R. H. (1972). Preimplantation differentiation in the mouse egg as revealed by micro-injection of vital markers. *Journal of Embryology and Experimental Morphology*, **27**, 467–79.

EXPLANATION OF PLATE

PLATE 1

(*a*) A 'Quartet' (see text) immediately after reassociation. The donor blastomeres are at the 8-cell stage and are in the centre of each group (see Fig. 1). The 'carrier' blastomeres are of the C57/J strain and may be distinguished by the coarse black granules in their cytoplasm.

(*b*) An 'Octet Pair' (see text). The blastomeres are arranged as in Fig. 2.

(*c*) Mice 1, 2 and 3 from donor embryo 1 in Table 2.

(*d*) Mice 2 and 3 from donor embryo 2 in Table 4.

DIFFERENTIATION IN THE TROPHECTODERM AND INNER CELL MASS

BY R. L. GARDNER AND V. E. PAPAIOANNOU

Department of Zoology, University of Oxford, Oxford OX1 3PS
and The Physiological Laboratory, University of Cambridge,
Cambridge CB2 3EG

It is evident from this volume that considerable effort has been devoted in recent years to the experimental investigation of mammalian development, despite the formidable technical difficulties involved. Obviously, it is hoped that such studies will lead to an understanding of human development and its manifold disorders. They may also prove relevant to a better general understanding of developmental phenomena. Certain processes such as protein synthesis, nervous conduction and the contraction of muscle, whose analysis has reached the molecular level, have been found to operate by similar mechanisms in very different organisms. This encourages the belief that developmental processes may also have an underlying unity in spite of the bewildering diversity that they display at a gross level. If this outlook is valid, broad-based comparative studies may be extremely valuable in elucidating mechanisms, by highlighting features common to different developmental systems.

We have been engaged for a number of years in developing and exploiting microsurgical procedures which enable investigation of lineage, interactions, determination and differentiation of cells in the early mammalian embryo. The mouse was chosen for this work because of its short gestation (approximately 20 days), availability of genetic markers, and the resilience of pre-implantation embryos to experimental abuse (Gardner, 1971). Recently, interspecific chimaeras between the mouse and rat have also been employed in these studies for reasons that will emerge later. However, before proceeding to a discussion of these experiments, a few prefatory remarks will be made about some of the aspects of development with which this paper is concerned.

Determination is familiar to all embryologists as the term used to describe the elusive processes whereby the potency of cells becomes restricted during embryogenesis. Whether or not cells are determined can only be established experimentally by testing their developmental potential in altered circumstances. Among the criteria employed are 'self-differentiation' following isolation or transplantation to atypical locations (Weiss,

1939), specific cell recognition in mixed aggregates, and inheritance of specific differentiative capacities through successive mitotic cycles in situations that promote division rather than differentiation (Gehring, 1972). Since these are by no means exhaustive, there always remains the possibility that circumstances might exist in which restriction of potency would not be sustained. Therefore it is perhaps desirable, as suggested by Needham (1942), to specify the way in which a particular determinative event has been defined. A feature of determination that is evident particularly in studies on amphibian, echinoderm and insect embryos is that it is progressive, proceeding stepwise from a more general to a more specific cellular commitment (Huxley & De Beer, 1934; Nöthiger, 1972; Horstadius, 1973). Thus, at a stage when groups of cells are determined for formation of a particular region or organ, the fate of individual cells appears, within certain limits, to be yet unspecified.

An accurate map of the normal fate of different regions of the early embryo is an essential prerequisite for detailed analysis of morphogenesis (Needham, 1942). The only satisfactory method of constructing such a map is to study the lineage relationships and deployment of cells marked in a way that avoids disturbing their spatial relations within the embryo. This limits the choice of markers to those produced by spontaneous and induced genetic or cytoplasmic mosaicism (Conklin, 1905; Sturtevant, 1929; Vogt, 1929; Davidson, 1968; Nöthiger, 1972). Since the early mammalian embryo has so far proved refractory to such analysis, knowledge of the fate of its constituent cells has, until recently, been based exclusively on examination of successive stages by standard histological procedures (Snell & Stevens, 1966). This is necessarily a most imprecise method because it lacks the continuity that cell markers provide. The discovery that combinations of early embryonic cells of different genotype can develop into viable offspring thus offers a potentially more satisfactory way of investigating cell lineage in mammals (Tarkowski, 1961; Mintz, 1962; Gardner, 1968). However, while affording a solution to the marker problem, induction of chimaerism inevitably entails disturbance of the relations between cells of the embryo. Therefore, one cannot necessarily expect the fate of a cell in chimaeric combination to be the same as if it had been left in its normal situation. Nevertheless, provided efforts are made to ensure that donor cells are correctly positioned in host embryos, it is reasonable to assume that they will at least approximate their normal fate in a chimaera. This is particularly so if they are already determined at the time of isolation. But how can one ascertain whether transplanted cells are already determined if there is no reliable alternative method of establishing their normal fate? The only available solution to this problem is to find out if they yield reproducible patterns of differentiation regardless of location in host embryos.

The apparent lack of precision in cleavage and the totipotency of blastomeres suggest that if normal cell lineages could be traced from an early stage they would vary between individual embryos of a given *mammalian* species. If this is indeed the case, they may be regarded as trivial in the sense that they would depend on the chance distribution of progeny of each cell. However, once the process of determination sets in, lineages will become progressively more predictable, and hence the fate of cells following transplantation will more closely resemble their fate in the intact embryo. Indeed, it is a basic tenet of vertebrate embryology that neuroblasts, neural crest, haemopoietic and other stem cells form mutually exclusive groups of mature cell types (e.g. Balinsky, 1970). This argues that cell lineages are no longer trivial once determination has begun, and suggests that particular differentiated cells are related in terms of the sequence of epigenetic programming through which their progenitors must pass (Holtzer, 1970).

This is not an appropriate place to attempt to define differentiation. Many authors have pointed out that a truly undifferentiated cell is an abstraction for which no counterpart in nature can be found (e.g. Holtzer, 1970). Thus the term will be employed in this paper in a relative sense to describe two phenomena. First, it will be used to denote an increase in morphological complexity of the embryo due to the appearance of discrete populations of cells. Second, it will be used in relation to cells, to denote the appearance of enduring physiological and or morphological characteristics which serve to distinguish them from other cells present at the same stage of development, and from their progenitors. Unfortunately, most cells of the mammalian embryo have not been sufficiently well-defined biochemically to enable them to be characterised in terms of luxury molecules (Holtzer & Abbott, 1968).

Experiments involving the isolation, transplantation and rearrangement of cells in cleaving mouse, rat and rabbit eggs, have demonstrated that at least some blastomeres retain totipotency until the 8-cell stage. Similar studies on the blastocyst suggest that the outer trophoblast, which will hereafter be referred to as the trophectoderm, and the inner cell mass (ICM), exist as two distinct determined populations of cells at $3\frac{1}{2}$ days *post coitum* (*p.c.*) in the mouse (Gardner, 1974*a*, *b*). Therefore, initial cellular commitment in the mouse embryo seems to occur between the 8- and approximately 64-cell stage. Furthermore, the relative position occupied by a cell during this period of development is evidently an important factor in deciding its fate. Cells placed in an outside position during later cleavage contribute most of their progeny to the trophoblast, while inside cells tend to colonise the ICM and its derivatives (Hillman, Sherman & Graham, 1972; Wilson, Bolton & Cuttler, 1972).

The various experiments on which the foregoing conclusions are based have been reviewed recently (Graham, 1973; Gardner, 1974*a*; Herbert & Graham, 1974) and are the subject of other contributions to this volume. They will therefore not be discussed further here. Instead, this article will deal with the subsequent development of the two primary tissues of the blastocyst in relation to the following questions.

(1) Do all trophoblast cells of the conceptus originate from the trophectoderm of the blastocyst, or are some derived from the ICM?
(2) What is the significance of the ICM in later development of the trophoblast?
(3) When do differences in developmental potential among ICM cells first arise?
(4) How is the differentiation of ICM cells affected by experimental changes in cell position?

LINEAGE RELATIONSHIPS OF TROPHOBLAST CELLS

A number of distinct types of trophoblast cells are found at the periphery of the mouse conceptus and constitute the immediate cellular boundary between embryonic and maternal tissues throughout most of pregnancy. Except in the chorio-allantoic placental region, this barrier is relatively simple, consisting of one or more layers of typically mononuclear giant cells. These cells, together with the underlying Reichert's membrane and distal endoderm, degenerate or disperse by the 15th day of gestation so that the proximal endoderm of the yolk sac splanchnopleure comes into direct contact with the uterine lumen thereafter (Amoroso, 1952).

In contrast, the mature chorio-allantoic placenta is an extremely complex organ in which a vast area of close approximation between foetal and maternal circulation is established. Analysis of the lineage of all the differentiated trophoblast cell populations that it contains would be impossible on the basis of histological studies alone. Clearly, this task can only be accomplished by following the fate of genetically marked cells. Nevertheless, the results of histological investigations can assist in defining the tissues of the early embryo which participate in the formation of later trophoblast. The following brief outline of the development of trophoblast in the mouse embryo is based on the work of several authors (Duval, 1892; Jenkinson, 1902; Snell & Stevens, 1966).

The mural trophectoderm cells which surround the blastocoel of the blastocyst cease dividing and transform into giant cells during early implantation. In contrast, the polar trophectoderm cells overlying the ICM continue to divide and thus form the ectoplacental cone which caps the mesometrial surface of the developing egg-cylinder. Giant cells, identical to the primary ones of the mural trophectoderm, are generated continuously in the periphery of the ectoplacental cone. These secondary giant cells eventually spread anti-mesometrially, so as to surround the rest of the conceptus. As noted earlier, they disappear before term.

As well as producing numerous giant cells, the ectoplacental cone acts as an invasive spearhead of tissue that penetrates the decidua basalis of the endometrium. The extra-embryonic ectoderm lining the ectoplacental cavity is incorporated into the developing ectoplacenta by mid-gestation as a result of fusion between the chorion and the base of the cone. This extra-embryonic ectoderm is believed later to form the barrier of the vast placental labyrinth (Amoroso, 1952), which consists of two cellular layers, and possibly a third syncytial layer, of trophoblast cells (Jollie, 1964; Enders, 1965; Kirby & Bradbury, 1965). The foetal vasculature of the placenta is provided by the allantois.

Hence, according to histological investigations, all trophoblast cells of the conceptus originate from the trophectoderm of the blastocyst and the extra-embryonic ectoderm of the slightly later egg-cylinder stage. Both a trophectodermal (Jenkinson, 1902) and an ICM origin (Rugh, 1968) have been claimed for the extra-embryonic ectoderm. Clearly, the former favours a unitary origin of all trophoblast cells, while the latter requires a dual origin. Several recent experiments employing genetically marked cells have helped to resolve this issue.

Experimental investigation of the origin of ectoplacental cone and giant trophoblastic cells

The precocious transformation of mural trophectoderm cells in the late blastocyst leaves little doubt that these are the progenitors of the primary giant cells (Alden, 1948; Dickson, 1966). This conclusion has been confirmed by the finding that typical giant cells are formed *in vitro* in outgrowths of trophectodermal vesicles in which development of the ICM has been completely suppressed (Ansell & Snow, 1974). The origin of the ectoplacental cone and secondary giant cells has been studied by experimentally combining trophectoderm and ICM tissue isolated from blastocysts homozygous for different alleles at the glucose phosphate isomerase locus (*Gpi-1*) (Gardner, Papaioannou & Barton, 1973). Nineteen conceptuses suitable for analysis were recovered six days after transplanting a series of 'reconstituted' blastocysts to the uteri of pseudopregnant mice. Each was dissected into a trophoblastic and an embryonic fraction prior to electrophoresis, as shown in Plate 1. The trophoblastic fraction of thirteen expressed only the isozyme of the trophectoderm donor genotype. Contamination with maternal and/or ICM-derived cells appeared to account for the presence of both isozymes in the remaining six. These results thus confirm histological studies in showing that the vast majority, if not all, ectoplacental cone and secondary giant cells originate, as do primary giant cells, from the trophectoderm of the blastocyst.

Evidence for trophectodermal origin of extra-embryonic ectoderm

The embryonic fractions of sixteen of the above conceptuses were sufficiently well-developed to enable them also to be analysed for glucose phosphate isomerase (GPI). Both trophectodermal and ICM donor isozymes were present in all except four of the sixteen. Furthermore, although it invariably formed the minor contribution, trophectodermal isozyme nevertheless represented a high proportion of the total activity in eight cases (see Fig. 2 in Gardner *et al.*, 1973). This suggested that progeny of trophectoderm cells may also give rise to part of the embryonic fraction of the conceptus.

Support for this idea comes from the analysis of post-implantation embryos made chimaeric by transplanting ICMs from 4½ day *p.c.* rat blastocysts into the blastocoel of 3½ day *p.c.* mouse blastocysts. The composite blastocysts can develop into viable foetuses, containing a substantial proportion of rat cells, after transfer to the uteri of recipient mice (Gardner & Johnson, 1973, 1974; Johnson & Gardner, 1974). The value of such chimaeras for developmental studies lies in the fact that all cells originating from either species can be identified in serially sectioned material by an indirect immunofluorescent technique, using antisera directed against species antigens (Gardner & Johnson, 1973). Hence the spatial distribution of the two populations of cells comprising the chimaera can be analysed with a degree of precision that is not possible at present in intra-specific chimaeras (Gardner & Johnson, 1974).

The donor rat ICM usually aggregates with that of the host mouse blastocyst so that a single embryo is formed. Rat cells are widely distributed in post-implantation stages, but are conspicuously absent from mural and ectoplacental trophoblast, and also from the extra-embryonic ectoderm.

Evidently, the ICMs do not always aggregate together because two separate egg-cylinders have been found within a single trophoblastic shell in three cases. Each of the three conceptuses showed the same pattern of distribution of rat and mouse cells by immune fluorescence. Thus, in each case the ectoplacental cones overlying both egg-cylinders, together with all other trophoblast cells, stained only with the anti-mouse serum. The larger cylinder was composed entirely of mouse cells. All tissues of the smaller twin egg-cylinder were rat, except the extra-embryonic ectoderm, which was composed exclusively of mouse cells (Gardner & Johnson, 1974).

Exchange between the two ICMs seems most unlikely to account for the observed distribution of rat and mouse cells. The results are most readily explained by supposing that the rat ICMs attached to the cells of the mural trophectoderm rather than the ICM of the host mouse blastocysts in these three instances, and thereby stimulated them to proliferate. As a result of local proliferation the trophectoderm cells are presumed to have pushed outwards to form a second ectoplacental cone, and also inwards, thereby displacing the underlying rat ICM into the blastocoel and forming

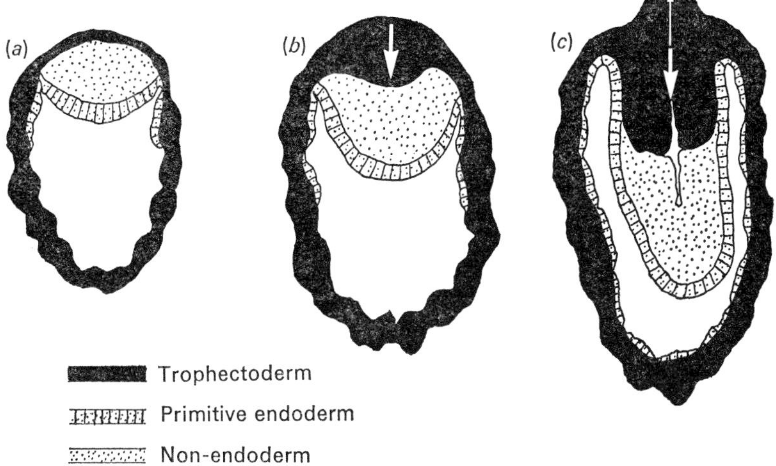

Fig. 1. Diagrams illustrating how the extra-embryonic ectoderm may develop from polar trophectoderm between $4\frac{1}{2}$ and $5\frac{1}{2}$ days *p.c.* (*a*) $4\frac{1}{2}$ day blastocyst before local proliferation of polar trophectoderm has begun. (*b*) Local proliferation results in accumulation of polar cells at the mesometrial (upper) pole of the blastocyst, which leads to invagination of the ICM and future extra-embryonic ectoderm into the blastocoel. (*c*) The latter pushes inside the sheath of primitive endoderm which is anchored mesometrially to adjacent mural trophectoderm cells. Invagination may be an active process, or the passive consequence of close investment of the blastocyst by uterine tissues. Once the blastocoelic space is thus filled, further proliferation of the polar trophectoderm leads to the mesometrial outgrowth of the ectoplacental cone into the space made available by degeneration of the uterine epithelium.

the extra-embryonic ectoderm. The primitive endoderm delaminates as a monolayer of cells on the blastocoelic surface of the ICM (Snell & Stevens, 1966; also see later). The most peripheral of these cells extend beyond the edge of the ICM and attach to the inner surface of adjacent non-dividing mural trophectoderm cells. Hence, one can readily visualise on this model how the trophoblast cells that grow inwards to form the extra-embryonic ectoderm come to be invested in a sheath of endoderm (Fig. 1).

The above hypothesis would also explain why trophectoderm cells appeared to make a significant cellular contribution to the embryonic fraction in most, rather than all, of the conceptuses which had developed from 'reconstituted' mouse blastocysts. It was assumed in this study that the extra-embryonic ectoderm lining the ectoplacental cavity was of ICM origin; therefore efforts were made to isolate it with the embryonic rather than the trophoblastic fraction (Plate 1). However, this was not possible in some of the more advanced embryos in which the chorion had already united with the base of the ectoplacental cone.

If the extra-embryonic ectoderm does indeed originate from the trophectoderm rather than the ICM, one might expect that its cells would have properties similar to those of the ectoplacental cone, at least in initial stages of its development. The following preliminary experiments indicate that this is so.

Similarities between extra-embryonic ectoderm and ectoplacental cone

Very early egg-cylinder stage embryos were recovered from the uterus at 5½ days *p.c.* and dissected as follows. First, Reichert's membrane was torn back to the base of the ectoplacental cone. The embryo was then divided in two just below this level with solid glass needles, held in Leitz micromanipulator units, so as to isolate the egg-cylinder from the as yet small cone. Finally, the egg-cylinder was itself cut in two across the junction between embryonic and extra-embryonic regions. The separated fragments were grafted individually under the testis capsule of adult male mice. Host organs were inspected for haemorrhagic sites five to seven days later, and then fixed and processed for histological examination.

Embryonic fragments formed compact grafts which had not caused any obvious erosion of adjacent testicular tissue. In contrast, extra-embryonic fragments resembled ectoplacental cones in producing haemorrhagic grafts in which cells of trophoblastic morphology predominated (Table 1 and Plate 2). Trophoblastic cells were in fact more numerous in grafts of the former than the latter, though clearly less advanced in giant transformation.

Grobstein (1950) grafted the embryonic regions of 6½ day *p.c.* mouse embryos into the anterior chamber of one eye of a series of adult mice.

Corresponding ectoplacental cones were placed in the opposite eyes. The latter almost invariably produced intra-ocular haemorrhage, whereas the former did not. He also grafted a limited number of extra-embryonic fragments which produced weak or doubtful haemorrhage in several cases.

Table 1. *Development of fragments of $5\frac{1}{2}$ day* p.c. *mouse egg-cylinders 5 to 7 days after transplantation under the testis capsule of adult male mice*

Type of fragment transplanted	Number transplanted	Number of grafts found	Number of compact non-haemorrhagic grafts lacking trophoblast cells	Number of invasive haemorrhagic grafts containing cells of trophoblastic morphology
Embryonic region	5	4	4	0
Extra-embryonic region*	5	4	0	4
Ectoplacental cone	4	3	0	3

* In two cases almost pure extra-embryonic ectoderm was transplanted because most of the surrounding endoderm was detached during isolation from the ectoplacental cone.

He did not describe the cellular composition of these grafts, but inclined towards the view that positive results were due to contamination with ectoplacental cone cells.

Care was taken in the present experiments to avoid contamination. The fact that extra-embryonic grafts were composed mainly of trophoblastic cells would be difficult to explain in this way. The reduced invasiveness of $6\frac{1}{2}$ day fragments in Grobstein's experiments may be due to the difference in graft site, or to changes in properties of the ectodermal cells as a consequence of differentiation. Certainly, grafts of later choric-allantoic placental tissue, whose trophoblast is believed to originate from extra-embryonic ectoderm, did not produce intra-ocular haemorrhage (Grobstein, 1950).

Experimental data thus support the view of Jenkinson (1902) that extra-embryonic ectoderm is derived from the trophectoderm rather than the ICM. All trophoblast cells of the mouse conceptus may therefore be mitotic descendents of the primary trophectoderm of the blastocyst. We are currently attempting to test this hypothesis by analysing the deployment of cells in conceptuses developing from blastocysts reconstituted from rat ICM and mouse trophectoderm.

ROLE OF THE ICM IN TROPHOBLAST PROLIFERATION AND DIFFERENTIATION

The fact that the trophectoderm of the mouse blastocyst exhibits several specific properties argues that its cells are already differentiated at $3\frac{1}{2}$ days *p.c.* (Gardner, 1971, 1972, 1974*a*, *b*). There is, however, no evidence to suggest that polar and mural cells are distinct populations at this stage of development (Gardner, 1974*b*). Twenty-four hours later this is no longer the case. Those of the mural trophectoderm are terminal cells by $4\frac{1}{2}$ days *p.c.* since, although they can continue to synthesise DNA (Barlow, Owen & Graham, 1972; Barlow & Sherman, 1972), they have lost the capacity to divide. Polar cells continue to divide actively and, on evidence presented earlier, probably generate all other trophoblast tissues.

Trophectodermal vesicles devoid of ICM cells can be produced either by microsurgery on the blastocyst (Gardner, 1971, 1972), or by culturing cleavage stages in the presence of [^{3}H]thymidine (Snow, 1973*a*, *b*). Though capable of implantation *in utero* and hatching and outgrowth *in vitro*, such vesicles do not show any further mitotic activity (Gardner, 1972; Ansell & Snow, 1974). Instead, their cells continue to synthesise DNA and transform into giant cells. Thus their subsequent development resembles that of mural rather than polar trophectoderm tissue. 'Half-blastocysts' yield extensive trophoblast following implantation in most cases, so lack of mitotic activity is not simply due to damage or reduction in the number of cells in the trophectodermal vesicles (Gardner, 1974*b*, *c*); rather, it would seem that all trophectoderm cells lose their intrinsic capacity to divide by the late blastocyst stage. This conclusion is supported by the fact that vesicles of mural trophectoderm show extensive proliferation *in utero* if ICM tissue is injected into them before giant cell formation begins (Gardner, 1971). The trophoblast produced in this way expresses the marker of the trophectoderm rather than ICM donor blastocysts (Gardner *et al.*, 1973). Therefore, the injected ICMs are not themselves producing the trophoblast, but stimulating mural trophectoderm cells to do so. The ICM thus appears to be acting like an inducing tissue (Spemann, 1938). As noted earlier, rat ICMs that develop separately in host mouse blastocysts also exhibit this property.

By extrapolation from these experimental results, it would appear that regional differentiation of the mural and polar trophectoderm is determined by the presence and position of the ICM in the intact blastocyst. We have argued that the trophectoderm cells to which the developing ICM attaches receive a mitotic stimulus essential for further proliferation, while the remainder transform into non-dividing giant cells through lack of inductive support. A detailed statement of this hypothesis, together with the evidence on which it is based, has been presented elsewhere (Gardner, 1971, 1972, 1974*a*, *b*; Gardner *et al.*, 1973).

Little can be said at present as to how the ICM exerts its effect. Giant transformation begins consistently in trophectoderm cells remote from the ICM, but eventually involves all except those immediately overlying it (Dickson, 1966). While the former observation might suggest that the ICM produces a diffusible substance, the latter clearly does not. The two observations may be reconciled if mural trophectoderm is being recruited by division of polar cells in the later blastocyst stage, in which case cells closer to the ICM would have passed through their last mitosis more recently than those that are further away. This possibility has yet to be investigated.

Dependence of trophoblast on the ICM for its continued division seems to differ from a classical inductive interaction in that it is evidently not transient, but extends into later stages of development. Cessation of cell division and giant transformation is the fate of all trophoblastic cells in the periphery of the ectoplacental cone, and in other parts of the conceptus remote from derivatives of the developing ICM. It is not a specific response to the proximity of maternal decidual cells, because it also occurs in ectopic grafts (Avery & Hunt, 1969), and is the ultimate fate of trophoblast cells in culture (Gwatkin, 1966; Koren & Behrman, 1968). Further evidence of the need for sustained interaction with the ICM comes from histological observations on the trophoblast of embryos developing abnormally *in utero* and in ectopic sites (Gardner, 1972, 1974*c*; Johnson, 1972). In general, the extent of trophoblast proliferation correlates well with the duration and degree of development of the ICM. Early death or arrest of ICM growth is accompanied by fewer trophoblast cells than later death. Furthermore, when the ICM dies or its tissues are walled off from the trophoblast by Reichert's membrane material, trophoblast cell division appears soon to cease, and all existing cells assume the giant form. From the time of implantation mitotic activity in the trophoblast is restricted to the ectoplacental cone and extra-embryonic ectoderm, and later to the chorio-allantoic placenta (Gardner, 1974*b*). The placenta contains allantoic tissue derived from the ICM (see later) in intimate association with trophoblast cells. However, it remains to be seen whether ICM-derived tissue is necessary to sustain proliferation of the trophoblast throughout development.

Effort is currently being devoted to finding an in-vitro system that will enable the nature of the interaction between trophoblast and ICM to be more precisely defined. We are also trying to determine the types of cells that can promote cell division in trophectodermal vesicles.

DEVELOPMENTAL POTENTIAL OF ICM CELLS

It was suggested earlier that the ectoplacental cone, trophoblastic giant cells and extra-embryonic ectoderm of the egg-cylinder originate from the trophectoderm of the blastocyst. Thus all other tissues presumably arise from the ICM. These are the embryonic ectoderm, the distal endoderm, and both the embryonic and extra-embryonic parts of the proximal endoderm (Snell & Stevens, 1966). The primitive endoderm can first be discerned at approximately 4½ days *p.c.*; the embryo attains the egg-cylinder stage 24 hours later. Several experimental approaches have been employed to establish when ICM cells are first committed to form these different tissues. The data presented here concern the status of ICM cells at 3½ and 4½ days *p.c.*

At 3½ days the ICM consists of a cluster of some 10–15 cells which are somewhat similar in properties to those of earlier cleavage stages (Gardner, 1972), though apparently unable to form trophoblast (Gardner, 1974*b*; Rossant, 1975*a*, *b*). The cells exhibit no special ultrastructural features, and are linked together by interdigitating membranes rather than defined, junctional complexes (Enders & Schlafke, 1965; Enders, 1971). By 4½ days *p.c.* a distinct cell monolayer can be seen extending across the blastocoelic surface of the ICM and adjacent mural trophectoderm in living embryos, by both bright-field and phase contrast microscopy. This primitive endoderm layer stains more deeply than other cells in sectioned embryos, and possesses certain special ultrastructural features. Adjacent cells are united by junctional complexes similar to those found between trophectoderm cells, and are separated from underlying ICM cells by a basement membrane (Enders, 1971). They are readily distinguished from both trophectoderm and other ICM cells by the fact that their cytoplasm is rich in rough endoplasmic reticulum (Enders, 1971).

Dalcq (1957) suggested on the basis of histochemical staining of rat blastocysts that the endoderm is formed by ingrowth of trophectoderm rather than delamination of cells from the blastocoelic surface of the ICM. Recent data show that this viewpoint is untenable. First, vesicles of mural trophectoderm do not form endoderm following transfer to the uterus (Gardner, 1971, 1972). Second, rat ICMs isolated prior to endoderm formation colonise the endoderm as well as other tissues of host mouse blastocysts (Gardner & Johnson, 1974). Finally, the outer cells of ICMs isolated from 3½ day *p.c.* mouse blastocysts and injected into empty zonae, exhibit the ultrastructural features characteristic of endoderm after one to two days in the oviduct (Rossant, 1975*b*).

Regulative capacity of $3\frac{1}{2}$ day ICM tissue

Live young have been obtained following destruction of some ICM cells in $3\frac{1}{2}$ day *p.c.* blastocysts (Lin, 1969). Histological analysis of serially sectioned conceptuses, recovered six to eight days after transplanting half-blastocysts to recipient uteri, demonstrated that even more radical reduction of the number of cells in $3\frac{1}{2}$ day ICMs is compatible with normal development (Gardner, 1974*b*, *c*). No specific embryonic defects or deficiencies were found in the latter experiments. Instead, a broad spectrum of developmental forms was encountered that ranged from morphologically normal foetuses, through retarded embryos, to conceptuses in which derivatives of the ICM were minimal or lacking altogether. The converse experiment can be performed by injecting disaggregated ICM cells or single entire ICMs removed from $3\frac{1}{2}$ day blastocysts into host blastocysts of the same age (Gardner, 1971, and unpublished data). This rarely leads to twinning. Donor cells typically aggregate with those of the host ICM to form normal-sized chimaeric foetuses and offspring.

These results show that the ICM possesses a considerable capacity for regulative development, and rule out the possibility that it is a strict mosaic of determined cells in the sense that each one is committed to form a different part of the later conceptus. They do not, however, exclude the existence of a few subpopulations of determined cells. This can only be explored by injecting single ICM cells into host blastocysts (Gardner & Lyon, 1971), and studying the distribution of their clonal descendants later in development. Once again, isozymes of GPI were used as markers.

Developmental potential of single $3\frac{1}{2}$ day ICM cells

Preliminary results obtained following injection of single $3\frac{1}{2}$ day cells into host blastocysts have been presented elsewhere (Gardner, 1974*b*). Briefly, eleven definite chimaeras were analysed six days after transplanting a series of injected blastocysts in to recipient uteri. Progeny of the donor cell were detected in the embryonic region only of three, and in the extra-embryonic region of a further four. The donor cell clone embraced both regions of the remaining four conceptuses.

A further six chimaeric conceptuses were recovered at an advanced foetal stage so as to permit separate analysis of the placenta, yolk sac splanchnopleure, amnion and umbilical cord, as well as a variety of organs derived from all three germ layers of the foetus. The donor cell had colonised only the yolk sac splanchnopleure and placenta in three cases, and only foetal organs in a further two. In the sixth, chimaerism was detected in the amnion and umbilical cord in addition to all parts of the foetus. These data show that at least some ICM cells are relatively unrestricted in developmental potential at 3½ days *p.c.* However, they are as yet insufficient to resolve the question of the presence of subpopulations of determined cells. Thus the findings in the six advanced conceptuses could be due to existence of two subpopulations, one determined for formation of ICM derivatives present only in the yolk sac and placenta, and the other committed to form all other ICM derivatives. They could also be explained by the chance distribution of progeny of equipotential cells. We will return to this problem after describing the results of experiments on the 4½ day ICM.

Developmental potential of 4½ day ICM cells

The fate of transplanted 4½ day ICM cells was investigated by injecting them into 3½ day blastocysts, because 4½ day blastocysts do not develop normally after recovery and transfer to the uteri of recipient mice (A. K. Tarkowski, personal communication). The use of such asynchronous combinations may be justified by the routine production of chimaeric offspring from 3½ day mouse blastocysts injected with whole 4½ day mouse ICMs (Gardner, 1971, and unpublished data).

Endodermal and non-endodermal fragments, each containing at least twenty cells, were isolated microsurgically from 4½ day $Gpi\text{-}1^b/Gpi\text{-}1^b$ blastocysts and injected individually into two series of 3½ day $Gpi\text{-}1^a/Gpi\text{-}1^a$ blastocysts. The host blastocysts were then transferred to the uteri of $Gpi\text{-}1^a/Gpi\text{-}1^a$ mice. All foetuses recovered 13–15 days later were separated into the parts listed in Table 2 before electrophoretic analysis. The chimaeras that had developed from blastocysts injected with endodermal tissue exhibited a high level of donor isozyme in the yolk sac splanchnopleure and usually a lower level in the placenta. Progeny of the donor cells appeared to be confined to these two extra-embryonic organs in all but a single case (Table 2). In contrast, transplanted non-endodermal tissue yielded widespread chimaerism in most instances. When compared for specific yolk sac splanchnopleure and placental colonisation versus more general chimaerism, the difference in fate of transplanted endodermal and non-endodermal cells is statistically significant ($P = <0.05 > 0.02$ using a χ^2 test with Yates' correction for continuity). Is it attributable to junctional complexes between endoderm cells imposing restraint on their movement in these tissue fragments, or does it reflect a more fundamental difference between cells of the two populations? Very recently, we have begun to examine this question by transplanting dissociated cells into host blastocysts.

Table 2. *GPI analysis of chimaeric conceptuses developing from* Gpi-*1*[a]/Gpi-*1*[a] *blastocysts injected with* Gpi-*1*[b]/Gpi-*1*[b] *endodermal and non-endodermal tissue*

Individual parts analysed	Conceptus no.: blastocysts injected with fragments of endodermal tissue										Conceptus no.: blastocysts injected with fragments of non-endodermal tissue													
	1	2	3	4	5	6	7	8	9	10	1	2	3	4	5	6	7	8	9	10	11	12	13	14
Placenta	■	■	■	■	■	■	■	■	■	■	■	■	■	■	■	■	■	■	■	■	■	■	■	■
Yolk sac	■	■	■	■	■	■	■	■	■	■	■	■	■	■	■	■	■	■	■	■	■	■	■	■
Amnion + umb. cord	□	□	□	□	□	□	□	□	□	■	■	■	■	■	■	■	■	■	■	■	□	□	□	□
Liver	□	□	□	□	□	□	□	□	□	■	■	■	■	■	■	■	■	■	■	□	□	□	□	□
Lungs	□	□	□	□	□	□	□	□	□	■	■	■	■	■	■	■	■	■	■	□	□	□	□	□
Heart	□	□	□	□	□	□	□	□	□	■	■	■	■	■	■	■	■	■	■	□	□	□	□	□
Brain	□	□	□	□	□	□	□	□	□	■	■	■	■	■	■	■	■	■	■	□	□	□	□	□
Remainder – foetus	□	□	□	□	□	□	□	□	□	■	■	■	■	■	■	■	■	■	■	□	□	□	□	□

■ Chimaeric

□ Host type

The surface of living cells obtained by disaggregation of endodermal tissue looks rougher than that of non-endodermal cells. The distinction is not as marked in $4\frac{1}{2}$ day ICMs as in the disaggregated embryonic region of $5\frac{1}{2}$ day egg-cylinders. Therefore, only the 'roughest' and 'smoothest' cells were selected from disaggregated $4\frac{1}{2}$ day ICMs for injection into host blastocysts. The results are summarised in Table 3. It will be seen that

Table 3. *GPI analysis of chimaeric conceptuses developing from* Gpi-1^{a}/Gpi-1^{a} *blastocysts injected with dissociated* Gpi-1^{b}/Gpi-1^{b} *endoderm or non-endoderm cells*

Individual parts analysed	Conceptus no.: blastocysts injected with 1 endoderm cell					Conceptus no.: blastocysts injected with 3 endoderm cells					Conceptus no.: blastocysts injected with 1 non-endoderm cell				
	1	2	3	4	5	1	2	3	4	5	1	2	3	4	5
Placenta	■	■	■	□	□	■	■	■	□	□	■	□	□	■	□
Yolk sac	■	■	□	■	□	■	■	■	■	■	■	□	□	□	□
Amnion	□	□	□	□	□	□	□	□	□	□	■	□	□	■	■
Umbilical cord	□	□	□	□	■	□	□	□	□	□	■	□	□	■	■
Gut	na	na	na	na	na	□	□	□	□	□	na	na	■	■	■
Liver	□	□	□	□	□	□	□	□	□	□	na	na	□	■	■
Heart	□	□	□	□	□	□	□	□	□	□	na	na	□	■	■
Lungs	□	□	□	□	□	□	□	□	□	□	na	na	□	■	■
Brain	□	□	□	□	□	□	□	□	□	□	na	na	□	■	■
Blood	na	na	na	na	na	□	□	□	na	na	na	na	■	□	■
Remainder – Foetus	□	□	□	□	□	□	□	□	□	□	na	na	□	■	■
Whole foetus	na	na	na	na	na	na	na	na	na	na	■	■	na	na	na

na, individual parts not analysed.

■ Chimaeric □ Host type

colonisation by disaggregated endoderm and non-endoderm cells is very similar to that obtained with the undissociated tissue fragments. The difference in fate of the two types of cells therefore depends on their individual properties rather than on whether or not they are coupled to similar cells by defined intercellular junctions.

Fate of the primitive endoderm

Both the yolk sac splanchnopleure and the placenta are composed of different tissues which it is impractical to attempt to separate prior to electrophoretic analysis. This is an obvious shortcoming of GPI, and indeed most other cell markers in the mouse (see Gardner & Johnson, 1974, for a more detailed discussion of this problem), which precludes precise localisation of the progeny of transplanted cells. Nevertheless, except in one case involving transplantation of a fragment which may have been contaminated with non-endoderm cells, the distribution of injected endoderm cells seems to coincide with the distribution of proximal endoderm in host embryos. Thus the placenta is known to incorporate proximal endoderm during its morphogenesis (Duval, 1892; Jenkinson, 1902). The fact that the umbilical cord was also weakly chimaeric in one case (Table 3) may be explained by the presence of this tissue at its point of insertion in the hilar region of the placenta (Duval, 1892; Jenkinson, 1902). The yolk sac splanchnopleure, which is composed of both extra-embryonic endoderm and mesoderm, usually showed the highest levels of donor isozyme. The mesodermal tissue of this organ is the source of haemopoietic stem cells (Snell & Stevens, 1966; Metcalf & Moore, 1971). However, foetal blood and organs rich in foetal blood cells were chimaeric in only one conceptus that had received endoderm cells (Table 2). This argues that the donor cells colonised the endoderm rather than mesoderm of the yolk sac splanchnopleure.

The parietal or distal endoderm had already degenerated by the stage at which the analyses were performed. It is at best a sparse population of small cells which are difficult to obtain free from trophoblastic giant cells. However, since it is formed by spreading of primitive endoderm cells round the mural trophectoderm of the late blastocyst (Snell & Stevens, 1966), it is most likely that the transplanted endoderm cells also colonised this tissue.

These data are thus consistent with the hypothesis that the primitive endoderm cells of the $4\frac{1}{2}$ day blastocyst form only the extra-embryonic endoderm of the later conceptus, and do not contribute progeny to endodermal or other organs of the foetus. It would perhaps be premature to state that these cells are already determined at $4\frac{1}{2}$ days *p.c.* since they were injected into the optimal site for finding their normal location in the host blastocysts. Nevertheless, their absence from all parts of host embryos except those in which extra-embryonic endoderm is known to occur is in accord with this view. The fact that in three cases single $3\frac{1}{2}$ day ICM cells yielded only yolk sac and placental colonisation raises the possibility that cells determined for endodermal differentiation may be present in the early blastocyst.

If the present conclusions regarding the fate of trophectoderm and primitive endoderm are correct, it follows that the non-endoderm cells of the late blastocyst must give rise only to the embryonic ectoderm of the egg-cylinder. One must further conclude that the embryonic ectoderm

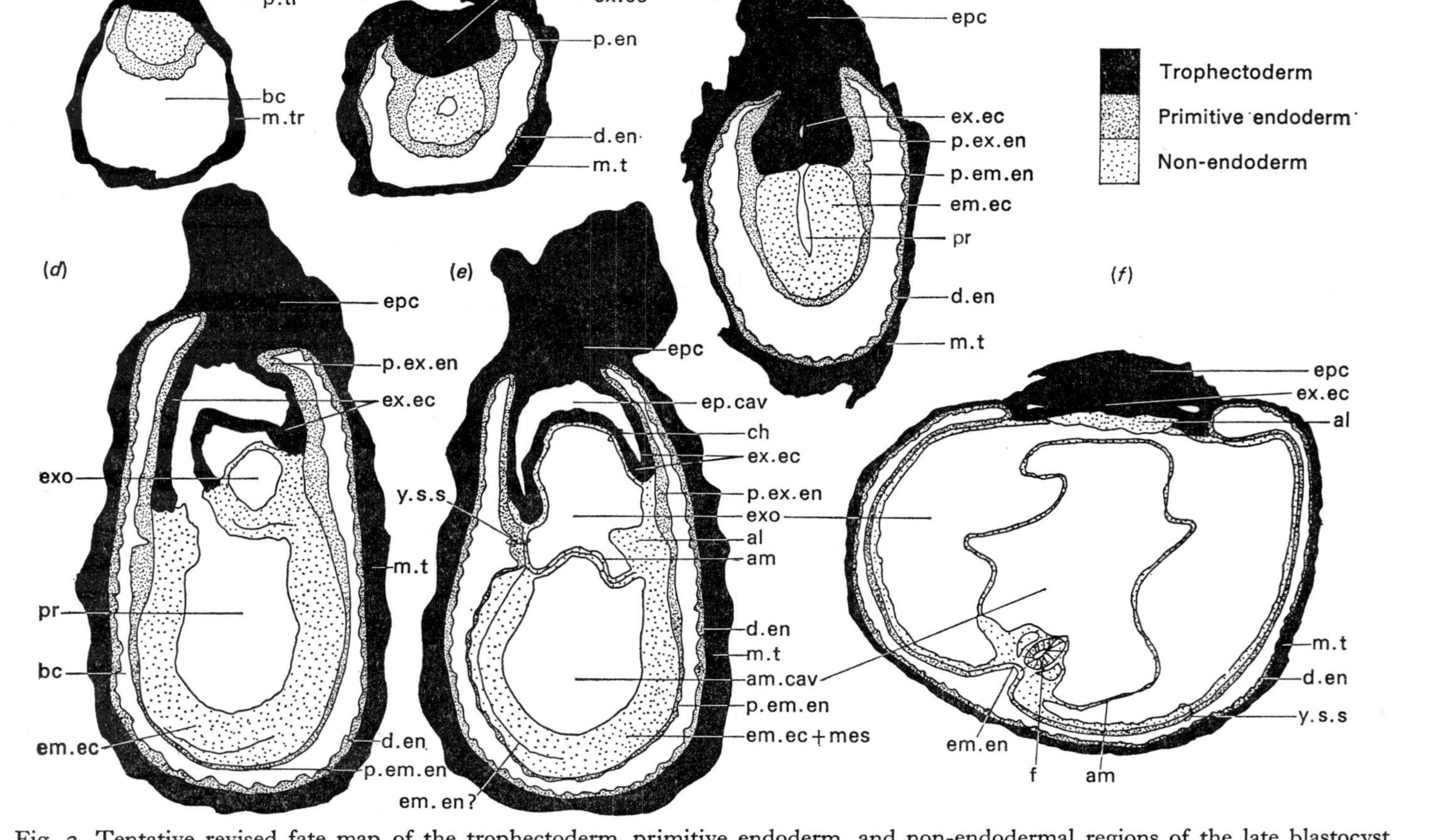

Fig. 2. Tentative revised fate map of the trophectoderm, primitive endoderm, and non-endodermal regions of the late blastocyst projected onto successive stages redrawn and modified from Snell & Stevens (1966). (*a*) 4 days 5 hours, (*b*) 5- or 6-day egg-cylinder, (*c*) 5 days 12 hours, (*d*) 7 days 1 hour, (*e*) 7 days 6 hours, (*f*) 8 days 11 hours. Key: al, allantois; am, amnion; am.cav, amniotic cavity; bc, blastocoel; ch, chorion; d.en, distal endoderm; em.ec, embryonic ectoderm; em.en, *definitive* embryonic endoderm; epc, ectoplacental cone; ep.cav, ectoplacental cavity; ex.ec, extra-embryonic ectoderm; exo, exocoelom; f, foetus; mes, mesoderm; m.t, mural trophoblast;

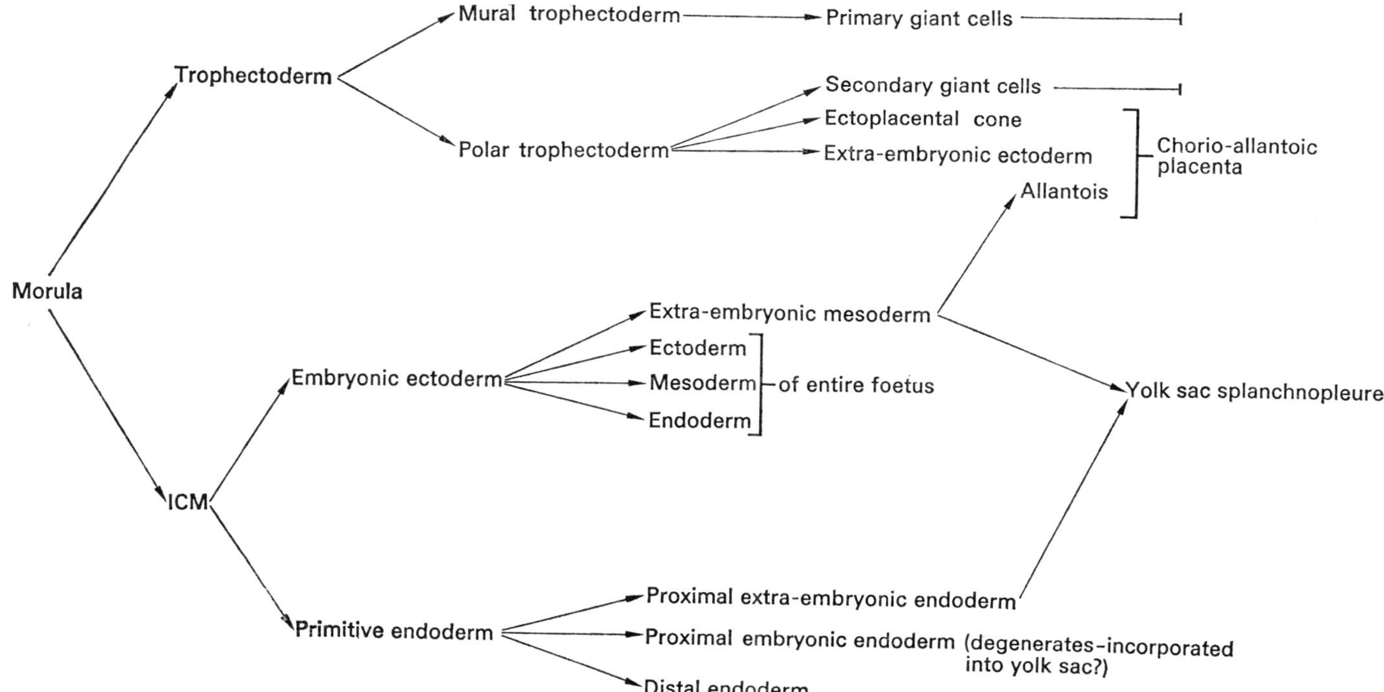

Fig. 3. Diagrammatic representation of the tentative revised cell lineages in the mouse embryo.

later forms the entire foetus and the extra-embryonic mesoderm of the allantois and exocoelom (Figs. 2 and 3). Earlier studies in which germ layers isolated from egg-cylinders or embryonic shields were grafted in ectopic sites endorse these conclusions. Grobstein (1952) mechanically dissected the primitive endoderm from 7th day mouse egg-cylinders and transplanted the remainder into the anterior chamber of the eye after a brief period in culture. The grafts formed many foetal tissues including those of the gut. These results have been confirmed recently by Levak-Svajger & Svajger (1971) who achieved clean separation of endoderm and ectoderm of 8-day rat embryonic shields by a combination of enzymatic treatment and mechanical dissection. Isolated endoderm showed no differentiation as renal homografts, while embryonic ectoderm consistently produced tissues characteristic of all three germ layers of the foetus. These authors suggested that the definitive endoderm may be formed in an analogous way to that of the chick, where it is produced by invagination of epiblast in the anterior part of the primitive streak (Nicolet, 1967).

Interspecific chimaeras between rat and mouse afford the most promising way of verifying the fate of primitive endoderm and non-endoderm cells because of the resolution of immunofluorescent analysis. Preliminary experiments have demonstrated that single rat ICM cells injected into the blastocoel of mouse blastocysts continue to divide following incorporation into the host ICM (Gardner & Johnson, 1974). Furthermore, dissociated rat endoderm and non-endoderm cells differ in appearance, as do those of the mouse. The problem of determination is being investigated by examining the fate of the two tissues after aggregation with morulae (Rossant, 1975*a*), injection into trophoblastic vesicles, or transplantation of cells directly into the depth of the host ICM.

EFFECT OF CELL POSITION ON DIFFERENTIATION OF ICM CELLS

Observations on the development of interspecific chimaeras and isolated ICMs indicate that cell position may be important in differentiation within the ICM, as appears to be the case with the earlier differentiation of trophectoderm versus ICM (Hillman *et al.*, 1972; Wilson *et al.*, 1972). Rat ICMs isolated before endoderm cells have differentiated morphologically form mainly endoderm in host mouse egg-cylinders (Gardner & Johnson, 1974). Their cells seem to spread out over the blastocoelic surface of the host ICM rather than penetrate into it, so that they occupy the position in which primitive endoderm normally develops. If, on the other hand, the rat ICM attaches to mural trophectoderm instead of the host ICM, it develops into an egg-cylinder. Hence, in one situation donor cells form largely endoderm, and in the other ectoderm and mesoderm as well.

None of the cells of $3\frac{1}{2}$ day mouse ICMs show endodermal features (Enders & Schlafke, 1965; Enders, 1965). However, when the ICMs are isolated, aggregated in pairs and injected into empty zonae, they form cells possessing such characteristics after 1–2 days in the oviduct (Rossant, 1975*b*). There is little or no increase in cell number in these spherical masses of ICM tissue. All surface cells show the ultrastructural features peculiar to endoderm. It is possible that a number of determined but cytoplasmically undifferentiated endoderm cells exist in the ICMs at the time of isolation. However, given variations in initial cell number due to differences in rate of development and damage during isolation of the ICMs, it seems unlikely that they would always be appropriate to form the whole surface and only the surface layer in the aggregates. Taken together, these two sets of observations suggest that exposure of ICM cells to outside conditions, whether provided by the blastocoel or oviduct fluid, is one factor promoting endodermal differentiation.

CONCLUSIONS

It would be unwise to draw firm conclusions at this stage since most of these lines of investigation are still in progress. Nevertheless, several points of interest emerge from these studies. Thus experiments involving transplantation and recombination of genetically marked cells have produced results that are at variance with observations on fixed and sectioned embryos concerning the fate of polar trophectoderm, primitive endoderm and embryonic ectoderm. The revised fate map, which requires confirmation by more detailed studies on interspecific chimaeras in particular, is presented in Figs. 2 and 3. The main points to note are the possibility of a unitary origin of all trophoblast cells, and the formation of the entire foetus and extra-embryonic mesoderm from the non-endoderm cells of the ICM.

Cells of the ICM and trophectoderm are evidently committed to mutually exclusive pathways of development by $3\frac{1}{2}$ days *p.c.* (Gardner, 1974*a*, *b*; Rossant, 1975*a*, *b*). Both of these tissues are further divided into two morphologically distinct subpopulations of cells during the next twenty-four hours. Initiation of giant transformation in mural trophectoderm cells is first evident at the abembryonic pole of the blastocyst at midnight on day 4, and is completed approximately twelve hours later. Therefore, determination of these cells may occur sequentially rather than simultaneously. Some mural cells are obviously not determined at noon on this day because they can be converted into polar cells by recombination with ICM tissue. Polar trophectoderm cells clearly retain the capacity to undergo giant transformation, but are evidently prevented from doing so by the underlying ICM. Regional differentiation of the trophectoderm thus appears to be dictated by the ICM during the second half of the 4th day of development.

Two populations of cells can be distinguished in the $4\frac{1}{2}$ day ICM both morphologically and in their fate following transplantation. The fact that endoderm and non-endoderm cells produce different patterns of chimaerism in host embryos suggests that they are determined by $4\frac{1}{2}$ days *p.c.* Indeed, cell injection data are consistent with there being even earlier determination well before the two types of cells can be identified morphologically. However, the ability of $3\frac{1}{2}$ day ICMs to vary the number of endoderm cells they form when aggregated with either rat or other isolated mouse ICMs, is more readily explained by supposing that determination depends on the position that cells occupy after this stage. It is hoped that the experimental approaches outlined earlier will provide a more precise estimate of the time of determination in ICM cells.

Finally, if the pattern of chimaerism produced by tranplanted primitive endoderm cells proves to be as specific as we suggest, it will provide a means of testing the hypothesis that primordial germ cells are of extra-embryonic endodermal origin (Brambell, 1956; Hamilton & Mossman, 1972).

We wish to thank Dr M. McBurney, Dr C. F. Graham and Dr M. H. Johnson for valuable discussion, and Mrs S. C. Barton, Mrs L. Ofer and Mrs V. N. Ovelowo for assistance. Our experimental work was supported by The Ford Foundation, The Medical Research Council, The Smith Kline and French Foundation, The Wellcome Trust and The World Health Organisation.

REFERENCES

Alden, R. H. (1948). Implantation of the rat egg. III. Origin and development of primary trophoblast giant cells. *American Journal of Anatomy*, **83**, 143–81.

Amoroso, E. C. (1952). Placentation. In *Marshall's Physiology of Reproduction*, ed. A. S. Parkes, vol. 2, pp. 127–311. London: Longmans, Green.

Ansell, J. D. & Snow, M. H. L. (1974). The development of trophoblast *in vitro* from blastocysts containing varying amounts of inner cell mass. *Journal of Embryology and Experimental Morphology*, in press.

AVERY, G. B. & HUNT, C. V. (1969). The differentiation of trophoblast giant cells in the mouse studied in kidney capsule grafts. *Transplantation Proceedings*, **1**, 61–6.

BALINSKY, B. I. (1970). *An Introduction to Embryology*, 3rd edition. Philadelphia: Saunders.

BARLOW, P. W., OWEN, D. & GRAHAM, C. F. (1972). DNA synthesis in the preimplantation mouse embryo. *Journal of Embryology and Experimental Morphology*, **27**, 431–45.

BARLOW, P. W. & SHERMAN, M. I. (1972). The biochemistry of differentiation of mouse trophoblast: studies on polyploidy. *Journal of Embryology and Experimental Morphology*, **27**, 447–65.

BRAMBELL, F. W. R. (1956). Ovarian changes. In *Marshall's Physiology of Reproduction*, ed. A. S. Parkes, vol. 1, part I, pp. 397–542. London: Longmans, Green.

CONKLIN, E. G. (1905). The organisation and cell lineage of the ascidian egg. *Journal of the Academy of Natural Sciences of Philadelphia*, **13**, 5–119.

DALCQ, A. M. (1957). *Introduction to General Embryology*. London: Oxford University Press.

DAVIDSON, E. H. (1968). *Gene Activity in Early Development*. New York & London: Academic Press.

DICKSON, A. D. (1966). The form of the mouse blastocyst. *Journal of Anatomy, London*, **100**, 335–48.

DUVAL, M. (1892). Le Placenta des Rongeurs. Extrait du *Journal de l'Anatomie et de la Physiologie*, Années 1889–1892, ed. F. Alcan. Paris: Ancienne Librairie Germer Baillière.

ENDERS, A. C. (1965). Comparative study of the fine structure of the trophoblast in several hemochorial placentas. *American Journal of Anatomy*, **116**, 29–68.

ENDERS, A. C. (1971). The fine structure of the blastocyst. In *The Biology of the Blastocyst*, ed. R. J. Blandau, pp. 71–94. Chicago: University of Chicago Press.

ENDERS, A. C. & SCHLAFKE, S. J. (1965). The fine structure of the blastocyst: some comparative studies. In *Preimplantation Stages of Pregnancy: A Ciba Foundation Symposium*, ed. G. E. W. Wolstenholme & M. O'Connor, pp. 29–54. London: J. & A. Churchill.

GARDNER, R. L. (1968). Mouse chimaeras obtained by the injection of cells into the blastocyst. *Nature, London*, **220**, 596–7.

GARDNER, R. L. (1971). Manipulations on the blastocyst. In *Schering Symposium on Intrinsic and Extrinsic Factors in Early Mammalian Development. Advances in the Biosciences*, vol. 6, ed. G. Raspé, pp. 279–96. Oxford: Pergamon Press.

GARDNER, R. L. (1972). An investigation of inner cell mass and trophoblast tissue following their isolation from the mouse blastocyst. *Journal of Embryology and Experimental Morphology*, **28**, 279–312.

GARDNER, R. L. (1974*a*). Origin and properties of trophoblast. In *The Immunobiology of Trophoblast*. London: Cambridge University Press. (In press.)

GARDNER, R. L. (1974*b*). Analysis of determination and differentiation in the early mammalian embryo using intra- and inter-specific chimaeras. In *The Developmental Biology of Reproduction: 33rd Symposium of the Society for Developmental Biology*, ed. C. L. Markert. New York & London: Academic Press. (In press.)

GARDNER, R. L. (1974*c*). Microsurgical approaches to the study of early mammalian development. In *Birth Defects and Fetal Development: Endocrine and Metabolic Factors*, ed. K. S. Moghissi, pp. 212–33. Springfield: C. C. Thomas.

GARDNER, R. L. & JOHNSON, M. H. (1973). Investigation of early mammalian development using interspecific chimaeras between rat and mouse. *Nature New Biology*, **246,** 86–9.

GARDNER, R. L. & JOHNSON, M. H. (1974). Investigation of cellular interaction and deployment in the early mammalian embryo using interspecific chimaeras between the rat and mouse. In *Ciba Foundation Symposium on Pattern Formation*. (In press.)

GARDNER, R. L. & LYON, M. F. (1971). X-Chromosome inactivation studied by injection of a single cell into the mouse blastocyst. *Nature, London*, **231,** 385–6.

GARDNER, R. L., PAPAIOANNOU, V. E. & BARTON, S. C. (1973). Origin of the ectoplacental cone and secondary giant cells in mouse blastocysts reconstituted from isolated trophoblast and inner cell mass. *Journal of Embryology and Experimental Morphology*, **30,** 561–72.

GEHRING, W. (1972). The stability of the determined state in cultures of imaginal disks in *Drosophila*. In *The Biology of Imaginal Disks*, ed. H. Ursprung & R. Nöthiger, pp. 35–58. Berlin: Springer-Verlag.

GRAHAM, C. F. (1973). The necessary conditions for gene expression during early mammalian development. In *Genetic Mechanisms of Development: 31st Symposium of The Society for Developmental Biology*, ed. F. H. Ruddle, pp. 201–24. New York & London: Academic Press.

GROBSTEIN, C. (1950). Production of intra-ocular hemorrhage by mouse trophoblast. *Journal of Experimental Zoology*, **114,** 359–73.

GROBSTEIN, C. (1952). Intra-ocular growth and differentiation of clusters of mouse embryonic shields cultured with and without primitive endoderm and in the presence of possible inductors. *Journal of Experimental Zoology*, **119,** 355–80.

GWATKIN, R. B. L. (1966). Amino acid requirements for attachment and outgrowth of mouse blastocysts *in vitro*. *Journal of Cellular Physiology*, **68,** 335–45.

HAMILTON, W. J. & MOSSMAN, H. W. (1972). *Hamilton, Boyd and Mossman's Human Embryology*, 4th edition. Cambridge: Heffer.

HERBERT, M. C. & GRAHAM, C. F. (1974). Cell determination and biochemical differentiation of the early mammalian embryo. In *Current Topics in Developmental Biology*, **8,** 151–78.

HILLMAN, N., SHERMAN, M. I. & GRAHAM, C. F. (1972). The effect of spatial arrangement on cell determination during mouse development. *Journal of Embryology and Experimental Morphology*, **28,** 263–78.

HOLTZER, H. (1970). Proliferative and quantal cell cycles in the differentiation of muscle, cartilage, and red blood cells. In *Control Mechanisms in the Expression of Cellular Phenotypes: Symposia of the International Society for Cell Biology*, vol. 9, ed. H. A. Padykula, pp. 69–88. New York & London: Academic Press.

HOLTZER, H. & ABBOTT, J. (1968). Oscillations of the chondrogenic phenotype *in vitro*. In *The Stability of the Differentiated State*, ed. H. Ursprung, pp. 1–16. Berlin: Springer-Verlag.

HORSTADIUS, S. (1973). *Experimental Embryology of Echinoderms*. Oxford: Clarendon Press.

HUXLEY, J. S. & DE BEER, G. R. (1934). *The Elements of Experimental Embryology*. London: Cambridge University Press.

JENKINSON, J. W. (1902). Observations on the histology and physiology of the placenta of the mouse. *Tijdschrift der Nederlandsche dierkundige Vereeniging*, **2,** 124–98.

JOHNSON, M. H. (1972). Relationship between inner cell mass derivatives and trophoblast proliferation in ectopic pregnancy. *Journal of Embryology and Experimental Morphology*, **28,** 306–12 (Appendix).

JOHNSON, M. H. & GARDNER, R. L. (1974). Analysis of rat : : mouse chimaeras by immunofluorescence: a preliminary report. In *Proceedings of The 1st International Congress on the Immunology of Obstetrics and Gynaecology*, Padua 1973, ed. N. Garrett. Amsterdam: Excerpta Medica. (In press.)

JOLLIE, W. P. (1964). Fine structural changes in placental labyrinth of the rat with increasing gestational age. *Journal of Ultrastructural Research*, **10**, 27–47.

KIRBY, D. R. S. & BRADBURY, S. (1965). The hemo-chorial mouse placenta. *Anatomical Record*, **152**, 279–82.

KOREN, Z. & BEHRMAN, S. J. (1968). Organ culture of pure mouse trophoblast. *American Journal of Obstetrics and Gynecology*, **100**, 576–81.

LEVAK-SVAJGER, B. & SVAJGER, A. (1971). Differentiation of endodermal tissues in homografts of primitive ectoderm from two-layered rat embryonic shields. *Experientia*, **27**, 683–4.

LIN, T. P. (1969). Microsurgery of the inner cell mass of mouse blastocysts. *Nature, London*, **222**, 480–1.

METCALF, D. & MOORE, M. A. S. (1971). *Haemopoietic Cells*. Amsterdam: North-Holland.

MINTZ, B. (1962). Experimental recombination of cells in the developing mouse egg: normal and lethal mutant genotypes. *American Zoologist*, **2**, abstract 145.

NEEDHAM, J. (1942). *Biochemistry and Morphogenesis*. London: Cambridge University Press.

NICOLET, G. (1967). La chronologie d'invagination chez le poulet: étude à l'aide de la thymidine tritiée. *Experientia*, **23**, 576–7.

NÖTHIGER, R. (1972). The larval development of imaginal disks. In *The Biology of Imaginal Disks*, ed. H. Ursprung & R. Nöthiger, pp. 1–34. Berlin: Springer-Verlag.

ROSSANT, J. (1975*a*). Investigation of the determinative state of the mouse inner cell mass. I. Aggregation of isolated inner cell masses with morulae. *Journal of Embryology and Experimental Morphology*, in press.

ROSSANT, J. (1975*b*). Investigation of the determinative state of the mouse inner cell mass. II. The fate of isolated inner cell masses transferred to the oviduct. *Journal of Embryology and Experimental Morphology*, in press.

RUGH, R. (1968). *The Mouse: Its Reproduction and Development*. Minneapolis: Burgess.

SNELL, G. D. & STEVENS, L. C. (1966). Early embryology. In *Biology of The Laboratory Mouse*, 2nd edition, ed. E. L. Green, pp. 205–45. New York: McGraw-Hill.

SNOW, M. H. L. (1973*a*). The differential effect of [^{3}H]thymidine upon two populations of cells in pre-implantation mouse embryos. In *The Cell Cycle in Development and Differentiation*, ed. M. Balls & F. S. Billett, pp. 311–24. London: Cambridge University Press.

SNOW, M. H. L. (1973*b*). Abnormal development of pre-implantation mouse embryos grown *in vitro* with [^{3}H]thymidine. *Journal of Experimental Embryology and Morphology*, **29**, 601–15.

SPEMANN, H. (1938). *Embryonic Development and Induction*. New Haven, Conn.: Yale University Press.

STURTEVANT, A. H. (1929). The *claret* mutant type of *Drosophila simulans*: a study of chromosome elimination and of cell-lineage. *Zeitschrift für wissenschaftliche Zoologie*, **135**, 324–55.

TARKOWSKI, A. K. (1961). Mouse chimaeras developed from fused eggs. *Nature, London*, **190**, 857–60.

VOGT, W. (1929). Gestaltungsanalyse am Amphibienkeim mit örtlicher Vitalfärbung. II. Gastrulation und Mesodermbilding bei Urodelen und Anuren. *Wilhelm Roux' Archiv für Entwicklungsmechanik der Organismen*, **120**, 385–706.

WEISS, P. (1939). *Principles of Development.* New York: Holt.

WILSON, I. B., BOLTON, E. & CUTTLER, R. H. (1972). Preimplantation differentiation in the mouse egg as revealed by micro-injection of vital markers. *Journal of Embryology and Experimental Morphology*, **27**, 467–79.

EXPLANATION OF PLATES

PLATE 1

Photograph of a partially dissected early somite embryo (*a*) with outline drawing, (*b*) indicating the line of dissection into trophoblastic (T) and embryonic (E) fractions: al, allantois; am, amnion; emb, embryo proper; epc, ectoplacental cone; mt, mural trophoblast with Reichert's membrane and attached distal endoderm. From Gardner *et al.* (1973).

PLATE 2

(*a*) Compact non-invasive graft formed by the embryonic region of a $5\frac{1}{2}$ day *p.c.* egg-cylinder 5 days after transplantation under the testis capsule.

(*b*) Corresponding extra-embryonic region which has yielded an invasive haemorrhagic graft containing mainly trophoblastic cells.

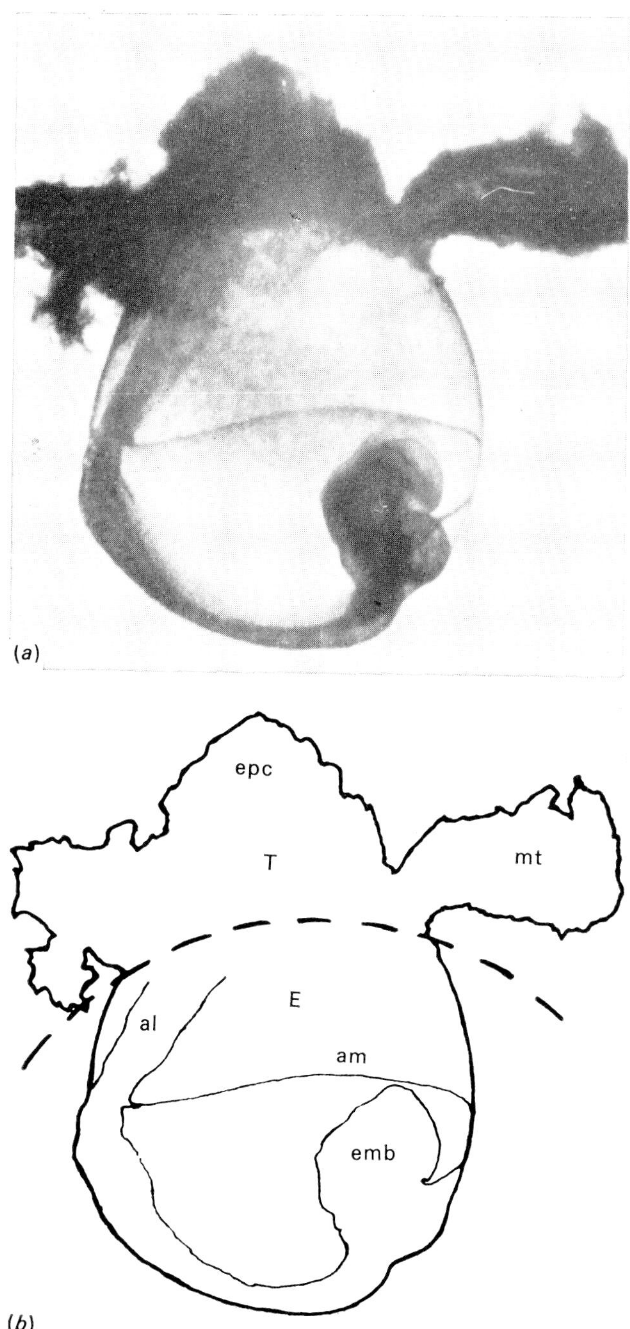

(Facing p. 132)

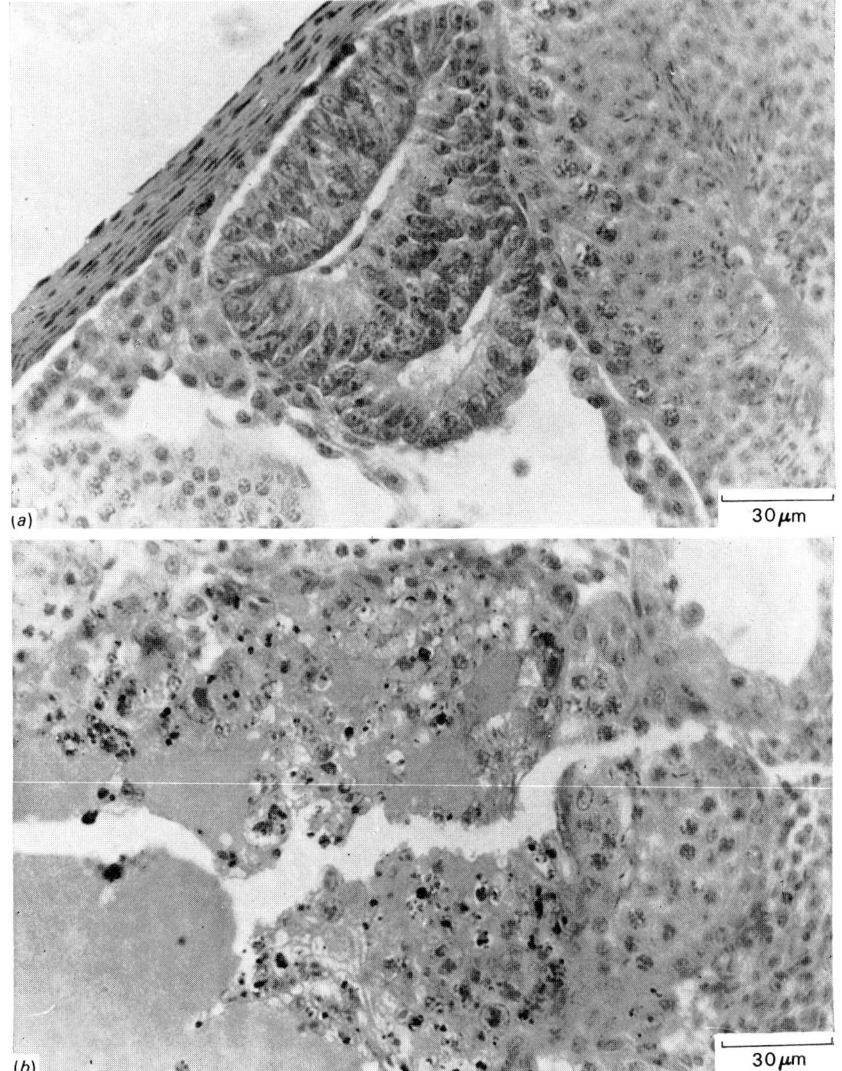
(a)
30 μm
(b)
30 μm

THE DIFFERENTIATION AND DEVELOPMENT OF MOUSE TROPHOBLAST

BY J. D. ANSELL

ARC Unit of Reproductive Biology and Chemistry,
Animal Research Station, Cambridge CB3 0JQ

During early cleavage of mouse embryos two distinct cell populations arise, which by the blastocyst stage have morphologically and functionally differentiated into an outermost layer, the trophoblast, enclosing the blastocoel cavity, and an inner cell mass (ICM). In the mouse regional differences appear in the trophoblast cell layer at implantation. The abembryonic or mural trophoblast, i.e. those cells constituting the walls of the blastocyst, give rise to highly invasive primary giant cells, which invade the maternal epithelium and serve to anchor the blastocyst to the uterine wall (Amoroso, 1952; Kirby & Cowell, 1968). The cells overlying the ICM, the polar trophoblast or Räuber's layer, unlike the mural trophoblast remain diploid and proliferate until about the 8th day of pregnancy, developing into the ectoplacental cone or träger. Peripheral trophoblast cells of the träger give rise to secondary giant cells which are again highly invasive, eroding into the newly formed decidual tissue and eventually penetrating maternal blood spaces (Amoroso, 1952; Snell & Stevens, 1966; Zybina, 1970). The experiments of Gardner, Papaioannou & Barton (1973) involving the reconstruction of blastocysts from microsurgically dissected ICM and trophoblastic fragments containing different biochemical markers, have very elegantly demonstrated that the ectoplacental cone is derived from trophoblast cells and not as a result of proliferation of the cells of the ICM.

Thus these early stages of trophoblast development represent a series of differentiation steps, from the separation of trophoblast and ICM cells, which is the first major morphogenetic event in mammalian embryogenesis, through the involvement of some of these trophoblast cells in the implantation process, to the formation of the ectoplacental cone.

In this paper three aspects of the early development of trophoblast tissue will be discussed:

(1) The factors governing the differentiation of trophoblast and ICM cells.
(2) The effect of the presence or absence of the ICM on the subsequent proliferation and development of trophoblast.
(3) The formation and chromosomal arrangement of trophoblast giant cells.

DIFFERENTIATION OF TROPHOBLAST AND INNER CELL MASS

The separation of trophoblastic and ICM elements in the early embryo is currently thought to be due to cell position during cleavage. Graham (1971) has reviewed elsewhere the early experiments concerning the lability of early embryonic cells and the effects of their position on subsequent development. I will, then, briefly review the more recent experiments on this topic.

Barlow, Owen & Graham (1972) demonstrated that by the 8–16-cell stage, on average one blastomere had become totally enclosed by the 'outside' blastomeres. This inside cell, and inside cells throughout development to the blastocyst stage, continued to have a higher labelling index after incubation with [^{3}H]thymidine, suggestive of a higher rate of cell division, than those on the outside. Thus the inside cells acquire this different characteristic from those on the outside from the 8–16-cell stage. Hillman,

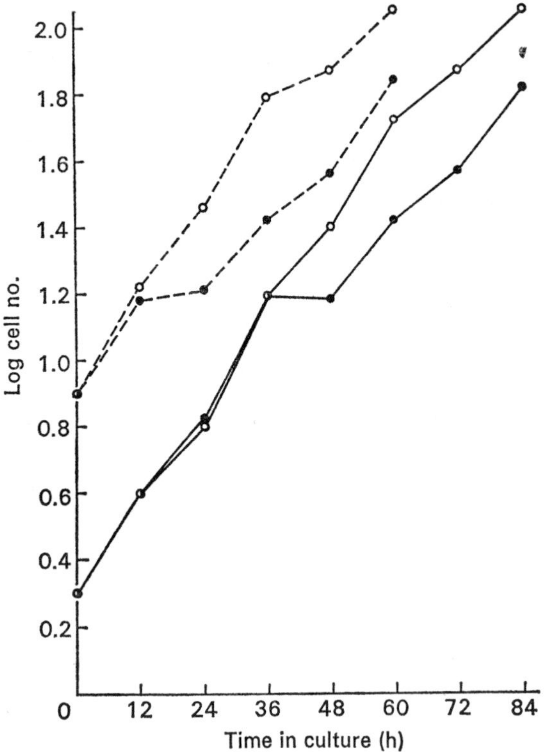

Fig. 1. The progress of 2-cell and 8-cell embryos to blastocysts when grown *in vitro* both with (●) and without (○) 0.05 μCi/ml [^{3}H]thymidine (methyl-T-TdR). ——, 2-cell embryos, ----, 8-cell embryos. Radiation damage is clearly restricted to the 16–32-cell stage. (From Snow, 1973*b*)

Sherman & Graham (1972) made reaggregation chimaeras from two cell populations differing in their expression of the two isozyme variants of the enzyme glucose phosphate isomerase. Blastomeres expressing one isozyme variant were arranged around the outside of blastomeres of the other genotype. The chimaeric foetuses that subsequently developed generally expressed the enzyme phenotype of the outside cells in the trophoblast.

Treatment of 2- or 8-cell stage embryos in culture with [^{3}H]thymidine (Snow, 1973*a*, *b*) selectively damaged the cells inside the 16-cell stage embryo (Fig. 1) whilst those cells on the outside continued cleaving normally such that the subsequent 'blastocyst' is composed entirely of trophoblast cells. Such blastocysts are termed trophoblast vesicles and are essentially similar to those produced by the microsurgical techniques of Gardner (1971). The evidence is strong therefore that the inside cells develop into the ICM and the outside cells into trophoblast. The labelling and disaggregation studies of Stern (1972) and Stern & Wilson (1972) suggest, however, that all cells are potentially labile even up until the late morula and early blastocyst stages and can still form either trophoblast or ICM.

EFFECTS OF PRESENCE OR ABSENCE OF ICM ON TROPHOBLAST DEVELOPMENT

Snow and I have used the trophoblast vesicles discussed in the previous section to study the effects of absence of the ICM on the subsequent development of trophoblast both in culture and after transfer to ectopic sites (Ansell & Snow, 1974). The effect of [^{3}H]thymidine is seen over a range of concentrations: at 0.01 μCi/ml blastocysts are produced with several remaining ICM cells, whilst at 0.025 μCi/ml the ICM is reduced to two or three cells in 50% of the treated embryos. The remainder of these and approximately 80% of those eggs treated at 0.05 μCi/ml show a complete absence of ICM.

Trophoblast vesicles produced at these concentrations were transferred into a Brinster's medium supplemented with 5% foetal calf serum to facilitate the outgrowth of trophoblast tissue as a monolayer of cells. The eggs were scored as having hatched, or attached and outgrown in culture. Hatching is the crucial stage in the outgrowth process, since all hatched blastocysts tend to outgrow. The appearance of a significant difference in the response to outgrowth on day 4 in culture (in spite of the 100% outgrowth of hatched blastocysts), between control and 0.05 μCi/ml treated eggs, suggested a difference in the rate of giant cell formation and outgrowth between the control blastocysts containing an ICM and the vesicles without *any* ICM cells. An analysis of the rate of outgrowth expressed as the number of hatched blastocysts that had outgrown per day of culture shows such an effect (Fig. 2). All of the vesicles in the 0.05 μCi/ml group had begun outgrowth by day 3 whilst the controls did not complete the attachment and outgrowth phase until day 4. Thus, *in vitro* at least, the presence of the ICM exerts some controlling influence over the

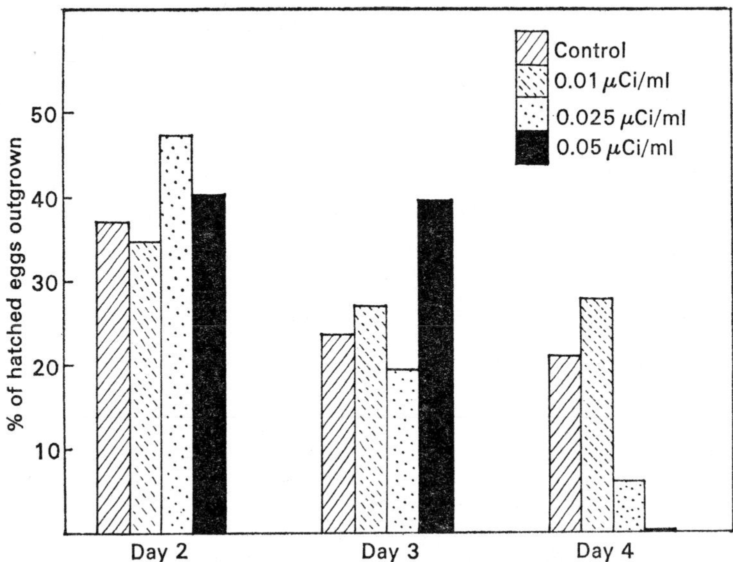

Fig. 2. The rate of outgrowth of control and [^{3}H]thymidine-treated blastocysts, expressed as the percentage of hatched blastocysts that have outgrown each day in culture.

timing of giant cell formation and outgrowth. It would be of interest to know whether this effect is shown *in utero*.

Several groups of control blastocysts, and vesicles grown at 0.05 μCi/ml [^{3}H]thymidine were outgrown on glass coverslips, fixed *in situ* and stained. Typical outgrowths from both groups are shown in Plate 1. The cellular content of vesicle outgrowths was trophoblast giant cells only, whose maximum nuclear sizes were comparable to those of control outgrowths. The latter contained as well as giant cells a central nodule of presumably diploid ICM cells and possibly proliferating trophoblast. It was possible to count the nuclei in these stained preparations and to differentiate those cells which were obviously giant from others. The counts are summarised in Table 1. This table includes some data from cell counts of tetraploid blastocysts and outgrowths. In the majority of tetraploid blastocysts the inner cell mass was drastically reduced or absent and these were therefore

Table 1. *Mean cell number* (± *1 standard error*) *in diploid and tetraploid blastocysts, vesicles and outgrowths of similarly treated embryos*

Treatment	Blastocysts: total cell no.	Outgrowths: total cell no.	Outgrowths: no. of giant cells
Controls	61.9 ±6.4	85.31 ±8.18	56.56 ± 6.53
0.05 μCi/ml	30.3 ±2.5	29.11 ±2.46	26.74 ± 2.29
Controls*	61.3 ±3.2	73.9 ±4.5	63.0 ± 4.0
Tetraploid*	18.11 ±1.4	20.2 ±1.7	17.7 ± 1.5
Controls with ICM	–	80.00 ±3.4	66.3 ± 3.8
Controls without ICM	–	–	52.00 ±13.1

* Unpublished data of M. H. L. Snow.

similar to the trophoblast vesicles produced by [^{3}H]thymidine treatment. It can be seen that the outgrowths from both types of vesicle contained only giant cells, in similar numbers to the vesicle cell number. Outgrowths from control blastocysts, however, showed considerable cellular proliferation. In a normal blastocyst containing 61 cells one would expect 52 of those cells to be trophoblast and 9 to be ICM cells. Therefore the outgrowths from control blastocysts showed trophoblast proliferation. The number of giant cells counted in control outgrowths will also be

Table 2. *The development of* [^{3}H]*thymidine-treated and control blastocysts in the cryptorchidised testis and the kidney.* T *signifies trophoblast development only, and* E *a range of embryonic development from egg-cylinder to small patches of embryonic tissue enclosed within proliferating trophoblast*

Ectopic site	[^{3}H]thymidine (μCi/ml)	No. of single (S) or multiple (M) transfers	No. showing some development	Type of development
Kidney	0	7M	7	T, E
Kidney	0.01	3M	2	T
Kidney	0.025	5M	2	T
Kidney	0.05	1M	0	–
Cryptorchidised testis	0	5M	5	T, E
Cryptorchidised testis	0	2S	1	T, E
Cryptorchidised testis	0.01	5M	4	T
Cryptorchidised testis	0.01	3S	0	–
Cryptorchidised testis	0.025	4M	0	–
Cryptorchidised testis	0.025	3S	0	–
Cryptorchidised testis	0.05	4M	0	–
Cryptorchidised testis	0.05	4S	0	–

an underestimate of those present since only *obviously* giant cells were counted.

From this data we postulate that the presence of the ICM is necessary for trophoblast proliferation. The results of the transfer of vesicles and blastocysts to ectopic sites (kidney and testis) support these conclusions (Table 2). Only control blastocysts showed embryonic as well as trophoblast proliferation in these sites. No growth or proliferation of vesicles from the 0.05 μCi/ml was observed in either the testis or the kidney but some proliferation of trophoblast was recorded after transfer of 0.01 μCi/ml and 0.025 μCi/ml vesicles, i.e. those containing some ICM cells, to both these sites.

Results obtained by Gardner & Johnson (1972) from experiments on microsurgically prepared trophoblast vesicles are supported by our observations and show that whilst the mural trophoblast of mouse blastocysts can implant and become giant irrespective of the presence of the ICM, the proliferation of trophoblast cells and hence the formation of the ectoplacental cone is specifically dependent upon the presence of ICM cells.

FORMATION AND CHROMOSOMAL ARRANGEMENT OF TROPHOBLAST GIANT CELLS

The formation of giant cells in trophoblast is a well-known characteristic of this tissue. It is associated with vast increases in the DNA content of trophoblast nuclei which in the placenta of the white rat has been shown by Nagl (1972) to reach 4096 times the haploid DNA content (C) and by other workers to be at least many hundreds of times the haploid DNA content in both the rat and the mouse (Zybina, 1961, 1963, 1970; Zybina & Mos'yan, 1967; Barlow & Sherman, 1972). Zybina (1970) considers that by the 18th day of gestation in the mouse, DNA levels of not less than 128C are found in all sections of the placenta.

The first evidence of nuclear enlargement in trophoblast has been observed at 117 hours post-ovulation in the mouse (Barlow, Owen & Graham, 1972) and this phenomenon occurs not only normally *in utero*, but after culture of eggs from the 2-cell stage (Fawcett *et al.*, 1947; Kirby, 1960; McLaren & Tarkowski, 1963; Hunt & Avery, 1971; Barlow & Sherman, 1972), in ectopic sites, and in the absence of the ICM (Gardner & Johnson, 1972; Snow, 1973*a*, *b*; Ansell & Snow, 1974). Hatching of the blastocyst from its zona pellucida is not a prerequisite for giant cell formation either (Barlow & Sherman, 1972) and these authors consider that the formation of the blastocoel cavity is the most likely event to initiate giant cell formation.

Originally three separate mechanisms were put forward to account for the increased amounts of DNA in giant trophoblast cells:

(1) Endomitosis with the formation of polytene or polyploid chromosomes (Zybina, 1961, 1963; Jollie, 1964).
(2) Direct inclusion of nuclei or macromolecular DNA from maternal (decidual) tissue into the trophoblast nucleus (Galassi, 1967; Avery & Hunt, 1969).
(3) Fusion of trophoblast nuclei following the formation of syncitial trophoblast (Saccoman, Morgan & Wells, 1967).

Trophoblast–trophoblast and trophoblast–host fusion as mechanisms for nuclear enlargement were shown by the experiments of Chapman, Ansell & McLaren (1972) and Gearhart & Mintz (1972) not to contribute significantly to placental formation and it was concluded that the increase in DNA content of trophoblast giant nuclei was by endoreduplication of DNA rather than by cell fusion. The fact that the ratio of satellite to main band DNA in trophoblast giant nuclei was similar to that found in normal diploid cells (Sherman, McLaren & Walker, 1972), indicated that the complete genome was replicated during DNA synthesis rather than parts of the genome having a disproportionate increase in DNA content during nuclear enlargement. These studies throw little light on the organisation of chromosomes within the trophoblast giant nucleus; in particular whether the nucleus is polyploid, containing many chromosome sets, or if it has the diploid number of chromosomes which are polytene such as those observed in Dipteran species. During the normal course of giant cell development in the mouse, trophoblast chromosomes do not condense enough to be properly visualised, although Zybina (1970) and Zybina, Kudryavtseva & Kudryavtsev (1973) claim to have seen polytene-like chromosomal structures in the larger giant nuclei of the rat and the rabbit.

We have reasoned that if mouse trophoblast nuclei are indeed polytene, they may have some characteristics in common with Dipteran polytene chromosomes in which there is an extension of chromatin into 'puffs' in regions active in DNA synthesis. Actinomycin D, a specific inhibitor of DNA-dependent RNA synthesis (Reich & Goldberg, 1964) is known to cause condensation of such 'puffs' on Dipteran chromosomes (Beerman, 1967) and also to cause a retraction of the DNA loops in the lampbrush chromosomes of amphibian oocytes, into their parent chromomeres, as well as a coalescence of neighbouring chromomeres (Izawa, Allfrey & Mirsky, 1963; Snow & Callan, 1969).

Treatment of trophoblast outgrowths from blastocysts and ectoplacental cones in culture with actinomycin D did cause a range of degrees of condensation of the nuclear contents (Plate 2). Where condensation was minimal the nuclei exhibited a fibrous structure; where condensation was sufficient to visualise chromosomes these exhibited a beaded appearance reminiscent of the polytene chromosomes of Diptera. At the maximum degree of condensation chromosome bodies appeared as discrete lumps of chromatin which were counted. In no cases did the number of these bodies exceed forty, the normal diploid number of mouse chromosomes. In the largest trophoblast cells, containing highly DNA-enriched nuclei, chromosomal condensation is less well resolved, the chromatin aggregating together in clumps. Barlow & Sherman (1974) in a study of the centromeric heterochromatin of trophoblast nuclei have also concluded that some fusion of centromeres occurred, possibly by aggregation of the heterochromatin around the nucleolus. We have concluded that the giant nuclei of mouse trophoblast have the normal diploid number of chromosomes which show a considerable degree of polyteny (Snow & Ansell, 1974). This type of polyteny is markedly different from that found in *Drosophila*, however, trophoblast giant chromosomes having a full complement of satellite DNA and *not* showing a close association of homologous chromosomes.

Further studies of trophoblast chromosomes would be facilitated by the removal of the tough nuclear membrane and the more efficient spreading of the chromosomes, but so far this has not been achieved.

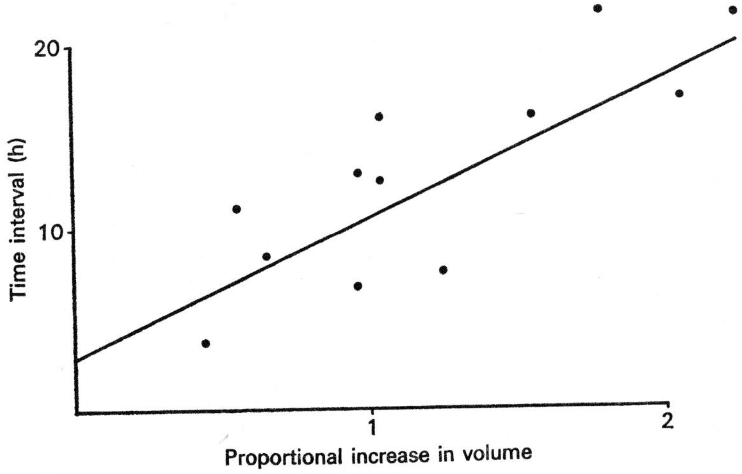

Fig. 3. The increase in volume of trophoblast giant nuclei continuously observed *in vitro* by time-lapse cinephotomicrography (shape of nuclei assumed to be oblate spheroid). From the regression of y on x ($P > 0.01$), the estimated volume doubling time of these nuclei was 8.6 h.

P. W. Barlow & V. M. Papaioannou (personal communication) have produced the first convincing evidence of an endomitotic cycle in mouse trophoblast nuclei and estimate that the total cell cycle time of the secondary giant nuclei is of the order of 12 hours. From a time-lapse cinephotomicrographic study of the hatching and outgrowth of mouse blastocysts, I have been able to make some measurements on the increase in size of primary trophoblast nuclei and to estimate their proportional increase in volume. Given that the increase in volume of these nuclei is directly proportional to increase in DNA content (Zybina & Mos'yan, 1967) these measurements indicate that the time taken for the nuclear volume (and therefore DNA content) to double is of the order of 8–9 hours (Fig. 3).

CONCLUSIONS

Some aspects of the early differentiation and development of trophoblast have been studied. Its determination as trophoblast tissue arises by the effect of cell position during cleavage and up to the blastocyst stage the ability of trophoblast to implant and become giant is to some degree an intrinsic property of this tissue. Proliferation and further development of trophoblast into the ectoplacental cone is, however, dependent upon the presence of ICM cells, which may also influence the rate of giant cell transformation. The accumulation of large amounts of DNA in giant trophoblast nuclei has been shown to be by the rapid endoreduplication cycles of trophoblast chromosomes, during which there is some attraction of sister chromatids and the formation of polytene chromosomes.

REFERENCES

AMOROSO, E. C. (1952). Placentation. In *Marshall's Physiology of Reproduction*, 3rd edition, ed. A. S. Parkes, vol. 2, pp. 127–311. London: Longmans, Green.

ANSELL, J. D. & SNOW, M. H. L. (1974). The development of trophoblast *in vitro* from blastocysts containing varying amounts of inner cell mass. *Journal of Embryology and Experimental Morphology*. (In press.)

AVERY, G. B. & HUNT, C. V. (1969). The differentiation of trophoblast giant cells in the mouse studied in kidney capsule grafts. *Transplantation Proceedings*, **1**, 61–6.

BARLOW, P. W., OWEN, D. A. J. & GRAHAM, C. F. (1972). DNA synthesis in the preimplantation mouse embryos. *Journal of Embryology and Experimental Morphology*, **27**, 431–45.

BARLOW, P. W. & SHERMAN, M. I. (1972). The biochemistry of differentiation of mouse trophoblast: studies on polyploidy. *Journal of Embryology and Experimental Morphology*, **27**, 447–65.

BARLOW, P. W. & SHERMAN, M. I. (1974). Cytological studies on the organisation of DNA in giant trophoblast nuclei in the mouse and the rat. *Chromosoma*, **47**, 119–31.

Beerman, W. (1967). Gene action at the level of the chromosome. In *Heritage from Mendel*, ed. R. A. Brink, pp. 179–201. Madison: University of Wisconsin Press.

Chapman, V. M., Ansell, J. D. & McLaren, A. (1972). Trophoblast differentiation in the mouse: expression of Glucose Phosphate Isomerase (GPI-1) electrophoretic variants in transferred and chimeric embryos. *Developmental Biology*, **29**, 48–54.

Fawcett, D. W., Wislocki, G. B. & Waldo, C. M. (1947). The development of mouse ova in the anterior chamber of the eye and in the abdominal cavity. *American Journal of Anatomy*, **81**, 413–43.

Galassi, L. (1967). Reutilization of maternal nuclear material by embryonic and trophoblastic cells in the rat for the synthesis of deoxyribonucleic acid. *Journal of Histochemistry and Cytochemistry*, **15**, 573–9.

Gardner, R. L. (1971). Manipulations on the blastocyst. In *Schering Symposium on Intrinsic and Extrinsic Factors in Early Mammalian Development, Advances in the Biosciences vol. 6*, ed. G. Raspé, pp. 279–96. Oxford: Pergamon.

Gardner, R. L. & Johnson, M. H. (1972). An investigation of inner cell mass and trophoblast tissues following their isolation from the mouse blastocyst. *Journal of Embryology and Experimental Morphology*, **28**, 279–312.

Gardner, R. L., Papaioannou, V. M. & Barton, S. C. (1973). Origin of the ectoplacental cone and secondary giant cells in mouse blastocysts reconstituted from isolated trophoblast and inner cell mass. *Journal of Embryology and Experimental Morphology*, **30**, 561–72.

Gearhart, J. D. & Mintz, B. (1972). Glucose phosphate Isomerase Subunit reassociation tests for maternal–fetal and fetal–fetal cell fusion in the mouse placenta. *Developmental Biology*, **29**, 55–64.

Graham, C. F. (1971). The design of the mouse blastocyst. In *Control Mechanisms of Growth and Differentiation*, Symposium of the Society for Experimental Biology No. 25, ed. D. D. Davies & M. Balls, pp. 371–8. London: Cambridge University Press.

Hillman, N., Sherman, M. I. & Graham, C. F. (1972). The effect of spatial arrangement on cell determination during mouse development. *Journal of Embryology and Experimental Morphology*, **28**, 263–78.

Hunt, C. V. & Avery, G. B. (1971). Increased levels of deoxyribonucleic acid during trophoblast giant cell formation in mice. *Journal of Reproduction and Fertility*, **25**, 85–94.

Izawa, M., Allfrey, N. G. & Mirsky, A. E. (1963). The relationship between DNA synthesis and loop structure in lampbrush chromosomes. *Proceedings of the National Academy of Sciences*, USA, **49**, 544–51.

Jollie, W. P. (1964). Radioautographic observations on variations in desoxyribonucleic acid synthesis in rat placenta with increasing gestational age. *American Journal of Anatomy*, **114**, 161–71.

Kirby, D. R. S. (1960). The development of mouse blastocysts beneath the kidney capsule. *Nature, London*, **187**, 707–8.

Kirby, D. R. S. & Cowell, T. P. (1968). Trophoblast–host interactions. In *Epithelial–mesenchymal interactions in carcinogenesis*, ed. R. Fleischmajer & R. E. Billingham, pp. 64–77. Baltimore: Williams & Wilkins.

McLaren, A. & Tarkowski, A. K. (1963). Implantation of mouse eggs in the peritoneal cavity. *Journal of Reproduction and Fertility*, **6**, 384–92.

Nagl, W. (1972). Giant cell chromatin in endopolyploid trophoblast nuclei of the rat. *Experientia*, **28**, 217–18.

REICH, E. & GOLDBERG, I. H. (1964). Actinomycin and nucleic acid function. *Progress in Nucleic Acid Research and Molecular Biology*, **3**, 183–234.

SACCOMAN, F. H., MORGAN, C. F. & WELLS, L. F. (1967). Radioautographic studies of DNA synthesis in the developing extraembryonic membranes of the mouse. *Anatomical Record*, **158**, 197–206.

SHERMAN, M. I., MCLAREN, A. & WALKER, P. M. B. (1972). Mechanism of accumulation of DNA in giant cells of mouse trophoblast. *Nature, London*, **238**, 175–6.

SNELL, C. D. & STEVENS, L. C. (1966). Early embryology. In *Biology of the Laboratory Mouse*, ed. E. L. Green, pp. 205–45. New York: McGraw-Hill.

SNOW, M. H. L. (1973*a*). Abnormal development of pre-implantation mouse embryos grown *in vitro* with H^3-thymidine. *Journal of Embryology and Experimental Morphology*, **29**, 601–15.

SNOW, M. H. L. (1973*b*). The differential effect of H^3-thymidine upon two populations of cells in pre-implantation mouse embryos. In *The Cell Cycle in Development and Differentiation*, ed. M. Balls & F. S. Billett, pp. 311–24. London: Cambridge University Press.

SNOW, M. H. L. & ANSELL, J. D. (1974). The chromosomes of the giant cells in mouse trophoblast. *Proceedings of the Royal Society, London*, Ser. B, **187**, 93–9.

SNOW, M. H. L. & CALLAN, H. G. (1969). Evidence for a polarized movement of newt lampbrush chromosomes during oogenesis. *Journal of Cell Science*, **5**, 1–25.

STERN, M. S. (1972). Experimental studies on the organisation of the preimplantation mouse embryo. II. Reaggregation of disaggregated embryos. *Journal of Embryology and Experimental Morphology*, **28**, 255–61.

STERN, M. S. & WILSON, I. B. (1972). Experimental studies on the organisation of the preimplantation mouse embryo. I. Fusion of asynchronously cleaving eggs. *Journal of Embryology and Experimental Morphology*, **28**, 247–54.

WILSON, I. B., BOLTON, E. & CUTTLER, R. H. (1972). Preimplantation differentiation in the mouse egg as revealed by micro-injection of vital markers. *Journal of Embryology and Experimental Morphology*, **27**, 467–79.

ZYBINA, E. V. (1961). Endomitosis and polyteny of trophoblast giant cells. *Doklady Akademia Nauk, SSSR*, **140**, 1177–80. (In Russian.)

ZYBINA, E. V. (1963). Cytophotometer determination of DNA content of nuclei of trophoblast giant cells. *Doklady Akademii Nauk, SSSR*, **153**, 1428–31. (In Russian.)

ZYBINA, E. V. (1970). Anomalies of polyploidisation of the cells of the trophoblast. *Tsitologiya*, **12**, 1081–93. (In Russian.)

ZYBINA, E. V., KUDRYAVTSEVA, M. V. & KUDRYAVTSEV, B. N. (1973). Morphological and cytophotometric study of giant cells of rabbit trophoblast. *Tsitologiya*, **15**, 833–44. (In Russian.)

ZYBINA, E. V. & MOS'YAN, I. A. (1967). Sex chromatin bodies during endomitotic polyploidisation of trophoblast. *Tsitologiya*, **2**, 265–72. (In Russian.)

EXPLANATION OF PLATES

PLATE 1

Outgrowths from control and [^{3}H]thymidine-treated blastocysts, showing the differences in cellular content.

(*a*) Control outgrowth composed of many cells of different sizes and the characteristic central nodule of presumed diploid inner cell mass cells.

(*b*) High power section of part of a control trophoblast outgrowth showing a giant trophoblast nucleus and a mitotic figure (arrowed).

(*c*) Trophoblast outgrowth from a [^{3}H]thymidine-treated blastocyst composed almost entirely of giant trophoblast cells. The central nodule of smaller cells is absent from such outgrowths.

(*d*) High power of a section of the outgrowth from (*c*). Note the comparable size of these two giant nuclei to those shown in (*b*).

PLATE 2

The range of condensation of nuclear contents after treatment of trophoblast giant nuclei with actinomycin D.

(*a*) Control untreated nuclei from an ectoplacental cone outgrowth. The range of sizes of nuclei present indicates that most of the nuclei observed after actinomycin D treatment were giant.

(*b*) The least contracted nuclear contents observed were a fine meshwork of thin chromatin threads.

(*c*) High power of the nucleus described in (*b*).

(*d*) A nucleus estimated, by virtue of its size, to be between 32 C and 64 C. Actinomycin D treatment has condensed the nuclear contents such that the better displayed chromatin bodies exhibit a beaded appearance.

(*e*) The maximum degree of condensation observed in trophoblast giant nuclei. Forty discrete chromatin bodies are visible.

(*f*) Diploid nuclei from an actinomycin D treated blastocyst. Again approximately forty discrete chromatin bodies are visible.

(*g*) A large (approx. 500 C) nucleus in which the chromatin has aggregated together in one part of the nucleus.

(*h*) A nucleus similar in size to that shown in (*g*) in which there is better resolution of nuclear material.

(*j*) A section of fixed and embedded actinomycin D treated ectoplacental cone, just glancing the periphery of a trophoblast giant nucleus. Beaded chromosome-like structures similar to those seen in (*d*) are visible just under the nuclear membrane.

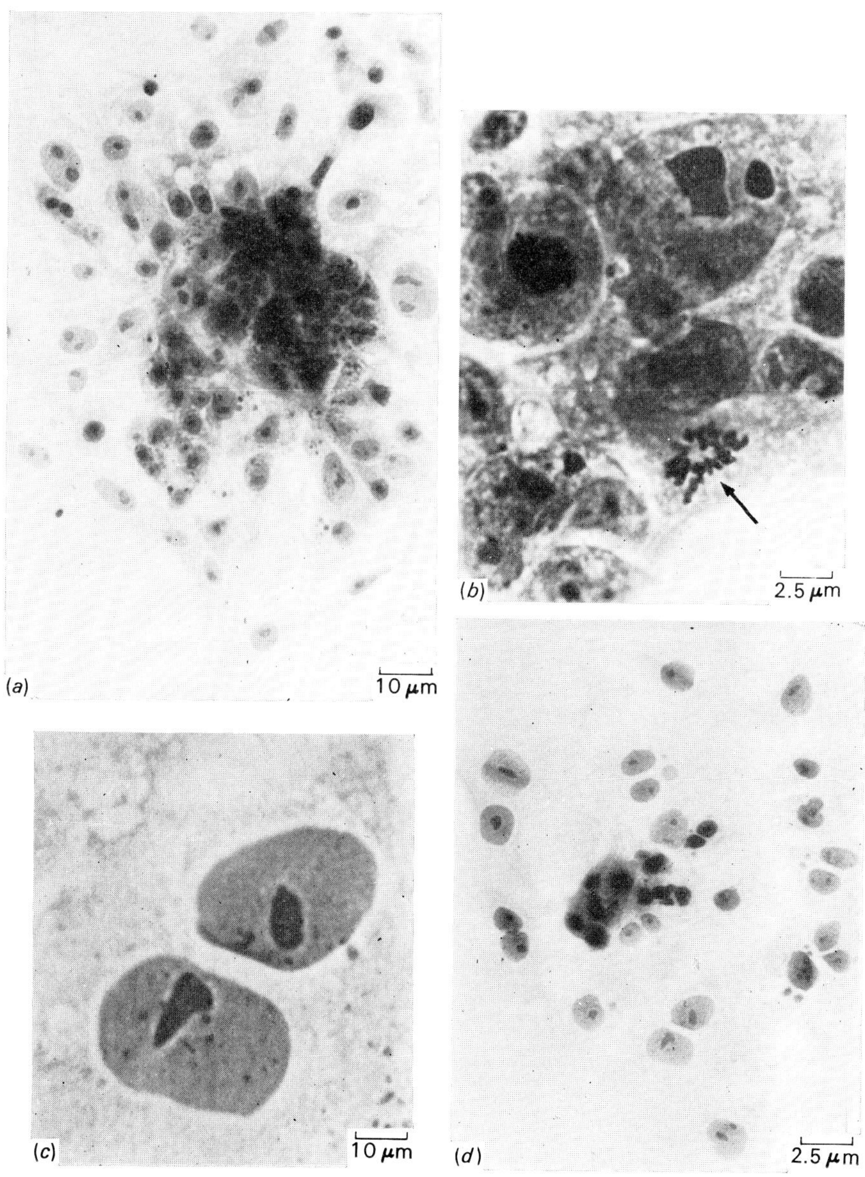

(Facing p. 144)

PLATE 2

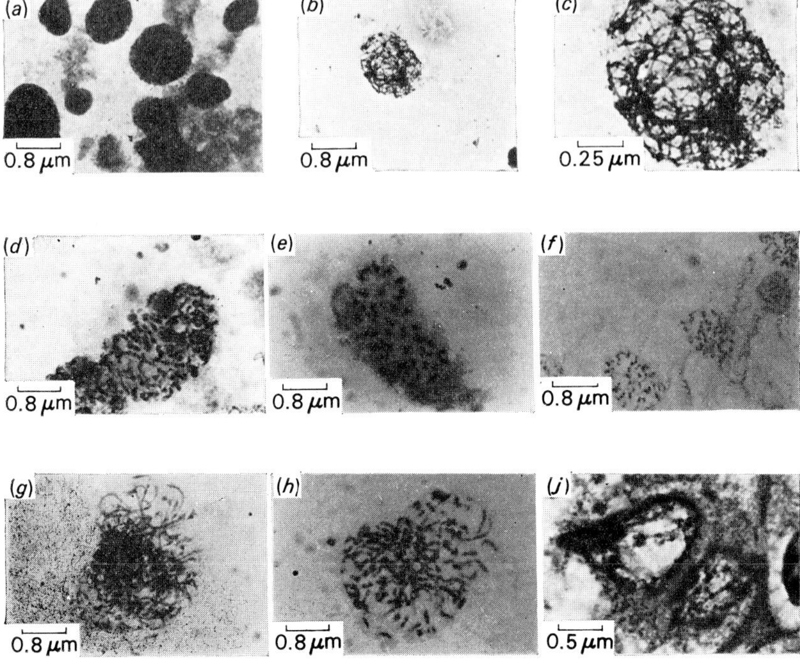

THE ROLE OF CELL–CELL INTERACTION DURING EARLY MOUSE EMBRYOGENESIS

BY M. I. SHERMAN

Department of Cell Biology, Roche Institute of Molecular Biology, Nutley, New Jersey 07110, USA

Mouse blastocysts consist only of two cell types: the trophectoderm, a continuous, single-celled layer and the inner cell mass (ICM), a clump of cells enclosed by the trophectoderm. The developmental fate of the two cell types is predictable; all the available evidence suggests that the trophectoderm forms the trophoblast layer of the placenta, while the ICM gives rise to the remainder of the extra-embryonic membranes, as well as the embryo proper (Duval, 1891; Jenkinson, 1913; Gardner, 1971; Hillman, Sherman & Graham, 1972; Gardner, Papaioannou & Barton, 1973). Recent developments in postblastocyst culture technique (Hsu, 1971, 1973; Hsu, Baskar, Stevens & Rash, 1974; Sherman, 1972*b*; Bell & Sherman, 1973; Sherman, 1974*a*) permit experimentation *in vitro*, outside the sphere of any possible maternal influences (see Sherman & Salomon, 1974); trophectoderm and ICM derivatives develop *in vitro* and can be distinguished both morphologically (Hsu, 1971; Hsu *et al.*, 1974; Sherman, 1974*a*) and biochemically (see Herbert & Graham, 1974, Sherman, 1974*b* and Sherman & Salomon, 1974 for reviews). Insofar as the blastocyst is a relatively simple structure, and since the fate of its component cell types can be followed in culture with ease, we are presented with a potentially useful system for the study of the role of cell–cell interaction during development.

In our efforts to determine the degree of dependence of the early embryo upon the uterine environment for normal development, we have removed blastocysts from the uterine milieu and studied the properties of the cell types (particularly those of trophoblast) which develop (Barlow & Sherman, 1972; Sherman, 1972*a*, *b*; Bell & Sherman, 1973; Chew & Sherman, 1975; Sherman, 1974*c*; Sherman & Salomon, 1974). In the course of these studies, we have found ample evidence that the differentiated properties of trophoblast and yolk sac can be initiated outside the maternal environment, and even that the time of appearance of these properties *in vitro* parallels the schedule *in vivo*. By the same line of reasoning, the degree of interdependence of ICM derivatives and trophoblast for normal development should become clear if we could place one of the cell types into the position of having to develop in the absence of the other. To date, a number of techniques have been developed for preparing trophoblast cells free of ICM derivatives. On the other hand, it is a more difficult matter to obtain pure ICMs (see Discussion). Consequently, this article will focus upon the degree of differentiation of trophectoderm to trophoblast cells in the absence of the ICM.

In our early postblastocyst culture studies (Barlow & Sherman, 1972; Sherman, 1972*a*), we found that all ICM cells commonly disappeared after about five days in culture, leaving a monolayer of trophoblast cells. As mentioned above, recently improved culture conditions do not select against ICM development. However, the addition to the culture medium of appropriate concentrations of antimetabolites, such as colcemid, cytosine arabinoside or bromodeoxyuridine (Sherman, unpublished data), as well as actinomycin D, cordycepin or cycloheximide (J. Rowinski, D. Solter & H. Koprowski, personal communication) specifically kills ICM cells within 48–72 hours, while there is little or no effect upon trophoblast cells. Gardner (1971; see this volume) has used his elegant microsurgical techniques to cut blastocysts so that one half contains only trophectoderm cells which round up to form 'trophoblastic vesicles'. When transferred to incubator mothers, these trophoblast vesicles give rise to a few giant trophoblast cells, but no ICM derivatives (Gardner, 1971, 1972).

The above techniques have been, and are being, used to study a number of properties of early mouse embryos. However, in each case, trophectoderm cells had initially been in contact with ICM. It would be more desirable in studying the ability of trophectoderm cells to differentiate independently if these cells had *never* been exposed to possible influences of the ICM. This has apparently been achieved by Snow (1973*a*, *b*), utilizing [^{3}H]thymidine. Two-cell embryos cultured in the presence of the isotopically-labeled DNA precursor develop in such a way that the majority appear not to have an ICM when the blastocyst stage is reached, i.e. trophoblastic vesicles are formed. Presumably, the precursor cells of the ICM are destroyed by radiation damage. Upon transfer to uteri of incubator mothers or to ectopic sites of adult hosts, the trophoblastic vesicles give rise only to giant trophoblast cells (Snow, 1973*b*). However, Snow (1973*a*) has pointed out that a number of the cells in these trophoblastic vesicles had abnormal nuclei and possessed chromosomal abnormalities. Misleading results might be obtained from a study of the developmental potential of trophoblastic vesicles if they contained many abnormal cells. Tarkowski & Wroblewska (1967) observed that single blastomeres disaggregated from 4-cell or 8-cell mouse embryos gave rise to trophoblastic vesicles, among other forms (see below), upon subsequent culture.

Since the trophectoderm cells in these structures had never encountered an ICM, and did not appear to be in any way abnormal, this system was chosen for further study.

METHODS

The embryos used in these studies were obtained by mating SJL/J males with previously superovulated (Runner & Palm, 1953) SWR/J females (Jackson Laboratories, Bar Harbor, Maine). The day of observation of the copulation plug is considered the first day of pregnancy. Unless otherwise indicated, 3rd day embryos were used. These were flushed from oviducts and uteri with the medium of Whitten & Biggers (1968), hereafter designated WB medium. A mixture of 4-cell and 8-cell embryos was commonly obtained. On some occasions, to serve as controls, blastocysts were flushed from uteri in phosphate-buffered saline (PBS; solution of Dulbecco & Vogt, 1954) on the 4th day of gestation. The blastocysts were cultured in NCTC–109 medium (Evans, Bryant, Kerr & Schilling, 1964; obtained from Microbiological Associates, Bethesda, Maryland), supplemented with 10 % heat-inactivated (56 °C for 15 min) fetal calf serum (Microbiological Associates) and antibiotics (100 Units/ml penicillin, 100 μg/ml streptomycin and kanamycin, all purchased from Gibco, Grand Island, New York). This mixture will be referred to as 'supplemented NCTC-109 medium'.

Embryos were disaggregated as described by Tarkowski & Wroblewska (1967), with some minor modifications. Briefly, embryos were washed with PBS, and treated with 0.5% pronase (Calbiochem, LaJolla, California) in PBS to remove the zona pellucida (Mintz, 1962). The embryos were then immediately returned to WB medium and sucked up and down in a siliconized micropipette with a flame-polished tip and an aperture slightly greater than that of the intact embryo. Most blastomeres were disaggregated by this time. More stubborn groups of blastomeres were disaggregated by sucking them up and down in a micropipette with a narrower bore. Treatment with ethylenediamine tetra-acetic acid (Tarkowski & Wroblewska, 1967) was found not to be necessary. The procedure used here resulted in very high survival rates (see Tables 1 and 2). After disaggregation, blastomeres were placed in wells of a Falcon microtest dish, type 3034 (Falcon Plastics, Oxnard, California), containing WB medium. Blastomere reaggregates were generated by nudging the cells together, and keeping them in contact with each other for a few seconds. Incubation was carried out in microtest dishes at 37 °C in 5 % CO_2/air.

After 48 hours, the type of structure formed by each blastomere was determined by the use of a Wild M40 inverted phase microscope. The structures were then treated in one of three ways: (1) placed individually into wells of a microtest dish containing supplemented NCTC-109 medium; (2) placed with other similar structures into a Petri dish containing a monolayer of primary uterine cells which had been cultured on the previous day (see Salomon & Sherman, 1974); or (3) placed with other similar structures into wells of a larger microtest dish (Falcon, type 3040) containing supplemented NCTC-109 medium (100 μl per well). After 72 hours of culture in the last case, the medium was changed to supplemented NCTC-109 containing 1 μg/ml pregnenolone (previously purified by crystallization). Every 48 hours, the medium was collected and frozen, while new medium was added to the cultures. At the end of the culture period, aliquots of the collected media were assayed for progesterone content by a highly sensitive and specific radioimmunoassay procedure (see Chew & Sherman, 1975). As a control, medium incubated for similar periods of time in wells containing no cells was also assayed for progesterone content. The background values obtained were subtracted from the experimental values.

TERMINOLOGY

The cells constituting the outer layer of the blastocyst have been referred to by some as 'trophectoderm' and by others as 'trophoblast'. The term 'trophoblast' is also used to describe the fetal moiety of the placenta, i.e. initially the giant cells and the ectoplacental cone, later the intermediate layers of the placenta (spongiotrophoblast and labyrinth). Since many of our studies have revealed marked morphological and biochemical differences between the cells constituting the outer layer of the blastocyst and those forming the fetal placenta, the former will be referred to as 'trophectoderm' cells, to distinguish them from the 'trophoblast', trophectoderm-derived cells which have implanted *in vivo*, or, by analogy, outgrown *in vitro*.

The remaining terminology used is generally in accordance with that of Tarkowski & Wroblewska (1967). Single blastomeres from the 4- and 8-cell stages will be referred to as $\frac{1}{4}$ or $\frac{1}{8}$ blastomeres, respectively. Pairs of blastomeres from 8-cell embryos will be referred to as $\frac{2}{8}$ blastomeres. Often, some of the embryos obtained were intermediate between the 4- and 8-cell stages, i.e. they contained 5–7 (most often 6) cells. In these embryos, the blastomeres which had divided for the third time could be distinguished from those which had only divided twice by their smaller size; the two could therefore be separated and used as $\frac{1}{8}$ and $\frac{1}{4}$ blastomeres, respectively.

The forms developing from disaggregated blastomeres in this study were the same as those described by Tarkowski & Wroblewska (1967). Trophoblastic vesicles did not contain any cells which were fully enclosed (Plate 1*c*, *f*). Miniblastocysts, although smaller than normal control blastocysts (cf. Plate 1*a* and *b*), appeared to contain an ICM. However, it is not always possible without sectioning to distinguish between true blastocysts and 'false' blastocysts, trophoblastic vesicles which, due to a thickening on one side, appear to contain enclosed cells (Tarkowski & Wroblewska, 1967). Consequently, the miniblastocyst group contains both true and false blastocysts. The two other structures observed were 'morulae', which did not cavitate (Plate 1*e*) and 'non-integrated forms', disorganized clusters of cells which often contained vacuolated cells (Plate 1*d*).

RESULTS

Fate of single and paired blastomeres

As Table 1 indicates, almost all cleavage stage embryos formed blastocysts following incubation for two days in WB medium. Disaggregation of the

Table 1. *Blastocyst formation by control 4-cell and 8-cell embryos or disaggregated and reaggregated 4-cell and 8-cell embyros* in vitro

	Number studied	Number forming blastocysts	%
Control embryos:			
(4-cell & 8-cell)	61	59	97
Reaggregated embryos:			
(4/4)	34	30	88
(8/8)	29	28	97

Embryos containing 4–8 cells were treated with pronase to remove the zona pellucida. Control embryos were placed in WB medium, while experimentals were first disaggregated and reaggregated as described in Methods. Embryos were inspected for blastocyst formation after 48 h in WB medium.

blastomeres did not damage them since high percentages of reaggregated embryos also formed blastocysts.

After 24 hours, the majority of $\frac{1}{4}$, $\frac{2}{8}$ and $\frac{1}{8}$ blastomeres divided, although at different frequencies (Table 2); almost all $\frac{1}{4}$ blastomeres divided, but only two-thirds of $\frac{1}{8}$ blastomeres did so. In 91% of $\frac{2}{8}$ blastomeres, at least one of the pair had divided. However, only in 64% of cases did *both* of the $\frac{2}{8}$ blastomeres divide (i.e. forming a total of four or more cells; see Table 2).

Blastomeres at the 4-cell stage showed no apparent dependence upon other cells for further division and development, since they progressed equally well as singles or in reaggregates (cf. Tables 1 and 2). On the other hand, cell–cell interaction appears to play a role in the development of $\frac{1}{8}$ blastomeres. While 97 % of $\frac{8}{8}$ reaggregates formed blastocysts, only 68 % of isolated $\frac{1}{8}$ blastomeres showed the ability to divide. The beneficial effect of 8-cell blastomeres upon each other is further evidenced by a comparison of the capacity of $\frac{1}{8}$ and $\frac{2}{8}$ blastomeres to undergo division: if each of the blastomeres in the $\frac{2}{8}$ pair possessed a 68 % chance of division (the case for isolated $\frac{1}{8}$ blastomeres), it can be calculated that only 46 % of $\frac{2}{8}$ blastomeres would be expected to contain four or more cells. The observed value of 64% is substantially higher. The dependence upon cell–cell interaction at the morula stage is even more striking. While we have been able to repeat Stern's (1972) observation that reaggregates of morula cells can form blastocysts, very few disaggregated and isolated cells showed the ability to divide even once (Bell & Sherman, unpublished observations).

After 48 hours of incubation, the blastomeres were again examined (Table 2). In all cases, the majority of blastomeres had formed either miniblastocysts, trophoblastic vesicles, 'morulae' or non-integrated forms. In fact, although only 91% of $\frac{2}{8}$ blastomeres contained three or more cells after 24 hours, 98 % of the blastomeres formed one of the four structures by 48 hours. This discrepancy is due to delayed division of the blastomeres as well as to formation of structures, particularly trophoblastic vesicles, by doublets which had failed to divide (e.g. see Plate 1(*f*) for a 2-celled trophoblastic vesicle).

The incidence of the various structures agrees well with that found by Tarkowski & Wroblewska (1967). The basic observation is that the majority of $\frac{1}{4}$ blastomeres form miniblastocysts while $\frac{1}{8}$ blastomeres are more likely to form trophoblastic vesicles, and $\frac{2}{8}$ blastomeres are intermediate. There are minor differences, however: first, the incidence of 'morulae' was less in this study than in Tarkowski & Wroblewska's; second, in the present study, a marked difference was not observed between the frequency of non-integrated forms depending upon the initial stage of the blastomeres (Tarkowski & Wroblewska found an incidence for non-

Table 2. *Fate of isolated and paired blastomeres* in vitro

Number and stage of blastomeres	Number studied	Cell number after 24 h						Forms observed after 48 h							
		2 or more		3 or more		4 or more		MBc		TV		Mor		NIF	
		No.	%	No.	%	No.	%	No.	%	No.	%	No.	%	No.	%
$\frac{1}{8}$	489	334	68	–		–		69	14	119	24	11	2	66	14
$\frac{2}{8}$	333	–		302	91	212	64	146	44	99	34	11	3	58	17
$\frac{1}{4}$	331	313	95	–		–		173	52	55	17	16	5	55	17

Embryos were pronased to remove the zona pellucida and disaggregated as described in Methods. Single 4- and 8-cell blastomeres as well as 8-cell pairs were cultured individually in WB medium. After 24 h, cell numbers were determined. After a further 24 h, cultures were inspected for the formation of miniblastocysts, true or false (MBc), trophoblastic vesicles (TV), 'morulae' (Mor) and non-integrated forms (NIF).

integrated forms of 11.9, 21.3 and 32.9% for $\frac{1}{4}$, $\frac{2}{8}$ and $\frac{1}{8}$ blastomeres, respectively).

Tarkowski & Wroblewska (1967) noted that trophoblastic vesicles contained fewer cells than blastocysts. Based on this observation, these authors proposed that an ICM would be present only if there were an adequate number of blastomeres in the structure so that one or more cells could be completely enclosed by the others at the time of formation of the blastocoel cavity. Such a proposal would in turn imply that cavitation is a response to gestation age, or some other parameter, rather than cell number. In support of this is the observation that $\frac{2}{8}$ blastomeres can fail to divide (or a $\frac{1}{8}$ blastomere can divide only once), and yet, a blastocoel is induced at the expected time, resulting in a trophoblastic vesicle (Plate 1*f*). The actual trigger for blastocoel formation remains unknown.

Developmental capacity of trophoblastic vesicles

To determine whether trophoblast cells can differentiate from trophoblastic vesicles in a normal manner, the developmental capacity of these structures was monitored and compared with that of control blastocysts as well as miniblastocysts. For this purpose, control and experimental forms were placed in supplemented NCTC-109 medium. Within 96 hours, virtually all control blastocysts as well as blastocysts developing from reaggregated 4- and 8-cell embryos had attached to the culture dish, and trophoblast outgrowth had begun (Table 3). Furthermore, in 94% of cases, the ICM could be clearly seen as a clump of cells sitting atop the trophoblast monolayer (see e.g. Plate 2*a* and *c*). Trophoblastic vesicles as well as miniblastocysts also attached to the culture dish and showed trophoblast outgrowth at high frequencies (Table 3). As would be expected, almost none of the trophoblastic vesicles contained a clump of cells above the trophoblast monolayers, as did the controls; of the two exceptional cases, only one definitely contained an ICM (the other possessed a small cluster of cells, which may have been abnormal trophoblast since the cells did not proliferate, but disappeared within a few days). The structure had obviously been incorrectly categorized as a trophoblastic vesicle.

It was somewhat surprising that miniblastocysts showed such a low incidence of ICMs (Table 3). Possible explanations are that many miniblastocysts might in fact have been false blastocysts, and that there may be a critical number of cells required for the maintenance of the ICM as a clump of cells under our culture conditions.

The number of trophoblast cells in outgrowths of trophoblastic vesicles

Table 3. *Outgrowth and presence of inner cell mass in control and experimental embryos in supplemented NCTC-109 medium*

Form	Number studied	Number outgrown	%	Number with ICM	%
Control blastocysts	105	101	96.2	99	94.3
Reaggregated embryos	17	17	100	16	94.1
Miniblastocysts	186	158	84.9	19	10.2
Trophoblastic vesicles	104	92	90.4	2	1.9

Control blastocysts were removed from the uterus on the 4th day of gestation and placed directly into supplemented NCTC-109 medium. Reaggregated embryos were those forming blastocysts from 4- or 8-cell reaggregates after 48 h in WB medium. Miniblastocysts and trophoblastic vesicles were generated from $\frac{1}{4}$ or $\frac{2}{8}$ embryos after 48 h in WB medium. Cultures were inspected for outgrowth and presence of ICM after 48 and 96 h in supplemented NCTC-109 medium.

and miniblastocysts was estimated by counting the number of nuclei present (Fig. 1). There is not an exact 1 : 1 correspondence because a small number of cells were binucleate (see e.g. Plate 2*d*). The range of nuclei present in trophoblastic vesicle outgrowths derived from $\frac{1}{4}$ and $\frac{2}{8}$ blastomeres varied from 2–21, with a mean of 9.5. These values are in good agreement with cell numbers determined by Tarkowski & Wroblewska (1967) on sectioned trophoblastic vesicles (range of 3–25, with a combined mean for both $\frac{1}{4}$ and $\frac{2}{8}$ blastomeres of 10.7). Since the two sets of data agree so well with each other, this would suggest that most cells in the trophoblastic vesicles were viable, and capable of outgrowth.

It was interesting to note (Fig. 1) that the number of trophoblast cells growing out from miniblastocysts was very similar to that of trophoblastic vesicles (range 2–21, mean 9.9). These values did not agree well with those obtained by Tarkowski & Wroblewska for true blastocysts derived from $\frac{1}{4}$ and $\frac{2}{8}$ blastomeres (range 8–28, mean 15.3), but were closer to those for false blastocysts (range 5–20, mean 12.2). These factors taken together suggest that trophoblastic vesicles and miniblastocysts originally have the same number of trophectoderm cells, and that the greater number of total cells in true blastocysts is due to the presence of ICM cells. The ICM either degenerates or remains as a clump in culture (nuclei cannot be seen in these clumps by phase optics, so they are not counted), but does not outgrow along the dish within 96 hours of culture. Consequently, only

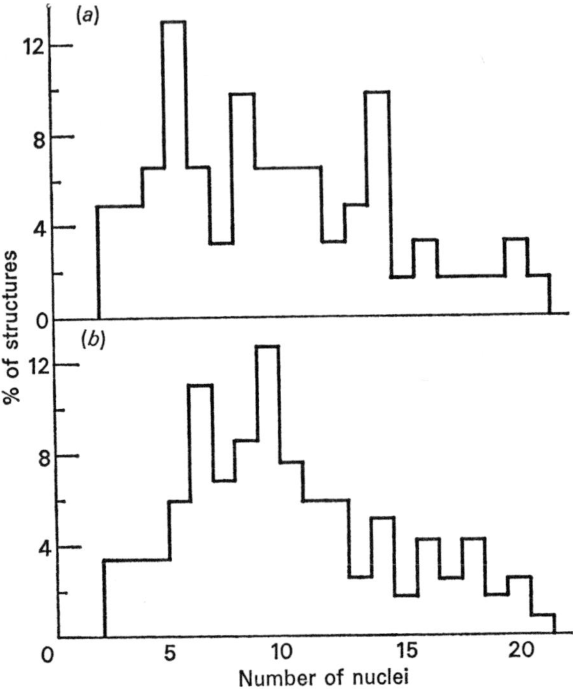

Fig. 1. Nuclear numbers in outgrowths of trophoblastic vesicles (*a*) and miniblastocysts (*b*). The structures were derived from $\frac{1}{4}$ or $\frac{2}{8}$ blastomeres cultured for 2 days in WB medium, and were then individually transferred to wells of a microtest dish containing supplemented NCTC-109 medium. Nuclei were counted after 2 and 4 days in the latter medium, and the larger of the two numbers was used. The number of structures analyzed was 62 for trophoblastic vesicles and 118 for miniblastocysts.

trophoblast nuclei were counted in both trophoblastic vesicle and miniblastocyst cultures, and, as is then to be expected, the numbers were the same.

Morphology of trophoblast cells from trophoblastic vesicles

The morphological properties of trophoblast cells during their development in postblastocyst cultures has been described elsewhere (Sherman, 1974*a*, *b*; Sherman & Salomon, 1974). Trophoblast cells outgrowing from trophoblastic vesicles were inspected for three morphological properties characteristic of most trophoblast cells developing from normal blastocysts: (1) a high degree of vacuolation during the initial stages of outgrowth (Plate 2*a*); (2) the presence of cytoplasmic granules concentrated in the perinuclear region after about three or four days in culture (Plate 2*c*); and (3) the presence of giant nuclei (Fig. 2*a*). Plate 2(*b*) demonstrates that after 48 hours in supplemented NCTC-109 medium the trophoblast cells of trophoblastic vesicles closely resemble those of control or reaggregated embryos: there is a high degree of vacuolation. After a further 48 hours, however, most of the vacuoles had disappeared. At this time, perinuclear granules could be seen in many trophoblast cells (Plate 2*d*), just as in the controls. Not unexpectedly, the trophoblast cells growing out from mini-blastocysts had the same properties (not shown).

The nature and function of these cytoplasmic inclusions is not clear. Because Dickson (1969) has reported that trophectoderm cells contain conspicuous lipid droplets just prior to implantation, outgrowing trophoblast cells were stained with sudan black III (Dickson, 1969). The vacuoles did not take up the stain. It is possible that the vacuoles are evidence of residual pinocytotic activity related to the pumping of trophectoderm cells during cavitation and hatching from the zona pellucida. Unlike the vacuoles, the granulation of the cytoplasm remains and, in fact, intensifies throughout the life of the cell. These granules may perform a storage or secretory role, or alternatively, they may be swarms of lysosomes. Further studies are in progress to clarify the nature of both vacuoles and granules.

By comparison of Plate 2(*a*) with (*c*) and of 2(*b*) with (*d*), it can be seen that there is an increase in nuclear size of trophoblast cells with time spent in culture. We have shown previously that trophoblast cells undergo polyploidization both *in vivo* and *in vitro*, and that there is a proportionality between DNA content and nuclear diameter (Barlow & Sherman, 1972). After 96 hours in supplemented NCTC-109 medium, diameters of trophoblast nuclei in control blastocysts ranged from 10–57 μm (Fig. 2*a*). The diameters of 96% of a sample of nuclei from blastocyst-derived diploid cell lines (Sherman, 1974*a*) were found to be less than 21 μm. Consequently, it is assumed here that any nuclei with diameters greater than 21 μm ('giant nuclei') are likely to be polyploid. By this criterion, 98 % of the nuclei measured in monolayers of control blastocysts were giant.

Sample miniblastocysts and trophoblastic vesicles also contained giant nuclei (Fig. 2*b* and *c*). In both cases, about 85 % of the nuclei had diameters ranging from 24–63 μm. It must, therefore, be concluded that interaction with ICM cells is not necessary for the formation of giant nuclei and polyploidization of trophoblast cells. Indeed, the available evidence suggests that trophoblast cells lose the ability to divide and normally become polyploid only after loss of contact with the ICM (Barlow & Sherman, 1972; Gardner, 1972; Gardner *et al.*, 1973). It would therefore be expected that trophoblastic vesicles, since they lack an ICM, would contain only polyploid cells. As Table 4 indicates, the majority of trophoblastic vesicles possessed only cells with giant nuclei, while almost

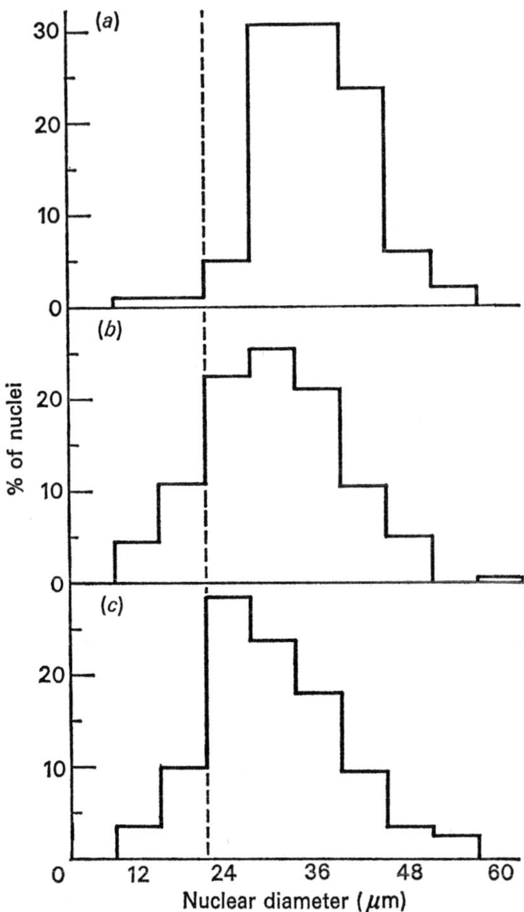

Fig. 2. Diameter of trophoblast nuclei in outgrowths from control blastocysts (*a*), miniblastocysts (*b*), and trophoblastic vesicles (*c*). Control blastocysts were derived from 4- and 8-cell embryos cultured for 2 days in WB medium; miniblastocysts and trophoblastic vesicles were derived from $\frac{1}{4}$ or $\frac{2}{8}$ blastomeres cultured in the same way. The structures were then transferred to supplemented NCTC-109 medium, and trophoblast nuclei were measured after 4 days of outgrowth. The values shown are the averages of the largest and smallest diameter of each nucleus. Values to the right of the dotted lines represent giant nuclei. The mean nuclear diameters are (*a*) 37.2 μm, (*b*) 31.2 μm and (*c*) 31.2 μm. The number of nuclei measured in each case are (*a*) 101, (*b*) 242 and (*c*) 168.

half of the miniblastocysts did so as well. Many miniblastocysts would be expected to contain only giant cells since, as has been mentioned, some are false blastocysts, and the ICM in true miniblastocysts often degenerates. Some trophoblast cells in control blastocysts, trophoblastic vesicles, and miniblastocysts contained two nuclei, each of which fell within the diploid nucleus size range, but, when taken together, fell within the giant nucleus size range. About 70% of both trophoblastic vesicles and miniblastocysts contained cells which contained only giant cells and binucleates (Table 4). Since there was no sign of cell division in any of the trophoblastic vesicle outgrowths, it is likely that the small proportion of cells which did not contain giant nuclei were inactive. Such cells also occur *in utero*; they have been observed in sections of post-implantation trophoblast (R. L. Gardner, personal communication).

Table 4. *Proportions of miniblastocysts and trophoblastic vesicles containing only giant nuclei after 4 days in culture medium*

Form	Number studied	Giant nuclei only	%	Giant nuclei and binucleates only	%	Total %
Miniblastocysts	45	21	46.7	10	22.2	68.9
Trophoblastic vesicles	45	27	60.0	5	11.1	71.1

Nuclear diameters were measured after 96 h in supplemented NCTC-109 medium (see Fig. 2). Only structures with 4 or more cells were considered.

Functional capacity of trophoblast cells from trophoblastic vesicles

Gardner (1971) has shown that trophoblastic vesicles generated by cutting blastocysts in two can induce a decidual response when placed in the uterus of an incubator mother. Snow (1973*b*) also noted that trophoblast vesicles resulting from [^{3}H]thymidine treatment of cleavage embryos could induce a decidual response, but did not implant normally. Trophoblast cells from these structures did develop normally in ectopic sites and showed invasive properties characteristic of trophoblast cells from normal blastocysts.

We have devised an in-vitro system for testing the functional capacity of trophoblast cells (Salomon & Sherman, 1974; Sherman & Salomon, 1974). This involves the attachment of blastocysts to a monolayer of uterine cells and the subsequent invasive outgrowth of trophoblast cells into the uterine monolayer. As Plate 3(*a*) illustrates, trophoblast cells from control blastocysts had the ability to displace uterine cells as they grew out along the culture dish. Trophoblastic vesicles (Plate 3*b*) and miniblastocysts (not shown) also attached to a uterine monolayer, and the developing trophoblast cells showed the same invasive properties as did those of control blastocysts. Preliminary observations suggested that a higher proportion of miniblastocysts than trophoblastic vesicles were capable of 'implanting' *in vitro*, but controlled quantitative studies have yet to be done.

Progesterone production by trophoblast cells from trophoblastic vesicles

By the use of a sensitive radioimmunoassay procedure, we have shown that homogenates of trophoblast cells possess the enzyme Δ^5,3β-hydroxysteroid dehydrogenase (3β-HSD). This enzyme, actually a coupled enzyme system, functions in the conversion of pregnenolone to progesterone, and thereby establishes trophoblast as hormone-producing cells. In our studies, we found that trophoblast homogenates possessed 3β-HSD activity by the 9th day of gestation. Specific enzyme activity increased to a maximum on the 11th day and then fell (Chew & Sherman, 1973, 1975).

Homogenates from trophoblast cells developing in postblastocyst cultures possessed 3β-HSD activity (Chew & Sherman, 1975). Progesterone production has also been measured in living trophoblast cells by adding pregnenolone to the culture medium; conveniently, more than 95% of the progesterone produced by the cell is secreted and can be detected in the culture medium by radioimmunoassay (Chew & Sherman, 1975). Accordingly, the pattern of progesterone production during trophoblast development in a given culture can be followed merely by daily collection and assay of the culture medium (Sherman & Salomon, 1974). Initially, we found the temporal pattern of progesterone production by trophoblast cells developing from control blastocysts to be unpredictable. We later discovered this to be due to the fact that although cells derived from the ICM could not themselves produce progesterone, they were capable of further metabolizing the progesterone synthesized and secreted by trophoblast cells (Sherman, unpublished observations). More recently, assays have been carried out to determine progesterone production by trophoblast cells in the absence of ICM derivatives, either by dissecting out postimplantation trophoblast (Sherman & Salomon, 1974), or by specifically killing ICM cells of blastocysts with antimetabolites (Sherman, in preparation), as mentioned at the beginning of this article. The results are clear and consistent: progesterone production begins between the 6th and 7th days (of equivalent gestation age, i.e. the age of the embryos had they been left *in utero*), but not earlier, rises to a peak between the 10th and 12th days, and falls thereafter. In other words, the pattern of 3β-HSD activity by trophoblast cells in culture faithfully follows that observed in homogenates of uterine trophoblast. (Progesterone production by cultured trophoblast is probably detected at an earlier time than in uterine trophoblast homogenates because living cells are analyzed and because very much longer incubation periods are used in the former case.)

Trophoblast cells outgrowing from trophoblastic vesicles were assayed for progesterone production during development in culture (Fig. 3). Progesterone synthesis was clearly detectable. It is difficult to compare this activity meaningfully with cultured control blastocysts since the latter originally possessed four times as many cells compared to the trophoblastic vesicles and also contained ICM derivatives. Nevertheless, in this experiment, a control blastocyst culture assayed simultaneously produced about ten times as much progesterone as the trophoblastic vesicle culture between

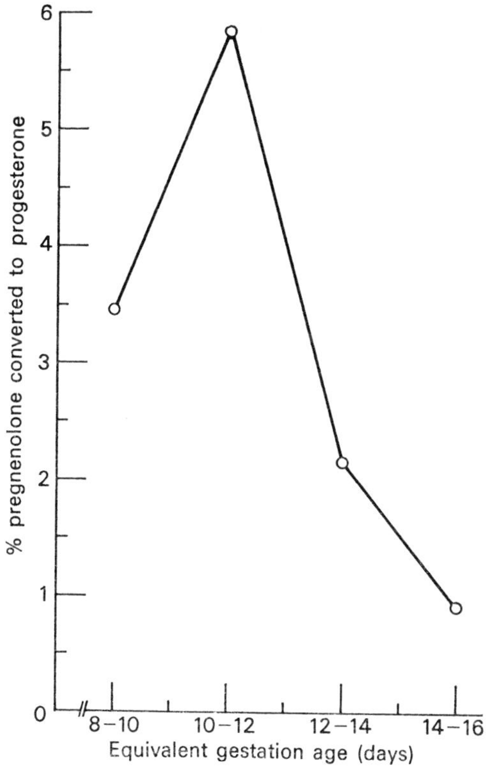

Fig. 3. Progesterone production by trophoblastic vesicle outgrowths. Trophoblastic vesicles were generated from $\frac{1}{4}$ or $\frac{2}{8}$ blastomeres during 2 days of culture in WB medium. The structures were then transferred to supplemented NCTC-109 medium and assayed for progesterone production during 48-hour intervals as described in Methods. The absolute amount of progesterone formed during the peak of activity (10–12 days equivalent gestation age) was 5.85 ng.

the 10th and 12th equivalent gestation day (1.52 fmoles/min per blastocyst as against 0.16 fmoles/min per trophoblastic vesicle). Most impressively, the temporal pattern of progesterone production by trophoblastic vesicles (Fig. 3) closely followed that of uterine trophoblast homogenates.

DISCUSSION AND CONCLUSIONS

Taken together, the work of Mintz, Tarkowski, Graham, Wilson, Gardner and their coworkers provides convincing evidence that blastomere fate is determined on the basis of cell position (see Graham, 1971, and Herbert & Graham, 1974, for reviews). Factors which induce the cells of the early mouse embryo to undergo differentiation after they have become determined are less well understood. The results of a number of experiments have led to the conclusion that while the maternal environment might provide the best possible conditions for embryonic development, uterine factors do not directly trigger differentiation of trophectoderm and ICM cells (see Sherman, 1974*b*; Sherman & Salomon, 1974). The present report describes the beginning of the next step in the search for factors influencing differentiation of early embryonic cell types, namely an investigation of the interaction of cells within the embryo. I have specifically considered here the ability of trophectoderm cells to differentiate in the absence of an ICM. Earlier studies (Barlow & Sherman, 1972; Sherman, 1972*a*; Gardner, 1971, 1972; Snow, 1973*b*) provided reason to expect that trophoblast cells could develop normally under these conditions. For the reasons presented in the introductory section, however, trophoblastic vesicles developing from disaggregated blastomeres could best be used to attack the problem. The present studies have shown clearly by morphological, functional and biochemical criteria that factors triggering trophoblast differentiation do not reside in the ICM.

The data have, however, provided some evidence for an effect of cell–cell interaction upon early embryos which may be unrelated to differentiation. By the 8-cell stage, blastomeres appear to show some dependence upon others for division, and by the morula stage, the dependence is almost absolute. Even though ICM cells at the blastocyst stage quickly become acclimatized to growth and proliferation *in vitro*, several months of culture are required before these cells can be successfully cloned (Sherman, 1974*a* and unpublished observations).

A count of nuclei revealed that almost 60% of the trophoblastic vesicles and 70% of miniblastocyst outgrowths showed a decrease in trophoblast cell number between 48 and 96 hours in supplemented NCTC-109 medium (Sherman, unpublished observations). This does not appear to be the case with control blastocysts, although it is difficult to be certain because the ICM often covers a number of trophoblast cells, preventing an accurate count. It may, therefore, be that there is a cell–cell 'feeding' effect in culture, and that control blastocysts, with many more cells than trophoblastic vesicles or miniblastocysts, can survive better.

Snow (1973*b*) noted that trophoblast cells derived from [^{3}H]thymidine-induced trophoblastic vesicles did not seem to be as invasive *in utero* as trophoblast cells from control blastocysts; on the other hand, he observed that trophoblast nuclei in ectopic implants appeared to be larger when derived from trophoblastic vesicles than from control blastocysts. I have not noted any differences in the invasive capacity of control and experimental trophoblast cells on uterine monolayers (see Plate 3), although, as mentioned, the initial attachment of trophoblastic vesicles may be at a somewhat lower frequency than miniblastocysts and control blastocysts. Also, our in-vitro 'implantation' system is less demanding than the situation *in vivo* (Sherman & Salomon, 1974). As far as nuclear size is concerned, there is no indication from the data in Fig. 2 that trophoblast nuclei from trophoblastic vesicles are larger than those from control blastocysts; if anything, they are, on average, slightly smaller.

In-vitro systems have been used in this study to test the hypothesis that trophoblast cells developing from trophoblastic vesicles should become polyploid since there is no contact with ICM cells. The first evidence in support of this hypothesis stemmed from our earlier studies wherein blastocysts were cultured under conditions hostile to ICM development. ICM cells could not be detected by the 5th day of culture; by the 8th day, almost all the trophoblast cells were polyploid as measured by microspectrophotometry (Barlow & Sherman, 1972). Almost simultaneously, Gardner (1972) reported that trophoblastic vesicles from fragmented blastocysts formed only cells with giant nuclei when implanted to uteri of foster mothers. Snow (1973*b*) made similar observations when [^{3}H] thymidine-treated embryos were implanted in ectopic sites. The present results, which support the hypothesis that trophoblast cells not in contact with ICM undergo polyploidization, are perhaps more definitive than the previous studies, since diploid trophoblast as well as ICM cells might have been selected against in the culture system used in the Barlow & Sherman experiments, and since some diploid cells might have been either missed or mistaken as host cells during inspection of sections of the uterine or ectopic transplants carried out by Gardner and by Snow. Nevertheless, proof of polyploidization must be satisfied by direct measurement of DNA content, not nuclear size, and such studies are now in progress.

The fate of the 'morulae' and non-integrated forms generated from disaggregated blastomeres has also been followed. In some cases, especially with the former, these forms, if cultured long enough, will in turn generate miniblastocysts or trophoblastic vesicles. The 'morulae' and non-integrated forms which are not precursors for these other structures often do not survive when transferred to supplemented NCTC-109 medium. However, some do survive and attach to the culture dish. The result is the outgrowth of cells indistinguishable from trophoblast cells by the criteria discussed here. In fact, 'morulae' and non-integrated forms are even capable of producing small, but detectable, amounts of progesterone (Sherman, unpublished observations). It is remarkable that cells from non-integrated forms, which had never been exposed to any obvious organization or order, would nevertheless differentiate along the predictable and precise pathway that is characteristic of trophoblast cells. This observation reinforces the implication from the work of Hillman, Sherman & Graham (1972) that cells, unless purposefully and completely enclosed by other cells by the blastocyst stage, will differentiate as trophoblast cells.

The next logical step in the study of the role of cell–cell interaction during early embryogenesis is to determine the effect of the trophectoderm upon ICM differentiation. As mentioned in the introductory section, this is less easily done since techniques are not available for generating ICM cells that have never come into contact with trophectoderm. However, it would be instructive to determine the fate in culture of the ICMs that Gardner (1972) is able to separate from the trophectoderm at the blastocyst stage. In two markedly atypical cases in my culture studies, the particular lots of media used were found to support development of ICM derivatives but were deleterious to trophoblast cells, which did not become giant and died within a few days of culture. The ICMs dislodged from the degenerated trophoblast cells and within a week, large multicellular vesicles were observed. These structures were morphologically very similar to the yolk sac vesicles formed by blastocysts under normal culture conditions (Bell & Sherman, 1973; Sherman, 1974*a*, *c*). Analyses did not reveal any errors in the amino acid concentrations of the atypical media. The development of culture conditions selective for ICM development would be very useful in pursuing these preliminary observations which suggest that ICM derivatives may be able to differentiate independently, in the same way as trophoblast cells.

The author wishes to thank Mr Jose M. Marcal and Ms Sui Bi Atienza for assistance with these experiments.

REFERENCES

Barlow, P. W. & Sherman, M. I. (1972). The biochemistry of differentiation of mouse trophoblast: studies on polyploidy. *Journal of Embryology and Experimental Morphology*, **27**, 447–65.

BELL, K. R. & SHERMAN, M. I. (1973). Enzyme markers of mouse yolk sac differentiation. *Developmental Biology*, **33**, 38–47.

CHEW, N. J. & SHERMAN, M. I. (1973). Δ^5,3β-Hydroxysteroid dehydrogenase activity in mouse giant trophoblast cells *in vivo* and *in vitro*. *Biology of Reproduction*, **9**, 79.

CHEW, N. J. & SHERMAN, M. I. (1975). Biochemistry of differentiation of mouse trophoblast: Δ^5,3β-hydroxysteroid dehydrogenase. *Biology of Reproduction*, in press.

DICKSON, A. D. (1969). Cytoplasmic changes during the trophoblastic giant cell transformation of blastocysts from normal and ovariectomized mice. *Journal of Anatomy*, **105**, 371–80.

DULBECCO, R. & VOGT, M. (1954). Plaque formation and isolation of pure lines with poliomyelitis virus. *Journal of Experimental Medicine*, **99**, 167–82.

DUVAL, M. (1891). Le placenta des rongeurs: le placenta de la souris et du rat. *Journal de l'Anatomie et de la Physiologie normales et pathologiques de l'Homme et des Animaux*, **27**, 24–106.

EVANS, V. J., BRYANT, J. C., KERR, H. A. & SCHILLING, E. L. (1964). Chemically defined media for cultivation of long-term cell strains from four mammalian species. *Experimental Cell Research*, **36**, 439–74.

GARDNER, R. L. (1971). Manipulations of the blastocyst. *Advances in the Biosciences*, **6**, 279–96.

GARDNER, R. L. (1972). An investigation of inner cell mass and trophoblast tissues following their isolation from the mouse blastocyst. *Journal of Embryology and Experimental Morphology*, **28**, 279–312.

GARDNER, R. L., PAPAIOANNOU, V. E. & BARTON, S. C. (1973). Origin of the ectoplacental cone and secondary giant cells in mouse blastocysts reconstituted from isolated trophoblast and inner cell mass. *Journal of Embryology and Experimental Morphology*, **30**, 561–72.

GRAHAM, C. F. (1971). The design of the mouse blastocyst. *Symposium of the Society for Experimental Biology*, **25**, 371–8.

HERBERT, M. C. & GRAHAM, C. F. (1974). Cell determination and biochemical differentiation of the early mammalian embryo. *Current Topics in Developmental Biology*, **8**, 151–78.

HILLMAN, N., SHERMAN, M. I. & GRAHAM, C. F. (1972). The effect of spatial arrangement on cell determination during mouse development. *Journal of Embryology and Experimental Morphology*, **28**, 263–78.

HSU, Y.-C. (1971). Post-blastocyst differentiation *in vitro*. *Nature, London*, **231**, 100–2.

HSU, Y.-C. (1973). Differentiation *in vitro* of mouse embryos to the stage of early somite. *Developmental Biology*, **33**, 403–11.

HSU, Y.-C., BASKAR, J., STEVENS, L. C. & RASH, J. E. (1974). Development *in vitro* of mouse embryos from the two-cell stage to the early somite stage. *Journal of Embryology and Experimental Morphology*, **31**, 235–45.

JENKINSON, J. W. (1913). *Vertebrate Embryology*. London: Oxford University Press.

MINTZ, B. (1962). Experimental studies of the developing egg: removal of the zona pellucida. *Science*, **138**, 594–5.

RUNNER, M. N. & PALM, J. (1953). Transplantation and survival of unfertilized ova of the mouse in relation to postovulatory age. *Journal of Experimental Zoology*, **124**, 303–16.

SALOMON, D. S. & SHERMAN, M. I. (1974). Implantation and invasiveness of mouse blastocysts on uterine monolayers. *Experimental Cell Research*. (In press.)

SHERMAN, M. I. (1972*a*). The biochemistry of differentiation of mouse trophoblast: alkaline phosphatase. *Developmental Biology*, **27**, 337–50.

SHERMAN, M. I. (1972*b*). Biochemistry of differentiation of mouse trophoblast: esterase. *Experimental Cell Research*, **75**, 449–59.

SHERMAN, M. I. (1974*a*). Long term culture of cells derived from mouse blastocysts. *Differentiation*. (In press.)

SHERMAN, M. I. (1974*b*). *In vivo* and *in vitro* differentiation during early mammalian embryogenesis. *Frontiers in Radiation Therapy and Oncology*, **9**, 28–41.

SHERMAN, M. I. (1974*c*). Esterase isozymes during mouse embryonic development *in vivo* and *in vitro*. In *Third International Conference on Isozymes*, ed. C. L. Markert. New York & London: Academic Press. (In press.)

SHERMAN, M. I. & SALOMON, D. S. (1974). The relationships between the early mouse embryo and its environment. *Symposium of the Society for Developmental Biology*, vol. 33, ed. C. L. Markert. New York & London: Academic Press. (In press.)

SNOW, M. H. L. (1973*a*). Abnormal development of pre-implantation mouse embryos grown *in vitro* with [^{3}H]thymidine. *Journal of Embryology and Experimental Morphology*, **29**, 601–15.

SNOW, M. H. L. (1973*b*). The differential effect of [^{3}H]thymidine upon two populations of cells in pre-implantation mouse embryos. In *The Cell Cycle in Development and Differentiation*, ed. M. Balls & F. S. Billett, pp. 311–24. London: Cambridge University Press.

STERN, M. S. (1972). Experimental studies on the organization of the pre-implantation mouse embryo. II. Reaggregation of disaggregated embryos. *Journal of Embryology and Experimental Morphology*, **28**, 255–61.

TARKOWSKI, A. K. & WROBLEWSKA, J. (1967). Development of blastomeres of mouse eggs isolated at the 4- and 8-cell stage. *Journal of Embryology and Experimental Morphology*, **18**, 155–80.

WHITTEN, W. K. & BIGGERS, J. D. (1968). Complete development *in vitro* of the preimplantation stages of the mouse in a simple chemically defined medium. *Journal of Reproduction and Fertility*, **17**, 399–401.

EXPLANATION OF PLATES

PLATE 1

(*a*) Blastocyst derived from 8-cell embryo cultured in WB medium for 2 days.

(*b*) Miniblastocyst derived from $\frac{2}{8}$ blastomeres after 2 days of culture in WB medium.

(*c*) Trophoblastic vesicle derived from $\frac{1}{8}$ blastomere after 2 days of culture in WB medium.

(*d*) Non-integrated form derived from $\frac{1}{8}$ blastomere after 2 days culture in WB medium. Note large intracellular vacuoles.

(*e*) 'Morula' derived from $\frac{1}{8}$ blastomere after 2 days culture in WB medium.

(*f*) Two-celled trophoblastic vesicle derived from $\frac{1}{8}$ blastomere after 2 days culture in WB medium.

Photographs were taken at the same magnification using phase optics.

PLATE 2

(*a*) Outgrowth of a blastocyst derived from $\frac{8}{8}$ blastomeres after 2 days culture in supplemented NCTC-109 medium. The ICM (I) is the large clump in the middle.

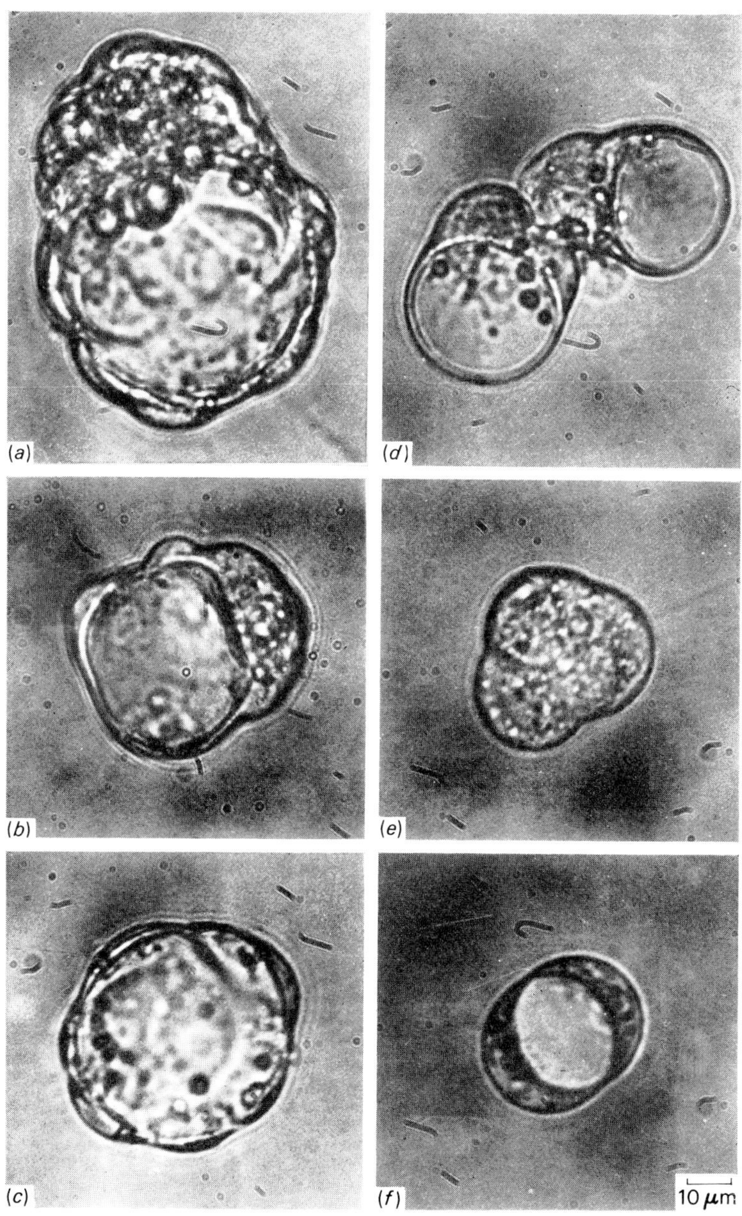

(*Facing p. 164*)

PLATE 2

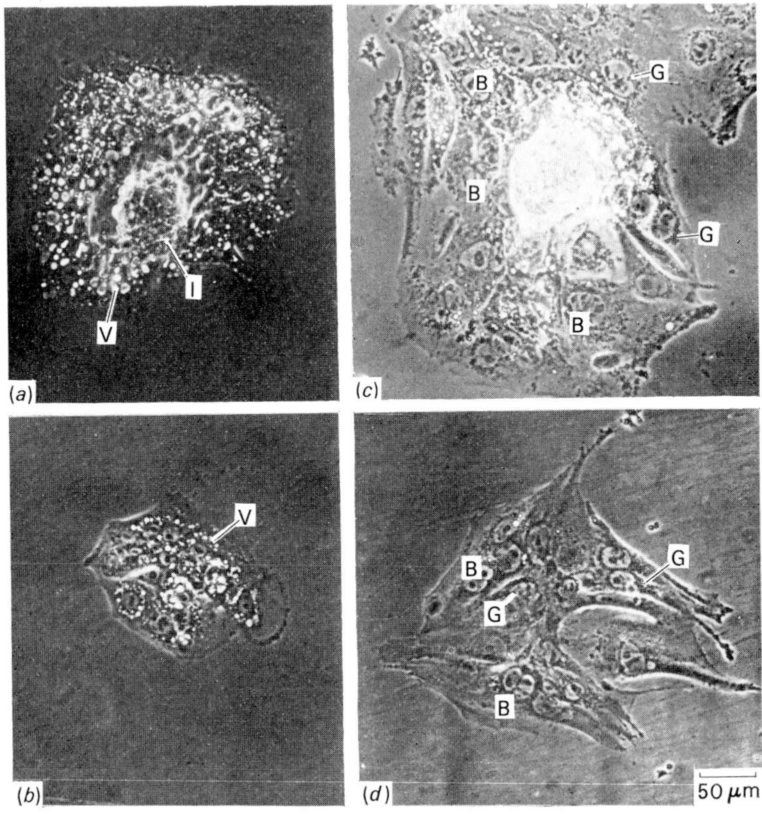

PLATE 3

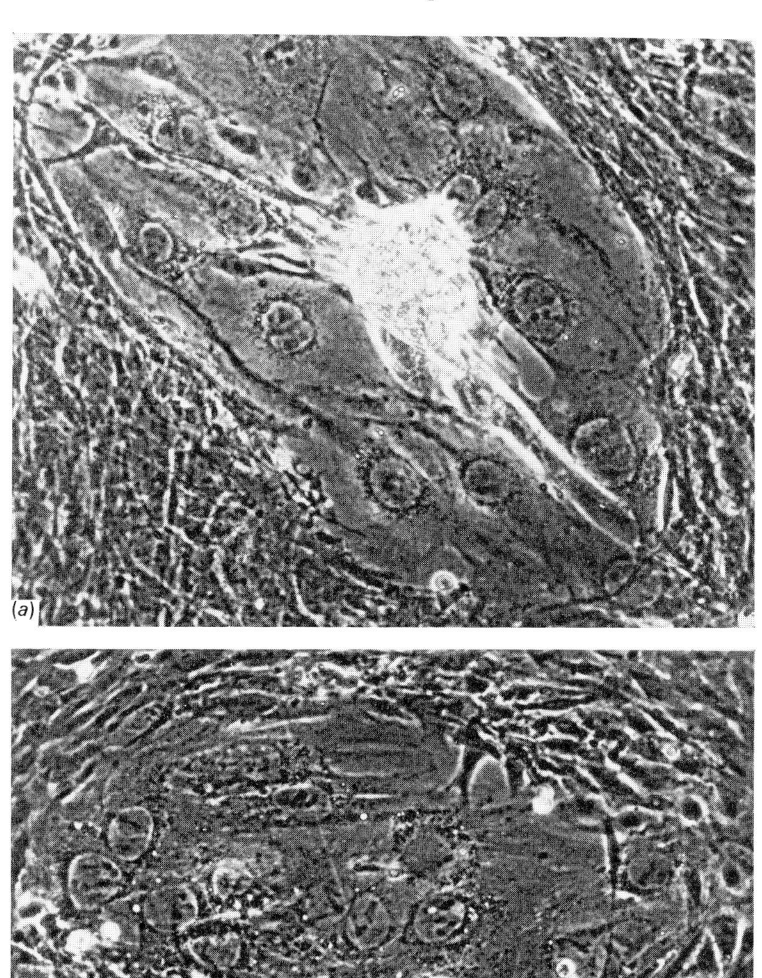

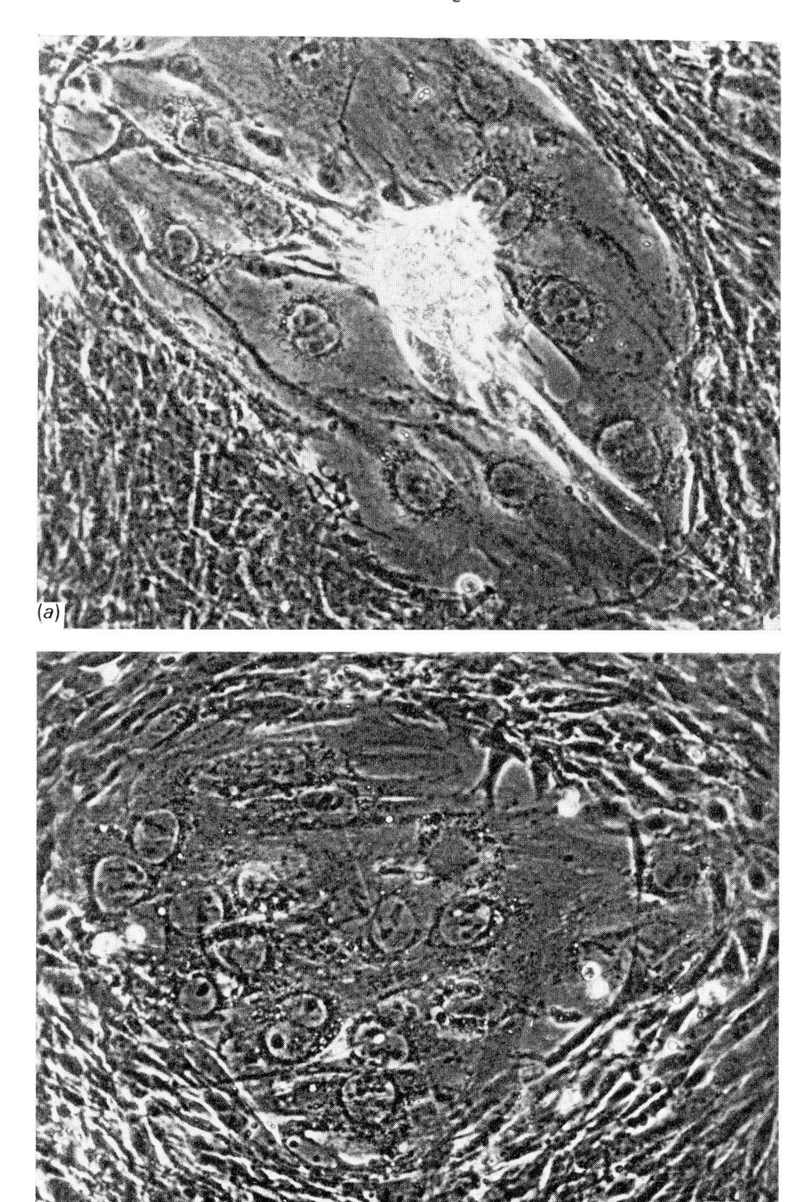

There are at least 37 trophoblast nuclei in the structure. V, vacuole.

(*b*) Outgrowth of a trophoblastic vesicle, derived from $\frac{1}{4}$ or $\frac{2}{8}$ blastomeres, after 2 days culture in supplemented NCTC-109 medium. The structure contains 20 nuclei. V, vacuole.

(*c*) Outgrowth of a control blastocyst removed from the uterus on the 4th day of gestation and cultured for a further 4 days in supplemented NCTC-109 medium. The structure contains at least 33 trophoblast nuclei. G, granules in the perinuclear region of the cytoplasm; B, binucleate cells.

(*d*) Outgrowth of a trophoblastic vesicle, derived from $\frac{1}{4}$ or $\frac{2}{8}$ blastomeres, after 4 days in supplemented NCTC-109 medium. The structure contains 18 nuclei. G, granules in the perinuclear region of the cytoplasm; B, binucleate cells.

Photographs were taken at the same magnification using phase optics.

PLATE 3

(*a*) Outgrowth of a control blastocyst removed from the uterus on the 4th day of gestation and cultured on a uterine monolayer in supplemented NCTC-109 medium for 5 days. The clump in the center of the outgrowth is derived from the ICM.

(*b*) Outgrowth of a trophoblastic vesicle on a uterine monolayer in supplemented NCTC-109 medium after 5 days in culture. The outgrowth contains 18 nuclei. Note perinuclear granules and binucleate cells.

Photographs were taken at the same magnification using phase optics.

EXPRESSION OF MITOCHONDRIAL AND NUCLEAR GENES DURING EARLY DEVELOPMENT

BY L. PIKÓ

Developmental Biology Laboratory, Veterans Administration Hospital, Sepulveda, California 91343 and Division of Biology, California Institute of Technology, Pasadena, California 91125, USA

This review considers some of the macromolecular and fine-structural aspects of differentiation in the pre-implantation mouse embryo. In recent years a considerable amount of data has accumulated on both the synthesis of various macromolecules and the fine-structural changes which occur during this period and has provided us with insights into the pattern of genetic activity underlying these events. Much of this information has been obtained in embryos developing *in vitro*. There is sufficient evidence to indicate that these data also apply to embryos developing *in vivo*.

The following topics will be discussed:

(1) The DNA content of the unfertilized mouse egg, with special reference to the amount and properties of mitochondrial DNA, and some aspects of DNA synthesis during early development.

(2) Expression of the nuclear genome and its role in pre-implantation development. This subject has been dealt with in several recent reviews (e.g. Graham, 1973; Church & Schultz, 1974) and here a brief account of the main features of nuclear genetic activity is given.

(3) Expression of the mitochondrial genome and its role in the early embryo. Recent work in the mouse indicates that the mitochondrial genome is transcribed during cleavage but that the contribution of the mitochondrial system does not become essential until the blastocyst stage.

(4) Expression of 'virus-like' particles. Recent observations in several laboratories have identified up to four morphologically distinct particles in the early mouse embryo. These particles are formed in a developmental-stage-dependent fashion and seem to be a general feature of early mouse development.

MOUSE EGG DNA

The problem of an excess amount of DNA in animal eggs has been satisfactorily settled recently by the demonstration that most or all of the cytoplasmic DNA represents mitochondrial DNA (cf. Dawid, 1972). The amount of mitochondrial DNA depends on the number of mitochondria stored, that is, roughly on the size of the egg as shown by studies in amphibians (Dawid, 1966), sea urchins (Pikó, Tyler & Vinograd, 1967) and the echiuroid worm (Dawid & Brown, 1970). The cytoplasmic DNA in a mammalian egg has not been characterized.

The ovulated unfertilized mouse egg is at the second maturation metaphase, containing the diploid amount of nuclear DNA (Alfert, 1950). In a recent study in our laboratory (Matsumoto & Pikó, unpublished observations) we have isolated the total egg DNA from 6600 unfertilized mouse eggs by buoyant density centrifugation in an ethidium bromide/caesium chloride gradient. The eggs were treated previously with hyaluronidase and subtilisin (bacterial protease) to remove cumulus cells, the zona pellucida and the remnants of the first polar body. A fluorometric assay performed on an aliquot of the purified DNA gave a DNA content per egg of about 8 pg. From this value and from the relative distribution of DNA in the lower band (representing mitochondrial DNA) and the upper band (representing nuclear DNA) in the gradient we estimate that the mitochondrial DNA amounts to 2–3 pg per egg (or approximately equal to the haploid amount of nuclear DNA). Electron microscopic examination of the egg DNA by the formamide–protein monolayer technique showed long linear duplex molecules and 5 μm circular (mitochondrial) DNA. About 70% of the mitochondrial DNA was in the form of clean duplex circles and about 30% contained a small single-stranded displacement loop (D-loop). (D-loop mitochondrial DNA was shown to be an early replicative intermediate which is usually present at a high frequency and serves as a pool of primed molecules for further synthesis; Kasamatsu *et al.*, 1973.) No larger replicating forms were observed, indicating an absence of mitochondrial DNA replication in the mature oocyte. The significance of why some molecules are stored as clean circles and others as molecules arrested at the D-loop stage is not known. In two other animal groups that have been so far examined, the mature frog oocyte contains at least 76% of the mitochondrial DNA in the D-loop form (Hallberg, 1974) while in the mature sea urchin egg all the mitochondrial DNA occurs in the form of clean duplex circles (Matsumoto, Kasamatsu, Pikó & Vinograd, 1974). Therefore, the mouse egg resembles the frog egg as to the structural form of its mitochondrial DNA. It remains to be determined whether the presence of D-loops is generally true for vertebrate eggs and whether it is related to the nuclear and cytoplasmic maturation of these eggs.

The fertilizing spermatozoon contributes the haploid amount of paternal nuclear DNA to the zygote but the sperm mitochondrial DNA is probably eliminated. A strictly maternal inheritance of mitochondrial DNA has been demonstrated in *Xenopus* (Dawid & Blackler, 1972); fine structural evidence suggests that the sperm mitochondria disintegrate in the egg cytoplasm in the rat (Szollosi, 1965).

During cleavage the nuclear DNA replicates early in the cell cycle (Graham, 1973) and can be labeled to a high specific activity with radioactive thymidine provided in the culture medium (Pikó, 1970). The analog, 5-bromodeoxyuridine, is incorporated into DNA with about the same efficiency as is thymidine. It was found that uniformly labeled [^{14}C] thymidine is incorporated both into DNA and into a polysaccharide component of mouse blastocysts. The sugar moiety is most likely responsible for the latter since labeling of the polysaccharide is absent when the embryos are cultured in the presence of *methyl*-[^{3}H]thymidine (Pikó, 1970, and unpublished observations).

Mitochondrial DNA synthesis in the mouse embryo is probably absent, or at a low level, up to the blastocyst stage (Pikó, 1970). Similarly, no significant replication of mitochondrial DNA has been found in the amphibian embryo up to the swimming tadpole (Chase & Dawid, 1972) and in the sea urchin up to the feeding pluteus (Matsumoto *et al.*, 1974). These observations suggest that the number of mitochondria remains constant during early development and mitochondrial replication is initiated when the embryo begins to grow in mass.

EXPRESSION OF THE NUCLEAR GENOME

A striking feature of mouse embryo development is the early stage at which true nucleoli appear (cf. Mintz, 1964). In recent years the fine-structure and biochemical correlates of nucleolar differentiation have been studied in detail.

The pronuclei of the fertilized egg and the nuclei of the early 2-cell embryo contain a number of small fibrillar primary nucleoli. A peripheral granular zone in some of the nucleoli appears at the late 2-cell stage and increases in amount in the nucleoli of the 4-cell embryo. Two definitive nucleoli with an advanced fibrillogranular structure are usually present in each cell of the 8-cell embryo and granularity is abundant in the nucleoli of the morula (Hillman & Tasca, 1969).

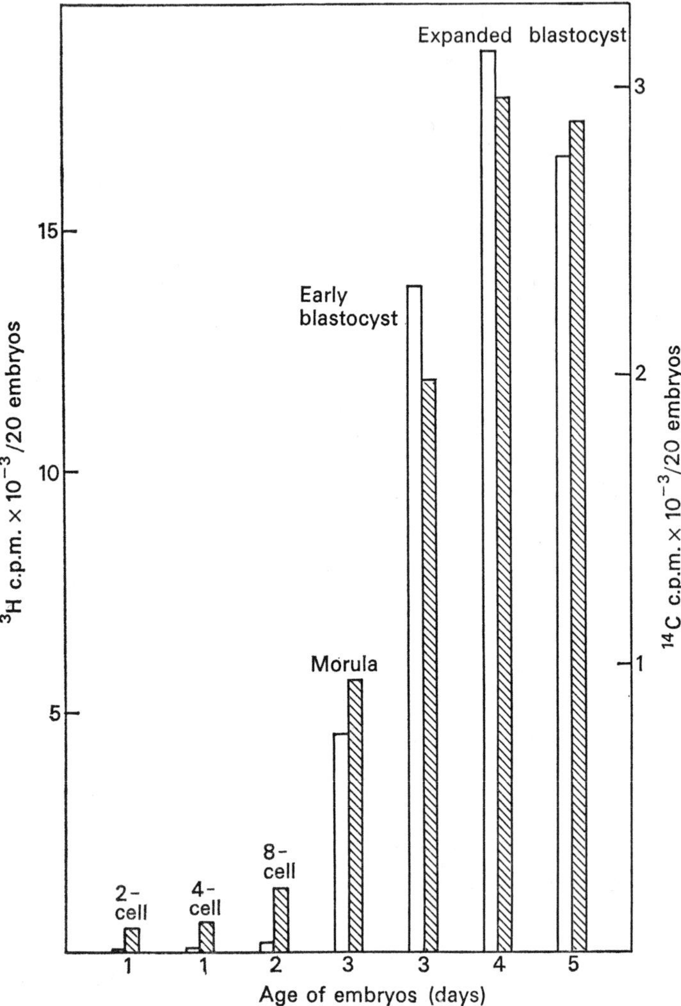

Fig. 1. Rate of incorporation of radioactive precursors into total cellular RNA and protein at different stages of development in mouse embryos cultured *in vitro* from the 2-cell stage (day 1) to the blastocyst (up to day 5). In the study of RNA synthesis □, approximately 60 embryos per 100 μl culture drop were incubated with 200 μCi/ml of [^{3}H]uridine (15.3 mCi/μmole) for 2 h and processed to obtain cold trichloroacetic acid (TCA) precipitable, alkali-resistant counts. In the study of protein synthesis ▧, a similar number of embryos was incubated for 1 h with a mixture (1 μCi/ml each) of three uniformly labeled [^{14}C]amino acids: L-aspartic acid, 167 μCi/μmole, L-lysine, 223 μCi/μmole, and L-valine, 190 μCi/μmole. Incorporation represents cold TCA-precipitable, hot TCA-insoluble counts. The counting efficiencies were about 7% for ^{3}H and 60% for ^{14}C. For procedural detail see Pikó (1970) and Pikó & Chase (1973).

The synthesis of mature 18 S and 28 S ribosomal RNA species, as well as 4 S transfer RNA, is clearly detectable in the 4-cell mouse embryo (Woodland & Graham, 1969; Knowland & Graham, 1972). The rate of incorporation of uridine into stable RNA products rises sharply from the 8-cell stage onwards to a high level in the blastocyst (Fig. 1). Analysis by methylated albumin kieselguhr chromatography (Ellem & Gwatkin, 1968), sucrose gradient centrifugation (Pikó, 1970) and acrylamide gel electrophoresis (Knowland & Graham, 1972) shows that the bulk of the RNA accumulated during this period is ribosomal RNA and transfer RNA. Fig. 2 illustrates the preponderance of label in these RNA species when mouse blastocysts are incubated with [^{3}H]uridine for a period of 2 hours; however, with shorter (0.5 hour) incubation, labile heterogeneous nuclear RNA and 39 S ribosomal precursor RNA become labeled primarily (Pikó, 1970).

The actual rate of increase of RNA synthesis on a per cell basis during cleavage is uncertain because the uptake of uridine is limited at the early

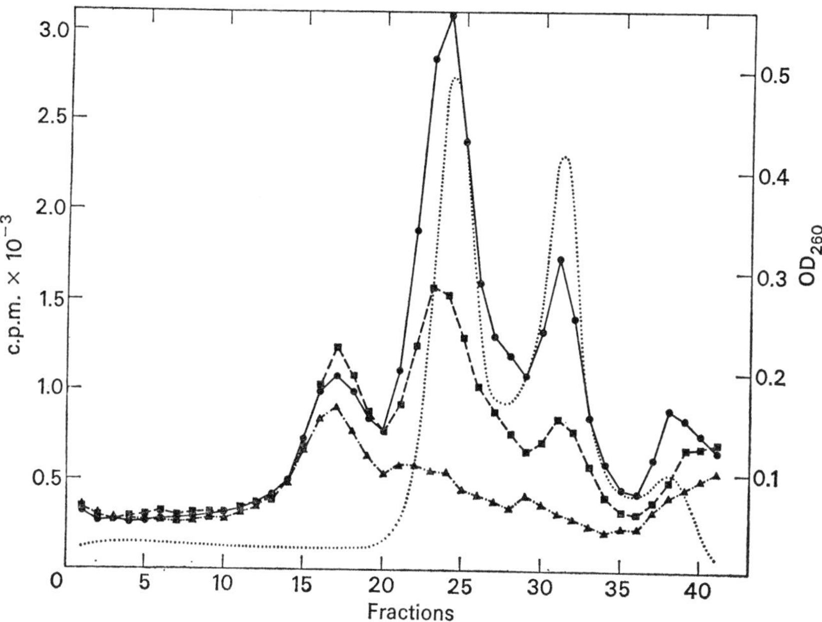

Fig. 2. Sedimentation pattern of total cellular RNA extracted from day 4 mouse blastocysts after different periods of labeling with [^{3}H]uridine. The embryos, which were cultured *in vitro* from the 2-cell stage onwards were exposed to [^{3}H]uridine (200 μCi/ml, 15.3 mCi/μmole) within the same experiment for 0.5 h (-•-▲-•-), 1 h (-■-) and 2 h (-●-), respectively. The RNA was extracted and centrifuged in a 5–20% (w/v) sucrose gradient in a Beckman SW41 rotor at 40 krpm at 20 °C for 140 min. The dotted line is the OD_{260} recording of carrier RNA from mouse liver ribosomes; it shows (from right to left) peaks of 4 S, 18 S and 28 S RNA (after Pikó, 1970).

stages and the size of the endogenous precursor pool is not known (Daentl & Epstein, 1971). The presence of fully developed nucleoli, as well as the rate of incorporation of uridine in the morula, indicates that a high rate of ribosomal RNA synthesis per nucleus is reached by this stage. A quantitative evaluation of ribosomal RNA synthesis in the blastocyst shows that new ribosomes are produced at a rate of at least 10^4 ribosomes per minute per cell, which is comparable to the rate of ribosome synthesis in HeLa cells (Pikó, 1970). The relative amount of form I (nucleolar) RNA polymerase to the form II (nucleoplasmic) enzyme appears to be greatly enhanced in mouse blastocysts as compared with adult liver nuclei, suggesting a possible mechanism for the regulation of RNA synthesis in the early mouse embryo (Warner & Versteegh, 1974).

The massive production of ribosomes results in a sharp increase in the ribosome concentration in the cytoplasm between the 8-cell stage and the blastocyst as indicated by cytochemical studies (Alfert, 1950; Austin & Bishop, 1959; cf. Mintz, 1965) electron microscopy (Enders & Schlafke, 1965; Calarco & Brown, 1969; Hillman & Tasca, 1969; Szollosi, 1972; Chase & Pikó, 1973) and the increase of the RNA content of the embryo (Olds *et al.*, 1973). The total cytoplasmic mass does not increase appreciably during this period (Brinster, 1967). By the above criteria, the unfertilized mouse egg and the mouse embryo up to the 8-cell stage are characterized by a sparsity of cytoplasmic ribosomes and a low RNA content. The mouse embryo evidently inherits a very limited protein synthetic machinery as a result of oogenesis and a major aspect of pre-implantation differentiation is the gradual build-up of an active protein synthesizing system. The newly formed ribosomes appear to be immediately utilized because the rate of amino acid incorporation increases markedly from the 8-cell stage onwards in a manner roughly parallel to the increase in RNA synthesis (Mintz, 1964; Monesi & Salfi, 1967; Brinster, 1971; Epstein & Smith, 1973; see Fig. 1). By the same token it is clear that the low level of protein synthesis which is observed up to the 4-cell stage is carried out primarily with maternally-inherited components.

A similar general pattern of RNA synthesis has been described in the early rabbit embryo, but there are differences in the details. In the rabbit a low level of incorporation of [^{3}H]uridine into 28 S and 18 S ribosomal RNA is first detectable at the 16-cell stage (Church & Schultz, 1974) but full production of ribosomal RNA, and increase in total embryo RNA, begins in the late morula–early blastocyst containing about 100–130 cells (Manes, 1969, 1971; Karp, Manes & Hahn, 1973; Church & Schultz, 1974). On the other hand, the synthesis of 4 S RNA and heterogeneous RNA is detectable from the 2-cell stage onwards (Manes, 1971).

Several lines of evidence indicate that messenger RNA is produced and translated early in the mouse embryo. Large molecular weight heterogeneous RNA is synthesised in the 2-cell embryo (Knowland & Graham, 1972) and RNA species with a DNA-like base composition and with the properties of 'heterogeneous nuclear RNA' have been detected from the 8-cell stage onwards (Ellem & Gwatkin, 1968; Pikó, 1970). In the rabbit, poly-A-containing heterogeneous nuclear RNA is synthesized at least from the 16-cell stage onwards through the blastocyst stage; during the same period labeled poly-A-containing heterogeneous RNA is found in association with polysomes in the cytoplasm (Schultz, Manes & Hahn, 1973). The diversity of transcription has been estimated in DNA–RNA hybridization experiments. In the pre-implantation mouse embryo the RNA transcripts hybridize with about 1% of the non-repeated portion of the mouse genome and with about 2–5% of the repeated sequences; in the rabbit embryo the corresponding figures are 1.8 and 4%. In both species, the proportion of unique DNA that is transcribed increases after implantation (Church & Schultz, 1974; Schultz *et al.*, 1973).

The marked sensitivity of early mouse embryos to very low doses of actinomycin (Mintz, 1964; Thomson & Biggers, 1966; Monesi, Molinaro, Spaletta & Davioli, 1970; cf. Graham, 1973) and to α-amanitin (Golbus, Calarco & Epstein, 1973; Warner & Versteegh, 1974) suggests that new RNA synthesis is required in each interphase to maintain normal development from the 2-cell stage onwards, and possibly as early as the 1-cell stage (Golbus *et al.*, 1973). Incorporation of [^{3}H]uridine into both pronuclei in the 1-cell embryo has been shown by autoradiography (Mintz, 1964), but the RNA products at this stage have not been characterized. The observation that protein synthesis during cleavage continues at a relatively high level in the presence of actinomycin for up to 24 hours (Mintz, 1964; Tasca & Hillman, 1970; Monesi *et al.*, 1970; Molinaro, Siracusa & Monesi, 1972) indicates that the bulk of the messenger RNA is quite stable in the mouse embryo. A relatively long lifetime, of the order of one cell generation, also appears to be the rule for most messenger RNA (with the notable exception of histone messenger RNA) in rapidly proliferating tissue culture cells (cf. Perry & Kelley, 1973).

Clearcut evidence for the early utilization of embryonic messenger RNA is provided by the expression of the paternal variant of glucose phosphate isomerase-1 in hybrid embryos. In an earlier study, the paternal isozyme variant was first observed in the blastocyst (Chapman, Whitten & Ruddle, 1971) but recently, as a result of using a larger number of embryos for assay, it could be detected as early as the 8-cell stage (Brinster, 1973). The regulation of the activity of the X-linked enzyme, hypoxanthine–

guanine phosphoribosyl transferase, in embryos of XO females, suggests that the paternal allele for this enzyme is also expressed before the blastocyst stage (Epstein, 1972). In a recent study Epstein & Smith (1974) found that a major shift occurs in the polyacrylamide gel pattern of the proteins synthesized at the 8–16-cell stage as compared with the 2-cell stage. This observation is consistent with the hypothesis that new kinds of embryonic messenger RNA for major structural proteins are also being produced and translated at this time. For example, it might be expected that the synthesis of ribosomal proteins becomes quantitatively important in the 8–16-cell embryo.

In summary, it is clear that nuclear genetic activity is initiated early in the mouse, possibly at the 1-cell stage, and that this activity is essential for normal development and differentiation during cleavage and blastocyst formation. It appears that the mouse egg stores a limited amount of maternal genetic information in its cytoplasm and has a low capacity for protein synthesis. The early protein synthesis may serve primarily a regulatory function for the activation of the nuclear genome. The differentiation of true nucleoli in the 2–4-cell embryo signals the beginning of rapidly increasing transcriptional and translational activity. Among the main products of RNA- and protein synthesis are the components of the cytoplasmic protein synthetic apparatus which is built up rapidly from the 8-cell stage onwards. Undoubtedly, many other aspects of early differentiation, such as changes in transport, enzyme activities and energy metabolism are also dependent, directly or indirectly, on nuclear genetic expression; their biochemical details remain to be elucidated. The biochemical evidence and the early appearance of some paternally inherited enzymes suggest that the paternal genome is active during early development.

EXPRESSION OF THE MITOCHONDRIAL GENOME

The increase in the rate of macromolecular synthesis in the pre-implantation mouse embryo is accompanied by a corresponding increase in energy metabolism. The rate of O_2 consumption, which is low in the 1–4-cell embryo, increases 3.5 times between the 8-cell and blastocyst stage (Mills & Brinster, 1967). Changes in the activities of several mitochondrial enzymes (Brinster, 1970), an increase in the number of utilizable substrates (Whitten, 1957; Brinster & Thomson, 1966; Barbehenn, Wales & Lowry, 1974) and an increased transport of TCA-cycle intermediates (Wales & Biggers, 1968; Kramen & Biggers, 1971) were also observed at the 8-cell stage. The metabolism of ATP increases during cleavage and blastocyst formation and ATP synthesis becomes progressively more sensitive to cyanide inhibition, indicating increased aerobic metabolism in the pre-implantation embryo (Ginsberg & Hillman, 1973). The mitochondria undergo marked changes in their fine structure, beginning with the 4–8-cell stage both in embryos developing *in vivo* (Calarco & Brown, 1969; Maraldi & Monesi, 1970; Stern, Biggers & Anderson, 1971) and *in vitro* (Hillman & Tasca, 1969); the possible significance of these changes to the metabolic state of the mitochondria has been discussed (Stern *et al.*, 1971; Ginsberg & Hillman, 1973).

We have been interested in the role of the mitochondrial genome in mitochondrial differentiation and in the development and differentiation of the embryo as a whole. We have used the drugs ethidium bromide and chloramphenicol to inhibit specifically mitochondrial DNA-, RNA- and protein synthesis in mouse embryos developing *in vitro* from the 2–4-cell stage (day 1 of culture) to the blastocyst (days 4 and 5). The effects of these inhibitors, and the corresponding parameters in control embryos were monitored by electron microscopy and biochemical analysis of RNA- and protein synthesis. An abbreviated account of the results is presented below (cf. Pikó & Chase, 1973).

Before describing the data obtained with mouse embryos it is appropriate to summarize briefly the available information on the mitochondrial genetic system in animal cells. The 5 μm mitochondrial DNA codes for 12 S and 16 S mitochondrial RNA and 4 S transfer RNA (Borst & Grivell, 1971; Borst, 1972) and, probably, for mitochondrial messenger RNA (Hirsch & Penman, 1973; Ojala & Attardi, 1974). Mitochondrial protein synthesis is carried out on unique 60 S ribosomes; the products are highly insoluble polypeptides (Costantino & Attardi, 1973) and their likely function is to contribute to the structural and functional integrity of mitochondrial cristae. However, all or most of the soluble mitochondrial proteins and enzymes, including those involved in mitochondrial protein synthesis as well as in mitochondrial DNA replication and transcription, are synthesized in the cytoplasm and are transported into the mitochondria (cf. Borst, 1972).

Plate 1 illustrates the fine structural changes in the mitochondria of control mouse embryos. The mitochondria of the 2-cell embryo (Plate 1*a*) are small, vacuolated, with a very high matrix density and few, often concentrically arranged, cristae. The early morphological change which occurs between the late 4-cell stage and the morula consists mainly of a progressive enlargement of the mitochondria; at the same time the matrix density decreases but there is little or no change in the number of cristae

(Plate 1*b* and *c*). The late morphological change, from early to expanded blastocyst, is characterized by an increase in matrix density and the number of cristae and considerable heteromorphy with regard to the shape and size of mitochondria (Plate 1*d*). During early development the mitochondria are typically found in close association with rough endoplasmic reticulum, suggesting an active functional relationship between these organelles (cf. Szollosi, 1972).

The electron microscopic study also revealed that the number of mitochondrial ribosomes is very low up to the morula stage but increases approximately 15-fold between the morula stage and the expanded blastocyst (Table 1). The presence of a low concentration of ethidium bromide (0.1 μg/ml), which preferentially inhibits mitochondrial DNA- and RNA-synthesis, prevented the increase in the number of mitochondrial ribosomes; however, mitochondrial ribosome formation was reduced only slightly in the presence of chloramphenicol (31.2 μg/ml) which inhibits mitochondrial protein synthesis (Table 1). This result is in agreement with studies in other systems showing that probably all mitochondrial ribosomal proteins are synthesized in the cytoplasm (Brega & Baglioni, 1971; Lizardi & Luck, 1972).

Table 1. *Mitochondrial ribosomes in control and inhibitor-treated embryos*

Stage of development	Inhibitor	Ribosomes/μm^2 (Average ± SE)
Day 3 morula	Control	5.7 ± 0.7
Day 4 blastocyst	Control	91.4 ± 5.4
Day 4 blastocyst	EB*	4.8 ± 0.9
Day 4 blastocyst	CAP*	56.0 ± 3.3

* Embryos were grown from the 2–4-cell stage onwards (day 1) in the presence of 0.1 μg/ml ethidium bromide (EB) and 31.2 μg/ml chloramphenicol (CAP), respectively. (After Pikó & Chase, 1973.)

Biochemical analysis of mitochondrial RNA- and protein synthesis provided the following information of the functioning of the mitochondrial genome in control embryos. Long-term (24 h) incubation of the embryos in the presence of [^{3}H]uridine, from the morula stage to the expanded blastocyst, resulted in the labeling of three major RNA species, with sedimentation coefficients of 16 S, 12 S and 4 S in the mitochondrial fraction (Fig. 3*a*). These RNA species have been shown to represent mitochondrial ribosomal RNA and transfer RNA (Borst & Grivell, 1971; Borst, 1972). Their mitochondrial origin in the mouse embryo is indicated by their sensitivity to inhibition by a low concentration of ethidium bromide (Fig. 3*b*). When embryos were labeled at different early stages of development, the synthesis of 12 S and 16 S ribosomal RNA (but not of 4 S RNA) was first detected in the late 8-cell embryo (Fig. 3*d*). All three species of RNA (16 S, 12 S and 4 S) appear to be synthesized in the morula (Fig. 3*c*). The

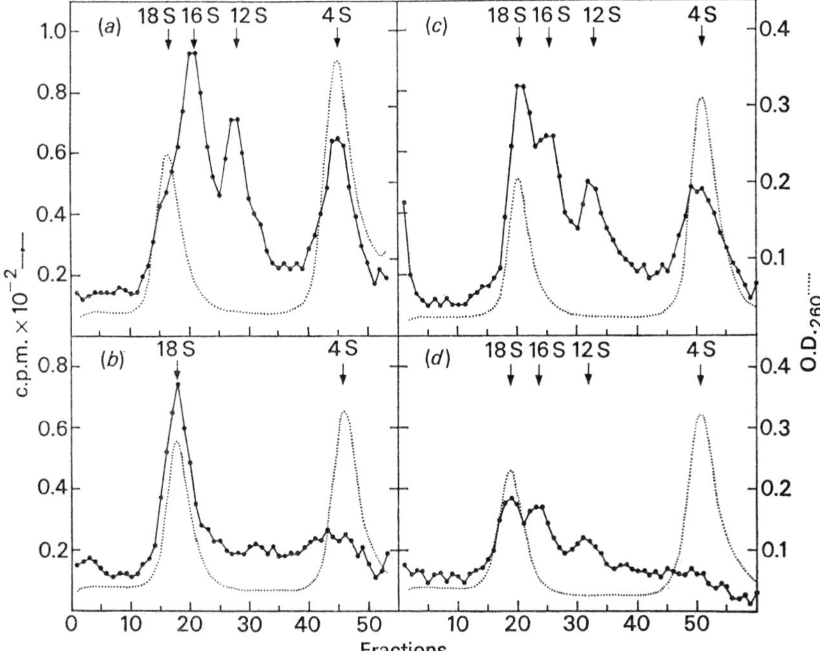

Fig. 3. Sedimentation pattern of RNA extracted from the mitochondrial fraction of mouse embryos incubated with [^{3}H]uridine at different stages of development. From 150 to 200 embryos per group were cultured in the presence of 5-[^{3}H] uridine (200 μCi/ml, 22.7 mCi/μmole) for the following periods: (*a*) 24 h labeling from the late morula (day 3) to the expanded blastocyst (day 4); (*b*) experimental conditions as in (*a*) except for the presence of 0.1 μg/ml ethidium bromide during (and 30 min preceding) incubation with [^{3}H]uridine; (*c*) 24 h labeling from the mid-8-cell stage (day 2) to the late morula (day 3); (*d*) 10 h labeling from the mid-8-cell to the late-8-cell stage (day 2). Centrifugation was in a SW41 rotor at 40 krpm at 20 °C, for about 6 h. The OD_{260} peaks (dotted line) derive from 18 S and 4 S carrier RNA isolated from mouse liver. (After Pikó & Chase, 1973.)

analysis of mitochondrial protein synthesis indicated that it is at a low level (less than 1% of the total cellular protein synthesis) in the blastocyst stage embryo.

When the embryos were cultured from the 2–4-cell stage onwards in the presence of low concentrations of ethidium bromide and chloramphenicol (which inhibited mitochondrial RNA- and protein synthesis but did not affect the nuclear and ctyoplasmic systems), development proceeded normally, albeit with a slight delay, to the blastocyst stage. Electron microscopic examination showed normal cellular differentiation. Mitochondrial morphogenesis also occurred normally up to the early blastocyst stage; however, abnormal mitochondrial cristae (dilated and vesicular rather than flattened saccular cristae) were evident in mid-to-late blastocysts. These changes are similar to those observed in the mitochondria of cultured tissue cells under the effects of these drugs (e.g. Soslau & Nass, 1971; King, Godman & King, 1972) and are also consistent with the evidence from other systems that mitochondrial protein synthesis is required for the normal structural and functional organization of the cristae (Borst, 1972). For example, mitochondrial contribution is essential for the formation of active cytochrome *c* oxidase (Schatz *et al.*, 1972) and oligomycin-sensitive ATPase (Tzagoloff, Akai & Sierra, 1972).

The blastocysts exposed to the effects of ethidium bromide and chloramphenicol apparently fully recover and develop into normal fetuses when transferred into the uteri of foster mothers.

These observations suggest that mitochondrial differentiation in the early mouse embryo involves the transformation of the pre-existing mitochondria and is controlled to a large extent by the nuclear and cytoplasmic systems. For example, the fine-structural and metabolic changes from the late 4-cell stage to the early blastocyst appear to be entirely under nucleo-cytoplasmic control. Indirect evidence suggests that nuclear transcription and the synthesis of new proteins in the cytoplasm are required for these changes, since inhibition of cellular RNA- and protein synthesis also arrests the mitochondrial transformation (Pikó & Chase, 1973).

The mitochondrial genome of the mouse embryo is transcribed from the 8–16-cell stage onwards and contributes to the establishment of a mitochondrial protein synthesizing system which is very limited in the early stages. Mitochondrial protein synthesis is needed for the normal growth and organization of the cristae in blastocyst mitochondria. However, embryo development and cellular differentiation up to the blastocyst stage are not dependent on the functioning of the mitochondrial genome. Evidently, the respiratory capacity of the mitochondria inherited from the mother is sufficient to support development through the blastocyst stage. These observations also argue against the possibility that mitochondria export specific morphogenetic substances in the form of RNA or protein into the cytoplasm during early development.

VIRUS-LIKE PARTICLES

Finally, I would like to discuss some recent reports on the occurrence of A- and C-type particles in the early mouse embryo. These observations are of interest because they provide direct evidence for the vertical transmission of viral genomes through the germ cells. In addition, the production of A and C particles appears to be associated with specific developmental stages of the embryo, which raises questions as to the cellular control of viral expression and the possible role of these particles in development. First, I will describe briefly the general characteristics of A- and C-type particles.

Intracisternal A-type particles are spherical structures 70–100 nm in diameter and consist of two concentric electron-dense shells surrounding a lucent core (Anon., 1966; Dalton, 1972). They are present in large amounts in a number of mouse tumours and also occur much less frequently in normal adult and embryonic mouse tissues (Wivel & Smith, 1971; Vernon, 1974). They contain 60–70 S RNA and have reverse transcriptase activity (Wilson & Kuff, 1972; Yang & Wivel, 1973, 1974) but are distinct from oncogenic RNA viruses in structure and antigenic properties (Kuff, Lueders, Ozer & Wivel, 1972; Wivel, Lueders & Kuff, 1973). Intracisternal A particles have no demonstrated infectivity and their biological function is unknown.

C-type particles are 90–110 nm in size and bud extra-cellularly from the plasma membrane. Their fine structure varies depending on the state of maturity of the particle and, to some extent, on the fixation used (Anon., 1966; Dalton, 1972). They possess 60–70 S RNA and reverse transcriptase activity, share common group-specific antigens and may be tumorigenic (Nowinski, Old, Sarkar & Moore, 1970). The structural genes of C-type viruses appear to be present in all mouse cells (Todaro & Huebner, 1972; Rowe, Lowy, Teich & Hartley, 1972) covalently linked to cellular DNA (Gelb, Milstien, Martin & Aaronson, 1973). However, different cells may carry different viral DNA sequences (Chattopadhyay *et al.*, 1974) and a single cell may harbor several populations of C-type virus identifiable on the basis of antigenic properties (Aoki *et al.*, 1974). The group-specific antigen has been detected during embryogenesis (Huebner *et al.*, 1970) and budding C-type particles have been found with some frequency in fetal mouse tissues of inbred and non-inbred strains (Vernon, Lane & Huebner, 1973).

Enders & Schlafke (1965) appear to be the first to have noted the presence of virus-like particles in an early mammalian embryo (the guinea-pig blastocyst). Calarco & Brown (1969) described doughnut-shaped bodies in the endoplasmic reticulum of cleaving mouse embryos. Recently,

Calarco & Szollosi (1973) have observed similar particles in early embryos of several mouse strains and identified them as intracisternal A particles; they also found them in dictyate oocytes but not in oocytes which had reached metaphase II. Biczysko, Pienkowski, Solter & Koprowski (1973) and Chase & Pikó (1973) have also recently reported on the occurrence and distribution of intracisternal A-type particles during early mouse development and noted the occasional presence of C-type particles at the blastocyst and later stages. In addition, the latter authors have described a small 'dense-cored vesicle' associated with endoplasmic reticulum; the nature and significance of this structure is not known.

Plate 2 presents the morphology of the particles that have been observed in early mouse embryos, and is from the work of Chase & Pikó (1973). These authors differentiated two types of intracisternal A particles on the basis of morphological criteria. 'Large' A particles (Plate 2*e* and *f*) are 85–100 nm in diameter and are structurally similar to the intracisternal A particles observed in normal adult and tumor cells; they have a large lucent core and their surface is covered by short radially oriented spines. 'Small' A particles (Plate 2*c* and *d*) are about 80 nm in size and have a small relatively electron-lucent center surrounded by an irregular dense shell. Delicate projections radiate from the surface membrane of small A particles. Dense-cored vesicles and tubules (Plate 2*a* and *b*) are about 50 nm in diameter and contain a granular or fibrous core; it has not been ascertained whether they actually exist as free intracisternal particles. Budding C-type particles (Plate 2*g* and *h*) have a lucent center surrounded by an electron-dense shell while mature C particles (Plate 2*j*) contain an irregular electron-dense core surrounded by a loosely contoured membrane; their morphology is similar to that of murine leukemia virus (Feller *et al.*, 1971).

Fig. 4 illustrates the distribution of the four types of particles at different stages of embryo development (Chase & Pikó, 1973). Small A particles are by far the most numerous class and their frequency shows a marked dependence on developmental stage. They seem to be absent in the 1-cell fertilized egg but clusters of budding particles (up to several hundred particles per equatorial section) are seen frequently in late 2-cell and 4-cell embryos; their numbers diminish from the 8-cell stage onwards. The other three particles appear later in development but are found consistently, albeit in low numbers, in the blastocyst.

Since all the studies reported so far, involving about a dozen different strains of mice, show the same general pattern of the occurrence and frequency of small A particles, the cycle of A particle production in the early embryo is most likely a general feature of mouse embryo de-

Type	Structure	Size (nm)	Stage of embryo development				
			1-cell	2–4-cell	8-cell	Morula	Blasto-cyst
?		50	0	±	±	±	++
Small A	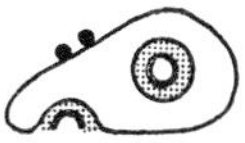	75–85	0	++++	+++	++	+
Large A	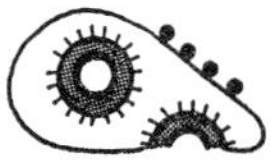	85–100	0	0	±	+	++
C		90–110	0	0	0	0	+

Fig. 4. Types of particles observed and their distribution at different stages of development in embryos of Swiss albino mice. (After Chase & Pikó, 1973.) Incidence is scored as 0, ± to + + + + scale.

velopment. However, careful quantitative comparisons between different strains have not yet been made. Biczysko *et al.* (1973) reported that the number of A and C particles seemed to be higher in blastocysts of AKR mice, a high leukemic strain.

The mechanism of formation of the small A particles – for example, whether it requires cellular RNA- and protein synthesis – is not known. Their sudden massive appearance at a time when cytoplasmic protein synthesis is of a very low level suggests that they are assembled, at least in part, from pre-formed components stored in the egg cytoplasm. Biczysko, Solter, Graham & Koprowski, (1974), studying parthenogenetically stimulated mouse eggs, observed dense granular structures, which may have represented precursors of A-type particles, in the nucleoplasm and between the duplicated inner leaflets of the nuclear membrane. Their study shows that fertilization and the paternal genome are not required for A particle expression. A similar conclusion can be drawn from the observed presence of A particles in the first polar body at the late 2-cell stage (Calarco & Szollosi, 1973).

The fate and possible function of the small A particles, as well as the other types of particles observed in the early embryo, are as yet unknown. The disappearance of the small A particles during oocyte maturation and the progressive reduction in their numbers from the 8-cell stage onwards suggest that the particles are turning over and/or being utilized. A crystalloid material which increases in amount from the 2-cell stage onwards is often found in association with endoplasmic reticulum containing A-type particles (Calarco & Brown, 1969; Calarco & Szollosi, 1973; Chase & Pikó, 1973) suggesting a possible role of these structures in the formation of the crystalloid material. However, the function of the crystalloid material itself in development remains to be elucidated.

SUMMARY

The mature mouse oocyte contains about 2–3 pg of mitochondrial DNA in the form of clean circles and D-loop molecules. There is no evidence of mitochondrial DNA replication during pre-implantation development.

Nuclear DNA transcription is required for the normal development of the mouse embryo, possibly as early as the 1-cell stage. From the 4-cell stage onwards the synthesis of ribosomal and transfer RNA dominates and contributes to a rapid build-up of protein synthesizing capacity, which is at a low level in the fertilized egg. There is evidence for the early synthesis and utilization of messenger RNA transcribed, in part, from the paternal genome.

The mitochondrial genome is transcribed from the 8–16-cell stage onwards and contributes, through the synthesis of mitochondrial ribosomal and transfer RNA, to the establishment of a mitochondrial protein synthesizing system. Mitochondrial protein synthesis is required for the normal growth and organization of mitochondrial cristae in the blastocyst. However, the development and differentiation of the mouse embryo up to the blastocyst stage is not dependent on mitochondrial genetic activity.

Several virus-like particles appear to be produced regularly in the early mouse embryo: two types of intracisternal A particles, a C-type particle and an endoplasmic reticulum-associated 'dense-cored vesicle'. A small intracisternal A particle is produced in particularly large numbers in the 2–4-cell embryo and may be involved in the formation of a crystalloid structure in the cytoplasm. The role, if any, of these particles, and the crystalloid material itself, in embryo development is unknown.

REFERENCES

ALFERT, M. (1950). A cytochemical study of oogenesis and cleavage in the mouse. *Journal of Cellular and Comparative Physiology*, **36**, 381–409.

Anon. (1966). Suggestions for the classification of oncogenic RNA viruses. *Journal of the National Cancer Institute*, **37**, 395–7.

AOKI, T., HUEBNER, R. J., CHANG, K. S. S., STURM, M. M. & LIU, M. (1974). Diversity of envelope antigens on murine type-C RNA viruses. *Journal of the National Cancer Institute*, **52**, 1189–97.

AUSTIN, C. R. & BISHOP, M. W. H. (1959). Differential fluorescence in living rat eggs treated with acridine orange. *Experimental Cell Research*, **17**, 35–43.

BARBEHENN, E. K., WALES, R. G. & LOWRY, O. H. (1974). The explanation for the blockade of glycolysis in early mouse embryos. *Proceedings of the National Academy of Sciences, USA*, **71**, 1056–60.

BICZYSKO, W., PIENKOWSKI, M., SOLTER, D. & KOPROWSKI, H. (1973). Virus particles in early mouse embryos. *Journal of the National Cancer Institute*, **51**, 1041–50.

BICZYSKO, W., SOLTER, D., GRAHAM, C. & KOPROWSKI, H. (1974). Synthesis of endogenous type-A virus particles in parthenogenetically stimulated mouse eggs. *Journal of the National Cancer Institute*, **52**, 483–9.

BORST, P. (1972). Mitochondrial nucleic acids. *Annual Review of Biochemistry*, **41**, 333–76.

BORST, P. & GRIVELL, L. A. (1971). Mitochondrial ribosomes. *FEBS Letters*, **13**, 73–88.

BREGA, A. & BAGLIONI, C. (1971). A study of mitochondrial protein synthesis in intact HeLa cells. *European Journal of Biochemistry*, **22**, 415–22.

BRINSTER, R. L. (1967). Protein content of the mouse embryo during the first five days of development. *Journal of Reproduction and Fertility*, **13**, 413–20.

BRINSTER, R. L. (1970). Metabolism of the ovum between conception and nidation. In *Mammalian Reproduction*, ed. H. Gibian & E. J. Plotz, pp. 229–63. New York: Springer Verlag.

BRINSTER, R. L. (1971). Uptake and incorporation of amino acids by the preimplantation mouse embryo. *Journal of Reproduction and Fertility*, **27**, 329–338.

BRINSTER, R. L. (1973). Parental glucose phosphate isomerase activity in three-day mouse embryos. *Biochemical Genetics*, **9**, 187–91.

BRINSTER, R. L. & THOMSON, J. L. (1966). Development of eight-cell mouse embryos *in vitro*. *Experimental Cell Research*, **42**, 308–15.

CALARCO, P. G. & BROWN, E. H. (1969). An ultrastructural and cytological study of preimplantation development of the mouse. *Journal of Experimental Zoology*, **171**, 253–84.

CALARCO, P. G. & SZOLLOSI, D. (1973). Intracisternal A particles in ova and preimplantation stages of the mouse. *Nature New Biology*, **243**, 91–3.

CHAPMAN, V. M., WHITTEN, W. K. & RUDDLE, F. H. (1971). Expression of paternal glucose phosphate isomerase-1 (*gpi*-1) in preimplantation stages of mouse embryos. *Developmental Biology*, **26**, 153–8.

CHASE, D. G. & PIKÓ, L. (1973). Expression of A- and C-type particles in early mouse embryos. *Journal of the National Cancer Institute*, **51**, 1971–5.

CHASE, J. W. & DAWID, I. B. (1972). Biogenesis of mitochondria during *Xenopus laevis* development. *Developmental Biology*, **27**, 504–18.

CHATTOPADHYAY, S. K., LOWY, D. R., TEICH, N. M., LEVINE, A. S. & ROWE, W. P. (1974). Evidence that the AKR murine-leukemia-virus genome is complete in DNA of the high-virus AKR mouse and incomplete in the DNA of the 'virus-negative' NIH mouse. *Proceedings of the National Academy of Sciences, USA*, **71**, 167–71.

CHURCH, R. B. & SCHULTZ, G. A. (1974). Differential gene activity in the pre- and postimplantation mammalian embryo. In *Current Topics in Developmental Biology*, ed. A. A. Moscona & A. Monroy, vol. 8, pp. 179–202, New York & London: Academic Press.

COSTANTINO, P. & ATTARDI, G. (1973). Atypical pattern of utilization of amino acids for mitochondrial protein synthesis in HeLa cells. *Proceedings of the National Academy of Sciences, USA*, **70**, 1490–4.

DAENTL, D. L. & EPSTEIN, C. J. (1971). Developmental interrelationships of uridine uptake. Nucleotide formation and incorporation into RNA by early mammalian embryos. *Developmental Biology*, **24**, 428–42.

DALTON, A. J. (1972). RNA-tumor viruses – terminology and ultrastructural aspects of virion morphology and replication. *Journal of the National Cancer Institute*, **49**, 323–7.

DAWID, I. B. (1966). Evidence for the mitochondrial origin of frog egg cytoplasmic DNA. *Proceedings of the National Academy of Sciences, USA*, **56**, 269–76.

DAWID, I. B. (1972). Cytoplasmic DNA. In *Oogenesis*, ed. J. D. Biggers & A. W. Schuetz, pp. 215–26. Baltimore: University Park Press.

DAWID, I. B. & BLACKLER, A. W. (1972). Maternal and cytoplasmic inheritance of mitochondrial DNA in *Xenopus*. *Developmental Biology*, **29**, 152–61.

DAWID, I. B. & BROWN, D. D. (1970). The mitochondrial and ribosomal DNA components of oocytes of *Urechis caupo*. *Developmental Biology*, **22**, 1–14.

ELLEM, K. A. O. & GWATKIN, R. B. L. (1968). Patterns of nucleic acid synthesis in the early mouse embryo. *Developmental Biology*, **18**, 311–30.

ENDERS, A. C. & SCHLAFKE, S. J. (1965). The fine structure of the blastocyst: some comparative studies. In *Preimplantation Stages of Pregnancy*, ed. G. E. W. Wolstenholme & M. O'Connor, pp. 29–54. Boston: Little, Brown.

EPSTEIN, C. J. (1972). Expression of the mammalian X-chromosome before and after fertilization. *Science*, **175**, 1467–8.

EPSTEIN, C. J. & SMITH, S. A. (1973). Amino acid uptake and protein synthesis in preimplantation mouse embryos. *Developmental Biology*, **33**, 171–84.

EPSTEIN, C. J. & SMITH, S. A. (1974). Electrophoretic analysis of proteins synthesized by preimplantation mouse embryos. *Developmental Biology*, **40**, 233–44.

FELLER, U., DOUGHERTY, R. M. & DI STEFANO, H. S. (1971). Comparative morphology of avian and murine leukemia viruses. *Journal of the National Cancer Institute*, **47**, 1289–98.

GELB, L. D., MILSTIEN, J. B., MARTIN, M. A. & AARONSON, S. A. (1973). Characterization of murine leukaemia virus-specific DNA present in normal mouse cells. *Nature New Biology*, **244**, 76–9.

GINSBERG, L. & HILLMAN, N. (1973). ATP metabolism in cleavage-staged mouse embryos. *Journal of Embryology and Experimental Morphology*, **30**, 267–82.

GOLBUS, M. S., CALARCO, P. G. & EPSTEIN, C. J. (1973). The effects of inhibitors of RNA synthesis (α-amanitin and actinomycin D) on preimplantation mouse embryogenesis. *Journal of Experimental Zoology*, **186**, 207–16.

GRAHAM, C. F. (1973). The necessary conditions for gene expression during early mammalian development. In *Genetic Mechanisms of Development*, ed. F. H. Ruddle, pp. 201–24. New York & London: Academic Press.

HALLBERG, R. L. (1974). Mitochondrial DNA in *Xenopus laevis* oocytes. I. Displacement loop occurrence. *Developmental Biology*, **38**, 346–55.

HILLMAN, N. & TASCA, R. J. (1969). Ultrastructural and autoradiographic studies of mouse cleavage stage. *American Journal of Anatomy*, **26**, 151–73.

HIRSCH, M. & PENMAN, S. (1973). Mitochondrial polyadenylic acid-containing RNA: localization and characterization. *Journal of Molecular Biology*, **80**, 379–391.

HUEBNER, R. J., KELLOFF, G. J., SARMA, P. S., LANE, W. T., TURNER, H. C., GILDEN, R. V., OROSZLAN, S., MEIER, H., MYERS, D. D. & PETERS, R. L. (1970). Group specific antigen expression during embryogenesis of the genome of the C-type RNA tumor virus: implications for ontogenesis and oncogenesis. *Proceedings of the National Academy of Sciences, USA*, **67**, 366–76.

KARP, G., MANES, C. & HAHN, W. E. (1973). RNA synthesis in the preimplantation rabbit embryo: radioautographic analysis. *Developmental Biology*, **31**, 404–8.

KASAMATSU, H., GROSSMAN, L. I., ROBBERSON, D. L., WATSON, R. & VINOGRAD, J. (1973). The replication and structure of mitochondrial DNA in animal cells. *Cold Spring Harbor Symposia on Quantitative Biology*, **38**, 281–8.

KING, M. E., GODMAN, G. C. & KING, D. W. (1972). Respiratory enzymes and mitochondrial morphology of HeLa and L cells treated with chloramphenicol and ethidium bromide. *Journal of Cell Biology*, **53**, 127–42.

KNOWLAND, J. S. & GRAHAM, C. F. (1972). RNA synthesis at the two-cell stage of mouse development. *Journal of Embryology and Experimental Morphology*, **27**, 167–76.

KRAMEN, M. A. & BIGGERS, J. D. (1971). Uptake of tricarboxylic acid cycle intermediates by preimplantation mouse embryos *in vitro*. *Proceedings of the National Academy of Sciences, USA*, **68**, 2656–9.

KUFF, E. L., LUEDERS, K. K., OZER, H. L. & WIVEL, N. A. (1972). Some structural and antigenic properties of intracisternal A particles occurring in mouse tumors. *Proceedings of the National Academy of Sciences, USA*, **69**, 218–22.

LIZARDI, P. M. & LUCK, D. J. L. (1972). The intracellular site of synthesis of mitochondrial ribosomal proteins in *Neurospora crassa*. *Journal of Cell Biology*, **54**, 56–74.

MANES, C. (1969). Nucleic acid synthesis in preimplantation rabbit embryos. *Journal of Experimental Zoology*, **172**, 303–10.

MANES, C. (1971). Nucleic acid synthesis in preimplantation rabbit embryos. II. Delayed synthesis of ribosomal RNA. *Journal of Experimental Zoology*, **176**, 87–96.

MARALDI, N. M. & MONESI, V. (1970). Ultrastructural changes from fertilization to blastulation in the mouse. *Archives d'Anatomie Microscopique et de Morphologie Expérimentale*, **59**, 361–82.

MATSUMOTO, L., KASAMATSU, H., PIKÓ, L. & VINOGRAD, J. (1974). Mitochondrial DNA replication in sea urchin oocytes. *Journal of Cell Biology*, **63**, 146–59.

MILLS, R. M., Jr, & BRINSTER, R. L. (1967). Oxygen consumption of preimplantation mouse embryos. *Experimental Cell Research*, **47**, 337–44.

MINTZ, B. (1964). Synthetic processes and early development in the mammalian egg. *Journal of Experimental Zoology*, **157**, 85–100.

MINTZ, B. (1965). Nucleic acid and protein synthesis in the developing mouse embryo. In *Preimplantation Stages of Pregnancy*, ed. G. E. W. Wolstenholme & M. O'Connor, pp. 145–55. Boston: Little, Brown.

MOLINARO, M., SIRACUSA, G. & MONESI, V. (1972). Differential effects of metabolic inhibitors on early development in the mouse embryo, at various stages of the cell cycle. *Experimental Cell Research*, **71**, 261–4.

MONESI, V., MOLINARO, M., SPALETTA, E. & DAVIOLI, C. (1970). Effect of metabolic inhibitors on macromolecular synthesis and early development in the mouse embryo. *Experimental Cell Research*, **59**, 197–206.

MONESI, V. & SALFI, V. (1967). Macromolecular synthesis during early development in the mouse embryo. *Experimental Cell Research*, **46**, 632–5.

NOWINSKI, R. C., OLD, L. J., SARKAR, N. H. & MOORE, D. H. (1970). Common properties of the oncogenic RNA viruses (oncornaviruses). *Virology*, **42**, 1152–7.

OJALA, D. & ATTARDI, G. (1974). Identification and partial characterization of multiple discrete polyadenylic acid-containing RNA components coded for by HeLa cell mitochondrial DNA. *Journal of Molecular Biology*, **88**, 205–19.

OLDS, P. J., STERN, S. & BIGGERS, J. D. (1973). Chemical estimates of the RNA and DNA contents of the early mouse embryo. *Journal of Experimental Zoology*, **186**, 39–45.

PERRY, R. P. & KELLEY, D. E. (1973). Messenger RNA turnover in mouse L cells. *Journal of Molecular Biology*, **79**, 681–96.

PIKÓ, L. (1970). Synthesis of macromolecules in early mouse embryos cultured *in vitro*: RNA, DNA and a polysaccharide component. *Developmental Biology*, **21**, 257–79.

PIKÓ, L. & CHASE, D. G. (1973). Role of the mitochondrial genome during early development in mice. Effects of ethidium bromide and chloramphenicol. *Journal of Cell Biology*, **58**, 357–78.

PIKÓ, L., TYLER, A. & VINOGRAD, J. (1967). Amount, location, priming capacity, circularity and other properties of cytoplasmic DNA in sea urchin eggs. *Biological Bulletin, Wood's Hole*, **132**, 68–90.

ROWE, W. P., LOWY, D. R., TEICH, N. & HARTLEY, J. W. (1972). Some implications of the activation of murine leukemia virus by halogenated pyrimidines. *Proceedings of the National Academy of Sciences, USA*, **69**, 1033–5.

SCHATZ, G., GROOT, G. S. P., MASON, T., ROUSLIN, W., WHARTON, D. C. & SALTZGABER, J. (1972). Biogenesis of mitochondrial inner membranes in baker's yeast. *Federation Proceedings*, **31**, 21–9.

SCHULTZ, G. A., MANES, C. & HAHN, W. E. (1973). Synthesis of RNA containing polyadenylic acid sequences in preimplantation rabbit embryos. *Developmental Biology*, **30**, 418–26.

SOSLAU, G. & NASS, M. M. K. (1971). Effects of ethidium bromide on the cytochrome content and ultrastructure of L cell mitochondria. *Journal of Cell Biology*, **51**, 514–24.

STERN, S., BIGGERS, J. D. & ANDERSON, E. (1971). Mitochondria and early development of the mouse. *Journal of Experimental Zoology*, **176**, 179–92.

SZOLLOSI, D. (1965). The fate of sperm middle-piece mitochondria in the rat egg. *Journal of Experimental Zoology*, **159**, 367–78.

SZOLLOSI, D. (1972). Changes in some cell organelles during oogenesis in mammals. In *Oogenesis*, ed. J. D. Biggers & A. W. Schuetz, pp. 47–64. Baltimore: University Park Press.

TASCA, R. J. & HILLMAN, N. (1970). Effects of actinomycin D and cycloheximide on RNA and protein synthesis in cleavage stage mouse embryos. *Nature, London*, **225**, 1022–5.

THOMSON, J. L. & BIGGERS, J. D. (1966). Effects of inhibitors of protein synthesis on the development of preimplantation mouse embryos. *Experimental Cell Research*, **41**, 411–27.

TODARO, G. J. & HUEBNER, R. J. (1972). The viral oncogene hypothesis: new evidence. *Proceedings of the National Academy of Sciences, USA*, **69**, 1009–1015.

TZAGOLOFF, A., AKAI, A. & SIERRA, M. F. (1972). Assembly of the mitochondrial membrane system. VII. Synthesis and integration of F_1 subunits into the rutamycin-sensitive adenosine triphosphatase. *Journal of Biological Chemistry*, **247**, 6511–16.

VERNON, M. L. (1974). Frequency distribution of intracisternal type A particles in selected tissues of embryonic and newborn mice. *Cancer Research*, in press.

VERNON, M. L., LANE, W. T. & HUEBNER, R. J. (1973). Prevalence of type-C particles in visceral tissues of embryonic and newborn mice. *Journal of the National Cancer Institute*, **51**, 1171–5.

WALES, R. G. & BIGGERS, J. D. (1968). The permeability of two- and eight-cell

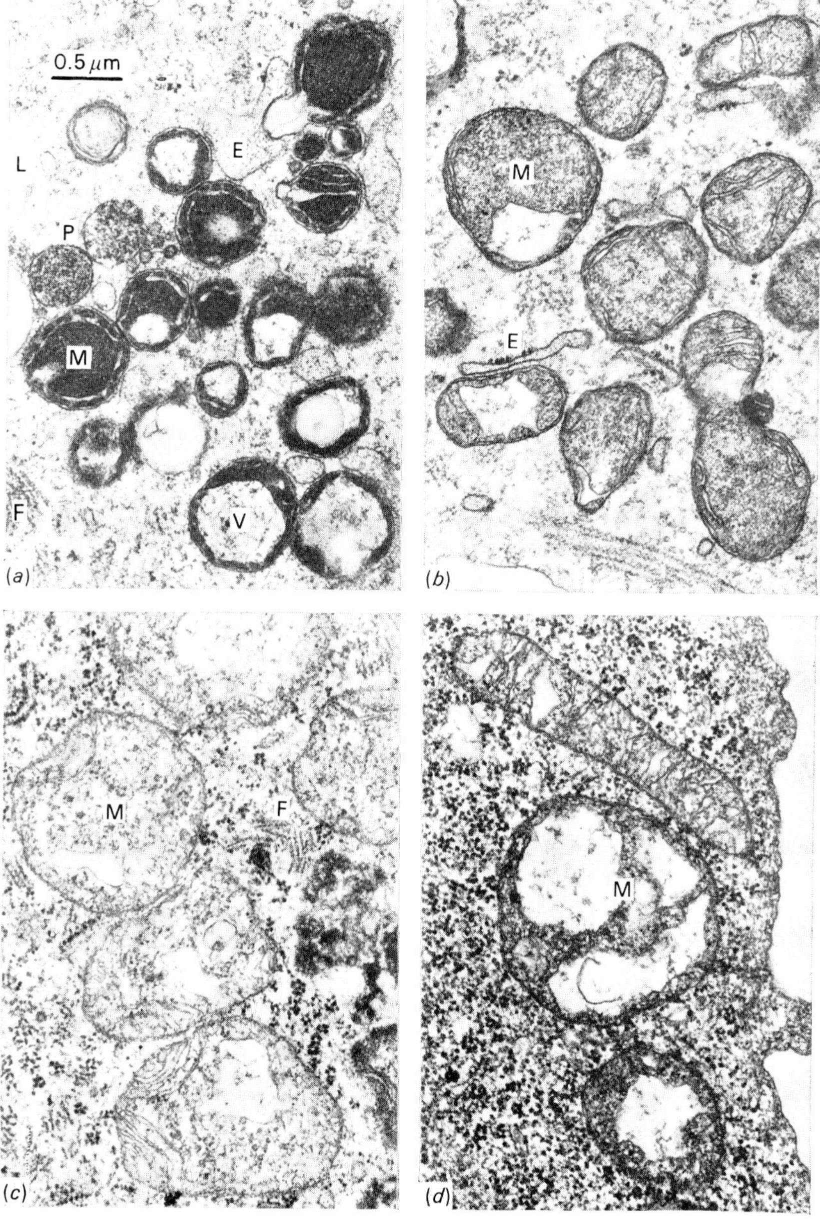

(*Facing p. 186*)

PLATE 2

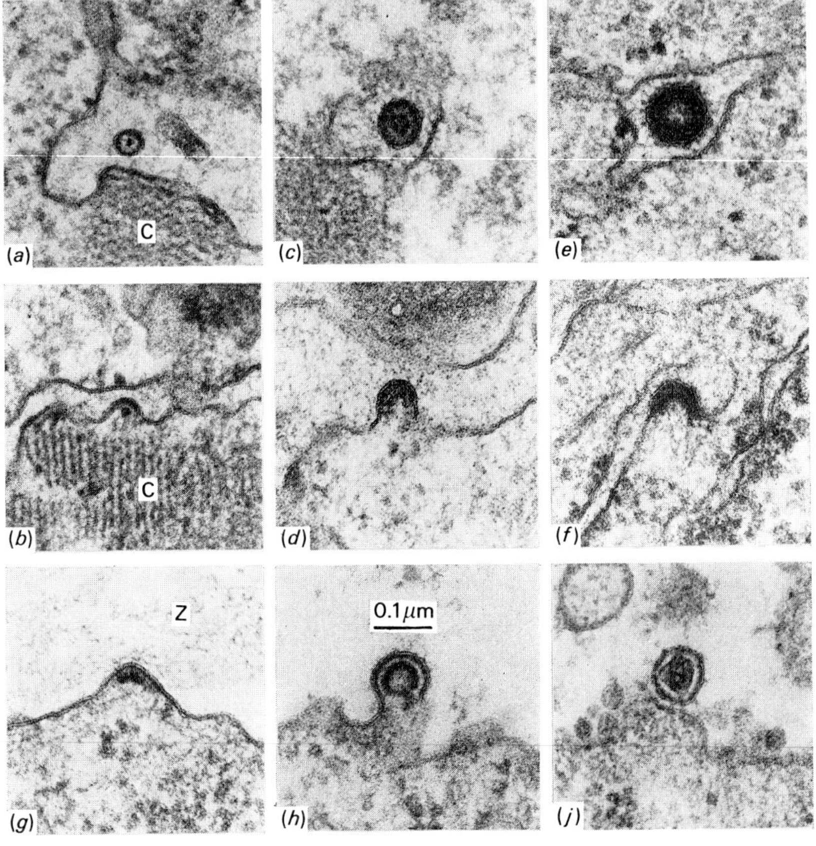

mouse embryos to L-malic acid. *Journal of Reproduction and Fertility*, **15**, 103–11.

WARNER, C. M. & VERSTEEGH, L. R. (1974). *In vivo* and *in vitro* effect of α-amanitin on preimplantation mouse embryo RNA polymerase. *Nature, London*, **248**, 678–80.

WHITTEN, W. K. (1957). Culture of tubal ova. *Nature, London*, **179**, 1081–2.

WILSON, S. H. & KUFF, E. L. (1972). A novel DNA polymerase activity found in association with intracisternal A-type particles. *Proceedings of the National Academy of Sciences, USA*, **69**, 1531–6.

WIVEL, N. A., LUEDERS, K. K. & KUFF, E. L. (1973). Structural organization of murine intracisternal A particles. *Journal of Virology*, **11**, 329–34.

WIVEL, N. A. & SMITH, G. H. (1971). Distribution of intracisternal A particles in a variety of normal and neoplastic mouse tissues. *International Journal of Cancer*, **7**, 167–75.

WOODLAND, H. R. & GRAHAM, C. F. (1969). RNA synthesis during early development of the mouse. *Nature, London*, **221**, 327–32.

YANG, S. S. & WIVEL, N. A. (1973). Analysis of high-molecular-weight ribonucleic acid associated with intracisternal A particles. *Journal of Virology*, **11**, 287–98.

YANG, S. S. & WIVEL, N. A. (1974). Characterization of an endogenous RNA-dependent DNA polymerase associated with murine intracisternal A particles. *Journal of Virology*, **13**, 712–20.

EXPLANATION OF PLATES

PLATE 1

The fine-structural changes in the mitochondria of mouse embryos cultured *in vitro* from the 2–4-cell stage to the blastocyst. M, mitochondrion; V, vacuole; E, endoplasmic reticulum; F, fibrous yolk; L, lipid; and P, peroxisome-like granule. The magnification is the same for all pictures. For further explanation see text. (*a*) Mitochondria of 2-cell embryo; (*b*) mitochondria of 8-cell embryo; (*c*) mitochondria of early blastocyst; (*d*) mitochondria from a trophoblast cell of an expanded blastocyst. (After Pikó & Chase, 1973.)

PLATE 2

Electron micrographs illustrating the morphology of free and budding particles in early mouse embryos. C, crystalloid material; Z, zona pellucida. The magnification is the same for all pictures. For further details see text. (*a*) and (*b*), dense-cored vesicle; (*c*) and (*d*), small A particle; (*e*) and (*f*), large A particle; (*g*) and (*h*), C-type particles in different stages of budding from the plasma membrane; (*j*), mature C particle. (After Chase & Pikó, 1973.)

STUDIES OF THE *T*-LOCUS

BY N. HILLMAN

Department of Biology, Temple University,
Philadelphia, Pennsylvania 19122, USA

The *T*-locus in the house mouse (*Mus musculus*) is located on linkage group IX (chromosome 17) and includes a wild-type allele, designated + or T^+, a dominant mutant allele, *T*, and a series of recessive t^n alleles. The first spontaneous mutation at this locus resulted in a mouse with a short-tail phenotype (Dobrovolskaia-Zavadskaia, 1927). This short-tailed mouse, when outcrossed to normal-tailed mice, gave rise to both short-tailed and normal-tailed progeny. Dobrovolskaia-Zavadskaia suggested that the short-tailed phenotype was the expression of a locus heterozygous for a wild-type allele (+) and a mutant dominant allele, which was designated *T*. When mice from the original short-tailed line (+/*T*) were outcrossed to normal-tailed, wild-type mice, three of the outcrosses produced tailless progeny. In *inter se* matings, the tailless progeny were found to breed true, producing only tailless offspring. Dobrovolskaia-Zavadskaia (1932) hypothesized that these tailless mice resulted from crosses between animals which carried balanced lethal mutations. The heterozygous condition would be responsible for the tailless phenotype and both the homozygous dominant and recessive mutant alleles would be lethal to the embryo.

A later investigation (Chesley, 1932) of the embryological effects of the dominant mutant allele (*T*) showed that this allele, in a homozygous condition, was indeed lethal, the embryos dying at 10½ to 11 days of gestation. Chesley & Dunn (1936) later proved that one of the three tailless lines contained, in addition to the dominant mutant allele *T*, a recessive lethal allele t^0, which in a homozygous condition was lethal during early embryogenesis. Since these original genetic and phenotypic studies, numerous other recessive *t*-alleles have arisen as spontaneous mutations both in laboratory stocks (t^n) and in wild populations (t^{wn}). Many of these t^n and t^{wn} alleles, when homozygous, produce a specific sequence of embryonic and fetal abnormalities which result in lethality during development. A number of these alleles are also meiotically driven. The recessive lethal alleles can be placed into five complementation groups according to their ability to partially complement with each other. Each complementation group is named in accordance with a designated prototype t^n allele which has a specific syndrome of embryonic effects. These prototypes are

t^{12}, t^0, t^{w5}, t^9 and t^{w1} (Bennett & Dunn, 1964). The members of each complementation group exhibit homozygous lethality at the same time during development, a similar syndrome of abnormal development and a similar increase in the frequency of transmission of *t*-bearing spermatozoa. The phenotypic similarities and differences both within and between complementation groups make the *t*-alleles unique tools for studying both mammalian embryogenesis and reproduction.

The original studies of the t^n/t^n embryos resulted in the publication of two hypotheses to explain the primary effect of the mutation on embryonic development. The first of these, which was based on light microscopic observations of homozygous *t* embryos which died following implantation (i.e. members of the t^0, t^{w5}, t^9 and t^{w1} complementation groups), proposed that the gene product of the *T*-locus controlled the ectodermal–mesodermal organization and differentiation of the embryo and that the *t*-alleles affected the normal interaction of ectoderm and mesoderm cells, resulting ultimately in the death of the embryo (for a review, see Bennett, 1964). This hypothesis, although offering a possible explanation for the later acting embryonic *t* lethal alleles, could not be used to explain the phenotypic changes and the lethality observed in those mutant *t* homozygotes which died prior to implantation (i.e. members of the t^{12} complementation group). An additional problem with this hypothesis is that it has now been shown that a member of the t^0 complementation group, t^6, which was originally believed to be a mesoderm mutation, dies prior to mesoderm formation (Nadijcka & Hillman, 1973, 1975).

The second hypothesis based on light microscopy, on histochemical observations and on low resolution autoradiographic studies suggested that t^{12}/t^{12} embryos, which belong to the t^{12} complementation group, died at the late morula stage as a result of aberrant nucleolar development and deficient RNA synthesis (Smith, 1956; Mintz, 1964*a*, *b*, *c*). The investigators (Smith, using Azure-B stain for the photometric determination of cellular RNA content, and Mintz, using labelled RNA precursors followed by low resolution autoradiography) reported that they could not distinguish between homozygous mutant embryos and their wild-type littermates either by total RNA content or by rates of RNA synthesis prior to the late morula stage. At the late morula stage, however, the homozygotes were arrested in development while the normal littermates formed blastocoels. Both techniques, histochemistry and autoradiography, showed that the noncavitating homozygotes had less total RNA and were synthesizing less RNA than were the embryos which were advancing into the blastocyst stage.

These combined studies defined the differences, therefore, between the mutant embryos (t^{12}/t^{12}) and their normal littermates as (1) the homozygous lethal embryos die at the late morula stage, (2) the nucleoli of the mutant morulae remain round while those of their normal cavitating littermates become irregularly long and (3) RNA synthesis in the mutant morulae is reduced compared with that of the normals. The observations indicated to these investigators that the primary action of the mutation was related to defective nucleolar development and function (synthesis of ribosomal RNA). It should be noted that all the above studies were qualitative rather than quantitative in nature and that comparative measurements of RNA synthetic activity were made between normally developing embryos and mutant morulae which were developmentally arrested.

Our first observation on t^{12} homozygotes was that some t^{12}/t^{12} embryos die as early as the 8–12-cell stage while others begin to form blastocoels and then die as late as the early blastocyst stage (Hillman, Hillman & Wileman, 1970). Although most homozygotes do die at the late morula stage, death is not limited to this developmental period. Secondly, we found that the t^{12}/t^{12} genotype is a cell lethal and embryonic lethality is dependent upon developmental arrest, which in turn is dependent upon the proportion of viable cells within the embryo.

We have also determined that the t^{12}/t^{12} homozygotes do not differ from their normal littermates in:

(i) *Nucleolar development.* The nucleoli develop normally up to the stage at which embryos are arrested in development. Those embryos which die as late morulae have elongated nucleoli which become round only when pycnosis is apparent in other parts of the cell (Hillman *et al.*, 1970).

(ii) *RNA labelling patterns.* Comparisons between high resolution autoradiographs of phenotypically mutant embryos and their phenotypically wild-type littermates after incubation in [^{3}H]uridine have revealed no significant difference in either the amount or distribution of the labelled precursors between these two groups of embryos. The labelling patterns of phenotypically wild-type late morula embryos and of t^{12}/t^{12} late morulae, at developmental arrest but prior to degeneration, are the same (Hillman, 1972).

(iii) *Ribosomal RNA synthesis.* At each developmental stage, mutant and wild-type embryos demonstrate the same capability to take up labelled precursors and to incorporate them into newly synthesized ribosomal RNA (rRNA). Sucrose density gradient analyses show no differences in the pattern of incorporation of labelled precursor into the various RNA peaks. Nor do they show differences in the percentage distribution of the label into the various peaks at any developmental stage, including the late morula stage (Hillman & Tasca, 1973).

(iv) *Polyribosomes and rough endoplasmic reticulum* (*RER*). In both normal and t^{12}/t^{12} mouse embryos large numbers of polyribosomes are first found in the cytoplasm at the 8-cell stage. This is correlated with a substantial increase in the amount of RER. There is a progressive increase in the quantity of both polyribosomes and RER as the mutant and wild-type embryos continue development through the later cleavage stages. This similarity continues through developmental arrest until the time of degeneration of the homozygous t^{12} embryos (Hillman & Tasca, 1969; Hillman *et al.*, 1970).

(v) *Rates of protein synthesis.* There are no differences in the uptake of [^{14}C]leucine nor the rates of incorporation of this precursor into PCA-precipitable material between t^{12}/t^{12} embryos and correspondingly staged wild-type embryos (Figs. 1 and 2).

These combined studies, therefore, do not support the earlier suggestions that the cause of lethality in homozygous t^{12} embryos is related to

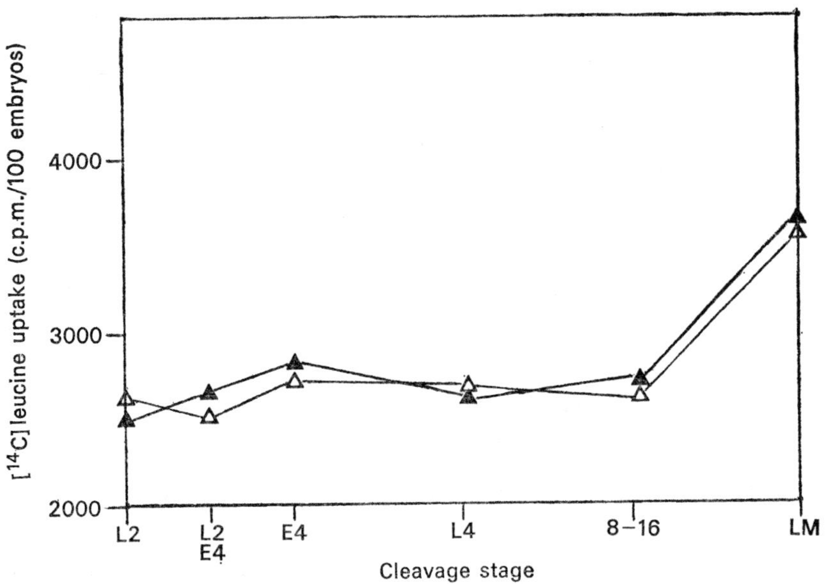

Fig. 1. The uptake of [^{14}C]leucine (c.p.m./100 embryos) by T^+/T^+ wild-type (▲—▲) and t^{12}/t^{12} mutant (△—△) 2-, 4-, 8–16-cell and late morula embryos. The samples labelled t^{12}/t^{12} are pooled litters from T^+/t^{12} *inter se* matings and are composed of approximately 40 % homozygous t^{12} embryos. The embryos were incubated at each cleavage stage in 0.6 μCi/ml [^{14}C]leucine for a 3 h period. (For protocol see Tasca & Hillman, 1970.) E, early; L, late; LM, late morula.

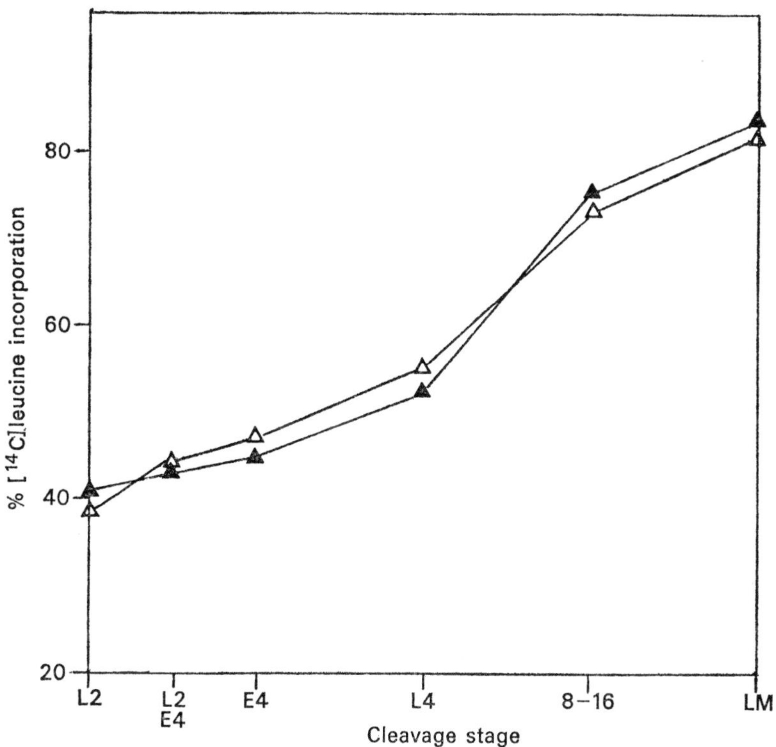

Fig. 2. Comparative rates of protein synthesis (incorporation of [^{14}C]leucine into PCA-precipitable, RNase-resistant material; for protocol see Tasca & Hillman, 1970). The embryos were incubated for 3 h at each cleavage stage in 0.6 μCi/ml [^{14}C]leucine. Approximately 40 % of each sample, designated t^{12}/t^{12}, are homozygous t^{12} embryos. Graph symbols as in Fig. 1. E, early; L, late; LM, late morula.

defective nucleolar development or function. Supportive evidence for our findings comes from a series of studies. Bennett (1965), who did karyotype analyses on various t^n genotypes (including T^+/t^{12}), found that (1) there are six secondary constrictions (nucleolar organizers) in the T^+/T^+ karyotype, (2) none of these constrictions are located on the ninth linkage group (the location of the *T*-locus) and (3) the karyotype of the heterozygote, T^+/t^{12}, does not differ from that of the wild-type. Ivanyi (1971) also noted that the numbers of nucleoli in mouse cells were not determined by a single major gene and Flaherty, Bennett & Graef (1972) have corroborated this report. Finally Calarco & Brown (1968) have reported that they could find no difference in the intensity of Azure-B stain when recently blocked late morula t^{12}/t^{12} embryos were compared with late morula T^+/T^+ embryos. These observations, taken together, further support the assumption that nucleolar abnormalities are not the cause of lethality of the t^{12} homozygotes.

One major shortcoming of the two original hypotheses which were used to explain the primary developmental effect of the t^n alleles and the resultant lethality of t^n homozygous embryos was that neither was able to be used to explain the second separable major phenotypic characteristic of these lethal t^n alleles, that of meiotic drive. Meiotic drive is a phenomenon which may occur either as a result of an alteration of normal meiosis (Sandler & Novitski, 1957) or as a result of a transmission anomaly (Lewontin, 1962). Both of these aberrations may, in turn, result in an increased frequency of fertilization by spermatozoa carrying one gene over those carrying its allele. The increased transmission frequency of the driven allele is independent of any deleterious effect of this allele. In the case of the *T*-locus, males heterozygous for the dominant mutant allele (T^+/T) and all heterozygous females (T^+/T; T^+/t^n) transmit the alleles in a 1 : 1 ratio. Conversely, males heterozygous for certain of the t^n alleles (e.g. t^{12}, t^{w32}, t^6) transmit t^n with a frequency greater than 50 %. There is no evidence that this changed male transmission frequency is a result of extra postmeiotic mitoses of t^n allele-bearing spermatozoa or that t^n spermatozoa are present in significantly higher numbers in seminal fluid (Bryson, 1944).

Recently, two hypotheses have been advanced to explain both the embryonic lethality and the meiotic drive of the t^n-bearing spermatozoa. The first suggests that the T^+ locus controls a specific cell surface antigen, that the lethal t^n alleles result in the production of altered but t^n specific antigens, and that these altered antigens affect both phenotypic characteristics (D. Bennett, personal communication and this volume). The second hypothesis, based on ultrastructural and intermediary metabolism studies of both t^n/t^n embryos and 'driven' t^n spermatozoa, proposes that the t^n alleles affect both embryonic lethality and meiotic drive by eliciting a metabolic error which results in nonphysiological levels of aerobic metabolism.

In support of this latter hypothesis we have specified ultrastructural differences between t^n/t^n embryos and their wild-type littermates and have attempted to correlate these ultrastructural differences with changes in the developmental physiology of lethal and viable embryos. Through our ultrastructural studies we have determined the phenocritical and phenolethal periods of t^{12}, t^{w32} and t^6 homozygotes and t^{12}/t^{w32} heterozygotes and have delimited those phenotypic expressions which are specific to each genotype as well as those which are characteristic of more than one genotype.

The first of the lethal recessive alleles studied was t^{12}. Although t^{12}/t^{12} embryos cannot be distinguished at the ultrastructural level from their littermates on the basis of nucleolar structure or function, they can be distinguished as early as the 2-cell stage by the presence of nuclear lipid droplets, nuclear fibrillo-granular bodies and by excessive cytoplasmic lipid. At the later cleavage stages, binucleated cells are an additional phenotypic expression of the mutant genotype (Hillman *et al.*, 1970).

A second recessive allele assigned to the t^{12} complementation group is t^{w32} (Bennett & Dunn, 1964). This assignment resulted from studies showing that t^{12} and t^{w32} do not complement each other and that the t^{w32} homozygotes as well as the heterozygous t^{12}/t^{w32} embryos die at the late morula stage. This is the same stage at which t^{12}/t^{12} embryos were reported to be blocked in development (Smith, 1956). Also, Bennett & Dunn (1964) reported that, at the light microscope level, both t^{w32} homozygotes and t^{12}/t^{w32} embryos show the same syndrome of phenotypic expression as previously described by Smith (1956) for the t^{12}/t^{12} genome. A more recent ultrastructural study has shown, however, that t^{w32}/t^{w32} embryos do differ from t^{12}/t^{12} embryos both in their phenolethal period and in their ultrastructure (Hillman & Hillman, 1975). An additional study has shown that the t^{12}/t^{w32} embryos are like t^{w32}/t^{w32} embryos in their phenolethal period but differ from both t^{12} and t^{w32} homozygotes in phenotypic expressions (see below).

As mentioned earlier, the t^{12}/t^{12} genotype can result in embryonic lethality as early as the 8-cell stage or as late as the early blastocyst stage; however, most of the t^{12}/t^{12} embryos are developmentally arrested as late morulae. Although the t^{w32}/t^{w32} genotype also exhibits a range of lethal periods, this range is from the 8-cell to the late morula stage, and unlike t^{12}/t^{12} embryos, the highest attrition occurs in the early morula. It should be emphasized that a distinction between early and late morulae can only be achieved by an ultrastructural examination of mouse embryos (Hillman & Tasca, 1969). This, in turn, probably accounts for the difference in the timing of the phenolethal periods for t^{w32}/t^{w32} embryos as reported by Bennett & Dunn (1964) and that reported by Hillman & Hillman (1975).

Additional morphological differences which distinguish t^{w32} homozygotes from t^{12} homozygotes can only be resolved at the ultrastructural level. For example, nuclear fibrillo-granular bodies which distinguish t^{12}/t^{12} embryos from their phenotypically normal littermates are not found in t^{w32} homozygotes and can therefore be used as a phenotypic characteristic to distinguish the two genotypes. Also, the mitochondria of t^{12}/t^{12} embryos undergo the normal sequential development which has been described for mouse cleavage staged embryos (Hillman & Tasca, 1969). Conversely, the mitochondria of some t^{w32}/t^{w32} embryos differ from those of their littermates as early as the 8-cell stage. In these mutant homozygotes, the mitochondria do not undergo the normal structural transition which typifies late 4- and early 8-cell mouse embryos (Hillman & Tasca,

1969). Many of the 8-cell and older t^{w32} homozygous embryos also contain mitochondria with crystal depositions. Crystal-containing mitochondria have not been seen in t^{12}/t^{12} embryos.

Therefore, the homozygous t^{12}/t^{12} and t^{w32}/t^{w32} genotypes are alike in that they result in (1) cell lethality, (2) pre-implantation lethality, (3) the presence of nuclear lipid droplets and excessive cytoplasmic lipid which distinguish the homozygotes as early as the 2-cell stage, (4) the appearance of binucleated cells and (5) the same chronology of degenerative changes. Embryos having the two homozygous genotypes differ from each other in (1) their phenolethal periods, (2) their nuclear inclusions, and (3) the presence or absence of mitochondrial variants. The two genotypes thus elicit both similar and dissimilar phenotypic expressions. Although the two alleles are each, in a homozygous condition, pre-implantation lethals, the evidence suggests that they are separate alleles.

A further test of the allelism between t^{12} and t^{w32} is the study of the phenotype of the heterozygous t^{12}/t^{w32} genotype. Recent ultrastructural studies (Hillman, unpublished observations) have shown that the heterozygous genotype results in characteristics which are common to both t^{12} and t^{w32} homozygotes. The phenolethal period of the heterozygous t^{12}/t^{w32} embryos is the same as that observed in t^{w32} homozygotes: death can occur between the 8-cell and the late morula stage, most being blocked as early morulae. Like the t^{12} homozygotes, the heterozygotes contain nuclear fibrillo-granular bodies (Plate 1*a*) and like both homozygotes, they can be distinguished by the presence of nuclear lipid droplets (Plate 1*b*) and excessive cytoplasmic lipid (Plate 1*c*). The later cleavage-staged heterozygotes also contain binucleated cells, a characteristic found in both homozygous genotypes. Unlike the homozygotes, however, the heterozygotes contain a variant form of mitochondria as early as the 2-cell stage. In wild-type embryos, 2-cell mitochondria are ovoid, contain few cristae and have a dense matrix (Hillman & Tasca, 1969; Pikó & Chase, 1973). Conversely, the mitochondria of the heterozygotes appear swollen and contain a less dense matrix (cf. Plate 2*a* and *b*). Variant mitochondrial ultrastructure distinguishes the heterozygotes from their phenotypically wild-type littermates at each cleavage stage.

A study of the phenotypic expression of the t^6/t^6 genotype has also been completed (Nadijcka & Hillman, 1973, 1975). The t^6 allele has been assigned to the t^0 complementation group (Bennett & Dunn, 1964). Gluecksohn-Schoenheimer (1940) had previously described the phenotypic expression of the lethal t^0/t^0 genotype, and Dunn & Gluecksohn-Schoenheimer (1950) reported that t^6 was a recurrence of t^0. In her light microscopic study, Gluecksohn-Schoenheimer found that the homozygous

t^0 embryos could not be distinguished from their littermates prior to developmental arrest which occurred at $5\frac{1}{4}$ days (beginning of the egg-cylinder) and that resorption of the necrotic embryos took place on day 7, at the time mesoderm begins to form in the phenotypically normal littermates. On the basis of these observations, she suggested that a possible cause of the lethality was the lack of mesoderm formation in the homozygous t^0 embryos.

Ultrastructural and light microscopic observations of t^6/t^6 embryos have shown, however, that these embryos, like t^{12}/t^{12}, t^{w32}/t^{w32} and t^{12}/t^{w32} embryos, have a range of phenocritical and phenolethal periods. The phenocritical period of the t^6/t^6 genome extends from the late blastocyst substages through the elongated egg-cylinder stage. The phenolethal period begins at the short egg-cylinder stage and extends through the elongated egg-cylinder stage, with most dying during the short egg-cylinder stage. Over one-half of the t^6/t^6 embryos can be identified as early as the late blastocyst substages by the presence of large electron-dense lipid droplets and by individual cell death. The viable short egg-cylinder-staged t^6/t^6 embryos can be distinguished from their phenotypically wild-type littermates at both the light and electron microscopic levels. Distinguishing characteristics of these embryos are aberrantly arranged endodermal cells, excessive cytoplasmic lipid and crystal-containing mitochondria. Such features are also characteristic of those homozygous t^6 embryos which are developmentally arrested at both the short and elongated egg-cylinder stages. All of the t^6/t^6 embryos die before mesoderm formation; and the lack of mesoderm formation, therefore, is the result rather than the cause of death (Nadijcka & Hillman, 1975).

These ultrastructural studies of t^n/t^n and t^n/t^{wn} embryos have shown that there are several characteristics which are common to all of the mutant embryos. Each of the genotypes results in cell death, the consequences of which are that the phenolethal periods cover a range of developmental stages. Also, each of the genotypes is characterized by excessive cytoplasmic lipid prior to blocked development. In addition, t^{w32}/t^{w32}, t^{12}/t^{w32} and t^6/t^6 embryos contain abnormal mitochondria prior to cell lethality.

It has been noted from studies on wild-type embryos that the ultrastructural configurational changes of the mitochondria which occur during the cleavage stages can be correlated with levels of aerobic metabolism (Ginsberg & Hillman, 1973*a*). The appearance in the mutant embryos, therefore, of both excessive lipid and variant forms of mitochondria suggests that one effect of the t^n alleles is on aerobic metabolism. This hypothesis is based on studies which show that the configuration of mitochondria can be correlated with the metabolic steady-state of the cell

(Hackenbrock, 1966, 1968, 1972; Hackenbrock, Rehn, Weinbach & LeMasters, 1971) and that excessive lipid synthesis and mitochondrial crystal deposition (characteristics of the mutant embryos) can occur either as a result of, or together with, excessive levels of total ATP and ATP metabolism (Atkinson, 1965; Brierley & Slautterback, 1964; Greenawalt, Rossi & Lehninger, 1964; Lehninger, Carafoli & Rossi, 1967; Trump, Croker & Mergner, 1971; Bonucci, Derenzini & Marinozzi, 1973; Newsholme & Start, 1973). Furthermore, it has also been shown that increased spermatozoan capacitation and fertilization can result from increased aerobic metabolism (Hamner & Williams, 1963). Since the meiotic drive of the t^n-bearing spermatozoa is not a result of extra postmeiotic mitoses of these spermatozoa (Bryson, 1944) and since it has been suggested that the drive is a result of a physiological advantage occurring between ejaculation and fertilization (Braden, 1958), we postulated that this increased transmission frequency could occur as a result of increased aerobic metabolism. The increased metabolism could result in either increased motility, viability, capacitation and/or ability to fertilize and could cause the t^n-bearing spermatozoa to be meiotically driven. On the basis of this reasoning, two series of experiments, the first on t^n/t^n embryos and the second on t^n spermatozoa, were undertaken in order to determine if t^n/t^n embryos and t^n-bearing spermatozoa differed from their wild-type counterparts in their levels of aerobic metabolism and if these differences could be attributable to the t^n alleles.

The first of these series of studies has shown that the homozygous t^{12}/t^{12} and t^{w32}/t^{w32} embryos have nonphysiological rates of ATP metabolism prior to developmental arrest. Two-cell t^n litters are characterized by lower ATP/ADP ratios, higher levels of total ATP and higher rates of net synthesis and turnover of ATP compared with correspondingly staged wild-type litters (Ginsberg & Hillman, 1972, 1973*a*, 1975). The increased levels of ATP metabolism continue until the 8-cell stage in the t^{w32}-containing litters and until the early morula stage in t^{12}-containing litters. At these respective stages, the levels of ATP metabolism in the mutant litters fall below control levels and remain low until the death of the homozygous t^n embryos. Following the death of the t^n/t^n embryos, the ATP metabolism of the mutant litters (then composed of only T^+/T^+ and T^+/t^n embryos) returns to control levels. The decline in the rates of ATP metabolism at different cleavage stages in t^{w32} and t^{12} litters can be correlated with the observation that these two t^n/t^n genotypes cause death at different cleavage stages. The decreases in ATP metabolism occur one stage prior to the respective phenolethal periods of most t^{w32}/t^{w32} (early morula) and t^{12}/t^{12} (late morula) embryos.

The increased levels of ATP metabolism can be correlated with the excessive lipid which characterizes the t^n homozygous embryos. It has been noted, in other systems, that acetyl CoA, in the presence of excessive total ATP, is diverted to fatty acid and lipid synthesis (Atkinson, 1965; Newsholme & Start, 1973). In the mutant embryos, the temporal coincidence of increased ATP metabolism and the presence of excessive lipid suggest the occurrence of a similar divergence of acetyl CoA. This hypothesis is supported by two recent studies on t^6/t^6 embryos. In the first of these, the initial characterization of the t^6 homozygotes is at the late blastocyst substages; at earlier stages, they cannot be distinguished from their littermates. At the late blastocyst, however, the homozygous embryos can be identified by the presence of excessive cytoplasmic lipid (Nadijcka & Hillman, 1975). In the second study, Ginsberg & Hillman (unpublished data) found that litters containing t^6/t^6 embryos do not differ from control litters in ATP metabolism prior to these late blastocyst substages. At these substages, however, t^6 litters exhibit excessive ATP metabolism. Therefore, the first distinctive phenotypic characteristic (excessive lipid) of the t^6/t^6 genome occurs at a time coincident with increased ATP metabolism.

The second series of experiments to determine the effect of the t^n and t^{wn} alleles on aerobic metabolism utilized epididymal spermatozoa from T^+/t^{w32}, T^+/t^6, T^+/T^+ and T^+/T males (Ginsberg & Hillman, 1973*b*, 1974). In addition, the t^{w32} allele was placed on a *BALB/c* background and backcrossed to *BALB/c* for five generations. These crosses yielded both control $BALB/cT^+//BALB/cT^+$ and experimental $BALB/cT^+//BALB/ct^{w32}$ males. These males are 97% isogenic for all loci except those on chromosome 17, which itself should be isogenic for loci greater than ± 20 map units from the *T*-locus (Crow & Kimura, 1970). Two energy parameters were studied: comparative NADH/NAD ratios and O_2 uptake (ZO_2) using defined media supplemented with specific substrates.

Our findings show that there is an inverse relationship between the NADH/NAD ratios and the frequency of spermatozoan transmission (Table 1). In each case, the NADH/NAD ratio observed in pooled spermatozoa from T^+/t^n males is lower than that in pooled spermatozoa from T^+/T^+ or T^+/T males. Both of the latter samples have equivalent ratios and these are consistent with ratios reported for other mammalian spermatozoa (Brooks & Mann, 1971). The T^+/t^6 sample has the lowest NADH/NAD ratio of those tested and the t^6 allele is transmitted in the highest frequency. The lowered ratio in spermatozoa from T^+/t^n males is, in each case, a result of a lower NADH and a higher NAD concentration than is present in either the T^+/T^+ samples or in samples containing the nonmeiotically driven T-allele. These findings suggest an increased rate of energy metabolism in the pooled samples containing t^n spermatozoa.

The oxygen uptake studies also support the hypothesis of an increased rate of energy metabolism. The data show that both T^+/T^+ and T^+/T samples utilize substrates equally, whereas the t^n-containing samples show a ZO_2 increase over comparable T^+/T^+ and T^+/T samples in the presence of fructose, lactate, pyruvate and succinate (Table 2). The T^+/T^+ and T^+/T samples have a higher ZO_2 than the samples containing recessive mutants only in the presence of glucose. Table 2 also contains the results of the O_2 uptake studies on the spermatozoa from the progeny of back-crossed $BALB/ct^{w32}$ mice. The data show that the ZO_2 levels of the spermatozoa from $BALB/cT^+//BALB/ct^{w32}$ and the spermatozoa from T^+/t^{w32} males are the same in the presence of either lactate or pyruvate (the two substrates tested). Similarly, the spermatozoa from the back-crossed $BALB/cT^+//BALB/cT^+$ littermates do not differ significantly from either T^+/T^+ or T^+/T pooled spermatozoa in their utilization of these same substrates. The results from these studies show, therefore, that the t^n-containing spermatozoan samples have an increased rate of aerobic metabolism in the presence of specific substrates and strongly indicate that the increased metabolism is a result of the presence of the meiotically driven allele.

As noted, it has been found that the abilities of spermatozoa to become capacitated and to fertilize are proportionally related to the levels of aerobic respiration. These levels are, in turn, dependent upon the length of time spermatozoa remain in the female reproductive tract prior to fertilization (Hamner & Williams, 1963). For a higher rate of respiration to be effective in increasing capacitation and fertilization *in vivo*, spermatozoa must first be incubated in the female reproductive tract for a period longer than two hours and less than seven hours. Optimum capacity is reached after a six-hour incubation period.

Several reports state that delayed matings significantly decrease the transmission frequency of t-bearing spermatozoa which are normally driven in spontaneous matings (Braden, 1958; Yanagisawa, Dunn & Bennett, 1961; Erickson, 1973). In these experiments the spermatozoa resided in the female for as little as one hour before fertilization (Braden, 1958). As a consequence of such a short time within the female reproductive tract, the increased respiration by the mutant-bearing spermatozoa over the T^+-bearing spermatozoa would not be effective in promoting non-random fertilizations. Under these conditions, the t-bearing spermatozoa would not be driven and ova would be fertilized at random by T^+ and t^n

Table 1. *A comparison of the transmission frequencies and NADH/NAD ratio of normal and mutant epididymal mouse spermatozoa*

Genotype	Number of trials	Frequency of transmission	μg NADH/10^8 sperm	μg NAD/10^8 sperm	NADH/NAD ratio
T^+/T^+	8	0.50:0.50	17.9 ± 0.47	21.6 ± 0.18	0.80 ± 0.01
T^+/T	7	0.50:0.50	17.9 ± 0.26	21.8 ± 0.12	0.81 ± 0.01
T^+/t^6	7	0.10:0.90	15.1 ± 0.27	23.9 ± 0.52	$0.61^* \pm 0.01$
T^+/t^{w32}	7	0.22:0.78	16.5 ± 0.15	22.4 ± 0.11	$0.68^* \pm 0.02$

* Significant difference from normal controls at 5 % level.
From Ginsberg & Hillman (1974).

Table 2. *A comparison of the oxygen uptake (ZO_2)† by pooled mouse spermatozoa in the presence of different subtrates*

Substrate (10 mM)	T^+/T^+	T^+/T	T^+/t^6	T^+/t^{w32}	$BALB/cT^+$ $BALB/ct^{w32}$	$BALB/cT^+$ $BALB/cT^+$
Glucose	11.94 ± 0.42	11.15 ± 0.18	$7.13 \pm 0.42^*$	$8.58 \pm 0.27^*$	–	–
Fructose	12.46 ± 0.54	12.40 ± 0.21	$14.46 \pm 0.10^*$	$14.20 \pm 0.32^*$	–	–
Lactate	15.25 ± 0.44	15.73 ± 0.79	$18.33 \pm 0.65^*$	$18.30 \pm 0.53^*$	$18.15 \pm 0.46^*$	15.71 ± 0.21
Pyruvate	10.00 ± 0.28	9.00 ± 0.40	$14.44 \pm 0.64^*$	$13.37 \pm 0.55^*$	$13.33 \pm 0.61^*$	9.65 ± 0.35
Succinate	16.11 ± 2.70	17.86 ± 3.20	$30.47 \pm 1.30^*$	$30.85 \pm 2.10^*$	–	–

* Significant difference from control at 5% level.
† Values for ZO_2 are measured in terms of nmol O_2/h · 10^6 spermatozoa.
From Ginsberg & Hillman (1974).

spermatozoa. Under conditions of spontaneous mating, during which the spermatozoa remain in the uterus and oviduct for several hours, the *t*-bearing spermatozoa could attain comparatively higher levels of respiration. These higher levels of respiration would result from the t^n-bearing spermatozoa's increased ability to utilize specific available substrates.

In support of our hypothesis is evidence showing that epididymal spermatozoa can be capacitated and can fertilize mouse ova *in vitro*, and that the percentage of capacitation and fertilization is substrate-dependent (Miyamoto & Chang, 1973). Although glucose is present in oviducal fluid (Bishop, 1957; Hamner & Williams, 1965; Holmdahl & Mastroianni, 1965) and even though our studies show that this substrate supports a higher rate of respiration in pooled spermatozoa from T^+/T^+ and T^+/T males than from pooled spermatozoa from T^+/t^n males, glucose alone does not support capacitation and fertilization *in vitro* (Miyamoto & Chang, 1973). This indicates that oviducal glucose is not normally used by spermatozoa as a metabolic substrate. Lactate and pyruvate, however, are both present in oviducal fluid (Bishop, 1957; Hamner & Williams, 1965; Holmdahl & Mastroianni, 1965; David, Brackett, Garcia & Mastroianni, 1969) and have been shown to promote capacitation and fertilization *in vitro* (Miyamoto & Chang, 1973). Both of these substrates increase the rate of respiration of the mutant pooled spermatozoa significantly above that of pooled control spermatozoa. Fructose, present in seminal fluid (Mann, 1946), causes a slight increase in O_2 uptake in spermatozoan samples from T^+/t^n males, while the most significant increase is found when succinate is the sole energy source. This latter substrate, however, has not been reported in oviducal fluid.

A possible explanation for the excessive levels of ATP metabolism in early cleavage stage t^n/t^n embryos can be found in these latter studies on t^n-bearing spermatozoa. Embryonic ATP metabolism was assayed using embryos developing *in vitro* in Brinster's medium (Brinster, 1963). This medium contains both pyruvate and lactate, two of the energy sources which support the continued development of 2-cell mouse embryos *in vitro* (Brinster, 1965). As noted above, these substrates are also found in mouse oviducts. If t^n/t^n embryos, like t^n-bearing spermatozoa, utilize pyruvate and lactate, both *in vivo* and *in vitro*, more efficiently than do correspondingly staged T^+/T^+ embryos, the increased utilization could result in the increased ATP metabolism found in the t^{12}/t^{12} and t^{w32}/t^{w32} embryos. The metabolic imbalance caused by this excessive carbohydrate oxidation would result in the excessive levels of ATP metabolism. Additional studies are now in progress to delimit the primary effect of the mutant t^n genome on aerobic metabolism.

This work was supported by United States Public Health Research Grant No. HD 00827. I would like to thank Marie Morris and Geraldine Wileman for their technical assistance.

REFERENCES

Atkinson, D. E. (1965). Biological feedback control at the molecular level. *Science*, **150**, 851–7.

Bennett, D. (1964). Abnormalities associated with a chromosome region in the mouse. II. Embryological effects of lethal alleles in the *t*-region. *Science*, **144**, 263–7.

Bennett, D. (1965). The karyotype of the mouse with identification of a translocation. *Proceedings of the National Academy of Sciences, USA*, **53**, 730–7.

Bennett, D. & Dunn, L. C. (1964). Repeated occurrences in the mouse of lethal alleles of the same complementation group. *Genetics*, **49**, 949–58.

Bishop, D. W. (1957). Metabolic conditions within the oviduct of the rabbit. *International Journal of Fertility*, **2**, 11–22.

Bonucci, E., Derenzini, M. & Marinozzi, V. (1973). The organic–inorganic relationship in calcified mitochondria. *Journal of Cell Biology*, **59**, 185–211.

Braden, A. W. H. (1958). Influence of time of mating on the segregation ratio of alleles at the *T* locus in the house mouse. *Nature, London*, **181**, 786–7.

Brierley, G. P. & Slautterback, D. B. (1964). Studies on ion transport. IV. An electron microscope study of the accumulation of Ca^{2+} and inorganic phosphate by heart mitochondria. *Biochimica et Biophysica Acta*, **82**, 183–6.

Brinster, R. L. (1963). A method for *in vitro* cultivation of mouse ova from two-cell to blastocyst. *Experimental Cell Research*, **32**, 205–7.

Brinster, R. L. (1965). Studies on the development of mouse embryos *in vitro*. II. The effect of energy source. *Journal of Experimental Zoology*, **158**, 59–68.

Brooks, D. E. & Mann, T. (1971). NAD in the metabolism of motile spermatozoa. *Nature, London*, **234**, 301–2.

Bryson, V. (1944). Spermatogenesis and fertility in *Mus musculus* as affected by factors at the *T* locus. *Journal of Morphology*, **74**, 131–79.

Calarco, P. G. & Brown, E. H. (1968). Cytological and ultrastructural comparisons of t^{12}/t^{12} and normal mouse morulae. *Journal of Experimental Zoology*, **168**, 169–86.

Chesley, P. (1932). Lethal action in the short-tailed mutation in the house mouse. *Proceedings of the Society for Experimental Biology and Medicine*, **29**, 437–8.

Chesley, P. & Dunn, L. C. (1936). The inheritance of taillessness (anury) in the house mouse. *Genetics*, **21**, 525–36.

Crow, J. F. & Kimura, M. (1970). *An Introduction to Population Genetics Theory*. New York: Harper & Row.

David, A., Brackett, B. G., Garcia, C.-R. & Mastroianni, L., Jr (1969). Composition of rabbit oviduct fluid in ligated segments of the Fallopian tube. *Journal of Reproduction and Fertility*, **19**, 285–9.

Dobrovolskaia-Zavadskaia, N. (1927). Sur la mortification spontanée de la queue chez la souris nouveau-née et sur l'existence d'un caractère (facteur) héréditaire 'non-viable'. *Comptes rendus des Séances de la Société de Biologie, Paris*, **97**, 114–16.

Dobrovolskaia-Zavadskaia, N. (1932). A morphological study of the short-tail in mice. *Proceedings of the VI International Congress of Genetics*, **2**, 43–4 (abstract).

Dunn, L. C. & Gluecksohn-Schoenheimer, S. (1950). Repeated mutations in one area of a mouse chromosome. *Genetics*, **36**, 233–7.

Erickson, R. P. (1973). Haploid gene expression versus meiotic drive: the relevance of intercellular bridges during spermatogenesis. *Nature, London*, **243**, 210–12.

Flaherty, L., Bennett, D. & Graef, S. (1972). Genetic control of nucleolar number in mouse. *Experimental Cell Research*, **70**, 13–16.

Ginsberg, L. & Hillman, N. (1972). ATP synthesis in normal and mutant cleavage-staged mouse embryos. *Genetics*, **71**, S20.

Ginsberg, L. & Hillman, N. (1973*a*). ATP metabolism in cleavage-staged mouse embryos. *Journal of Embryology and Experimental Morphology*, **30**, 267–82.

Ginsberg, L. & Hillman, N. (1973*b*). Energy levels in *t*-mutant and wild-type mouse sperm. *Genetics*, **74**, S94.

Ginsberg, L. & Hillman, N. (1974). Meiotic drive in t^n-bearing mouse spermatozoa: a relationship between aerobic respiration and transmission frequency. *Journal of Reproduction and Fertility*, **38**, 157–63.

Ginsberg, L. & Hillman, N. (1975). ATP metabolism in t^n/t^n mouse embryos. *Journal of Embryology and Experimental Morphology*, in press.

Gluecksohn-Schoenheimer, S. (1940). The effect of an early lethal (t^0) in the house mouse. *Genetics*, **25**, 391–400.

Greenawalt, J. W., Rossi, C. S. & Lehninger, A. L. (1964). Effect of active accumulation of calcium and phosphate ions on the structure of rat liver mitochondria. *Journal of Cell Biology*, **23**, 21–38.

Hackenbrock, C. R. (1966). Ultrastructural bases for metabolically linked mechanical activity in mitochondria. I. Reversible ultrastructural changes with change in metabolic steady state in isolated liver mitochondria. *Journal of Cell Biology*, **30**, 269–97.

Hackenbrock, C. R. (1968). Ultrastructural bases for metabolically linked mechanical activity in mitochondria. II. Electron transport-linked ultrastructural transformations in mitochondria. *Journal of Cell Biology*, **37**, 345–369.

Hackenbrock, C. R. (1972). Energy-linked ultrastructural transformations in isolated liver mitochondria and mitoplasts. *Journal of Cell Biology*, **53**, 450–65.

Hackenbrock, C. R., Rehn, T. G., Weinbach, E. C. & LeMasters, J. J. (1971). Oxidative phosphorylation and ultrastructural transformation in mitochondria in the intact ascites tumor cell. *Journal of Cell Biology*, **51**, 123–37.

Hamner, C. E. & Williams, W. L. (1963). Effect of the female reproductive tract on sperm metabolism in the rabbit and fowl. *Journal of Reproduction and Fertility*, **5**, 143–50.

Hamner, C. E. & Williams, W. L. (1965). Composition of rabbit oviduct secretions. *Fertility and Sterility*, **16**, 170–6.

Hillman, N. (1972). Autoradiographic studies of t^{12}/t^{12} mouse embryos. *American Journal of Anatomy*, **134**, 411–24.

Hillman, N. & Hillman, R. (1975). Ultrastructural studies of t^{w32}/t^{w32} mouse embryos. *Journal of Embryology and Experimental Morphology*, in press.

Hillman, N., Hillman, R. & Wileman, G. (1970). Ultrastructural studies of cleavage stage t^{12}/t^{12} mouse embryos. *American Journal of Anatomy*, **128**, 311–40.

Hillman, N. & Tasca, R. J. (1969). Ultrastructural and autoradiographic studies of mouse cleavage stages. *American Journal of Anatomy*, **126**, 151–74.

Hillman, N. & Tasca, R. J. (1973). Synthesis of RNA in t^{12}/t^{12} mouse embryos. *Journal of Reproduction and Fertility*, **33**, 501–6.

Holmdahl, T. H. & Mastroianni, L., Jr (1965). Continuous collection of rabbit oviduct secretions at low temperatures. *Fertility and Sterility*, **16**, 587–95.

IVANYI, D. (1971). Genetic studies on mouse nucleoli. *Experimental Cell Research*, **64**, 240–2.

LEHNINGER, A. L., CARAFOLI, E. & ROSSI, C. S. (1967). Energy-linked ion movements in mitochondrial systems. In *Advances in Enzymology*, ed. F. F. Nord, pp. 259–320. New York: John Wiley & Sons.

LEWONTIN, R. C. (1962). Interdeme selection controlling a polymorphism in the house mouse. *American Naturalist*, **96**, 65–78

MANN, T. (1946). Studies on the metabolism of semen. III. Fructose as a normal constituent of seminal plasma. Site of formation and function of fructose in semen. *Biochemical Journal*, **40**, 481–91.

MINTZ, B. (1964*a*). Gene expression in the morula stage of mouse embryos, as observed during development of t^{12}/t^{12} lethal mutants *in vitro*. *Journal of Experimental Zoology*, **157**, 267–72.

MINTZ, B. (1964*b*). Formation of genetically mosaic mouse embryos, and early development of 'lethal ($t^{12}-t^{12}$)–normal' mosaics. *Journal of Experimental Zoology*, **157**, 273–92.

MINTZ, B. (1964*c*). Synthetic processes and early development in the mammalian egg. *Journal of Experimental Zoology*, **157**, 85–100.

MIYAMOTO, H. & CHANG, M. C. (1973). The importance of serum albumen and metabolic intermediates for capacitation of spermatozoa and fertilization of mouse eggs *in vitro*. *Journal of Reproduction and Fertility*, **32**, 193–205.

NADIJCKA, M. & HILLMAN, N. (1973). The lethal expression of the t^6/t^6 genotype in mouse embryos. *Genetics*, **74**, S189.

NADIJCKA, M. & HILLMAN, N. (1975). Studies of t^6/t^6 mouse embryos. *Journal of Embryology and Experimental Morphology*, in press.

NEWSHOLME, E. A. & START, C. (1973). *Regulation in Metabolism*. New York: John Wiley & Sons.

PIKÓ, L. & CHASE, D. G. (1973). Role of the mitochondrial genome during early development in mice. Effects of ethidium bromide and chloramphenicol. *Journal of Cell Biology*, **58**, 357–78.

SANDLER, L. & NOVITSKI, E. (1957). Meiotic drive as an evolutionary force. *American Naturalist*, **91**, 105–10.

SMITH, L. J. (1956). A morphological and histochemical investigation of a pre-implantation lethal (t^{12}) in the house mouse. *Journal of Experimental Zoology*, **157**, 267–72.

TASCA, R. J. & HILLMAN, N. (1970). Effects of actinomycin D and cycloheximide on RNA and protein synthesis in cleavage stage mouse embryos. *Nature, London*, **225**, 1022–5.

TRUMP, B. F., CROKER, B. P., Jr & MERGNER, W. J. (1971). The role of energy metabolism, ion, and water shifts in the pathogenesis of cell injury. In *Cell Membranes: Biological and Pathological Aspects*, ed. G. W. Richter & D. G. Scarpelli, pp. 84–128. Baltimore: Williams & Wilkins.

YANAGISAWA, K., DUNN, L. C. & BENNETT, D. (1961). On the mechanism of abnormal transmission ratios at *T* locus in the house mouse. *Genetics*, **46**, 1635–44.

EXPLANATION OF PLATES

PLATE I

(*a*) A portion of a cell from an 8-cell t^{12}/t^{w32} embryo. Note the presence of the two nuclear fibrillo-granular bodies (arrows).

(*b*) A portion of a cell from an early morula t^{12}/t^{w32} embryo. This section is throu the nucleus which contains lipid droplets (arrows). Nuclear lipid droplets d tinguish the heterozygotes from their littermates at each cleavage stage.
(*c*) A developmentally arrested t^{12}/t^{w32} early morula. Note the presence of t single large, electron-dense lipid droplets and the clusters of lipid droplets with the cytoplasm of the blastomeres.

PLATE 2

(*a*) This micrograph shows mitochondria (arrows) of a T^{+}/T^{+} 2-cell embr The mitochondria contain a condensed matrix and few cristae.
(*b*) The mitochondria (arrows) of 2-cell t^{12}/t^{w32} embryos contain a matrix whi is less electron-dense than the matrix of 2-cell T^{+}/T^{+} mitochondria (cf. Plate 2 Variant forms of mitochondria distinguish the t^{12}/t^{w32} embryos at each cleava stage.

PLATE I

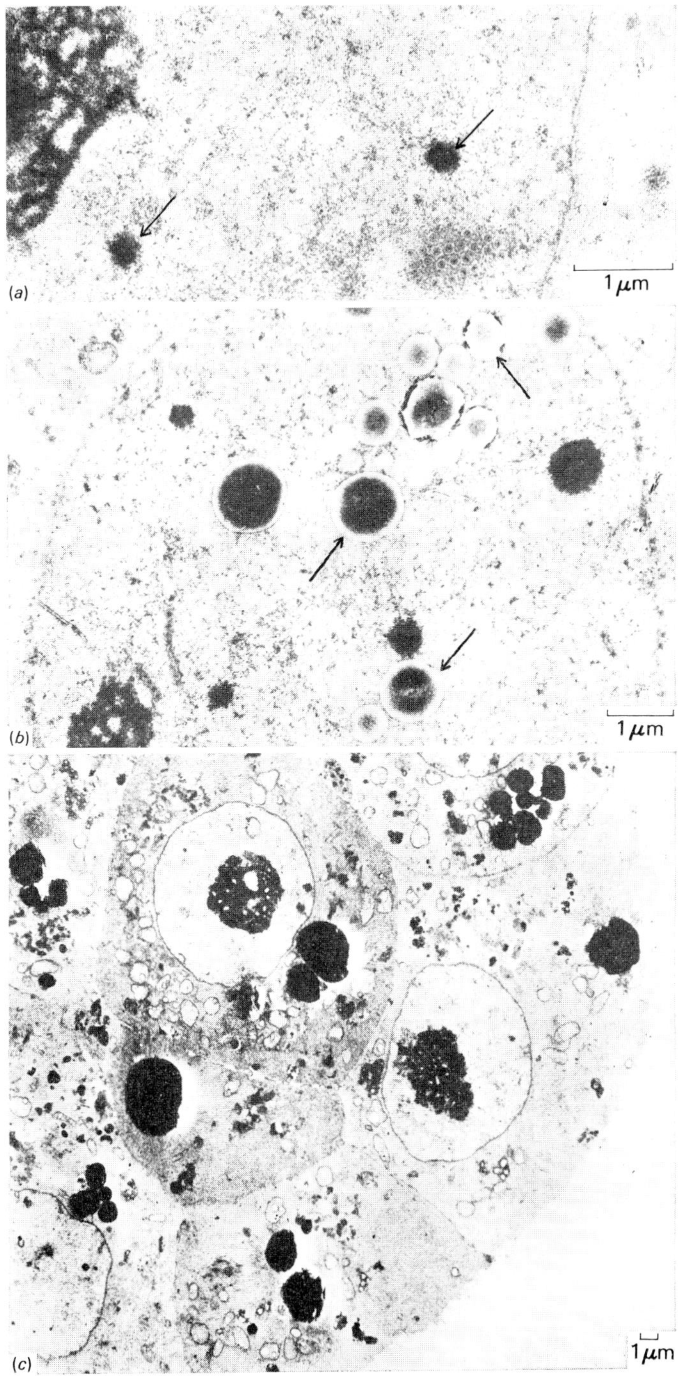

(*Facing p. 206*)

PLATE 2

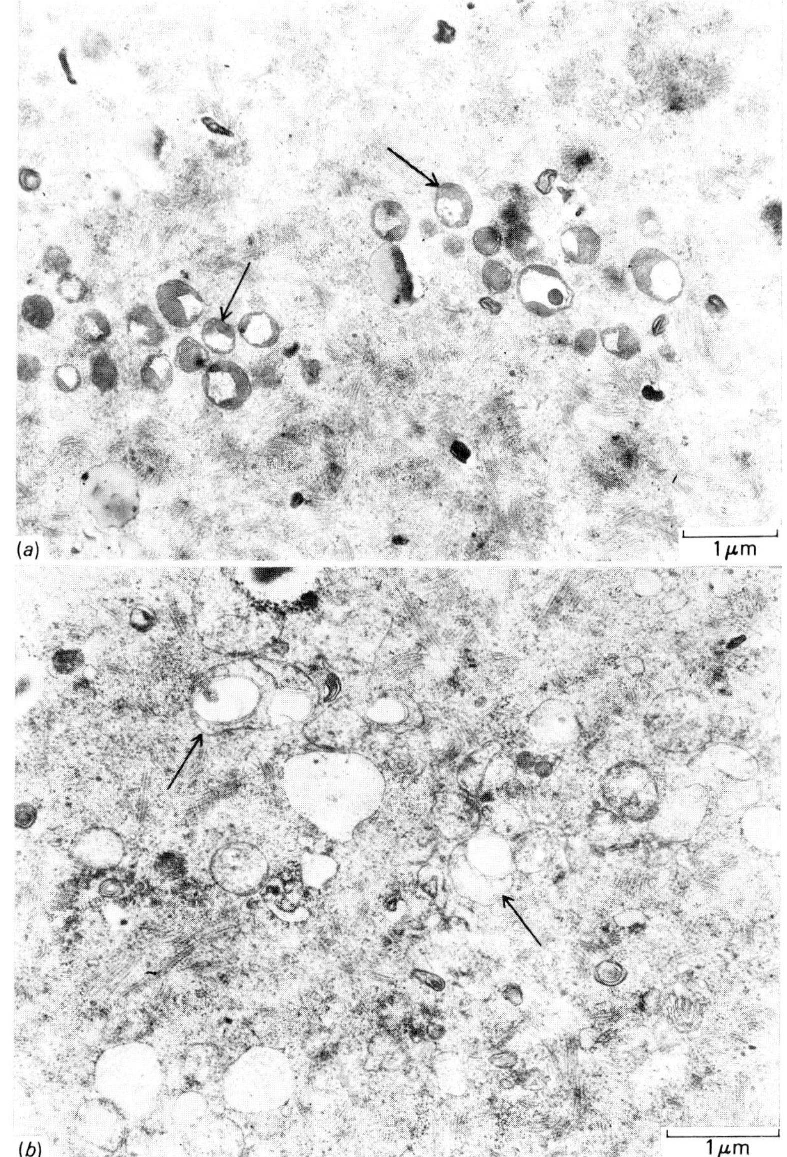

T-LOCUS MUTANTS: SUGGESTIONS FOR THE CONTROL OF EARLY EMBRYONIC ORGANIZATION THROUGH CELL SURFACE COMPONENTS

BY D. BENNETT

Cornell University Medical College,
New York, NY 10021, USA

One approach to the analysis of early embryonic development and cellular differentiation in mammals has been the study of mutant genes which impair specific processes or morphogenetic events during embryogenesis. The theoretical advantage to such systems is that, at least initially, each mutant gene presumably produces a unitary and ultimately definable biochemical defect. The work of many investigators over the past forty years (see Bennett, 1964; Gluecksohn-Waelsch & Erickson, 1970, for review) has shown that a restricted region within one chromosome of the mouse – the complex *T*-locus – appears to contain an important center controlling (*a*) fundamental organizational events essential for the establishment and further differentiation of the three germ layers of the early embryo prior to definitive organogenesis, (*b*) production and function of spermatozoa, and (*c*) crossing over and reorganization of genetical properties within the region. Furthermore, variant alleles at this locus have been found in most of the many populations of wild mice which have been tested, so it is clear that polymorphism at this locus plays some, as yet undefined, role in population dynamics, and that this polymorphism may have an important physiological role (Dunn, 1964).

The *T*-locus is defined by a dominant mutation, *T*, which shortens the tail of heterozygotes, and is lethal in homozygotes at about mid-gestation. *T* serves as diagnostic marker for identifying recessive mutant alleles (*t*) at the locus, because *t*-alleles interact with *T* to produce a distinctive tailless phenotype. This interaction has led to the detection of a number of different recessive mutations; the phenotype of the homozygote can be used to broadly categorize the recessive alleles as lethals, semilethals, and viables. The lethal alleles which have so far been well studied fall into five complementation groups designated t^0, t^9, t^{12}, t^{w1}, t^{w5}. (Members of different complementation groups are easily defined when tailless parents are crossed ($T/t^x \times T/t^y$); if t^x and t^y are members of the same complementation group, only tailless progeny result; if t^x and t^y differ, normal-tailed

(t^x/t^y) progeny are obtained.) Mutants in any one complementation group produce similar abnormalities in the homozygous condition with comparable lethal periods, and these effects are different from those of any other complementation group.

A detailed discussion of the morphogenetic defects produced by these mutants will not be given here (see Bennett, 1964, for review). Briefly, the abnormalities produced by members of each complementation group can be characterized as resulting from interference with apparent switch points that occur as derivatives of the ectoderm and mesoderm diverge along separate pathways during early embryonic development. Furthermore, the embryological defect produced by each mutant suggests an inability of particular groups of cells either to reach their normal location or to maintain their viability once they have attained their proper destinations. Thus, each of these mutants appears to affect early processes of organization upon which subsequent morphogenesis is dependent, and it appears that the *T*-locus must be considered a major region controlling the first determinative steps in the embryo. Our tentative interpretation is that genes in the *T*-region may switch on sequentially as development proceeds (Bennett, Boyse & Old, 1972).

At the same time it is clear that the orderly events of embryonic morphogenesis are governed at least in part by interactions between cells and/or between cells and their immediate environment. Much fruitful work has demonstrated that this kind of interaction is mediated through specific components on the cell surface that either provide recognition devices permitting cells to identify one another as similar or different, or serve as receptors for extracellular substances which modify or stabilize their behavior (See Moscona, 1973, for review). The homozygous effects of lethal *t*-alleles can be interpreted as being caused by improper cell–cell recognition and were therefore used to test the contention that the *t*-region specifies cell surface components which are important mediators of cellular interrelationships during embryogenesis. We have so far obtained both serological and morphological data which support this hypothesis.

The specific experimental approach leading into the serological analysis of *T*-locus mutants in terms of cell surface components was dictated by the unique effects of *t*-alleles on spermatozoa, since these mutants have clearly defined effects on both the function and production of spermatozoa.

Male (but not female) mice heterozygous for most *t*-alleles transmit the *t*-allele to their progeny in proportions which do not conform to Mendel's rules. In general, lethal and semilethal alleles are transmitted in higher than normal proportions – ranging from 75–99% depending on the allele concerned – whereas viable *t*-mutants are often transmitted in lower than expected proportions, or may show no distortion at all. In all cases, however, normal numbers of spermatozoa of normal gross morphology are produced. Therefore, it appears that meiosis followed by haploid gene expression during spermiogenesis in *t*-heterozygotes produces two populations of spermatozoa which are genetically and phenotypically unlike and possess different potentials for fertilization (Yanagisawa, Dunn & Bennett, 1961). It has been suggested that the difference between these two classes of spermatozoa may reside in disparities in genetically controlled cell surface components, since the spermatozoal membrane is, in the last analysis, presumably the essential element in effecting fertilization.

Thus, the effects of *t*-alleles on both embryogenesis and spermatogenesis may be taken to imply that the role of these genes is to specify cell surface components essential for the appropriate organization of the embryo and for the function of spermatozoa.

This idea has been tested. The initial goal was to determine simply if mutant alleles at the *T*-locus did in fact determine cell surface components that were serologically detectable on spermatozoa. To this end, allogeneic immunizations with spermatozoa carrying *T* and four unique *t*-alleles (t^0, t^{12}, t^{w1} and t^{w5}) into $+^t/+^t$ recipients were performed. Antisera cytotoxic for spermatozoa were obtained in all cases and after appropriate absorption with spermatozoa of recipient ($+^t/+^t$) type to remove non-specific spermatozoa auto-antibody these antisera were shown to be specific for spermatozoa carrying the immunizing *t*-allele (Bennett, Goldberg, Dunn & Boyse, 1972; Yanagisawa *et al.*, 1974). Further tests bore out our original hypothesis that only spermatozoa and embryonic cells carry the relevant antigens, since neither by direct cytotoxicity test nor by absorption procedures have any cells of the adult (except spermatozoa) been shown to carry antigens specified by mutant genes at the *T*-locus. Thus, *T*-locus antigens fall into the category of auto-antigens.

The apparent auto-antigenic nature of *t*-antigens led to an attempt at syngeneic immunizations with spermatozoa from genetically normal mice on the theory that a product of a '+' allele at the *T*-locus could be recognized by the same means that had been used to define the antigens specified by mutant alleles. And in fact serum from mice immunized in this way, after absorption with spermatozoa from a battery of T/t^x animals, proved to have specificity only for genotypes carrying at least one wild-type allele.

Thus we can recognize on sperm antigens specified by both '*T*' and '*t*' mutants, and by a wild-type allele at the *T*-locus. We suspect that similar antigens are displayed by cells in the embryo at the critical periods that are marked by the effects of mutant *t*-alleles in homozygous condition; and we suspect further that these antigens exist transiently – at discrete and appropriate stages during embryogenesis – since they are not present in the adult. The obvious next step is to attempt the serological demonstration of mutant cell components on the specific normal cell types in defective homozygous embryos. A variety of technical reasons have so far prevented us from achieving this goal, the greatest of which is that visual marking techniques applied to tissue sections are essential for this purpose and these techniques are at best capricious and at worst ineffectual.

Nevertheless, other types of experiments have shown that in one instance at least *T*-locus antigens can be detected on cells of what we consider to be the appropriate embryonic stage. This approach has involved the use of teratocarcinoma cell lines. Although the properties of teratocarcinomas and their use in developmental studies are discussed in detail elsewhere (Jacob, this volume), it may be worthwhile to reiterate some points salient to the argument being presented here. Naturally occurring teratocarcinomas are embryonic tumours which usually arise from abnormal germ cells. For our purposes their main advantage is that although at origin they contain a proliferating multipotential primitive stem cell component, they also generate cells which prove capable of further differentiation. In some cases the differentiated products appear to lose the element of malignancy, and therefore form benign histologically-normal teratomas which cease to proliferate, and likewise cease to be of interest to us. In other cases, however, cells differentiating from the primitive stem cell pool appear to become blocked in their capacity for further differentiation after reaching only a limited and still primitive stage of specialization. Lines of cells like this maintain their malignancy and their proliferative ability and can therefore be propagated indefinitely. In addition, some lines of originally primitive and multipotential cells stabilize as primitive cell lines which appear to have lost their ability to give rise to any differentiated derivatives. Thus, appropriate passages and manipulation either *in vivo* or *in vitro* of these various derivatives of teratocarcinomas can provide relatively homogeneous populations of embryonic cells at particular stages of differentiation (see Stevens, 1967, for review).

We have shown that antisera prepared against a primitive teratocarcinoma cell line (F9) in syngeneic hosts were highly cytotoxic for the immunizing cell type. This fact alone indicates the presence of an embryonic antigen on F9 cells, since the syngeneic adult is able to respond with antibody production against cells which are genetically identical; this corroborates our idea of transient expression of antigens during embryonic development. Furthermore, anti-F9 antiserum was cytotoxic as well for other primitive cell lines but completely negative for partially differentiated cell lines derived from the same teratocarcinoma (Artzt *et al.*, 1973). This again fits with our ideas of the transient and sequential expression of antigens that may play roles in embryonic morphogenesis.

Furthermore, normal mouse embryos at various cleavage stages (2-, 4-, 8-cells, and morula) have also been found to react with anti-F9 antiserum, thus demonstrating that F9 cells share, as would be suspected from their primitive and undifferentiated nature, an antigen that characterizes cleavage embryos. Interestingly, however, anti-F9 antiserum is negative in cytotoxicity tests on intact blastocysts (J. Caldwell, unpublished observations). This suggests that the temporal constraints on the expression of antigens during differentiation may be very narrow indeed, since the transition from morula to blastocyst occurs over only a few hours. We do not yet have information on the question whether the multipotential inner cell mass cells of the blastocyst display, as our reasoning would lead us to believe, the F9 antigen.

Reasons have been given above for considering genes at the *T*-locus to be of major importance in controlling events of early differentiation. It seemed necessary, therefore, to examine the possibility that the antigen shared by primitive teratocarcinoma cells and cleaving embryos was controlled by an allele at this locus. Since we knew *T*-locus antigens to be present on spermatozoa this possibility could be broadly tested by examining whether anti-F9 antiserum reacted with spermatozoa. This proved to be the case; in absorption tests wild-type spermatozoa were capable of removing all activity from anti-F9 antiserum.

Thus the antigen present on primitive cells was also present on spermatozoa; and this distribution coincided with the distribution known (in the case of spermatozoa) and postulated (in the case of embryos) to be characteristic of *T*-locus antigens. These findings were compatible with an hypothesis that the F9 antigen was controlled by a '+' gene at the *T*-locus. An obvious possibility for testing this hypothesis presented itself because of two known facts. First of all, one of the recessive *t*-mutants (t^{12}) curtails development at the morula stage, and therefore the '+' allele of t^{12} could be considered an outstanding candidate for the genetic source of F9 antigen. Second, it was known that, as is the usual case in genetically controlled antigens, the expression of '+' and '*t*' alleles on spermatozoa occurs in a co-dominant fashion, so that spermatozoa from $+^t/+^t$ males carry twice as much $+^t$ antigen as do spermatozoa from $+/t^{12}$ males. Therefore it was clear that since spermatozoa from $+^t/+^t$ males were capable of absorbing out all activity from anti-F9 antiserum, if our hypothesis were correct spermatozoa from $+/t^{12}$ males should be only half as effective in removing activity. Absorptions done in a quantitative way, with measured numbers of spermatozoa of the two genotypes, with anti-F9 antiserum of known titer tested under standard conditions showed that this is in fact the case. This evidence indicates that the F9 antigen is coded for by '$+^{t12}$' gene at the *T*-locus. This finding vindicates our original hypothesis and lends assurance that other antigens specified by other *t*-genes will be detected and found operative at the expected locations at the appropriate embryonic stages (Artzt, Bennett & Jacob, 1974).

Although it seems obvious from the work discussed above, and from many other uncited experiments, that the cell surface is crucial in differentiation, the actual way in which cell surface components play a role in morphogenesis is not clear at present. On theoretical grounds – but with mechanisms unknown – membrane elements may serve as receptor sites for humoral, cellular or 'micro-environmental' factors which may either elicit new responses from the genome, evoke pre-programmed genetic responses, or simply passively control the behavior of cells by governing their ability to recognize these or still other factors in their environment. Since the biochemistry of these no doubt important events in cell interactions is, although much explored, so little understood, we have attempted to bypass this arena for the moment, and look instead for ultrastructural evidence of membrane-mediated interplay between cells in developing embryos and for interactions between membranes and intracellular organelles in differentiating cells. We have chosen to do this by the study of developmental anomalies produced by mutant genes at the *T*-locus.

One of the lethal *t*-alleles, t^9, produces a unique abnormality of gastrulation which seemed particularly well-suited for our purpose. Homozygotes for t^9 develop normally until the time of the primitive streak formation; however, once this process begins the primitive streak becomes progressively enlarged relative to the dimensions of the embryo, and conversely very few mesoderm cells migrate out to assume their typical location between endoderm and ectoderm. *In utero* these embryos die by about ten days gestation, presumably because their deficient supply of mesodermal elements prevents them from establishing an effective circulatory system (Bennett & Dunn, 1960). Furthermore, the 'mesoderm' that does appear to form in these embryos is apparently defective both in proliferative ability and in playing a role in epithelial–mesenchymal interactions, since when t^9/t^9 embryos are transplanted under conditions where they can continue to grow, they form teratoma-like growths (often resembling neuro-epithelial malignancies), in which can be detected neither tissue of mesodermal type nor organs known to be dependent on mesenchymal interaction (Artzt & Bennett, 1972).

It appeared to us that the enlarged primitive streak and deficient mesoderm seen in these mutants might result from the immobilization of primitive streak and 'mesoderm' cells during gastrulation, and that this paralysis might in turn result from the inability of these cells to recognize their immediate environment or make appropriate responses to it. Fine-structural studies with the electron microscope have provided morphological evidence for abnormalities in these cells which can hardly be interpreted other than to indicate membrane-controlled defects in cell–cell interaction. 'Mesodermal' cells of the 8-day mutant embryo are arranged in a sheet-like configuration, rather than the stellate network seen in normal embryos, and the abnormal cells usually display bulging lobate pseudopodia instead of the filiform pseudopodia typical of normal mesoderm cells. Furthermore, the lobopodia do not contain the subsurface microfilaments characteristic of filopodia. Most importantly, although cellular junctions of some kind are invariably found wherever filopodia from two different cells abut on one another, they are essentially never seen between adjacent lobopodia of mutant cells (Plate 1).

Therefore it appears that in the mesoderm of mutant embryos abnormalities of cell shape, cellular interaction and cell movement are related to the lack of surface adhesiveness and microfilament cytostructure. We speculate that the occurrence of cellular junctions as well as the formation of microfilaments at adjacent sites may depend on specific cell surface components which make possible appropriate cellular interactions at this time. It should be noted that the defects attributable to abnormal membrane components in this case appear to operate at both the inner and outer face of the membrane, and we have as yet no information on the interesting question of whether the inability of a cell to recognize contacts with another impedes filopodial development, junction formation and microfilament assembly, or whether the reverse sequence of events, or both, may be true (Spiegelman & Bennett, 1974).

Another strong morphological indication of a role for *T*-locus controlled membrane elements in the assembly and organization of cellular organelles comes from electron microscope studies on spermatozoa. Males that are homozygous for the semilethal allele t^{w2} have normal gross morphology and are hormonally normal, yet they are completely sterile and produce almost no spermatozoa (Bennett & Dunn, 1971). Studies of spermiogenesis in these mice have shown that spermatid development appears to proceed entirely normally until relatively late stages. By this time the manchette, which is an orderly longitudinal array of microtubules, has formed. These microtubules insert in flocculent dense material associated with the plasma membrane in a restricted region near the posterior end of the nucleus, the perinuclear ring. After this time, development of almost all cells goes awry, but in two strikingly dissimilar ways. In the majority of spermatids the microtubules of the manchette proliferate to become unusually numerous, and establish bizarre relationships with the plasma membrane. These excessive numbers of disorganized microtubules impinge on large expanses of the plasma membrane, thus often grossly distorting the nucleus. Interestingly though, wherever the abnormal arrays of microtubules approach the plasma membrane they insert in dense flocculent material that resembles the material confined to the perinuclear ring in normal cells (Plate 2). In a minority of spermatids quite another phenomenon occurs at the same stage; in these, the microtubules appear to undergo a sudden depolymerization which coincides with the disappearance, through vesiculization, of the nuclear envelope. The meaning of this is entirely unclear. It can be speculated that in normal spermatids a specialized region of the plasma membrane (the perinuclear ring) contains elements responsible for the assembly of the dense flocculent material which perhaps initiates and stabilizes the assembly of microtubules, and that this region is abnormally extensive and topographically distorted in most t^{w2}/t^{w2} spermatids, and unstable in others (Dooher & Bennett, 1974).

In summary, we have tried to test the hypothesis that early development, in regulative embryos at least, may be controlled largely by interactions mediated by genetically specified components at the cell surface. In analyzing a series of mutants at the *T*-locus, which for many reasons can be thought of as a genetic region containing genes crucial to early embryonic organization, we have found serological evidence that these genes do in fact specify cell surface components, and morphological evidence that these genes have effects interpretable as operating at the cell membrane.

Further serological studies on spermatozoa which are now in progress lead us to believe that *t*-haplotypes are constellations of mutant cell surface antigens which may prove to be as complex as H-2 antigenic specificities in the adult. Our original data (Yanagisawa *et al.*, 1974) reported observations of four different lethal *t*-alleles, and each of these proved to be serologically unique when compared with the others. In other words, within this group of four alleles tested, no cross-reacting specificities were detected. We have extended our tests now to a member of the fifth known complementation group (t^9) and to the semilethal allele t^{w2}. Cytotoxicity tests were done on spermatozoa according to the protocols outlined in Yanagisawa *et al.* (1974); the results shown in Table 1 demonstrate a pattern of cross-reactions amongst these alleles and t^0. Further analysis of these reactions has been carried out by absorption procedures. In this case, specific antisera of predetermined titer were absorbed with known numbers of sperm cells and subsequently used in standard cytotoxicity tests. The number of cells of any given phenotype used for absorption was determined by the number

Table 1. *Results of direct cytotoxicity tests indicating cross-reacting antigens produced by t^0, t^9 and t^{w2}*

Antiserum	Test-cell haplotype	Cytotoxic Index*
Anti-t^9	t^9	42, 41, 37
	t^0	29, 24
	t^{w2}	35
Anti-t^0	t^9	36
	t^0	55, 52, 51, 50, 49
	t^{w2}	55, 53
Anti-t^{w2}	t^9	26
	t^0	32, 24
	t^{w2}	36, 35, 28

* Each number represents a separate test; for each test negative controls were included and always had a cytotoxic index below 10.

necessary to remove all activity from antisera specific for that phenotype. These numbers varied greatly amongst the three genotypes studied here: to remove all activity from 0.05 ml of anti-t^9 antiserum diluted 1/12 required 50×10^6 spermatozoa; for the same amount and dilution of anti-t^{w2} antiserum approximately 36×10^6 sperm cells, and for anti-t^0 antiserum only 24×10^6 spermatozoa. These numbers were adhered to in absorptions reported in Table 2 or they were performed quantitatively using graded numbers of sperm. The results in Table 2 demonstrate that more than one specificity can be detected in both anti-t^0 and anti-t^{w2}

Table 2. *Results of absorption experiments indicating that t^0, t^9 and t^{w2} antigens each have a unique component as well as antigens shared with each of the others.*

	Antiserum								
	Anti-t^9			Anti-t^0			Anti-t^{w2}		
	Absorbing cell haplotype								
Test-cell haplotype	t^9	t^0	t^{w2}	t^9	t^0	t^{w2}	t^9	t^0	t^{w2}
t^9	–	+	+	–	nt	+	–	–	nt
t^0	nt*	–	+	+	–	+	+	–	–
t^{w2}	nt	–	–	+	–	–	+	+	–
	3†	2	2	1	2	2	3	3	3

* nt, not tested; † number of repeats.

antisera; one which appears to be common to both because of cross-reaction and one which is different since absorption of anti-t^0 antiserum with t^{w2} spermatozoa does not remove all activity for t^0 cells, and the reciprocal situation is also true. At least part of the specificity common to anti-t^0 and anti-t^{w2} appears to be the same as the t^9 antigens since the cytotoxic activity of anti-t^0 and anti-t^{w2} antisera for t^9 cells can be removed by absorption with t^9 sperm, but anti-t^9 probably contains some different specificities because both t^0 and t^{w2} are capable of absorbing out some but not all cytotoxic activity for t^9 cells from anti-t^9 antiserum.

We now need to obtain additional information on the antigenic arrangement and genetic specification of these membrane elements in the hope of understanding better the organization of the complex genetic t-region that controls them, and their physiological significance in differentiation.

This work was supported in part by a grant from the National Science Foundation and a contract with the National Institute of Child Health and Human Development.

REFERENCES

ARTZT, K. & BENNETT, D. (1972). A genetically caused embryonal ectodermal tumor in the mouse. *Journal of the National Cancer Institute*, **48**, 141–58.

ARTZT, K., BENNETT, D. & JACOB, F. (1974). Primitive teratocarcinoma cells express a differentiation antigen specified by a gene at the *T*-locus in the mouse. *Proceedings of the National Academy of Sciences, USA*, **71**, 811–13.

ARTZT, K., DUBOIS, P., BENNETT, D., CONDAMINE, H., BABINET, C. & JACOB, F. (1973). Surface antigens common to mouse primitive teratocarcinoma cells in culture and cleavage embryos. *Proceedings of the National Academy of Sciences, USA*, **70**, 2988–92.

BENNETT, D. (1964). Abnormalities associated with a chromosome region in the mouse. II. The embryological effects of lethal alleles at the *t*-region. *Science*, **144**, 263–7.

BENNETT, D., BOYSE, E. A. & OLD, L. J. (1972). Cell surface immunogenetics in the study of morphogenesis. In *Proceedings of the Third Lepetit Colloquium, Cell Interactions*, ed. L. G. Silvestri, pp. 247–63. Amsterdam: North Holland.

BENNETT, D. & DUNN, L. C. (1960). A lethal mutant (t^{w18}) in the house mouse showing partial duplication. *Journal of Experimental Zoology*, **143**, 203–19.

BENNETT, D. & DUNN, L. C. (1971). Transmission ratio distorting genes on chromosome IX and their interactions. In *Proceedings of Symposium on Immunogenetics of the H-2 System*, ed. A. Lengerova & M. Vojtiskova, pp. 90–103. Basel: Karger.

BENNETT, D., GOLDBERG, E., DUNN, L. C. & BOYSE, E. A. (1972). Serological detection of a cell antigen specified by the *T* (Brachyury) mutant gene in the house mouse. *Proceedings of the National Academy of Sciences, USA*, **69**, 2076–80.

DOOHER, G. B. & BENNETT, D. (1974). Abnormal microtubular systems in mouse spermatids associated with a mutant gene at the T-locus. *Journal of Embryology and Experimental Morphology*, in press.

DUNN, L. C. (1964). Abnormalities associated with a chromosome region in the mouse. I. Transmission and population genetics of the t-region. *Science*, **144**, 260–3.

GLUECKSOHN-WAELSCH, S. & ERICKSON, R. P. (1970). The T-locus of the mouse: implication for mechanisms of development. *Current Topics in Developmental Biology*, **5**, 281–315.

MOSCONA, A. A. (1973). Cell aggregation. In *Cell Biology in Medicine*, ed. E. E. Bittar, pp. 571–91. New York: Wiley.

SPIEGELMAN, M. & BENNETT, D. (1974). Fine structural study of cell migration in the early mesoderm of normal and mutant mouse embryos (T-locus: t^9/t^9), *Journal of Embryology and Experimental Morphology*, in press.

STEVENS, L. C. (1967). The biology of teratomas. *Advances in Morphogenesis*, **6**, 1–28.

YANAGISAWA, K., BENNETT, D., BOYSE, E. A., DUNN, L. C. & DIMEO, A. (1974). Serological identification of sperm antigens specified by lethal t-alleles in the mouse. *Immunogenetics*, **1**, 57–67.

YANAGISAWA, K., DUNN, L. C. & BENNETT, D. (1961). On the mechanism of abnormal transmission ratios at the *T*-locus in the house mouse. *Genetics*, **46**, 1635–44.

EXPLANATION OF PLATES

PLATE 1

(*a*) Electron micrograph of mesoderm cells of a normal 8-day mouse embryo. Cells are stellate-shaped with filamentous cytoplasmic processes terminating in junctions (arrowed) with similar processes of neighboring cells. Glutaraldehyde and osmium fixation, uranyl acetate and lead citrate stain. 7500 ×.

(*b*) Electron micrograph of 'mesoderm cells' of mutant (t^9/t^9) 8-day mouse embryo. Cells have lobose pseudopodia (arrowed) which, although in close apposition to neighboring cells, share only a few small junctions. Glutaraldehyde and osmium fixation, uranyl acetate and lead citrate stain. 7500 ×.

PLATE 2

(*a*) Electron micrograph of two normal late mouse spermatids. Within the nuclei (N) chromatin is undergoing condensation. Note that anteriorly the chromatin exhibits a relatively homogeneous, densely staining appearance whereas posteriorly (arrowed), the chromatin forms a dense network of anastomozing fibers. The microtubules (Mt) of the manchette ensheath the posterior half of the nucleus in an orderly array. Glutaraldehyde and osmium fixation, uranyl acetate and lead citrate stain. 25,000 ×.

(*b*) Electron micrograph of a spermatid from a mouse homozygous for t^{w2} which is at the same stage of differentiation as the normal spermatids shown in *a*. Although the nucleus of this cell is very distorted in shape, condensation of the chromatin exhibits the normal pattern found at this stage. Note that this section includes two disconnected portions of the same nucleus (N). The microtubules of the manchette (Mt) are very disorganized. However, where the microtubules approach the plasma membrane anteriorly (arrowed), dense material is associated with the plasma membrane and with the microtubular termini. Flagella (F) do not show abnormalities of microtubular organization. Glutaraldehyde and osmium fixation, uranyl acetate and lead citrate stain. 25,000 ×.

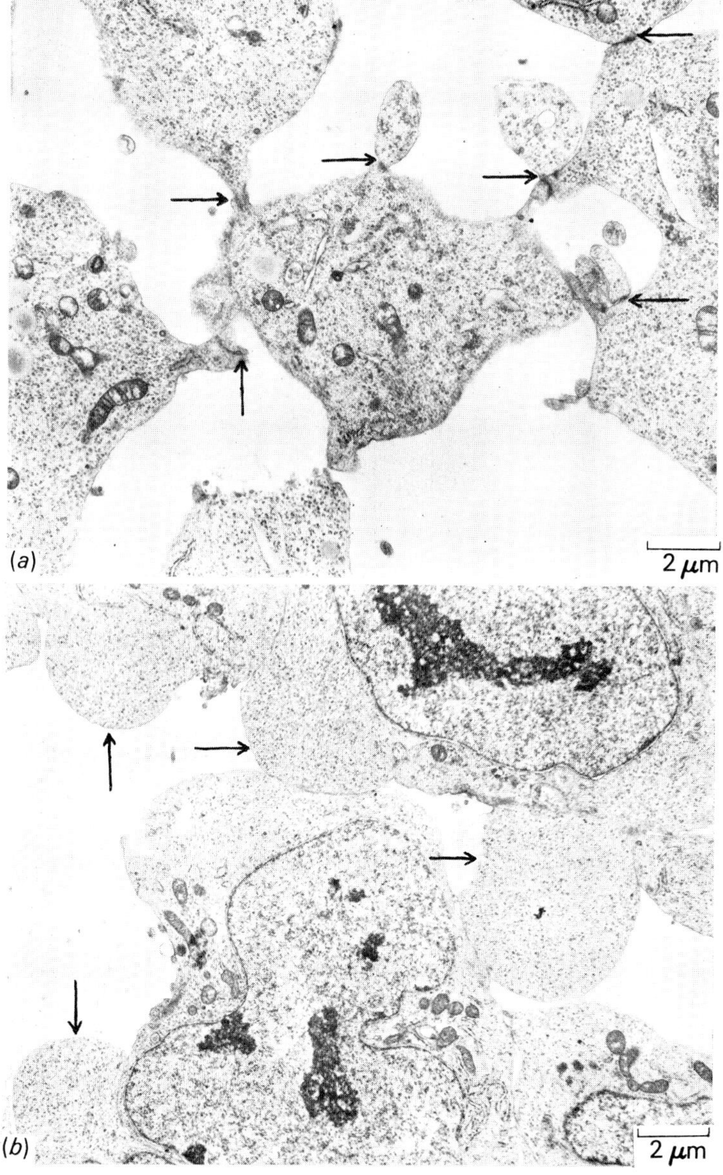

(Facing p. 218)

PLATE 2

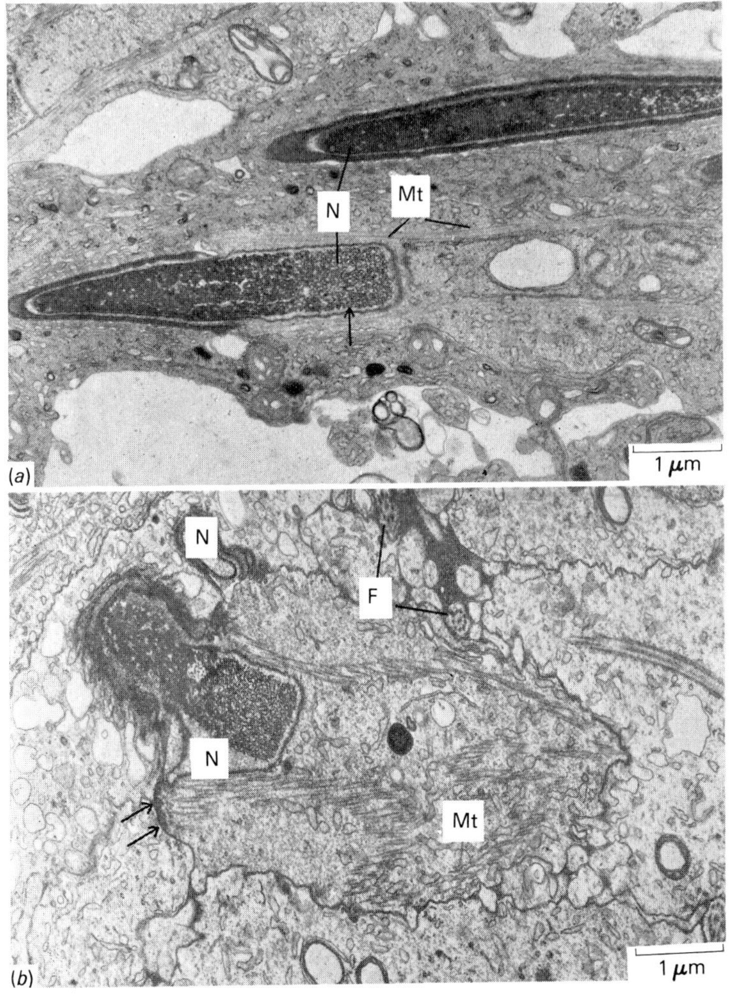

ANTIGEN EXPRESSION DURING EARLY MOUSE DEVELOPMENT

BY W. D. BILLINGTON AND E. J. JENKINSON

Reproductive Immunology Group, Department of Pathology, The Medical School, University of Bristol

Following fertilisation the mammalian embryo embarks upon a rapid programme of differentiation involving the appearance on the cell surface of macromolecules which endow the embryo with its antigenic properties. The identification of these antigens, which are the products of specific genetic loci, is of interest not only for an understanding of the immunological relationships between mother and foetus but also as a model of the pattern and control of differentiation at the cellular level. Evidence is accumulating that the establishment and maintenance of pregnancy depends upon a subtle balance between the antigens expressed on the embryo and the maternal immunological responses to them. The highly specific nature of these antigens and their surface location also make them especially useful as natural cell markers in the analysis of developmental processes. By comparison with biochemical and chromosomal markers presently employed in such studies, they possess the advantage that their detection need not necessitate disruption of the cell nor effect changes in its physiology. In addition, they may be detectable on all cells of the population, and at the precise time of the developmental event under investigation. Those natural markers, such as pigments and nuclear or nucleolar variants, that contributed significantly to the knowledge of morphogenetic movements in the infra-mammalian vertebrates, either arise relatively late in development or involve the use of techniques not feasible with the smaller mammalian embryo. It should also be noted that the presence of specific membrane antigens raises the possibility of using antisera to direct or inhibit the normal patterns of development.

There are both in-vivo and in-vitro techniques available for the detection of cell surface antigens. Transplantation of tissues may be carried out to assay for the presence of specific antigenic markers by their ability either to accelerate the rejection of subsequent grafts bearing those determinants or to reduce graft survival time in specifically pre-immunised hosts. In-vitro methods include visualisation of antigen by the binding of fluorescent or radiolabelled antisera or antibody-coated erythrocytes, quantitative absorption of antisera, and assessment of target cell susceptibility to specific lysis by immune cells or antibody. Three main antigen systems may be considered of potential value as indices of the antigenic status of the embryo; these are discussed below.

Xenogeneic antigens

These antigens are characteristic of a species, and facilitate analysis of cellular interactions where the embryo is an interspecific chimaera, as in the rat–mouse situation described by Gardner & Johnson (1973). The detection of these antigens could also be used to assess gene expression in interspecific hybrids, such as those occurring naturally in equine species and produced experimentally in ovine species.

Histocompatibility antigens

The existence of individual, or inbred strain, specific histocompatibility antigens offers a particularly wide application for the study of differentiation in normal and experimental interstrain hybrid embryos. The H-2, Ag-B and HL-A antigen complexes of mouse, rat and man respectively are becoming increasingly well characterised, and monospecific antisera for the detection of a number of different antigenic specificities within these can be obtained.

Foetal antigens

Some antigens normally appear to be expressed only during embryonic life; these are referred to as phase-specific or foetal antigens. Although it has been assumed that they must somehow be involved in differentiation and organogenesis (e.g. Volkova & Maysky, 1969), certain of these have become prominent because of their reappearance on human and experimental animal tumour cells. Abelev and his colleagues (1963) demonstrated that mouse liver tumours possess high concentrations of alpha-foetoprotein (AFP), a substance usually found in significant quantities only in foetal tissues. Tumours of the human gastrointestinal tract possess a substance known as carcinoembryonic (CEA) antigen, because of its presence otherwise only in the foetus (Gold & Freedman, 1965). Both of these foetal antigens have been well characterised (Laurence & Neville, 1972). In rodents, spontaneous and chemically induced tumours express embryonic antigens at the cell surface (Baldwin, Embleton, Price & Vose, 1974), and similar antigens are displayed on many virus-transformed cells (e.g. Coggin, Ambrose & Anderson, 1970). It is also of interest that an antigen present on a primitive teratocarcinoma has been identified on early mouse embryos (Edidin, Patthey, McGuire & Sheffield, 1971; Artzt *et al.*, 1973).

It is clear that a study of the tumour-associated embryonic antigens should provide a fruitful approach to an appreciation of the regulatory mechanisms involved in both ontogenetic and oncogenic processes. One example of this is AFP, which has already been used as a marker to study the differentiation of the hepatocyte in normal liver and in hepatomas (Abelev, 1974). The present discussion will, however, deal mainly with the histocompatibility (H) antigens and the evidence for their expression on the pre- and post-implantation stages of development in the mouse. It should perhaps be stated at the outset that despite the increasing attention given to this problem, there is as yet very little information on the precise nature and time of appearance of these antigens, although they are known to elicit both cellular and humoral immune reactions in the maternal organism. (See Maroni & Parrott, 1973; Kaliss, 1973.)

ANTIGENS OF THE PRE-IMPLANTATION EMBRYO

Apart from the so-called 'egg antigen', identified by the cytotoxic activity of xenogeneic antisera raised in guinea-pigs against mouse oocytes (Moskalewski & Koprowski, 1972), there is no evidence for antigen expression on unfertilised mouse eggs. In any case, the relationship between this antigen and any of those controlled by the histocompatibility genes is at present unknown. As far as xenogeneic antigens are concerned, recent evidence indicates that these are present at least on the 8-cell (Sellens & Jenkinson, 1975) and early blastocyst stages (Hakansson, 1973).

A number of studies have been carried out to detect H antigens on cleavage-stage embryos. Inhibition of development of oviducal and uterine eggs following ectopic transfer to the kidney of pre-immunised allogeneic hosts has provided evidence for the presence of histocompatibility antigens as early as the 2–8-cell stages (Simmons & Russell, 1965; Kirby, Billington & James, 1966). However, the strain combinations used in these investigations involved combined differences at both the major (H-2) and minor (non-H-2) histocompatibility loci. A claim for antigens on the 2-cell stage using a mixed antiglobulin method also did not distinguish between H-2 and non-H-2 systems (Olds, 1968). More recent attempts aimed at the specific detection of H-2 antigens using well-characterised serological reagents have not led to their identification on embryos of any stage from the 2-cell to the blastocyst, in either mixed antiglobulin, mixed agglutination (Gardner, Johnson & Edwards, 1973) or immunofluorescence tests (Palm, Heyner & Brinster, 1971). In contrast, cytotoxicity (Heyner, Brinster & Palm, 1969) and indirect immunofluorescence tests (Palm *et al.*,

1971) have indicated that at least some non-H-2 antigens are present.

The results of the in-vitro studies indicated that the inhibition of development of ectopically transplanted eggs might be mediated solely via immunity to non-H-2 antigens. The situation was therefore investigated using transfers between various combinations of inbred strains differing at H-2, non-H-2, or combined loci (Table 1). Blastocyst development was

Table 1. *Detection of histocompatibility antigens on the mouse blastocyst by ectopic transplantation to pre-immunised hosts**

Donor–host histocompatibility difference	Blastocyst transfer	Survival rate	% survival
Combined H-2 and non-H-2	Allogeneic	9/103	9
	Control	32/37	86
Non-H-2	Allogeneic	0/29	0
	Control	9/11	82
H-2	Allogeneic	45/52	86
	Control	7/7	100

* Data modified from Searle *et al.* (1974).

suppressed only when combined or non-H-2 differences were involved, thus supporting the evidence that H-2 antigens are not expressed on the pre-implantation embryo (Searle *et al.*, 1974). In contrast to the findings *in vivo*, however, it has been demonstrated that zona-free blastocysts and the trophoblastic outgrowths that they produce *in vitro* are not susceptible to attack by lymphocytes from animals immunised against the combined H-2 and non-H-2 antigens of the blastocyst donor strain (Jenkinson & Billington, 1974*a*; Table 2). Since it is known that this test system is capable of detecting non-H-2 antigens on other tissues (e.g. fibroblasts and tumour cells), some explanation must be sought for this discrepancy. It is now believed that trophoblast proliferation, which is used as an indication of ectopic blastocyst development, may depend upon an inductive influence from the inner cell mass (Gardner, 1971). Thus, as suggested by Billington (1973) and Gardner *et al.* (1973), inhibition of development could be explained by the presence of antigens on the inner cell mass, or embryonic tissues derived from it, rather than the outer trophectoderm which, as indicated by its essentially non-proliferative outgrowth and survival in the presence of immune cells *in vitro*, may have no effective antigen expression. There is a pointer in this direction from the report of Edidin and his colleagues (1971) that a teratoma antigen can be located on the inner cell mass but not on the trophectoderm. It has also recently been found that a teratoma antigen co-caps with H-2 antigens on certain cultured cell lines, indicating a close physical association between these molecules (Gooding & Edidin, 1974).

A final consideration concerns the time of expression of the paternal genome. In the ectopic transplantation system, F1 hybrid blastocysts were significantly suppressed regardless of the direction of mating (Searle *et al.*, 1974). This indicates that both paternally and maternally derived non-H-2 antigens are present by this time, a finding in accord with other evidence for the early expression of paternal genes based upon isoenzyme analyses (Chapman, Whitten & Ruddle, 1971). Unfortunately, in none of the serological studies used to demonstrate non-H-2 antigens were F1 embryos examined.

It would seem at present that only the weaker, non-H-2, antigens are expressed on the pre-implantation mouse embryo, and that by the blastocyst stage the majority of these may well be restricted to the inner cell mass. More detailed mapping of the distribution of the antigens is clearly required. This is now being attempted using refined immunofluorescence techniques with highly specific antisera (M. H. Johnson, personal communication). The reason why the embryo does not succumb to maternal immune attack may lie in this weak antigenicity and in the protective effect of the uterine epithelium, which may prevent access of effective levels of lymphocytes and of antibody and/or complement to the lumen.

ANTIGENS OF THE EARLY POST-IMPLANTATION EMBRYO

Around the time of implantation the trophectoderm transforms into polyploid giant cells which become phagocytic and undergo changes in their surface properties (Jenkinson & Wilson, 1973). This precocious differentiation may provide the embryo with the means whereby it can attach itself to the maternal tissues and at the same time avoid the immunological rejection that such an allograft might elicit. Whether the absence or reduced expression of antigens on the trophectoderm and its resistance to immune attack are in any way associated with the acquisition of these properties is as yet uncertain. It is clear, however, that the trophectoderm from a very early stage is committed to a pathway quite different from that of the inner cell mass, which goes on to acquire an increasingly complex array of antigens as embryonic differentiation proceeds.

Evidence on the antigenic status of the immediate post-implantation stages is extremely limited and the only studies reported to date involve the use of an in-vitro model. Under suitable conditions the mouse blastocyst will attach to the culture vessel and undergo a process of outgrowth forming a flattened sheet of large trophoblast cells overlain by smaller cells, presumably of inner cell mass origin (Cole & Paul, 1965). It has been suggested that this process may be analogous to implantation *in utero* (Gwatkin, 1966). Using immunofluorescence techniques on such outgrowths, Heyner (1973) has demonstrated the appearance of H-2 antigens on the smaller cells, but not on the trophoblast, after a few days in culture. Allowing for the period *in vitro* required to obtain a positive result it would seem that this represents antigen appearance at a time equivalent to about seven days *in utero*.

A particularly convenient stage of embryonic development for analysis is the $7\frac{1}{2}$ day conceptus. By this time there is a marked separation of the trophoblast and inner cell mass derived components, into an ectoplacental cone (EPC) and embryonic sac. These can be separated by micromanipulation to provide material for use in a variety of in-vivo or in-vitro procedures for antigen detection. An early observation (Simmons & Russell, 1966) was that these two components show a different susceptibility to immune attack following transplantation to ectopic sites in pre-immunised allogeneic hosts. The implication was that H antigens were present on the sac but not on the trophoblast cone. In a recent investigation we have confirmed this finding using a modification of an in-vitro cell-mediated microcytotoxicity test (Jenkinson & Billington, 1974*a*). Outgrowths of cells derived from sacs were almost totally destroyed on incubation with spleen cells from mice immunised against target cell histocompatibility antigens. In contrast, the ectoplacental cone outgrowths were completely unaffected, as shown in a more limited study by Vandeputte & Sobis (1972). Trophoblast at this stage clearly does not effectively express those antigenic determinants recognised by immune lymphocytes on the embryonic cells of the same conceptus (Table 2).

Although the trophoblast can resist immune attack it may still possess antigens capable of inducing an immune response in a host animal. Evidence that EPC trophoblast possesses such immunogenic properties is controversial. The claim that a second challenge of ectopically transplanted EPC results in a diminished growth of the trophoblast (Hulka & Mohr, 1968) has not been substantiated (Billington, Elson, Jenkinson & Searle, 1974). In addition, serial ectopic transfers of EPCs to allogeneic hosts does not lead to either an accelerated rejection or an enhanced survival time of subsequent allogeneic skin grafts (Billington, 1973). Currie, Van Doorninck

Table 2. *Cytotoxic effect of immune and non-immune spleen cells on histoincompatible mouse embryonic tissues* in vitro

Embryonic tissue	Developmental stage (days)	No. of cultures showing cytolysis*: Non-immune cells (controls)	No. of cultures showing cytolysis*: Immune cells
Blastocyst	3½	1/24	1/25
Embryonic sac	7½	0/14	13/14
Trophoblast (untreated)	7½	0/8	0/9
Trophoblast (+neuraminidase)	7½	0/16	0/24
Fibroblasts	10–17	0/48	57/57
Kidney	14	0/5	5/5
Lung	14	0/15	15/15
Yolk sac (whole)	10–14	0/40	40/40
Yolk sac (endoderm)	14–16	0/7	7/7
Amnion	14	0/3	3/3
Placenta	12–13	0/25	0/25†
Placenta	16	0/10	10/10

* For methods of assessment see Jenkinson & Billington (1974*a*, *b*).
† Slight effect observed.

& Bagshawe (1968) carried out experiments to test the hypothesis that antigens may be present on trophoblast but masked by surface mucoproteins (Kirby, Billington, Bradbury & Goldstein, 1964; Currie & Bagshawe, 1967). These authors demonstrated an accelerated rejection of allogeneic skin grafts on mice that had been given an intraperitoneal injection of EPC trophoblast cells previously incubated with neuraminidase. By analogy with similar experiments on tumour cells, these results were interpreted as demonstrating the unmasking of histocompatibility antigens by the removal of cell surface neuraminic acid. In a later study, however, Simmons, Lipschultz, Rios & Ray (1971) were unable to confirm these findings. In our microcytotoxicity system, pre-incubation of EPC outgrowths in various concentrations of neuraminidase did not render the trophoblast susceptible to cell-mediated immune lysis, despite a demonstrable effect of the enzyme on the cell surface as visualised by electron histochemistry (Jenkinson & Billington, 1974*a*). Using a similar injection procedure to Currie and his colleagues, but assaying sensitisation by the cytotoxic effect of host lymphocytes on fibroblast monolayers, we have also been unable to demonstrate trophoblast immunogenicity (Searle &

Jenkinson, unpublished observations). Hulka (1971) reported that EPC trophoblast developing in the kidney of allogeneic mice induced a cell-mediated immune response as judged by inhibition of trophoblast growth in recipients passively transferred with lymphocytes from the original host. Since we have failed to detect any in-vitro cytotoxic activity of lymphocytes from mice bearing such EPC transplants (Searle & Jenkinson, unpublished observations) it seems possible that Hulka may have been demonstrating *tissue-specific* antigen recognition, although here again, the evidence for the existence of this type of antigen on trophoblast is at best controversial (Billington *et al.*, 1974).

It must be concluded that the early mouse trophoblast either lacks or has no effective expression of histocompatibility antigens, at least as detectable by the test systems employed so far. There are, however, mechanisms possible whereby this tissue could possess antigenic determinants incapable of being recognised under such conditions. These include peculiarities of trophoblast membrane kinetics and cell surface topography, as discussed in detail by Jenkinson & Billington (1974*a*). In addition, the use of phytohaemagglutinin (PHA) stimulated lymphocytes, which are known to be capable of producing non-specific lysis of target cells, has no effect on EPC trophoblast outgrowths (Elson & Jenkinson, unpublished observations). This implies that trophoblast either has no PHA receptors or that the cell surface is inimical to the close approach of lymphocytes necessary for cytolysis to be effected. Whatever the reason, the relative antigenic deficiency of the trophoblast is a convenient property for a tissue having a fundamental role as a barrier protecting the foetus from potentially deleterious maternal immunity. The only way in which we have been able to effect the immune destruction of trophoblast is by incubation with a highly cytotoxic xenogeneic antiserum, which would recognise antigenic determinants of species specificity.

From a developmental point of view, it is of interest that H genes appear to be switched on in the embryonic component of the early mouse embryo but not the trophoblast. With the use of congenic strains of mice differing only at the H-2 locus it is now known that the H-2 genes are expressed only in the later post-implantation stages. The previously described work of Heyner (1973) supports the findings of Patthey & Edidin (1973) that H-2 antigens are recognised on 7 day (but not 6 day) embryonic sacs following transplantation to pre-immunised hosts. Since the sacs developed for 2–3 days before evidence of host cellular infiltration, it is not possible to say with certainty that the antigens were present at the time of transfer. This could mean that they do not make their appearance *in utero* until day 9 or later. Mitomycin treatment or X-irradiation of cells before transplantation would inhibit their further development, and allow a more precise evaluation of the time of antigen expression.

ANTIGENS OF ORGANISED TISSUES

By day 9 of gestation the mouse embryo can conveniently be considered as having three principal components; the embryo proper, the foetal membranes and the placenta. It is clear from the studies on $7\frac{1}{2}$ day embryonic sacs that the embryonic tissues by this time do express H antigens. There is, however, rather little information on the relative degree of antigen expression or changes in the level of antigenicity of the various tissues and organs during development. We have found that cell preparations obtained from whole embryos between days 10 and 17 of pregnancy are susceptible to immune cell lysis when donor strains differing at combined H-2 and non-H-2 loci are employed. Similarly, cultures of 14 day embryonic material in a test situation involving differences only at non-H-2 loci show significant destruction, although this is less than in the combined H-2 and non-H-2 system, suggesting that the effect seen in the latter case also involves the recognition of H-2 antigen. Preparations from both the kidney and lung of 14 day embryos have also been found to show an equal susceptibility in these tests, indicating a comparable degree of antigen expression on the cells of these organs (Table 2). We have not used this technique to investigate possible quantitative differences in antigenicity of other tissues since it is unlikely to detect other than fairly gross variations. There is, however, evidence from other studies that the over-all antigenicity of the embryo increases during development and that a pattern of different antigenic activities appears in various organs (see review by Edidin, 1972).

In common with the embryo, the foetal membranes are derived from the inner cell mass and have now been shown to possess similar antigenic determinants. Cells of the yolk sac, and of its outer endodermal layer, appear to express both H-2 and non-H-2 antigens on the basis of their susceptibility to immune cell (Jenkinson & Billington, 1974*b*) and complement-dependent antibody lysis (Jenkinson, Billington & Elson, 1975). Preparations of amnion are also susceptible to immune cell lysis (Table 2). The yolk sac findings are particularly interesting in view of the role of this membrane in the transmission of immunoglobulin to the foetus and the direct exposure of its endodermal component to the maternal environment in the latter part of pregnancy. The possible reasons for the survival of the yolk sac despite its exposure and possession of antigens have been fully discussed (Jenkinson & Billington, 1974*b*), but it is evident that the yolk sac differs from EPC trophoblast, the survival of which appears to depend upon a lack of effective antigen expression.

The origin of the placental trophoblast is less clear. It would seem unlikely that the stem cells which proliferate to give rise to the EPC, and subsequently the placenta, are derived from the highly specialised polyploid giant cells of the trophectoderm. This must mean either that part of the trophectoderm remains in a diploid state or that there is a second wave of trophoblast differentiation from the inner cell mass. Available evidence indicates that the former is more likely to be the case. Both diploid and polyploid trophoblast appear to be present in the early EPC (Searle & Jenkinson, unpublished observations) and have been detected by DNA estimations in the definitive placenta (Barlow & Sherman, 1972).

Whatever the antigenic status of its precursors there have been various studies suggesting that the placental trophoblast is antigenic (see review by Beer & Billingham, 1971). In most of these, however, assessment has been based upon procedures involving whole placental fragments or homogenates. The possibility that the effects are due to non-trophoblastic placental elements cannot, therefore, be ruled out. In our own studies we have found that cell cultures derived from whole 12- or 13-day placentae are less susceptible to immune cell lysis than those from later stage placentae (Table 2). This may reflect either an increase in placental trophoblast antigenicity as gestation proceeds or a change in the relative proportions of antigenic to non-antigenic components in the cell preparations. These antigenic components could either be non-trophoblastic or could represent one of the different biological forms of trophoblast present in the mature placenta. In this context it is perhaps of significance that the placental labyrinth increases in size during the course of development, possibly indicating that it is the lower ploidy cells of the labyrinthine trophoblast which are antigenic. A complete analysis of the antigenicity of the different cell populations of the placenta does, however, depend upon the development and application of suitable procedures for their separation. Whatever their precise location in the placenta, it is interesting to speculate that the antigens present in this organ may act as a filter to prevent the ingress of potentially harmful antibody directed against histocompatibility antigens of the foetus.

CONCLUSION

The possibility that mammalian differentiation may be directed or influenced largely by the presence of macromolecules on the surface of embryonic cells has been recognised for many years (Weiss, 1953). Since that time information has gradually accumulated on the expression of antigens on the embryo and on the characterisation of the genes responsible for them. It remains to be seen to what extent these particular macromolecules are actually involved in the developmental process. Minor histocompatibility antigens, at least, are detectable on the pre-implantation stages of the embryo. The H-2 antigens are probably expressed in significant quantity soon after day 7, and appear to be present on a variety of tissues examined later in pregnancy, including the foetal membranes. The trophoblast provides a notable exception to this, and appears antigenically neutral in the early stages of its differentiation. Whether effective antigen expression occurs later in this tissue has yet to be convincingly demonstrated. The pregnant female responds to the antigenic challenge of these paternally-inherited histocompatibility gene products with both cellular and humoral immunity that may play a significant role in the maintenance of the foetal allograft.

We are grateful to our colleagues Joanna Elson and Roger Searle for allowing us to include unpublished findings from our collaborative studies, and for discussions on the manuscript. The Rockefeller Foundation provided generous financial support.

REFERENCES

ABELEV, G. I. (1974). α-Foetoprotein as a marker of embryo-specific differentiation in normal and tumour tissues. *Transplantation Reviews*, **20**, 3–37.

ABELEV, G. I., PEROVA, S. D., KHRAMKOVA, N. I., POSTNIKOVA, Z. A. & IRLIN, I. S. (1963). Production of embryonal α-globulin by transplantable mouse hepatomas. *Transplantation*, **1**, 174–80.

ARTZT, K., DUBOIS, P., BENNETT, D., CONDAMINE, H., BABINET, C. & JACOB, F. (1973). Surface antigens common to mouse primitive teratocarcinoma cells in culture and cleavage embryos. *Proceedings of the National Academy of Sciences, USA*, **70**, 2988–92.

BALDWIN, R. W., EMBLETON, M. J., PRICE, M. R. & VOSE, B. M. (1974). Embryonic antigen expression on experimental rat tumours. *Transplantation Reviews*, **20**, 77–99.

BARLOW, P. W. & SHERMAN, M. I. (1972). The biochemistry of differentiation of mouse trophoblast: studies on polyploidy. *Journal of Embryology and Experimental Morphology*, **27**, 447–65.

BEER, A. E. & BILLINGHAM, R. E. (1971). Immunobiology of mammalian reproduction. *Advances in Immunology*, **14**, 1–84.

BILLINGTON, W. D. (1973). Does trophoblast express tissue-specific antigens? In *Immunology of Reproduction*, ed. K. Bratanov, pp. 475–9. Sofia: Bulgarian Academy of Sciences Press.

BILLINGTON, W. D., ELSON, J., JENKINSON, E. J. & SEARLE, R. F. (1974). Antigenicity of the trophoblast. *Proceedings of the First International Congress on Immunology in Obstetrics and Gynaecology*, pp. 111–15. Amsterdam: Excerpta Medica.

Chapman, V. M., Whitten, W. K. & Ruddle, F. H. (1971). Expression of glucose phosphate isomerase-1 (*Gpi*-1) in pre-implantation stages of mouse embryos. *Developmental Biology*, **26**, 153–8.

Coggin, J. H., Ambrose, K. R. & Anderson, N. G. (1970). Foetal antigen capable of inducing transplantation immunity against SV40 hamster tumour cells. *Journal of Immunology*, **105**, 524–6.

Cole, R. J. & Paul, J. (1965). Properties of cultured preimplantation mouse and rabbit embryos, and cell strains derived from them. In *Preimplantation Stages of Pregnancy, Ciba Foundation Symposium*, ed. G. E. W. Wolstenholme & M. O'Connor, pp. 82–112. London: J. & A. Churchill.

Currie, G. A. & Bagshawe, K. D. (1967). The masking of antigens on trophoblast and cancer cells. *Lancet*, **i**, 708–10.

Currie, G. A., van Doorninck, W. & Bagshawe, K. D. (1968). Effect of neuraminidase on the immunogenicity of early mouse trophoblast. *Nature, London*, **219**, 191–2.

Edidin, M. (1972). Histocompatibility genes, transplantation antigens and pregnancy. In *Transplantation Antigens: Markers of Biological Individuality*, ed. B. D. Kahan & R. A. Reisfeld, pp. 75–114. New York & London: Academic Press.

Edidin, M., Patthey, H. L., McGuire, E. J. & Sheffield, W. D. (1971). An antiserum to 'embryoid body' tumour cells that reacts with normal mouse embryos. In *Proceedings of the First Conference and Workshop on Embryonic and Foetal Antigens in Cancer*, ed. N. G. Anderson & J. H. Coggin, p. 239. Oak Ridge, Tennessee: USAEC.

Gardner, R. L. (1971). Manipulations on the blastocyst. In *Schering Symposium on Intrinsic and Extrinsic Factors in Early Mammalian Development*, Advances in the Biosciences vol. 6, ed. G. Raspé, pp. 279–96. Oxford: Pergamon Press Vieweg.

Gardner, R. L. & Johnson, M. H. (1973). Investigation of early mammalian development using interspecific chimaeras between rat and mouse. *Nature New Biology*, **246**, 86–9.

Gardner, R. L., Johnson, M. H. & Edwards, R. G. (1973). Are H-2 antigens expressed in the preimplantation blastocyst? In *Immunology of Reproduction*, ed. K. Bratanov, pp. 480–6. Sofia: Bulgarian Academy of Sciences Press.

Gold, P. & Freedman, S. O. (1965). Specific carcinoembryonic antigens of the human digestive system. *Journal of Experimental Medicine*, **122**, 467–81.

Gooding, L. R. & Edidin, M. (1974). Cell surface antigens of a mouse testicular teratoma. Identification of an antigen physically associated with H-2 antigens on tumour cells. *Journal of Experimental Medicine*, **140**, 61–78.

Gwatkin, R. B. L. (1966). Defined media and development of mammalian eggs *in vitro*. *Annals of the New York Academy of Sciences*, **139**, 79–90.

Hakansson, S. (1973). Effects of xenoantiserum on the development *in vitro* of mouse blastocysts from normal pregnancy and from delay of implantation with and without oestradiol. *Contraception*, **8**, 327–42.

Heyner, S. (1973). Detection of H-2 antigens on the cells of the early mouse embryo. *Transplantation*, **16**, 675–8.

Heyner, S., Brinster, R. L. & Palm, J. (1969). Effect of iso-antibody on pre-implantation mouse embryos. *Nature, London*, **222**, 783–4.

Hulka, J. F. (1971). In Discussion. In *Schering Symposium on Intrinsic and Extrinsic Factors in Early Mammalian Development*, Advances in the Biosciences, vol. 6, ed. G. Raspé, pp. 415–16. Oxford: Pergamon Press Vieweg.

HULKA, J. F. & MOHR, M. (1968). Trophoblast antigenicity demonstrated by altered challenge graft survival. *Science*, **161**, 696–8.

JENKINSON, E. J. & BILLINGTON, W. D. (1974*a*). Differential susceptibility of mouse trophoblast and embryonic tissue to immune cell lysis. *Transplantation*, **18**, 286–9.

JENKINSON, E. J. & BILLINGTON, W. D. (1974*b*). Studies on the immunobiology of mouse foetal membranes: the effect of cell-mediated immunity on yolk sac cells *in vitro*. *Journal of Reproduction and Fertility*, **41**, 403–12.

JENKINSON, E. J., BILLINGTON, W. D. & ELSON, J. (1975). The effect of cellular and humoral immunity on the mouse yolk sac. In *Transmission of Immunoglobulins from Mother to Young*, ed. W. A. Hemmings. London: Cambridge University Press. (In press.)

JENKINSON, E. J. & WILSON, I. B. (1973). *In vitro* studies on the control of trophoblast outgrowth in the mouse. *Journal of Embryology and Experimental Morphology*, **30**, 21–30.

KALISS, N. (1973). Immune reactions of multiparous female mice to foetal H-2 alloantigens. In *Immunology of Reproduction*, ed. K. Bratanov, pp. 495–511. Sofia: Bulgarian Academy of Sciences Press.

KIRBY, D. R. S., BILLINGTON, W. D., BRADBURY, S. & GOLDSTEIN, D. J. (1964). Antigen barrier of the mouse placenta. *Nature, London*, **204**, 548–9.

KIRBY, D. R. S., BILLINGTON, W. D. & JAMES, D. A. (1966). Transplantation of eggs to the kidney and uterus of immunised mice. *Transplantation*, **4**, 713–18.

LAURENCE, D. J. R. & NEVILLE, A. M. (1972). Foetal antigens and their role in the diagnosis and clinical management of human neoplasms: a review. *British Journal of Cancer*, **26**, 335–55.

MARONI, E. S. & PARROTT, D. M. V. (1973). Progressive increase in cell-mediated immunity against paternal transplantation antigens in parous mice after multiple pregnancies. *Clinical and Experimental Immunology*, **13**, 253–62.

MOSKALEWSKI, S. & KOPROWSKI, H. (1972). Presence of egg antigen in immature oocytes and preimplantation embryos. *Nature, London*, **237**, 167.

OLDS, P. J. (1968). An attempt to detect H-2 antigens on mouse eggs. *Transplantation*, **6**, 478–9.

PALM, J., HEYNER, S. & BRINSTER, R. L. (1971). Differential immunofluorescence of fertilized mouse eggs with H-2 and non-H-2 antibody. *Journal of Experimental Medicine*, **133**, 1282–93.

PATTHEY, H. L. & EDIDIN, M. (1973). Evidence for the time of appearance of H-2 antigens in mouse development. *Transplantation*, **15**, 211–14.

SEARLE, R. F., JOHNSON, M. H., BILLINGTON, W. D., ELSON, J. & CLUTTERBUCK-JACKSON, S. (1974). Investigation of H-2 and non-H-2 antigens on the mouse blastocyst. *Transplantation*, **18**, 136–41.

SELLENS, M. H. & JENKINSON, E. J. (1975). Permeability of the mouse zona pellucida to immunoglobulin. *Journal of Reproduction and Fertility*, **42**, 153–7.

SIMMONS, R. L., LIPSCHULTZ, M. L., RIOS, A. & RAY, P. K. (1971). Failure of neuraminidase to unmask histocompatibility antigens on trophoblast. *Nature New Biology*, **231**, 111–12.

SIMMONS, R. L. & RUSSELL, P. S. (1965). Histocompatibility antigens in transplanted mouse eggs. *Nature, London*, **208**, 698–9.

SIMMONS, R. L. & RUSSELL, P. S. (1966). The histocompatibility antigens of fertilized mouse eggs and trophoblast. *Annals of the New York Academy of Sciences*, **129**, 35–45.

VANDEPUTTE, M. & SOBIS, H. (1972). Histocompatibility antigens on mouse blastocysts and ectoplacental cones. *Transplantation*, **14**, 331–8.

VOLKOVA, L. S. & MAYSKY, I. N. (1969). Immunological interaction between mother and embryo. In *Immunology and Reproduction*, ed. R. G. Edwards, pp. 211–30. London: International Planned Parenthood Federation.

WEISS, P. (1953). Some introductory remarks on the cellular basis of differentiation. *Journal of Embryology and Experimental Morphology*, **1**, 181–211.

MOUSE TERATOCARCINOMA AS A TOOL FOR THE STUDY OF THE MOUSE EMBRYO

BY F. JACOB

Service de Génétique Cellulaire
du Collège de France et de l'Institut Pasteur,
25 rue du Dr Roux, 75015 Paris, France

Biochemical and immunological study of the early stages of mammalian development is made difficult by the scarcity of material and by the rapid transition of the cells through successive states of differentiation. Some of these difficulties could be overcome either by analysis at single cell level or by the use of stable cell lines maintaining, in culture, the properties of some embryonic cells. Growing, *in vitro*, specific cell types derived from early embryos has not so far been very successful. Attempts were made, therefore, to use cell lines derived from mouse teratocarcinoma.

The properties of the experimental mouse teratocarcinoma have been defined mainly through the work of L. C. Stevens (1967) and G. B. Pierce (1961). The mouse strain 129 exhibits a high incidence of spontaneous testicular teratocarcinomas, which contain a large variety of differentiated tissues. A similar situation can be obtained by grafting either blastocysts, or genital ridges of 12-day 129 embryos, into the testes of adult 129 mice. Some of these tumours can be serially transplanted in syngeneic mice, either subcutaneously or intraperitoneally. In the latter case, the injected mice exhibit ascites containing 'embryoid bodies', so called because of their morula- or blastocyst-like shape. When placed in culture, such embryoid bodies give rise to primitive teratocarcinoma (PTC) cells, as well as to a large variety of differentiated cells, i.e. nerve, muscle, epithelia, cartilage, etc. As shown by Rosenthal, Wishnow & Sato (1970) and by Kahan & Ephrussi (1970), PTC cell lines can be established which, although remaining undifferentiated in cultures, give rise, when injected into 129 mice, to tumours containing a variety of tissues derived from the three germ layers.

Since PTC cells appear to be 'multipotential' like early embryo cells, and can be grown in large amounts in culture, we hoped to use PTC cells as a tool for the study of the latter. If these two types of cells can be shown to share some characteristic properties, then it becomes possible to perform immunological and biochemical analysis on PTC cells and to apply the knowledge gained for the study of embryo cells. First it was decided to use immunological methods with the aim of detecting the presence of an antigenic determinant on the surface of both PTC and early embryo cells. This decision was based on the following premises: (1) early embryonic cells may be expected to possess specific determinants which are involved in development but which disappear early enough from the organism to evoke a specific immunological response in the *syngeneic* adult animal; (2) if PTC cells in culture possess such determinants, their injection into syngeneic adult mice should elicit the formation of specific antibodies reacting, not only with PTC cells, but also with early embryonic cells.

TERATOMA CELL LINES

129/Sv mice were originally obtained from Dr L. C. Stevens (The Jackson Laboratory) as well as a tumour (OTT6050) obtained by the graft of a 129 blastocyst into the testis of a 129 mouse. This tumour can be serially transplanted in syngeneic mice. Ascites were harvested and cultured in modified Eagle's medium (Dulbecco & Freeman, 1959) containing 15% calf serum. Under these conditions, embryoid bodies will attach to a plate and give rise to PTC cells mixed with a series of differentiated cells. From such plates, some differentiated (DTC) and a series of primitive (PTC) cell lines were isolated, purified, cloned and established in culture (Jakob *et al.*, 1973; Boon, Buckingham, Dexter, Jakob & Jacob, 1974). It is worth mentioning that while the DTC cell lines obtained so far are invariably aneuploid, PTC cells exhibit sometimes only very slight, and most frequently no, detectable alteration in the number and banding pattern of the chromosomes when compared with those of the mouse (Guénet, Jakob, Nicolas & Jacob, 1974). Two of the PTC cell lines isolated have now been maintained in cultures by serial transfer for more than a year and a half. They appear to be rather stable; they can be cloned easily and a very large fraction of the clones, when injected into 129 mice, give rise to tumours containing derivatives of the three germ layers. The same applies to a series of mutants resistant to several drugs.

ANTIGENIC PROPERTIES OF PTC CELLS

129 mice are H-2^b with respect to the major histocompatibility antigen. The presence of H-2^b on PTC cells was investigated using two different anti-H-2^b specific antisera. While these antisera are very potent by cytotoxicity test on 129 lymphocytes, they exhibit no detectable activity against PTC cells. Furthermore, absorption of antisera with PTC cells does not

remove any activity against lymphocytes (Artzt & Jacob, 1974). Thus, the use of sensitive immunological tests does not allow the detection of H-2^{b} on PTC cells; a result which has to be considered in the light of previous reports that H-2^{b} is not serologically detectable on pre-implantation embryos, specifically eggs and morulae (Heyner, Brinster & Palm, 1969; Heyner, 1973).

For a study of their antigenic properties, PTC cells grown in culture and irradiated (to prevent tumour formation and differentiation) were injected into syngeneic male (to avoid H-Y antigen) 129 mice according to a hyper-immunisation schedule. Sera harvested from the same mice before immunisation were used as controls. All sera were heat-inactivated; absorbed rabbit serum was added as a source of complement when required.

Three tests were used: direct cytotoxicity test, absorption, and indirect test with sheep antibodies directed against mouse immunoglobulins coupled with peroxidase (peroxidase test) according to Avrameas & Ternynck (1971). All three tests gave similar results; after the fourth immunisation, the immune sera were found to react strongly with that line of PTC cells used for immunisation. In the direct cytotoxicity test, the serum killed more than 90% of the cells. The serum dilution killing 50% of the cells under the conditions used ranged from 1/800 to 1/3200 (Artzt *et al.*, 1973).

EFFECT OF ANTI-PTC SERUM ON OTHER CELL TYPES

With these sera in hand, it became possible to investigate the presence of the PTC antigen on various cell types. The results can be summarised as follows:

(1) All the PTC cell lines isolated react with the serum, although they appear to differ both in the fraction of susceptible cells as evidenced by direct cytotoxicity test and in the amount of antigen(s), detectable by absorption, on the surface of the cells. These differences, however, disappear after sonication or neuraminidase treatment of the cells; in all PTC cell lines, the presence of the antigen(s) can be detected in the quasi-totality of the population and the amount of cells required to absorb 50% of activity in a given serum dilution can be shown to be similar (Dexter, Buc-Caron, Jakob, Nicolas & Gachelin, unpublished data).

(2) A series of differentiated cell types derived from the teratocarcinoma, or from adult 129 mice, as well as a variety of cell lines carrying known tumour viruses (such as SV40, polyoma, leukaemia, MTV), were tested; none was found to react with anti-PTC serum, except cells from the testis

and spermatozoa. Spermatozoa harvested from the epididymis and washed, can absorb all the anti-PTC activity. This applies to spermatozoa produced not only by syngeneic 129 mice but also by various allogeneic strains, either inbred (such as C57/Bl, BalbC, DBA, A) or random bred (NCS) (Artzt *et al.*, 1973). Preliminary attempts to localise the antigen on the surface of the sperm cells by means of the indirect fluorescence test indicate that only a small fraction of the cells (15–20%) can be specifically labelled as a ring on the postacrosomal area (Fellous, Gachelin, Buc-Caron, Dubois & Jacob, unpublished data).

(3) The anti-PTC serum was tested on cells of early stage embryos. Preimplantation embryos of different ages were harvested (at various times after appearance of a plug) by flushing the oviducts. The zona pellucida was removed with pronase. The embryos were incubated for 2 h with 5% CO_2 and then tested, either by indirect peroxidase, or by direct cytotoxicity tests. The results obtained with both techniques are in agreement. At the fertilised 1-cell stage, no reaction can be detected by these methods. At the 2-cell stage, a weak but definite reaction is observed (with serum dilutions up to 1/200). At the 8-cell morula stage, the reaction is as strong as that found with PTC cells (up to a serum dilution of 1/1600). Again this applies to embryos from a variety of strains such as C57/Bl, DBA, BalbC and NCS (Artzt *et al.*, 1973).

These results obtained with anti-PTC serum show, therefore, the progressive appearance of a surface antigen during the very first stages of embryo development; undetectable after fertilisation, this antigen reaches its full expression at the 8–16-cell stage. It seems unlikely that the observed increase in antigen could be due to a destruction of antigen by pronase followed by a differential resynthesis, since identical results are found both immediately after pronase treatment or after a 4 h incubation period. Most probably the appearance of the antigen results either from a new synthesis or from its unmasking.

GENETIC DETERMINATION OF THE PTC ANTIGEN

The finding that male germ cells were the only adult cells tested so far which shared a surface antigen in common with PTC and normal morula cells, suggested a way of analysing the genetic origin of this antigen. For example, in the mouse, mutations at the *T*-locus are known on the one hand to alter antigens present on the surface of sperm cells and, on the other, to stop embryonic development in specific ways interpreted as being due to cell surface defects (Bennett, Boyse & Old, 1971). Since the surface antigen revealed on morula cells by means of anti-PTC sera is likely to play a role at very early stages of embryonic development, the best candidate appeared to be that gene of the *T*-locus whose mutation (t^{12}) is known, in homozygous condition, to act earliest and to block development at the morula stage. This hypothesis can be tested because mouse sperm cells have been shown to express antigens determined by both wild-type and mutant *t* alleles (Yanagisawa *et al.*, 1974).

In the absence of sperm from t^{12}/t^{12} homozygous mouse embryos (their development is blocked at the morula stage), one has to compare, quantitatively, absorption of anti-PTC activity by sperm cells derived from both $+/+$ homozygous and $+/t^{12}$ heterozygous animals. It was thus found that to remove the same amount of anti-PTC activity (50% of a certain serum dilution) required twice as many sperm cells from the heterozygous $+/t^{12}$ as from the homozygous $+/+$ animal. In contrast, other *t* mutations which are known to act later during embryonic development, do not alter the absorbing capacity of sperm cells. Although other explanations cannot as yet be completely excluded, the most likely conclusion that can be drawn from these results is that the antigen detected on PTC, sperm and morula cells is determined by the $+^{t12}$ allele at the *T*-locus (Artzt, Bennett & Jacob, 1974). This can be confirmed by showing that t^{12}/t^{12} homozygous morulae are devoid of PTC antigen. Experiments designed to check this point are now in progress.

PRESENCE OF THE MOUSE PTC ANTIGEN IN OTHER MAMMALS INCLUDING MAN

Since homozygosity for t^{12} is lethal, the antigen revealed by anti-PTC sera is likely to play an important role in embryonic development. It might, therefore, have been subjected to severe constraints during evolution. If this were correct, the PTC antigen, or very similar constituents, might be expected to be present on the sperm of various mammals, including man. The presence of the antigen in man was investigated by direct cytotoxicity test and by quantitative absorption of anti-PTC sera with human sperm. While the presence of the PTC antigen could not be detected on human lymphocytes or erythrocytes, in contrast, a fraction of human sperm cells was killed in cytotoxicity tests with anti-PTC sera (cytotoxic index 0.35). Furthermore, sperm cells could absorb all the activity of the anti-PTC serum and the number of human sperm required to remove a given anti-PTC activity was roughly the same as that for mouse sperm cells. When anti-PTC serum was absorbed with different cell types, it was found that the remaining activity against PTC cells and against human sperm varied in parallel (Buc-Caron, Gachelin, Hofnung & Jacob, 1974). The location of the antigen on the sperm cells was determined by means of the indirect fluorescence technique. Just as with mouse sperm, it was found that only a fraction (15–20%) of human sperm cells was specifically labelled by a ring on the postacrosomal region (Fellous, Gachelin, Buc-Caron, Dubois & Jacob, unpublished data). Finally, absorption experiments have shown that the PTC antigen is also present on sperm cells of the bull, but not on those of the cock (Buc-Caron & Gachelin, unpublished data).

Since an identical, or a very similar, constituent appears to be present on the surface of sperm cells of several mammals, it seems reasonable to assume that it plays the same role in other mammals as it does in the mouse at some very early stage of embryonic development. This would imply the existence, in other mammals including man, of certain genes with the same functions as those of the *T*-locus in the mouse. Although recessive lethals are more difficult to analyse in humans than in mice, there are some defects known in man which in a way are reminiscent of lesions resulting from *T*-locus defects in the mouse. It might be useful to investigate such human lesions in the light of what is known of the *T*-locus of the mouse.

CONCLUDING REMARKS

The results obtained so far clearly justify the original assumption that teratocarcinoma can provide suitable in-vitro material for a study of embryonic development. They also support the initial hypothesis that surface antigens must exist which play a role at some stages of embryonic development and then disappear from the organism; an hypothesis which is now supported by embryological work (Bennett, Boyse & Old, 1971). These results, however, raise many questions, only a few of which will be discussed here, namely those pertaining to the appearance and the disappearance of the PTC antigen in embryonic life.

Let us consider first the question of its disappearance. From the results reported above we know that the antigen is strongly expressed on the surface of blastomeres of the morula; we know also that it has to disappear rather early from the organism since it is not recognised as 'self' by the adult animal. We want, therefore, to know how the antigen is distributed during the first morphological differentiations and at what stage it disappears. This, however, requires detecting the surface PTC antigen, not only on whole morulae, but also on sections of embryos. The examination of the peroxidase reaction must then be carried out by electron-microscopy which, in this case, is a much more reliable technique than optical microscopy. Only recently have the difficulties of having both good antigen–antibody reaction and good fixation been solved. Preliminary results indicate that, in morulae, the PTC antigen is present on the entire surface of all cells and that in blastocysts, it is also present on the entire surface of all cells, whether belonging to the trophectoderm or to the inner cell mass. (Babinet, Condamine, Dubois & Ryter, unpublished data.) Investigation of post-implantation stages is now in progress.

The detection of the PTC antigen on both early embryos and sperm warrants discussion. Although a variety of cell surface antigens have been detected on human and murine sperm cells, only HL-A (Fellous & Dausset, 1973) and the PTC antigen have so far been located on the post-acrosomal area. Indeed the postacrosomal membrane is thought to remain after capacitation while other membranes are removed (Bedford, 1968). Furthermore, the postacrosomal membrane of the sperm appears to fuse with the membrane of the ovum at fertilisation (Yanagimachi, Nicholson, Noda & Fujimoto, 1973). If the ovum does not possess the PTC antigen, one would expect, after fertilisation, to detect the membrane components brought in by the sperm cell. Although the methods used so far do not allow the detection of the PTC antigen on the fertilised egg, more refined techniques must be used to detect minute amounts of the antigen both on the ovum and the fertilised egg. If, as might well be, one finds no trace of PTC antigen on the ovum and only that small amount carried in by the sperm after fertilisation, then this poses the question as to the function of this antigen. The same question also applies to the surprising finding that, not only H-2, but apparently all antigens specified by genes at the *T*-locus, are present on sperm (Yanagisawa *et al.*, 1974). One possibility is that in mammals the display of such antigens on the cell surface requires a small amount of these antigens to be already present and serve as a primer. The function of the minute amounts of antigens carried by the sperm to the membrane of the fertilised egg would then be to serve as primers when more antigen is to be produced during development. In addition, these primers are likely to be unequally distributed during the first cellular divisions. Only those cells receiving a particular primer would then be able to produce, or position correctly, that antigen. Such a system would, therefore, provide a mechanism for differential properties in the membrane of the embryonic cells.

The work done in the author's laboratory was supported by grants from the Centre National de la Recherche Scientifique, The National Institutes of Health and the André Meyer Foundation.

REFERENCES

Artzt, K., Bennett, D. & Jacob, F. (1974). Primitive teratocarcinoma cells express a differentiation antigen specified by a gene at the *T*-locus in the mouse. *Proceedings of the National Academy of Sciences, USA*, **71**, 811–14.

Artzt, K., Dubois, P., Bennett, D., Condamine, H., Babinet, C. & Jacob, F. (1973). Surface antigens common to mouse cleavage embryos and primitive teratocarcinoma cells in culture. *Proceedings of the National Academy of Sciences, USA*, **70**, 2988–92.

Artzt, K. & Jacob, F. (1974). Absence of serologically detectable H-2 on primitive teratocarcinoma cells in culture. *Transplantation*, **17**, 633–4.

Avrameas, S. & Ternynck, T. (1971). Peroxidase-labeled antibody and Fabconjugates with enhanced intracellular penetration. *Immunochemistry*, **8**, 1175–1179.

Bedford, J. M. (1968). Ultrastructural changes in the sperm head during fertilization in the rabbit. *American Journal of Anatomy*, **123**, 329,

Bennett, D., Boyse, E. A. & Old, L. J. (1971). Cell surface immunogenetics in the study of morphogenesis. In *Cell Interactions*, Lepetit Colloquium, ed. L. G. Silvestri, pp. 247–63. Amsterdam: North Holland.

Boon, T., Buckingham, M. E., Dexter, D. L., Jakob, H. & Jacob, F. (1974). Tératocarcinome de la souris: isolement et propriétés de deux lignées de myoblastes. *Annales microbiologiques de l'Institut Pasteur*, **125B**, 13–28.

Buc-Caron, M. H., Gachelin, G., Hofnung, M. & Jacob, F. (1974). Presence of a mouse embryonic antigen on human spermatozoa. *Proceedings of the National Academy of Sciences, USA*, **71**, 1730–3.

Dulbecco, R. & Freeman, G. (1959). Plaque production by the polyoma virus. *Virology*, **8**, 396–7.

Fellous, M. & Dausset, J. (1973). Histocompatibility antigens on human spermatozoa. In *Immunology of reproduction.* Second International Congress, Sofia, pp. 332–43. Sofia: Bulgarian Academy of Sciences Press.

Guénet, J. L., Jakob, H., Nicolas, J. F. & Jacob, F. (1974). Tératocarcinome de la souris, étude cytogénétique de cellules à potentialités multiples. *Annales microbiologiques de l'Institut Pasteur*, **125A**, 135–51.

Heyner, S. (1973). Detection of H-2 antigens on the cells of the early mouse embryo. *Transplantation*, **16**, 675–7.

Heyner, S., Brinster, R. L. & Palm, J. (1969). Effect of isoantibody on preimplantation mouse embryos. *Nature, London*, **222**, 783–4.

Jakob, H., Boon, T., Gaillard, J., Nicolas, J. F. & Jacob, F. (1973). Tératocarcinome de la souris: isolement, culture et propriétés de cellules à potentialités multiples. *Annales microbiologiques de l'Institut Pasteur*, **124B**, 269–82.

Kahan, B. W. & Ephrussi, B. (1970). Developmental potentialities of clonal *in vitro* cultures of mouse testicular teratoma. *Journal of the National Cancer Institute*, **44**, 1015–29.

Pierce, G. B. (1961). Teratocarcinomas, a problem in developmental biology. In *Canadian Cancer Conference*, vol. 4, pp. 119–37. New York & London: Academic Press.

Rosenthal, M. D., Wishnow, R. M. & Sato, G. H. (1970). *In vitro* growth and differentiation of clonal populations of multipotential mouse cells derived from a transplantable testicular teratocarcinoma. *Journal of the National Cancer Institute*, **44**, 1001–9.

Stevens, L. C. (1967). The biology of teratomas. *Advances in Morphogenesis*, **6**, 1–28.

Yanagimachi, R., Nicholson, G. L., Noda, Y. D. & Fujimoto, M. (1973). Electron microscopic observations of the distribution of acidic anionic residues on hamster spermatozoa and eggs, before and during fertilization. *Journal of Ultrastructure Research*, **43**, 344–53.

YANAGISAWA, K., BENNETT, D., BOYSE, E. A., DUNN, L. C. & DIMEO, A. (1974). Serological identification of sperm antigens specified by lethal *t*-alleles in the mouse. *Immunogenetics*, **1**, 57–67.

EMBRYO-DERIVED TERATOMA: A MODEL SYSTEM IN DEVELOPMENTAL AND TUMOR BIOLOGY

BY D. SOLTER, I. DAMJANOV AND H. KOPROWSKI

The Wistar Institute of Anatomy and Biology, 36th Street at Spruce, Philadelphia, Pennsylvania 19104, USA

and

Department of Pathology, University of Connecticut Health Center, Farmington, Connecticut 06032, USA

Teratomas are benign tumors, composed of several adult tissues, haphazardly mixed. *Teratocarcinomas* or *malignant teratomas* contain, in addition, undifferentiated cells called embryonal carcinoma cells (ECC) (Pierce, 1967), considered to be the stem cells of the tumor. Spontaneous teratomas are rare in the more commonly used experimental animals (Stevens, 1967*a*) and this rarity has presented a major obstacle to intensive investigations. The discovery of a mouse strain with a high incidence of spontaneous testicular teratoma (Stevens & Little, 1954), and the subsequent development of several methods for the experimental induction of teratomas, have made these tumors an easy and attractive experimental model. The possibility that tumors in general may be caused by an aberration in the normal process of differentiation (Pierce, 1967; Markert, 1968) made the study of teratoma especially relevant. During the last twenty years, teratomas have been intensively studied by developmental and tumor biologists and readers are referred to several reviews for more details (Stevens, 1967*a*; Pierce, 1967; Damjanov & Solter, 1974*a*).

Spontaneous testicular and ovarian teratomas in mice occur with a high frequency in mice of strain 129/ter Sv and LT respectively (Stevens, 1973; Stevens & Varnum, 1974). Testicular teratomas most probably originate from primordial germ cells (Stevens, 1967*b*), while ovarian teratomas develop from parthenogenetically activated ova (Stevens & Varnum, 1974). Both types of teratoma characteristically mimic normal embryonic development at the onset and then the embryo-like structures become disorganized, and the typical teratoma appears (Stevens, 1959; Stevens & Varnum, 1974).

Experimental teratoma can be produced in several ways. Stevens (1964) grafted genital ridges into the testes of adult mice and subsequently observed that a high incidence of teratoma developed from such grafts. Successful induction of teratoma in grafted genital ridges was dependent on several factors: strain of mouse used (Stevens, 1970*a*); age of the embryo from which the genital ridge was removed (Stevens, 1966); and numerous environmental influences related to the site of the graft (Stevens & Mackensen, 1961; Stevens, 1970*b*).

Teratomas can also be produced by grafting pre-implantation (Stevens, 1968) and early post-implantation (Stevens, 1970*c*; Solter, Škreb & Damjanov, 1970) mouse embryos or post-implantation rat embryos (Škreb, Švajger & Levak-Švajger, 1971; Škreb, Damjanov & Solter, 1972) to various extra-uterine sites, with the testis or subcapsular space of the kidney being the most common (for review see Damjanov & Solter, 1974*a*). It is also possible to produce teratomas by grafting older fetuses (Salaün, 1968) or even extra-embryonic membranes (Payne & Payne, 1961; Sobis & Vandeputte, 1974). It is important to mention that teratocarcinomas (i.e. retransplantable tumors) were never observed in the last two groups.

Early investigations primarily concerned the origin and histogenesis of teratoma and teratocarcinoma. It has now been fairly well established that the two proposed theories of origin, i.e. from primordial germ cells and from misplaced embryonic cells, are, in fact, complementary (Stevens, 1970*c*; Damjanov, Solter, Belicza & Škreb, 1971*a*; Damjanov & Solter, 1974*a*).

The stem cells of teratoma possess the ability not only to divide and to remain in an undifferentiated state, but also to differentiate into various adult tissues. The presence of such cells in the tumor is a pre-requisite for the successful retransplantation, and the ability of the tumor to grow rapidly and finally cause the death of the host, usually by emaciation. A single ECC has been repeatedly shown to give rise to a tumor composed of numerous somatic tissues (Kleinsmith & Pierce, 1964; Jami & Ritz, 1974). The mechanism of transition from undifferentiated stem cells into well-differentiated tissues is, however, still unclear.

In recent years, several new experimental approaches to the study of teratocarcinoma have been introduced and are discussed in detail elsewhere (see Jacob, this volume).

Numerous cell lines have been derived from transplantable mouse teratocarcinoma (originated in 129 mice) and even after a prolonged period in culture they have retained their capacity for differentiation when injected into adult mice (Finch & Ephrussi, 1967; Kahan & Ephrussi, 1970; Rosenthal, Wishnow & Sato, 1970; Evans, 1972; Jami, Failly & Ritz, 1973; Martin & Evans, 1974; Jami & Ritz, 1974). These cell lines can be used to investigate differentiation and the factors that regulate differentiation *in vitro*, and recent reports suggest that such an approach will be fruitful (Bernstine, Hooper, Grandchamp & Ephrussi, 1973; Gearhart & Mintz, 1974).

Another useful approach is that of investigating the antigenic properties of embryonal carcinoma cells, their relationship with embryonic and other tumor cells and changes in antigenic properties during differentiation. Exploration of these aspects is also well under way (Edidin, Patthey, McGuire & Sheffield, 1971; Artzt *et al.*, 1973; Artzt, Bennett & Jacob, 1974; Buc-Caron, Gachelin, Hofnung & Jacob, 1974; Gooding & Edidin, 1974; Jacob, this volume).

It is obvious that a model with the complexity of the experimental teratoma can be approached from numerous viewpoints. From the beginning, our interest was centered on the developmental aspects of embryo-derived teratoma and teratocarcinoma in order to try to elucidate some of the problems of normal development of the early mammalian embryo. After finding that some transplanted embryos develop into benign teratoma composed of mature somatic tissues and some develop into teratocarcinoma that possess, in addition, embryonal carcinoma cells (Solter *et al.*, 1970), we have been trying to determine what causes such different behavior. Therefore, we will limit our present discussion to the following topics: (1) factors that regulate the behavior of the transplanted embryo, both graft- and host-related, and (2) histogenesis of embryo-derived teratoma.

MATERIALS AND METHODS

Mice of strains C3H/HeJ, C57BL/6J and AKR/J were obtained from Jackson Laboratory and kept in our laboratory on standard mouse diet and water *ad libitum*. Virgin females were caged with males overnight and checked for vaginal plugs the following morning. Seven-day-old embryos (the day the plug was found was considered as day 0) were removed in the morning from decidual swellings, cleaned from membranes and cut above the amnion. Only the embryonic part was used for grafting (Solter *et al.*, 1970). The host animal was anesthetized with an intraperitoneal (i.p.) injection of sodium pentobarbital, the left kidney was exposed through a lateral incision and fixed with Desmarres chalazion forceps. A small pocket was made between the kidney and kidney capsule with watchmaker's forceps, the embryo was transferred to this pocket with a braking pipette, and the kidney was then released and skin closed with clips. If, because of bleeding, it was not possible to discern the presence of embryo under the kidney capsule, the animal was discarded.

Splenectomy was performed at the same time as the embryo transfer. Spleen vessels were ligated and spleen removed. Ligation of ovaries was done one week before embryo grafting; a ligature was placed between the ovaries and corresponding tubes so that the blood supply to the ovaries was cut off, but the ovaries were left in place. Rabbit anti-mouse thymocyte serum (ATS, from Microbiological Associates, Bethesda, Maryland), at a dosage of 0.25 ml i.p. was injected one day before embryo transfer, on the day of the operation, one and three days after the operation and then at weekly intervals until the animals were sacrificed.

Tumor-bearing animals were sacrificed two months after the operation; tumor, spleen and liver were removed and weighed. Specimens of spleen and liver and several pieces from different areas of the tumor were removed and fixed in 10% formalin, embedded in paraffin, sectioned, stained with hematoxylin and eosin and examined.

FACTORS THAT REGULATE THE DEVELOPMENT OF THE GRAFTED EMBRYO

The factors that influence the development of the transplanted embryo are enumerated in Table 1. The list is obviously limited by our insufficient knowledge and should be expanded as the investigation progresses.

Table 1. *Factors which can influence and/or direct the development of embryo transplanted to extra-uterine sites*

Embryo-related factors
Stage of development
Size of the transplant – whole embryo
– part of an embryo
Manipulation of the embryo (carcinogens, viruses)?
Species?
Host-related factors
Nonspecific
Mechanical – site of transplant
Nutritional – site of transplant
Serum effect?
Hormonal effect?
Specific – species and strain related
Genetic
Immunological

Embryo-related factors

Age of the embryo

As stated previously teratoma can develop from whole embryos, from various parts of the embryo and from some extra-embryonic membranes. Teratocarcinoma, on the other hand, can develop only from embryos at a specific developmental stage and/or from genital ridges of a definite age (Stevens, 1966; Damjanov, Solter & Škreb, 1971*b*). Stevens (1968; 1970*c*) described transplantable teratocarcinoma derived from cleavage and early post-implantation mouse embryos of 129 and A/He strains. Damjanov *et al.* (1971*b*) showed that 6- and 7-day-old C3H/H embryos can develop into teratocarcinoma when transplanted under the kidney capsule while 8-day-old or older embryos develop only into mature teratoma. From these results, it seems that with the formation of the organ primordia, the capacity of the embryonic cells to remain undifferentiated disappears. The most probable reason for this change is irreversible determination of embryonic cells. It is also possible that transplantation and the subsequent disruption of the normal cellular relationship within the embryo is no longer a sufficient stimulus to keep some embryonic cells in an undifferentiated state. Mouse egg-cylinders do not develop into teratocarcinoma in certain mouse strains (see later) but only into teratoma. It will be of interest to determine whether this strain-related 'resistance' is also present when pre-implantation embryos are transplanted.

Size of the embryo

Here, the concern is not with the actual size of the graft, but with which part of the embryo is necessary for subsequent development of teratoma and/or teratocarcinoma. So far there has been no attempt to transplant parts of pre-implantation embryos to extra-uterine sites, although successful micromanipulation of the blastocyst offers the possibility for such experiments (Gardner, 1971). Transplantation of isolated germ layers has been done only in rats and the results suggested that the morphological division of an egg-cylinder into endoderm, mesoderm and ectoderm should be distinguished from the real developmental capacities of isolated germ layers (Levak-Švajger, Švajger & Škreb, 1969; Levak-Švajger & Švajger, 1971). Since teratocarcinomas have been observed in rats (Škreb *et al.*, 1971), the experiments should be repeated in mice to see whether one or more germ layers contribute undifferentiated cells to the final tumor. Our results (Solter & Damjanov, 1973) have shown that the extra-embryonic part of mouse egg-cylinder develops into a very small tumor composed of cells resembling parietal yolk sac cells. Grafting of the whole egg-cylinder produced the same type of tumors as when only the embryonic part was grafted (Solter & Damjanov, unpublished observations).

Manipulation of embryos before grafting

Embryo-derived teratomas are potentially a very useful model for studying the effects of different agents on mammalian embryo development; the embryos can be directly exposed and the subsequent development observed. The most obvious agents to be tested in this way are the various carcinogens and oncogenic viruses. Such possibilities, however, have so far been largely ignored. We have examined the effect of short periods of exposure of egg-cylinders to urethan and found that tumors derived from embryos so treated did not differ from controls. Recently, we also exposed C57BL/6 egg-cylinders to simian virus 40 (SV40) and polyoma virus for 4 hours before transplantation. Tumours obtained from these embryos were, as far as we could tell, quite similar to the controls and we could not detect the presence of viruses in tumor cells (Solter, unpublished observations), although egg-cylinders exposed to SV40 or polyoma and examined ultrastructurally showed signs of viral uptake and replication (Biczysko, Solter, Pienkowski & Koprowski, 1973; Solter, Biczysko & Koprowski, 1974*a*). These negative results, however, should not discourage further explorations of this system.

Another aspect which might prove fruitful is the investigation of developmental capacities of embryos developed *in vitro*. Billington, Graham & McLaren (1968) transplanted blastocysts grown *in vitro* to extra-uterine sites to prove that the uterine factor is not necessary for normal embryonic development. The period between grafting and sacrificing the host was too short, however, to show differences between tumors derived from blastocysts grown *in vitro* and those from blastocysts matured *in utero*. Mouse blastocysts can develop *in vitro* (Hsu, 1972, 1973; Pienkowski, Solter & Koprowski, 1974) into structures which are similar to egg-cylinders (Solter, Biczysko, Pienkowski & Koprowski, 1974*b*) and it will be interesting to investigate the growth and differentiation of such egg-cylinders transplanted to extra-uterine sites.

Species and strain

As mentioned previously, rat egg-cylinders grafted into the anterior chamber of the eye or under the kidney capsule develop into teratoma (Levak-Švajger & Škreb, 1965; Škreb *et al.*, 1971) but they never form teratocarcinoma. Similar findings were observed in some mouse strains (Damjanov & Solter, 1974*b*). Although it now seems that the inability to form teratocarcinoma is a host-dependent characteristic in mice (see later), it is possible that this phenomenon in rats could be embryo-related. The use of several different rat strains and their hybrids should determine if the rat embryo is intrinsically unable to develop into teratocarcinoma.

Host-related factors

Effect of different extra-uterine sites

The selection of the extra-uterine sites for the embryonal graft is important to the assessment of the developmental capacities of an embryo. Although the uterine environment is the most favorable for the embryo, it is also probably the most exacting. This has been shown by the uniform failure of parthenogenetically stimulated embryos to develop *in utero* beyond the early somitic stage, compared with their ability to grow and develop into teratoma in extra-uterine sites (Stevens & Varnum, 1974; C. F. Graham, personal communication). Apparently, although an embryo might be abnormal and unable to survive *in utero*, its components might be sufficiently vigorous to thrive in a less demanding environment.

Embryos have been transplanted to various extra-uterine sites with the testis, subcapsular space of the kidney and anterior chamber of the eye most often used. The testicular graft is technically the easiest, but makes it difficult to observe the actual deposition of the embryo. Despite this drawback, it is the preferred site for pre-implantation embryos. The kidney graft is somewhat more difficult to perform, but post-implantation embryos can be seen quite clearly after grafting. This makes possible the simultaneous use and juxtapositioning of several embryos or several parts of the embryos. The anterior chamber of the eye is inconvenient since it requires very deep anesthesia, sterile conditions and is technically the most difficult.

Beside these purely technical aspects, some other points should be considered. Levak-Švajger & Škreb (1965) showed that rat embryos composed of ectoderm and endoderm only, rarely develop mesodermal derivatives when grafted into the anterior chamber of the eye. However, when these embryos were grafted under kidney capsule, the developing tumors invariably possessed mesodermal derivatives (Škreb *et al.*, 1971). These results suggested that full developmental capacities of an embryo cannot be exposed unless the grafting site is optimal. We now know that an embryo deposited in the anterior eye chamber floats for several days before settling on the iris and establishing contact with the host's circulatory system. In kidney grafts, such contact is established immediately and this clearly better nutritional condition could account for differences in development in the two graft sites. We have also shown (Damjanov &

Solter, 1974*b*) that embryo-derived tumors are heavier when embryos were transplanted to the kidney than into testes. At present, it is unclear whether this is due to such mechanical factors as pressure by the tunica albuginea, less efficient blood supply once the tumor is big enough to compress blood vessels along the ductus deferens, or to some specific nonmechanical factor present in the testis and/or in the kidney.

Site-related influences were also reported for teratomas derived from grafted genital ridges (Stevens & Mackensen, 1961; Stevens, 1970*b*). In this case, teratomas developed only in those sites that descended into the scrotum and not in others. It was hypothesized that the lower temperature in the scrotum promoted teratocarcinogenesis in grafted genital ridges.

Effect of serum and hormones

Size and character of embryo-derived teratomas differ between various mouse strains (see later). Although it is possible that this is due to some specific, possibly immunological, factor, nonspecific differences (in, for example, serum composition) cannot be ruled out. Mouse serum is toxic for syngeneic cells *in vitro* (Mishell & Dutton, 1967; Jejeebhoy, 1974) and it is possible that this toxic effect is also present for embryonic cells *in vivo* in some mouse strains; this could account for observed differences. Stevens (1970*c*) investigated the effect of male and female sexual hormones on the development of teratoma in grafted genital ridges and concluded that hormones were not involved. We observed a difference in the weight of embryo-derived teratomas between male and female C57BL mice although the incidence of tumors was the same. In addition, ligation of ovaries significantly reduced both the weight and the incidence of tumors. These results suggested that there might be some hormonal influence and indicate the need for further investigation.

Effect of species and strain on embryo-derived teratoma and teratocarcinoma

Even though investigations of the fate of grafted early embryos have been confined to rats and mice only, one profound difference has been found: embryo-derived tumors in rats are always benign, with limited potential for growth, and they do not endanger the life of the host even six months after transplantation (Škreb *et al.*, 1971). In mice, however, a certain percentage of embryos develop into teratocarcinomas which grow fast and in an uncontrolled fashion, and can kill the host two months after transplantation (Solter *et al.*, 1970). The reasons for such different behavior are not clear. It is worth mentioning that spontaneous teratoma and/or teratocarcinoma have also never been reported in rats. The paucity of data makes it unprofitable to speculate whether determination of rat embryos takes place much earlier than it does in mice or if rats as a species are resistant to the development of teratocarcinoma. As suggested earlier, transplantation of pre-implantation rat embryos could furnish new data. It would seem worthwhile also to extend transplantation experiments to other mammalian species, both to see how common are the phenomena we discerned in mice, and also to get a better insight into the process of differentiation in other species of mammalian embryos.

Even within the mouse species, considerable difference in the occurrence of spontaneous and induced teratoma and teratocarcinoma exists between different strains. Spontaneous teratomas have been discussed and we can note that their appearance is limited to certain specific mouse strains and is probably genetically regulated with only one or very few genes involved (Stevens, 1973; Stevens & Varnum, 1974). Successful induction of teratoma from grafted genital ridges could be accomplished in some mouse strains (Stevens, 1970*a*, *b*, *c*) and not in others. Transplantation of embryos has so far always resulted in the development of teratoma in all the mouse strains tested, technical failures notwithstanding. However, our previous results (Škreb *et al.*, 1972; Damjanov & Solter, 1974*b*) and results presented in Table 2 clearly show teratocarcinomas were observed only in

Table 2. *Incidence of teratoma and teratocarcinoma*

Host	No. of animals	No. of animals with teratoma†	No. of animals with teratocarcinoma†
C3H males	46	13 (41)‡	28
C3H females	41	13 (32)‡	19
C57BL males	48	37	0
C57BL males splenectomized	49	41	0
C57BL males treated with ATS*	39	32	0
C57BL females	42	27	0
		$P < 0.01$ (27 vs 3)	
C57BL females splenectomized	34	3	0
		$P < 0.025$ (27 vs 8)	
C57BL females ligation of ovaries	26	8	0
AKR males	35	32	0

* ATS, rabbit anti-mouse thymocyte serum.
† Chi-squared analysis showed significant difference only where indicated.
‡ Total number of animals with tumors is given in brackets.

specific strains. There are several criteria by which tumors are divided into teratoma and teratocarcinoma. Teratocarcinomas are usually much bigger (average weight approx. 6 g) than teratomas (average weight about 0.5–1.0 g). On histological examination, teratomas are composed exclusively of mature somatic tissues while teratocarcinomas invariably also possess embryonal carcinoma cells (Plate 1*d*). The most critical test is that of the retransplantability of the tumor since only teratocarcinoma can be retransplanted indefinitely due to the presence of immature stem cells. It is obvious that some teratocarcinoma might be mistaken for teratoma since it is practically impossible to examine or retransplant the whole tumor, and such errors could account for differences in the percentage of teratocarcinoma between different experimental series. We therefore examined embryo-derived tumors in teratocarcinoma-negative strains with special care and tried to retransplant as many as possible; so far they have all proven to be teratomas. We have found that in C3H, CBA and A strains about half of transplanted embryos develop into teratocarcinoma (Damjanov & Solter, 1974*b*; Table 2), while in C57BL and AKR strains all embryos develop into mature teratomas (Table 2). These results suggested two interdependent questions: first, why cannot embryo-derived teratocarcinoma be produced in some mouse strains, and second, in strains where they can be produced, why do only some embryos develop into teratocarcinoma whereas others do not? One possibility might be that in teratocarcinoma-negative strains, the embryos became determined much earlier and thus the undifferentiated or multipotential cells were not present at the time of implantation. Two lines of evidence tend to refute this possibility. When 7-day-old C57BL embryos were transplanted into (C3H × C57BL) F1 hybrids about half of them developed into teratocarcinomas which were retransplantable in hybrids but not in C57BL mice (Škreb *et al.*, 1972; Damjanov & Solter, 1974*b*). Secondly, when embryo-derived tumors were examined 15 days after grafting instead of after 2 months practically all grafts had undifferentiated embryonal carcinoma cells (Šćukanec, Solter & Damjanov, unpublished observations). These results were interpreted as showing that all embryos are about equal at the time of implantation and that they develop differently depending on the interaction with the host.

As a working hypothesis, we assumed that this interaction was mostly immunological in nature. This assumption is based on several sets of data. First, it is now generally accepted that embryonic cells express antigens that are not present or expressed in adult cells and that the adult animal can react against such antigens (for general discussion on this subject see Alexander, 1972; Anderson & Coggin, 1972). Second, several antigens present in teratocarcinoma cells were observed after xenogeneic (Edidin *et al.*, 1971; Gooding & Edidin, 1974) or syngeneic immunization (Artzt *et al.*, 1973, 1974) with teratocarcinoma cells grown *in vivo* and *in vitro*. Antibodies so produced cross-reacted with mouse embryos and some of them also with several unrelated mouse tumor lines. Finally, the fact is that the degree of immune response in mice is genetically controlled and that so-called weak and strong responding mouse strains have been repeatedly described (for details see Melchers, Rajewsky & Shreffler, 1973; Gasser & Silvers, 1974; Green, 1974). All these data suggested that the existence of teratocarcinoma-resistant and teratocarcinoma-susceptible mouse strains could be explained on an immunological basis and also that manipulations of the immune system of the host might influence the development of transplanted embryos.

As shown in Table 2, although splenectomy or treatment with rabbit anti-mouse thymocyte serum (ATS) did not change the incidence of teratoma in C57BL males, the weight of tumors was significantly reduced (Table 4). In C57BL females in which embryo-derived teratoma are considerably smaller than in males (Table 3) splenectomy practically eliminated tumors. Ligation of ovaries also significantly decreased both weight

Table 3. *Weight of tumor, spleen and liver in several mouse strains*

		No. of animals	Tumor	Spleen	Liver
C3H males with:	Tumor	41	3242 ± 426	385 ± 37*	1560 ± 34*
	Control	25	–	109 ± 3	1324 ± 48
C3H females with :	Tumor	32	4146 ± 697	389 ± 47*	1539 ± 50*
	Control	20	–	137 ± 14	1147 ± 74
C57BL males with:	Tumor	37	2505 ± 418	384 ± 43*	1567 ± 58*
	Control	20	–	122 ± 14	1262 ± 85
C57BL females with:	Tumor	27	830 ± 255†	253 ± 41*	1237 ± 37
	Control	20	–	131 ± 14	1395 ± 87
AKR males with:	Tumor	32	1887 ± 303‡	321 ± 36*	1766 ± 45*
	Control	20	–	75 ± 2	1156 ± 63

* Significantly different from control ($P < 0.01$).
† Significantly different from all other tumor weights ($P < 0.01$).
‡ Significantly different from C3H and C57BL females ($P < 0.01$) and C3H males ($P < 0.02$) but not from C57BL males.
Weight is given in mg +S.E. Tumors were derived from grafts of 7-day-old syngeneic embryos under the kidney capsule.

and incidence of embryo-derived teratomas in C57BL females (Tables 2 & 4). None of the described manipulations, however, induced teratocarcinoma in C57BL mice.

This rather limited amount of data is difficult to fit into the already complex and controversial problem of tumor immunity. Our first idea was that the presence and survival of embryonal carcinoma cells might be due to the presence of blocking antibodies that protect tumor cells from cellular

Table 4. *Effect of splenectomy, ATS*[a] *treatment and ligation of ovaries on the weight of tumor, spleen and liver*

Host	No. of animals with tumor	Tumor	Spleen	Liver
C57BL males	41	2505 ± 418	384 ± 43	1567 ± 58
C57BL males: splenectomized	41	247 ± 62^{b}	–	1592 ± 40
C57BL males treated with ATS	32	620 ± 160^{b}	249 ± 24^{c}	1683 ± 52
C57BL females	27	830 ± 253	253 ± 41	1237 ± 37
C57BL females: splenectomized	3	(10; 20; 25)[e]	–	1146 ± 22^{d}
C57BL females: Ligation of ovaries	8	46 ± 20^{b}	94 ± 3^{b}	1105 ± 31^{c}

[a] ATS, rabbit anti-mouse thymocyte serum. [b] [c] [d] Significantly different from weight in untreated host; [b] ($P < 0.01$); [c] ($P < 0.02$); [d] ($P < 0.05$). [e] Mean was not calculated. Tumors' weight.

Weight is given in mg +S.E. Tumors were derived from grafts of 7-day-old syngeneic embryos under the kidney capsule.

immunity (Sjögren, 1973; Hellström & Hellström, 1974). Our results with splenectomy would fit this hypothesis since the spleen as a potent source of blocking antibodies might protect tumors from cellular immunity and hence splenectomy would inhibit tumor growth (Ferrer, 1968; Prehn & Lappé, 1971). Splenectomy significantly reduced the weight of embryo-derived teratoma in C3H mice (Damjanov & Solter, 1974*c*), a fact that seems to corroborate such a hypothesis. If our hypothesis is correct that the presence or absence of embryonal carcinoma cells in the graft and its subsequent development depends on the interplay between blocking antibodies and cellular immunity (Damjanov & Solter, 1974*c*) one would expect that reduction of cellular immunity would be beneficial for tumor growth. However, treatment with ATS (Table 4) or TIR (thymectomy, lethal irradiation and bone marrow reconstitution) of C57BL mice prior to embryo transfer actually reduced the weight of tumors (Damjanov & Solter, unpublished observations). There are several possible explanations for such results, the most attractive of which is the recently proposed theory of immunostimulation of tumor development (Prehn & Lappé, 1971; Prehn, 1972*a*, *b*; Fidler, 1974; Jejeebhoy, 1974). According to this theory, a small degree of cellular immunity might actually stimulate tumor growth. Several results suggest that this might be the case in embryo-derived tumors. The weight of tumors from C3H embryos transplanted to (C3H × C57BL) F1 hybrids was significantly higher than the weight of tumors produced in a syngeneic host (Damjanov & Solter, 1974*c*). The same was true for C57BL embryos transplanted to (C3H × C57BL) F1 hybrids (Damjanov & Solter, 1974*b*). More importantly, in the latter case about half of the C57BL embryos developed into teratocarcinomas which were again retransplantable in hybrids but not in C57BL syngeneic animals (Damjanov & Solter, 1974*b*). It is possible, therefore, that a low degree of immune response is necessary for development of teratocarcinoma and that inhibition of cellular immunity by ATS or TIR was too complete. In our future work, we will try to modify interference with cellular immunity in such a way that an early, possibly beneficial, response would be preserved and the later, detrimental response would be abolished, allowing the tumor to 'sneak through'.

There are several possible explanations for the permissiveness versus nonpermissiveness for teratocarcinoma in different mouse strains. The phenomenon is probably host-related and not embryo-related (Damjanov & Solter, 1974*b*, *c*). In addition, judging from hybrid experiments, it is also probable that permissiveness is a dominant characteristic. In C57BL mice, which are resistant to teratocarcinogenesis, blocking antibodies may not be produced or cellular immunity may be very strong, whereas the opposite would be true for the C3H mouse strain. It is also possible that both mechanisms exist simultaneously and that the nature of the embryo-derived tumor depends on the interaction of humoral and cellular immunity. The recently described phenomenon that a low dose of tumor cells actually induced tolerance and prevented tumor immunity (Kolsch, Mengersen & Diller, 1973) may also be relevant in this context.

At present, it is difficult to determine whether the effect of the ligation of ovaries is due to an immunological or a hormonal mechanism. Coggin & Anderson (1972), using ligation of ovaries in hamsters, described the spontaneous appearance of cytotoxic lymphoid cells against SV40-transformed tumor cells. They also found that females with ligated ovaries were resistant to challenges with SV40-induced tumors. They suggested that the removal of ovarian hormones might result in the development of immunity to fetal antigens present in tumor cells.

Another possibility that must be investigated is that the slow destruction of ovarian tissue following ligation actually serves as a source of antigens common to both the ovarian and the tumor tissues. Coggin & Anderson's (1972) finding that complete removal of ovaries offered less protection against tumor cells more adequately fits this latter hypothesis. The relevance of this hypothesis for teratoma systems is especially important because of the finding that the ovaries are the only adult tissue that shares antigenic characteristics with teratocarcinoma cells (Gooding & Edidin, 1974).

At this point it must be stressed that investigating the effects of the immune response to embryo-derived tumors does not just add another poorly defined system to the already immense and confusing area of tumor immunology. We are much more interested in the question of whether the immune system and immune reactions play any part in the complex differentiation of the mammalian embryo. It has recently been suggested that such may be the case (Bennett, Boyse & Old, 1971; Artzt *et al.*, 1974), and, if so, embryo-derived teratocarcinoma would be the most suitable model for elucidating these fascinating questions.

HOST REACTION TOWARDS TERATOMA

Our recent finding that embryo-derived teratocarcinoma elicits splenomegaly in the syngeneic host prompted us to investigate in more detail host reactions toward transplanted embryos and tumors derived from them (Damjanov & Solter, 1974*c*). As presented in Table 3, embryo-derived tumors induced a two- or threefold increase in spleen weights in comparison to controls. Splenomegaly is present in all tumor-bearing animals and is much more pronounced in animals with big tumors, especially in animals with teratocarcinoma (Damjanov & Solter, 1974*c*; and unpublished observations). A direct correlation also exists between tumor and spleen weight, with the correlation coefficient for the different groups in Table 3 ranging from 0.48 to 0.68. When examined histologically, spleens from tumor-bearing animals showed essentially two types of hyperplasia: follicular and diffuse. Follicular hyperplasia is due to enlarged follicles with very active germinate centers. In diffuse hyperplasia normal spleen structure is disturbed, follicles are no longer visible and the sinusoids are full of lymphoid cells and cells of extramedullary hematopoiesis (Plate 1*a*). Splenomegaly is considered to be a sign of an immunological reaction of the host to the tumor (Woodruff & Symes, 1962; Blamey & Evans, 1971; McKhann & Jagarlamoody, 1971) but its exact nature and significance are not clear and further investigation of this problem is in progress.

We also observed (Table 3) that teratomas were accompanied by hepatomegaly. The weight of liver in tumor-bearing animals was significantly higher than in controls but there was no correlation between the weights of tumor and liver. Histological examination revealed the presence of numerous binucleate hepatocytes, an increased number of Kupffer's cells and small centers of extramedullary hematopoiesis (Plate 1*b*). Extramedullary hematopoiesis is of special interest since it can be due to the reaction of the liver (Friedell, Sherman & Sommers, 1960; Lee & Aleyassine, 1971) but it can also be caused by teratocarcinoma cells that colonize the host liver. The hematopoietic potential of teratocarcinoma cells has already been suggested (Auerbach, 1972). Further studies of these questions are in progress.

Finally, if our hypothesis that embryo-derived tumors cause immune cellular response is valid, we should be able to find signs of such a response in the tumor itself and in regional lymph nodes. Lymph nodes have not yet been examined in detail but we often observed infiltration in tumors of small round cells that resembled lymphoid cells. These infiltrations were mostly located around areas composed of undifferentiated cells or around areas of mature and immature neural tissue (Plate 1*c*). It is possible that neural tissue, which is normally an immunologically privileged site, could act as a strong immunogen in ectopic sites.

HISTOGENESIS OF EMBRYO-DERIVED TERATOMA

Several publications have dealt in detail with the development of both spontaneous and induced teratoma, from the histological point of view (Stevens, 1959, 1970*c*). Embryonal carcinoma cells (ECC) are both the hallmark and a necessary prerequisite for the existence of retransplantable teratocarcinoma. They are usually present in small and distinct nests in highly cellular, immature mesenchyme (Plate 1*d*). Although it is believed that they are the progenitors of all differentiated tissues in teratoma (Kleinsmith & Pierce, 1964; Jami & Ritz, 1974; Damjanov & Solter, 1974*a*), the exact mechanism of such transitions is not known. Although a single ECC can give rise to tumors composed of various differentiated tissues (Kleinsmith & Pierce, 1964; Jami & Ritz, 1974), we do not know whether all ECC in teratocarcinoma are multipotential all the time or if their multipotentiality is gradually reduced until they can be precursors of only few or one type of tissue. It is also not clear whether ECC directly change into differentiated tissues or if some intermediary stages are necessary. It is sometimes possible to observe the direct transition of ECC into neural tissue (Damjanov, Solter & Šerman, 1973) or serous glands (Plate 2*a*) but such examples are rare and, since it is sometimes difficult to recognize ECC with absolute certainty, are inconclusive.

The other, and in our opinion much more likely, possibility is that ECC change into some intermediate types before final differentiation takes place. We also believe that such a transition does not occur at the level of the single cell but that the whole group or nest of ECC change into a structure that resembles either the whole or a part of an early post-implantation embryo. In areas with ECC one often finds structures that resemble mouse egg-cylinders. Such structures are called embryoid bodies, and can be at different levels of development. Some embryoid bodies have a single layer of cells resembling either ectoderm or endoderm and are then called ectodermal or endodermal vesicles (Plate 2*b*). Embryoid bodies can also resemble whole embryos with two or three germ layers (Plate 2*c* and *d*). In our opinion these different classes of embryoid bodies are the immediate precursors of mature tissues, so that endodermal vesicles differentiate into different types of epithelial cysts, ectodermal vesicles into neural tubes or cysts covered with keratinous epithelium, and more complex embryoid bodies can give rise to all the types of adult tissues observed in teratocarcinoma. If our hypothesis is correct, the histogenesis of teratocarcinoma is actually a byproduct of repeated embryogenesis.

Embryonal carcinoma cells act as zygotes; they divide until a certain size is reached and then form an embryo-like structure. Since conditions within the tumor certainly do not favor the further development of such embryos, the proper relations between various parts of embryo are disrupted and histogenesis proceeds in a haphazard manner. One consequence of the proposed hypothesis would be that the presence of ECC in a tumor does not necessarily imply retransplantability and/or multipotentiality. Namely if ECC lose their ability to form embryo-like structures they might still be tumorigenic but the resulting tumors would be embryonal cell carcinoma without a trace of differentiated tissues. Such reduction of multipotentiality of ECC has often been described (Stevens, 1970*c*; Bernstine *et al.*, 1973). Evidence for the progression of ECC into embryo-like structures that, in turn, differentiate into mature tissues is circumstantial but suggestive.

Since histological evidence is essentially in the form of static pictures, it cannot prove the existence of suggested dynamic development, and therefore some other points should be considered. The egg-cylinders we are transplanting are cleared of extra-embryonic membranes and of extra-embryonic parts. If we assume that an irreversible determination exists at this time, the cells of the embryonic part of the egg-cylinder would not be able to form extra-embryonic membranes. Nevertheless, we always find in a certain percentage of the teratoma, cells that resemble cells of the parietal yolk sac to the extent that they produce a Reichert's membrane and eventually sometimes form typical yolk sac carcinoma (Damjanov & Solter, 1973). One can argue that, if extra-embryonic membranes can form differentiated adult tissues (Sobis & Vandeputte, 1974), the opposite might also be possible, so we decided to look for another example.

Moore & Metcalf (1970) proved that hematopoietic cells are derived from precursors found in yolk sac of the 8-day-old mouse embryo. These precursors are necessary for further colonization of the embryo and subsequent intra-embryonal hematopoiesis. Moore & Metcalf (1970) also showed that such precursors are not present in embryonic parts of 7-day-old embryos and our results are in agreement with such conclusions (Pienkowski *et al.*, 1974). So although the precursors for hematopoietic cells do not exist in the part of the embryo we are grafting, bone marrow with active hematopoiesis is regularly found in teratoma. One possibility is that bone marrow is colonized with hematopoietic cells from the host. Although unlikely, such a possibility should be carefully investigated. Another possibility is that embryos with active yolk sac appeared in teratoma and that those hematopoietic cells colonize the bone marrow; the embryoid bodies mentioned often contain red blood cells and cells which resemble hematopoietic cells (Plate 2*c* and *d*). If these are really hematopoietic cells, our concept of the histogenesis in teratocarcinoma is much more tenable and the value of the embryo-derived tumor as a model for investigating early embryonic and somatic tissue differentiation gains considerably. It is important to realize that in our hypothesis a structure of definite size and shape and not an amorphous mass of cells is required for the beginning of differentiation.

CONCLUSIONS

Mouse and rat embryos transplanted to extra-uterine sites develop into tumors composed of different somatic tissues. Whether these tumors are benign teratoma with a limited capacity for growth or fast growing, re-transplantable teratocarcinomas that possess in addition to somatic tissues embryonal carcinoma cells, is dependent on the species and strain of the host. The fate of the transplanted embryo, whether it will become teratoma or teratocarcinoma, is regulated by numerous embryo- and host-related factors. Probably the most important single factor is the immune reaction of the host, the exact nature of which is at present unclear. The potential importance of elucidating the exact effect of the immune reaction on the development and differentiation of the transplanted embryo is considerable if we wish to understand the role of the immune system in normal development.

It is proposed that histogenesis in embryo-derived teratocarcinoma follows a regular pattern of development. Embryonal carcinoma cells would divide and become a group of cells of a certain size. At this point, such a group would form an embryo-like structure, either an ectodermal or endodermal vesicle or a more complex embryoid body composed of two or three germ layers, and further histogenesis would then proceed.

We deem it very likely that future research on mammalian development will shift from the investigation of normal embryonic development, hampered by the lack of material, to analysis of embryo-derived teratoma, as a much easier and more rewarding model.

The authors' original work was supported in part by grants from the Yugoslav Federal Science Foundation No. 812/3, the Council for Scientific Affairs of SR Croatia, the National Institutes of Health, USA through PL 480 Research Agreement Grant No. 02-038-1, and in part by CA 04534 and CA 10815 from the National Cancer Institute. One of us (D.S.) was the recipient of a fellowship from Damon Runyon Memorial Cancer Fund (DRF-810). We thank Nancy Adams, Elsa Fernbach and Marguarite Solomon for excellent technical help.

REFERENCES

Alexander, P. (1972). Foetal 'antigens' in cancer. *Nature, London*, **235**, 137–40.

Anderson, N. G. & Coggin, J. H., Jr (1972). Retrogenesis: problems and prospects. In *Proceedings of the Second Conference on Embryonic and Fetal Antigens in Cancer*, ed. N. G. Anderson, J. H. Coggin, Jr, E. Cole & J. W. Holleman, pp. 361–8. Springfield, Virginia: National Technical Information Service.

Artzt, K., Bennett, D. & Jacob, F. (1974). Primitive teratocarcinoma cells express a differentiation antigen specified by a gene at the *T*-locus in the mouse. *Proceedings of the National Academy of Sciences, USA*, **71**, 811–14.

Artzt, K., Dubois, P., Bennett, D., Condamine, H., Babinet, C. & Jacob, F. (1973). Surface antigens common to mouse cleavage embryos and primitive teratocarcinoma cells in culture. *Proceedings of the National Academy of Sciences, USA*, **70**, 2988–92.

Auerbach, R. (1972). Controlled differentiation of teratoma cells. In *Cell Differentiation*, ed. R. Harris, P. Alin & D. Viza, pp. 119–23. Copenhagen: Munksgaard.

Bennett, D., Boyse, E. A. & Old, L. J. (1971). Cell surface immunogenetics in the study of morphogenesis. In *Proceedings of the Third Lepetit Colloquium*, ed. L. G. Silvestri, pp. 247–63. Amsterdam: North-Holland.

Bernstine, E. G., Hooper, M. L., Grandchamp, S. & Ephrussi, B. (1973). Alkaline phosphatase activity in mouse teratoma. *Proceedings of the National Academy of Sciences, USA*, **70**, 3899–903.

Biczysko, W., Solter, D., Pienkowski, M. & Koprowski, H. (1973). Interaction of early mouse embryos with oncogenic viruses – simian virus 40 and polyoma I. Ultrastructural studies. *Journal of the National Cancer Institute*, **51**, 1945–54.

Billington, W. D., Graham, C. F. & McLaren, A. (1968). Extra-uterine development of mouse blastocyst cultured *in vitro* from early cleavage stages. *Journal of Embryology and Experimental Morphology*, **20**, 391–400.

Blamey, R. W. & Evans, D. M. D. (1971). Spleen weight in rats during tumour growth and in homograft rejection. *British Journal of Cancer*, **25**, 527–32.

Buc-Caron, M.-M., Gachelin, G., Hofnung, M. & Jacob, F. (1974). Presence of mouse embryonic antigen on human spermatozoa. *Proceedings of the National Academy of Sciences, USA*, **71**, 1730–3.

Coggin, J. H., Jr & Anderson, N. G. (1972). Phase-specific autoantigens (fetal) in model tumor systems. In *Proceedings of the Second Conference on Embryonic and Fetal Antigens in Cancer*, ed. N. G. Anderson, J. H. Coggin, Jr, E. Cole & J. W. Holleman, pp. 91–102. Springfield, Virginia: National Technical Information Service.

Damjanov, I. & Solter, D. (1973). Yolk sac carcinoma grown from explanted mouse egg cylinder. *Archives of Pathology*, **95**, 182–4.

Damjanov, I. & Solter, D. (1974*a*). Experimental teratoma. *Current Topics in Pathology*, **59**, 69–130.

Damjanov, I. & Solter, D. (1974*b*). Host-related factors determine the outgrowth of teratocarcinomas from mouse egg-cylinders. *Zeitschrift für Krebsforschung und Klinische Onkologie*, **81**, 63–9.

Damjanov, I. & Solter, D. (1974*c*). Embryo-derived teratocarcinomas elicit splenomegaly in syngeneic host. *Nature, London*, **249**, 569–71.

Damjanov, I., Solter, D., Belicza, M. & Škreb, N. (1971*a*). Teratomas obtained through extrauterine growth of seven-day old mouse embryos. *Journal of the National Cancer Institute*, **46**, 471–80.

Damjanov, I., Solter, D. & Šerman, D. (1973). Teratocarcinoma with the capacity for differentiation restricted to neuro-ectodermal tissue. *Virchows Archiv, Abteilung B Zellpathologie*, **13**, 179–95.

Damjanov, I., Solter, D. & Škreb, N. (1971*b*). Teratocarcinogenesis as related to the age of embryos grafted under the kidney capsule. *Wilhelm Roux' Archiv für Entwicklungsmechanik der Organismen*, **167**, 288–90.

Edidin, M., Patthey, H. L., McGuire, E. J. & Sheffield, W. D. (1971). An antiserum to 'embryoid body' tumor cells that reacts with normal mouse embryos. In *Embryonic and Fetal Antigens in Cancer*, ed. N. G. Anderson & J. H. Coggin, Jr, pp. 239–48. Springfield, Virginia: National Technical Information Service.

Evans, M. J. (1972). The isolation and properties of a clonal tissue culture strain of pluripotent mouse teratoma cells. *Journal of Embryology and Experimental Morphology*, **28**, 163–73.

Ferrer, J. F. (1968). Role of the spleen in passive immunological enhancement. *Transplantation*, **6**, 167–72.

Fidler, I. J. (1974). Immune stimulation–inhibition of experimental cancer metastasis. *Cancer Research*, **34**, 491–8.

Finch, B. W. & Ephrussi, B. (1967). Retention of multiple developmental potentialities by cells of a mouse testicular teratocarcinoma during prolonged culture *in vitro* and their extinction upon hybridization with cells of permanent lines. *Proceedings of the National Academy of Sciences, USA*, **57**, 615–21.

Friedell, G. M., Sherman, J. D. & Sommers, S. C. (1960). Spleen and liver in the anemia of the tumor-bearing hamster. *Archives of Pathology*, **70**, 363–71.

Gardner, R. L. (1971). Manipulations on the blastocyst. *Advances in the Biosciences*, **6**, 279–301.

Gasser, D. L. & Silvers, W. K. (1974). Genetic determinants of immunological responsiveness. *Advances in Immunology*, **18**, 1–66.

Gearhart, J. D. & Mintz, B. (1974). Contact-mediated myogenesis and increased acetylcholinesterase activity in primary cultures of mouse teratocarcinoma cells. *Proceedings of the National Academy of Sciences, USA*, **71**, 1734–8.

GOODING, L. R. & EDIDIN, M. (1974). Cell surface antigens of a mouse testicular teratoma. Identification of an antigen physically associated with H-2 antigens on tumor cells. *Journal of Experimental Medicine*, **140**, 61–78.

GREEN, I. (1974). Genetic control of immune responses. *Immunogenetics*, **1**, 4–21.

HELLSTRÖM, K. E. & HELLSTRÖM, I. (1974). Lymphocyte-mediated cytotoxicity and blocking serum activity to tumor antigens. *Advances in Immunology*, **18**, 209–77.

HSU, Y.-C. (1972). Differentiation *in vitro* of mouse embryos beyond the implantation stage. *Nature, London*, **239**, 200–2.

HSU, Y.-C. (1973). Differentiation *in vitro* of mouse embryos to the stage of early somite. *Developmental Biology*, **33**, 403–11.

JAMI, J., FAILLY, C. & RITZ, E. (1973). Lack of expression of differentiation in mouse teratoma–fibroblast somatic cell hybrids. *Experimental Cell Research*, **76**, 191–9.

JAMI, J. & RITZ, E. (1974). Multipotentiality of single cells of transplantable teratocarcinomas derived from mouse embryo grafts. *Journal of the National Cancer Institute*, **52**, 1547–52.

JEJEEBHOY, M. F. (1974). Stimulation of tumor growth by the immune response. *International Journal of Cancer*, **13**, 665–78.

KAHAN, B. W. & EPHRUSSI, B. (1970). Developmental potentialities of clonal *in vitro* culture of mouse testicular teratomas. *Journal of the National Cancer Institute*, **44**, 1015–36.

KLEINSMITH, L. J. & PIERCE, G. B., Jr (1964). Multipotentiality of single embryonal carcinoma cells. *Cancer Research*, **24**, 1544–52.

KOLSCH, E., MENGERSEN, R. & DILLER, E. (1973). Low dose tolerance preventing tumor immunity. *European Journal of Cancer*, **9**, 879–82.

LEE, S. M. & ALEYASSINE, H. (1971). Morphologic changes in the liver of mice bearing Ehrlich ascites tumor. *Laboratory Investigation*, **24**, 513–22.

LEVAK-ŠVAJGER, B. & ŠKREB, B. (1965). Intraocular differentiation of rat egg cylinders. *Journal of Embryology and Experimental Morphology*, **13**, 243–53.

LEVAK-ŠVAJGER, B. & ŠVAJGER, A. (1971). Differentiation of endodermal tissues in homografts of primitive ectoderm from two-layered rat embryonic shields. *Experientia*, **27**, 683–4.

LEVAK-ŠVAJGER, B., ŠVAJGER, A. & ŠKREB, N. (1969). Separation of germ layers in presomite rat embryos. *Experientia*, **25**, 1311–12.

MCKHANN, C. F. & JAGARLAMOODY, S. M. (1971). Evidence for immune reactivity against neoplasms. *Transplantation Reviews*, **7**, 55–77.

MARKERT, C. L. (1968). Neoplasia: a disease of cell differentiation. *Cancer Research*, **28**, 1908–14.

MARTIN, G. R. & EVANS, M. J. (1974). The morphology and growth of a pluripotent teratocarcinoma cell line and its derivatives in tissue culture. *Cell*, **2**, 163–72.

MELCHERS, I., RAJEWSKY, K. & SHREFFLER, D. C. (1973). Ir-LDH$_B$: map position and functional analysis. *European Journal of Immunology*, **3**, 754–61.

MISHELL, J. & DUTTON, R. W. (1967). Immunization of dissociated spleen cell cultures from normal mice. *Journal of Experimental Medicine*, **126**, 423–42.

MOORE, M. A. S. & METCALF, M. (1970). Ontogeny of the haemopoietic system: yolk sac origin of *in vivo* and *in vitro* colony forming cells in the developing mouse embryo. *British Journal of Haematology*, **18**, 279–99.

PAYNE, J. M. & PAYNE, S. (1961). Placental grafts in rats. *Journal of Embryology and Experimental Morphology*, **9**, 106–16.

PIENKOWSKI, M., SOLTER, D. & KOPROWSKI, H. (1974). Early mouse embryos: growth and differentiation *in vitro*. *Experimental Cell Research*, **85**, 424–8.

PIERCE, G. B., Jr (1967). Teratocarcinoma: model for a developmental concept of cancer. *Current Topics in Developmental Biology*, **2**, 223–46.

PREHN, R. T. (1972*a*). Immunosurveillance, regeneration and oncogenesis. *Progress in Experimental Tumor Research*, **14**, 1–24.

PREHN, R. T. (1972*b*). The immune reaction as a stimulator of tumor growth. *Science*, **176**, 170–1.

PREHN, R. T. & LAPPÉ, M. A. (1971). An immunostimulation theory of tumor development. *Transplantation Reviews*, **7**, 26–54.

ROSENTHAL, M. D., WISHNOW, R. M. & SATO, G. H. (1970). *In vitro* growth and differentiation of clonal populations of multipotential mouse cells derived from a transplantable testicular teratocarcinoma. *Journal of the National Cancer Institute*, **44**, 1001–14.

SALAÜN, J. (1968). Sur la formation expérimentale de tératomes chez le poulet et le rat. *Archives d'Anatomie Microscopique et de Morphologie Expérimentale*, **57**, 11–34.

SJÖGREN, H. O. (1973). Blocking and unblocking of cell-mediated tumor immunity. *Methods in Cancer Research*, **10**, 19–34.

ŠKREB, N., DAMJANOV, I. & SOLTER, D. (1972). Teratomas and teratocarcinomas derived from rodent egg-shields. In *Cell Differentiation*, ed. R. Harris, P. Alin & D. Viza, pp. 151–5. Copenhagen: Munksgaard.

ŠKREB, N., ŠVAJGER, A. & LEVAK-ŠVAJGER, B. (1971). Growth and differentiation of rat egg-cylinders under the kidney capsule. *Journal of Embryology and Experimental Morphology*, **25**, 47–56.

SOBIS, M. & VANDEPUTTE, M. (1974). Development of teratomas from displaced visceral yolk sac. *International Journal of Cancer*, **13**, 444–53.

SOLTER, D., BICZYSKO, W. & KOPROWSKI, H. (1974*a*). Host–virus relationship at the embryonic level. In *Viruses, Evolution and Cancer*, ed. E. Kurstak & K. Maramorosch, pp. 3–30. New York & London: Academic Press.

SOLTER, D., BICZYSKO, W., PIENKOWSKI, M. & KOPROWSKI, H. (1974*b*). Ultrastructure of mouse egg cylinders developed *in vitro*. *Anatomical Record*, **180**, 263–80.

SOLTER, D. & DAMJANOV, I. (1973). Explantation of extraembryonic parts of 7-day-old mouse egg cylinders. *Experientia*, **29**, 701–2.

SOLTER, D., ŠKREB, N. & DAMJANOV, I. (1970). Extrauterine growth of mouse egg-cylinders results in malignant teratoma. *Nature, London*, **227**, 503–4.

STEVENS, L. C. (1959). Embryology of testicular teratomas in strain 129 mice. *Journal of the National Cancer Institute*, **23**, 1249–95.

STEVENS, L. C. (1964). Experimental production of testicular teratomas in mice. *Proceedings of the National Academy of Sciences, USA*, **52**, 654–61.

STEVENS, L. C. (1966). Development of resistance to teratocarcinogenesis by primordial germ cells in mice. *Journal of the National Cancer Institute*, **37**, 859–61.

STEVENS, L. C. (1967*a*). The biology of teratomas. *Advances in Morphogenesis*, **6**, 1–31.

STEVENS, L. C. (1967*b*). Origin of testicular teratomas from primordial germ cells in mice. *Journal of the National Cancer Institute*, **38**, 549–52.

STEVENS, L. C. (1968). The development of teratomas from intratesticular grafts of tubal mouse eggs. *Journal of Embryology and Experimental Morphology*, **20**, 329–41.

STEVENS, L. C. (1970*a*). Experimental production of testicular teratomas in mice of strains 129, A/He, and their F_1 hybrids. *Journal of the National Cancer Institute*, **44**, 929–32.

STEVENS, L. C. (1970*b*). Environmental influences on experimental teratocarcinogenesis in testes of mice. *Journal of Experimental Zoology*, **174**, 407–14.

STEVENS, L. C. (1970*c*). The development of transplantable teratocarcinomas from intratesticular grafts of pre- and postimplantation mouse embryos. *Developmental Biology*, **21**, 364–82.

STEVENS, L. C. (1973). A new inbred subline of mice (129/ter Sv) with a high incidence of spontaneous congenital testicular teratomas. *Journal of the National Cancer Institute*, **50**, 235–42.

STEVENS, L. C. & LITTLE, C. C. (1954). Spontaneous testicular tumors in an inbred strain of mice. *Proceedings of the National Academy of Sciences, USA*, **40**, 1080–7.

STEVENS, L. C. & MACKENSEN, J. A. (1961). Genetic and environmental influences on teratocarcinogenesis in mice. *Journal of the National Cancer Institute*, **27**, 443–53.

STEVENS, L. C. & VARNUM, D. S. (1974). The development of teratomas from parthenogenetically activated ovarian mouse eggs. *Developmental Biology*, **37**, 369–80.

WOODRUFF, M. F. & SYMES, M. O. (1972). The significance of splenomegaly in tumour-bearing mice. *British Journal of Cancer*, **16**, 120–30.

EXPLANATION OF PLATES

PLATE 1

(*a*) Diffuse hyperplasia of the spleen in tumor-bearing C3H mouse. Sinusoids are filled with cells of extramedullary hematopoiesis.

(*b*) Several small regions of extramedullary hematopoiesis in the liver of tumor-bearing AKR mice.

(*c*) Infiltration of small, round cells in the area of teratocarcinoma composed mostly of mature neural tissue.

(*d*) Several small nests of embryonal carcinoma cells surrounded by cellular mesenchyme. Notice the transition of embryonal carcinoma cells into more differentiated cells which resemble the cells of ectodermal layer (arrow).

PLATE 2

(*a*) Serous glands in teratocarcinoma. The cells have eosinophylic cytoplasm and small, round nuclei. Notice the direct contact between embryonal carcinoma cells and cells composing serous glands (arrows).

(*b*) Ectodermal (EC) and endodermal (EN) vesicle in teratocarcinoma.

(*c*) Large embryoid body in teratocarcinoma. Three germ layers, yolk sac cavity and parietal endoderm are clearly visible. The cavity of the embryoid body is filled with presumably hematopoietic cells.

(*d*) One large (*) and three small (arrows) embryoid bodies. Small embryoid bodies have a central part composed of embryonal carcinoma cells with a single layer of endodermal cells around. The cavity of large embryoid body is filled with hematopoietic cells.

It is probably these cells which give rise to teratoma when this is produced by ectopic implantation of an early mouse embryo, but spontaneous testicular teratomas and teratomas formed from implanted germinal ridges are formed from primordial germ cells, which are also pluripotential cells (discussed by Damjanov & Solter 1974*b*).

A teratoma is a tumour of pluripotent cells which comprise the proliferating stem cells of the tumour. Some of these cells become organised, [illegible]

[illegible] was used to study the chemical nature of the basement membrane (Pierce, 1970). Auerbach (1972) has shown that the immature neural tissues present in a teratoma will provide the inductive stimulus necessary for kidney tubule formation from metanephric mesoderm *in vitro*, and he suggested that embryologists might turn to teratomas as a large and ready source of inductive factors.

Teratoma tissues, and cells, have been isolated and cultured as organ cultures, and as tissue cultures respectively. This isolation removes exogenous host factors and also allows all the experimental manipulative advantages of the in-vitro systems. This discussion is concerned with in-vitro investigations of the mouse teratoma system as a model for the analysis of the cellular events of early embryonic development.

Pierce & Verney (1960) made plasma-clot cultures of explanted pieces of a teratoma and of embryoid bodies. They observed a gradual simplification of the tissues present in the cultures explanted from teratomas until they were composed largely of parietal yolk sac and embryonal carcinoma. The embryoid body explants in which there were fewer tissues to start with became similar. Pluripotency (tested by re-innoculation into mice) was retained for many months in culture. From some explants there was a cellular outgrowth of granular cells which secreted hyaline droplets. Cultures of these cells when re-injected into mice gave slow-growing tumours composed solely of parietal yolk sac and this provided an identification of these cells as parietal yolk sac cells.

Kleinsmith & Pierce (1964) laid a major foundation for the experimental study of teratomas when they demonstrated by in-vivo cloning that the embryonal carcinoma cells were indeed the pluripotent stem cells of the

STUDIES WITH TERATOMA CELLS *IN VITRO*

BY M. J. EVANS

Department of Anatomy and Embryology,
University College, Gower Street,
London WC1E 6BT

In the process of embryonic development the possible pathways of differentiation of particular cell lineages become progressively restricted. The zygote has complete potentiality for development – it is totipotent. At an early stage all embryos contain cells whose mitotic descendants are able to develop into a wide variety of diversely differentiated cell types. These cells with a wide, although not necessarily complete, range of capacities for differentiation are termed pluripotent cells. As development proceeds, the developmental potentiality of cell lineages becomes progressively restricted as the processes of cell determination take place. It is this process of cell determination which is the crux of the problem of the establishment of the cell diversity which is finally recognisable as cellular differentiation. Paul Weiss (1967) has amply argued that it is the process of 'strain differentiation' which underlies the cellular and molecular-biological problems of developmental biology. The study of committed 'single-tracked' cells during or after the completion of their differentiation may reveal the established restrictions of their genetic expression and may also provide information about the 'modulation' of this genetic expression but it does not provide information about the establishment of this selectivity of genomic expression which is the most important aspect of eukaryote development. In order to study this one needs cells still capable of becoming irreversibly determined in a variety of ways; a population of pluripotent cells.

Whether there is a proliferating population of pluripotent cells or whether they represent only a small transient stage in the development of a particular cell lineage depends upon the biology of the system involved. In embryos showing a mosaic type of development, the progenitor cells for each tissue are very few at the time of their determination and much cell determination already occurs during cleavage. In the more regulative type of development a larger proportion of cells is built up before determinative events occur. The pattern of vertebrate development with induction of the main axial tissues at the time of mesoderm formation leaves a population of ecto-mesodermal cells pluripotent until this time.

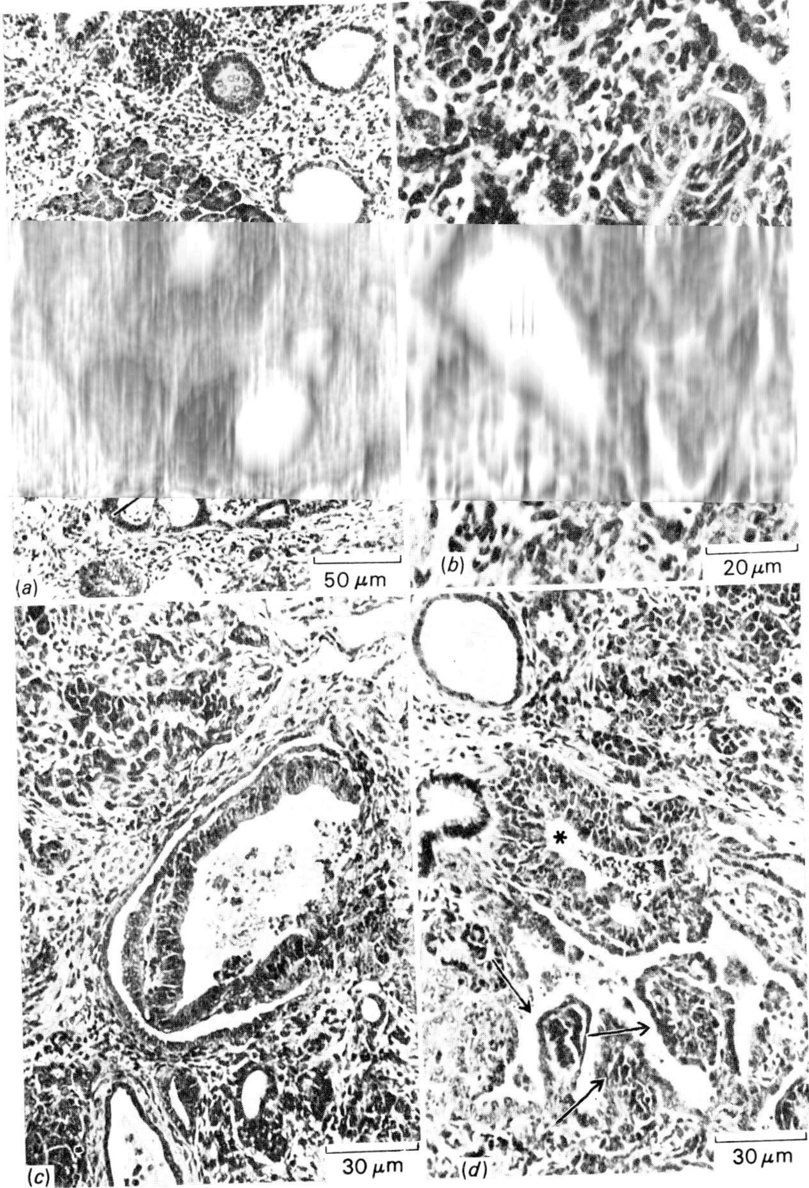
(a)
50 μm
(b)
20 μm
*
(c)
30 μm
(d)
30 μm

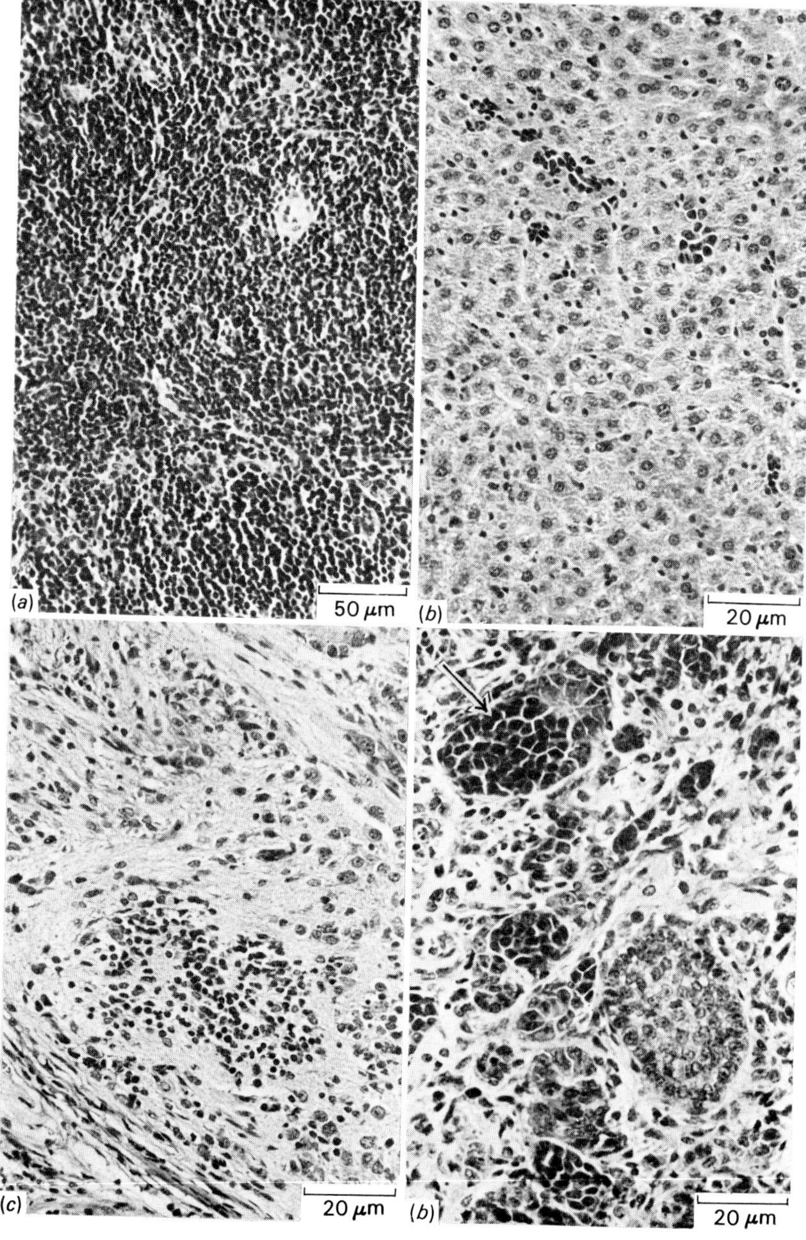

(*Facing p. 264*)

tumour, being able to give rise to all the other tissues. Since they were unsuccessful in their attempts at in-vitro cell cloning they took small embryoid bodies from the ascitic conversion of a diversely-differentiating teratoma. These small embryoid bodies, which had already been shown to be able to produce progressively growing teratomas after subcutaneous implantation (Pierce & Verney, 1960), consist of only two layers: parietal yolk sac and embryonal carcinoma. Kleinsmith & Pierce dissociated the [illegible]

[illegible] established cloning could be performed using a lethally irradiated feeder layer of fibroblastic cells.

Not all teratomas can be obtained in an ascites form and spontaneous teratomas are only available from particular strains of mice. The next line to be isolated, however (Evans, 1972), was obtained from the disaggregation of a solid tumour which had been derived from the implantation of a three-day blastocyst into the adult testis (Stevens, 1968). Subsequent lines have been derived from both the solid tumour (Jami & Ritz, 1974) and from the ascites conversion of an embryo-derived teratoma (Bernstine, Hooper, Grandchamp & Ephrussi, 1973; Jakob *et al.*, 1973). As it seems likely that the production of teratomas from ectopically implanted embryos is not restricted to particular strains, it may be possible to derive a pluripotent cell line from any genetic strain of mouse (Stevens, 1970; Damjanov, Solter, Belicza & Skreb, 1971). This may, however, be dependent upon the production of a transplantable, progressively growing teratoma (a teratocarcinoma) and many embryo-derived teratomas are not transplantable – although this property may be strain related (Damjanov & Solter, 1974*a*) – or indeed related to the immunological status of the histocompatible host. I should like to suggest that it may be quite feasible to obtain cultures of pluripotent cells directly from the embryo now that experience has been gained in handling such cells, and that the earlier results of Cole, Edwards & Paul (1966) with cultures from rabbit blastocysts should not necessarily inhibit further efforts in this direction.

Clonal and subclonal cultures of teratoma cells will give rise to diversely-differentiating teratomas on re-injection into a mouse (Rosenthal *et al.*,

1970; Kahan & Ephrussi, 1970; Evans, 1972; Jakob *et al.*, 1973; Jami & Ritz, 1974). This reconfirms the pluripotential stem cell origin of the various tissues of the teratoma and also demonstrates the maintenance of these cells in the tissue culture line. Kahan & Ephrussi (1970) found that one clone maintained its ability to produce a well-differentiated tumour during continuous culture for over six months and calculated the minimum number of cell generations to be eighty-eight. This must have been a very conservative estimate, as these cells are not contact-inhibited in growth (Martin & Evans, 1974) and in our hands have a generation time of about 9–12 h (Jakob *et al.*, 1973, report 10–15 h). Eighty-eight generations in six months represents a generation time of over two days. They did, however, notice that tumours formed by intra-peritoneal injection became progressively more poorly differentiated as the cultures were maintained through many generations. Subcutaneous inoculations, however, still produced well-differentiated tumours and they concluded that these represented a tumour from the entire inoculum whereas only a selected portion of the inoculum might be represented by one particular intraperitoneal tumour and this might be of more restricted cells. This was borne out by a variability in the extent of differentiation in the tumours of subclones. Other investigators have also noticed that the differentiation in tumours becomes gradually more restricted during a long-term in-vitro passage of the cells, and that there is a variability between subclones (Rosenthal *et al.*, 1970; Evans, 1972; Bernstine *et al.*, 1973). Nevertheless these cell lines display a remarkable stability *in vitro*; their pluripotency remains for very many generations and their karyotype shows a remarkable constancy at, or very near to, a diploid condition (Finch & Ephrussi, 1967; Kahan & Ephrussi, 1970; Rosenthal *et al.*, 1970; Evans, 1972; Bernstine *et al.*, 1973; Jakob *et al.*, 1973; Jami & Ritz, 1974). Kahan & Ephrussi (1970) made a careful analysis of the karyotypes of the original tumour and fourteen derived clones. All the clones contained one subtelocentric chromosome and two clones also had a metacentric chromosome. One clone which had been shown to produce a well-differentiated tumour was essentially tetraploid with a modal chromosome number of seventy-seven and this would indicate that diploidy is not an essential prerequisite of pluripotency.

Sit (1973) has examined the G banding pattern of the chromosomes of a teratoma cell line SIKR and of some non-pluripotent drug-resistant lines selected from it. The banding patterns of the SIKR cells are the same as the normal mouse cell, but the mutated lines show karyotypic alterations and banding changes. Lines resistant to 10^{-4}M bromodeoxy-uridine (BUDR) remained near-diploid (41 chromosomes) but showed altered chromosome banding patterns. Lines resistant to thioguanine showed grosser karyotypic alteration, becoming tetraploid and having some metacentric chromosomes. Guénét, Jakob, Nicholas & Jacob (1974) have also examined the G banding patterns of chromosomes of their cell lines and find normal or nearly-normal banding in the pluripotent lines but many alterations in F9 which is no longer pluripotent.

These studies have shown [illegible]

To simplify the discussions below I shall use the terminology which I have applied to cultures of the teratoma cell line SIKR and refer to the embryonal carcinoma cells in culture as C cells. Artzt *et al.* (1973) have called such cells PTC and others have referred to them as embryonal carcinoma or EC cells.

The availability of large numbers of cells analogous to early embryos has allowed their use as an immunogen. Edidin, Patthey, McGuire & Sheffield (1971) reported that antibodies raised in rabbits against mouse teratoma embryoid bodies cross-reacted with early embryo cells. Subsequently Artzt *et al.* (1973) have reported that embryonal carcinoma cells are strongly antigenic in syngeneic mice and an antibody is raised against a cell surface embryonic antigen. This antibody reacts specifically against the cell surface of teratoma stem cells, early cleavage cells and sperm. These observations strongly support the homology between the teratoma stem cells and early embryo cells as do their similarities of ultrastructure (Damjanov *et al.*, 1971). It is known that the recessive alleles of the *T*-locus in the mouse, which have an early lethal effect in the homozygous condition, affect the cell surface and are expressed on sperm (Bennett, Goldberg, Dunn & Boyse, 1972), and it has been demonstrated by comparison of absorption between normal and $t^{12}/+$ sperm that the teratoma antigen is the product of a normal allele of the *T*-locus (Artzt, Bennett & Jacob, 1974). As it is believed that the genes of the *t* series may act at the cell surface and hence affect the normal cellular interaction in embryogenesis, it is extremely interesting to find their normal counterparts expressed on the surface of pluripotent teratoma cells. It will be interesting to discover whether there is any causal relationship between the cell surface expression and cellular determination or whether the former will prove to be only one phenotypic expression of the cell's epigenetic state.

The large quantities of cells available would also allow biochemical studies which might then be related back to cells of the early embryo. As yet little has been done. A high alkaline phosphatase activity is characteristic of embryonal carcinoma in the tumours *in vivo* (Damjanov, Solter & Skreb, 1971) and also characterises the embryonal carcinoma cells *in vitro*, although different lines have very different levels of activity of the enzyme (Bernstine *et al.*, 1973; Evans & Braithwaite, unpublished observations). This enzyme activity is lost on differentiation to most other cell types. This may provide a marker for following cell behaviour *in vitro* and it may also be an interesting model in which to investigate the function of this enzyme which is often found associated with developing tissues (Moog, 1965). We have isolated this enzyme, which proves to be bound to a particulate fraction of the cell, and we are studying its properties (Braithwaite, unpublished observations).

Another foetal function found in teratomas which may prove to be a useful marker for cell differentiation is the production of α-foetoprotein; the presence of this protein may be used as a clinical diagnostic indication of the presence of either a hepatoma or a teratoma (Smith, 1970). It is produced by embryonic cells in the teratoma and possibly by the embryonal carcinoma cells themselves, as its production has been demonstrated in cultures of embryonal carcimona cells (Kahan & Levine, 1971). Engelhardt, Poltoranina & Yazova (1973), however, using indirect immunofluorescence found that the α-foetoprotein *in vivo* was mainly located in the endodermal epithelium. It could not be detected in the medium from the cultures of the teratocarcinoma cell line SIKR (J. B. Smith & M. J. Evans, unpublished observations). The discrepancies in these observations could well be re-investigated as α-foetoprotein could be a useful specific marker of cell function.

The availability of the pluripotent cells in monolayer tissue culture allows a variety of experimental manipulations. In the first place cloning and subcloning of these cells is feasible, although most authors have found that this is only possible using an X-irradiated feeder layer (Kahan & Ephrussi, 1970; Rosenthal *et al.*, 1970; Evans, 1972; Bernstine *et al.*, 1973; Jami & Ritz, 1974). Jakob *et al.* (1973) cloned in microwells without feeders in a medium containing 30% serum. Specific single cell cloning has not proved easy but has now been carried out successfully in at least two laboratories (Jami & Ritz, 1974; G. R. Martin, personal communication). The subcloning allows dissection of cell types in the cultures and should be able to provide evidence of any progressive restrictions which may take place. As yet, although various C cell subclones have been found to show different degrees of differentiation, the evidence has been taken more to show essential equivalence of pluripotency than to indicate restrictions. More extensive investigations and in particular comparisons with teratoma lines *in vivo* whose range of differentiation is restricted, either after extensive in-vivo passage (Stevens, 1958, 1970) or genetically (Artzt & Bennett, 1972), may indicate what significance these apparent restrictions of differentiative ability [illegible]

[illegible] teratomas but fibrosarcomatous. Similar results were reported by Jami, Failly & Ritz (1973). One possible explanation is that the fibroblastoma represents the differentiated epigenotype which is epistatic over the undetermined pluripotent teratoma cell. For this hypothesis to be substantiated more attention needs to be focussed upon the effect of hybridisation procedures on the C cells, and other, more demonstrably specific, differentiated phenotypic characteristics of the determined cell parent, need to be used. Some of these experiments have been undertaken in our laboratory (Sit, 1973).

Hybrid cells were made between the teratoma cell line SIKR and the mouse L cell line A9 (HGPRT$^-$) and a mouse neuroblastoma cell line (HGPRT$^-$) using the half-selective system (Davidson & Ephrussi, 1965) (HGPRT designates hypoxanthine-guanine phosphoribosyl transferase). Cell hybrid formation was stimulated by inactivated Sendai virus. A monolayer of the HGPRT$^-$ parent ($1-5 \times 10^6$ cells) was treated with Sendai virus at 4 °C, the unabsorbed virus removed, and 10^4 SIKR cells added and incubated at 4 °C for ten minutes followed by incubation at 37 °C for ten minutes. The dish was washed to remove unbound SIKR cells and then incubated in HAT medium. Colonies were picked and karyotypically analysed to identify hybrid cells. SIKR has only telocentric chromosomes and so hybrid cells were identifiable as HAT-resistant cells which contained the metacentric and submetacentric marker chromosomes of the other parent cells. HAT resistant revertants of the drug resistant parent cells were less than one in 10^7 cells. The results were consistent in that hybrids of SIKR cells, both with fibroblastic cells and with neuroblastoma cells, gave cultures which formed tumours in appropriate mice which were not teratomas but a fibrosarcomatous type (Table 1). The neuroblastoma and teratoma hybrids did not display the common available phenotype of neural tubule formations in the tumour, although *in vitro* formalin-induced fluorescence indicated that a high monoamine content, which was a characteristic only of the neuroblastoma parent, was retained in the hybrid cells. Two types of control for the effect of these hybridisation procedures on the pluripotency of SIKR cultures were performed; the effect of HAT selection, and of cell hybridisation *per se*.

SIKR cultures grown in HAT retain their pluripotency; however, a tenfold increase in HAT concentration destroys this pluripotency within five days. Sit (1973) examined the cause by scoring clones for the morphologically recognisable C cells after subcloning from treated cultures.

Table 1. *Hybrids of SIKR cells. From Sit (1973)*

Cell strain or hybrid	Expected karyotype	Actual karyotype	Tumour produced in syngeneic mice or F1 hybrid
SIKR(S)	41	41	Teratoma
SIKR(S) × SIKR(S)	82		
SIKR(S) × SIKR(S)		143	Fibrosarcoma
SIKR(S) × SIKR(S)		139	Fibrosarcoma
SIKR(S) × SIKR(S)		82	Fibrosarcoma
SIKR(S) × SIKR(S)		80	Fibrosarcoma
SIKR(S) × SIKR(S)		77	Fibrosarcoma
SIKR(S) × SIKR(S)		75	Fibrosarcoma
SIKR(S) × SIKR(S)		74	Fibrosarcoma
SIKR(S) × SIKR(S)		71	Fibrosarcoma
NS86*	86	86	Neuroblastoma
NS86 × SIKR(S)	127		
NS86 × SIKR(S)		165	Fibrosarcoma
NS86 × SIKR(S)		160	Fibrosarcoma
NS86 × SIKR(S)		152	Fibrosarcoma
NS86 × SIKR(S)		141	Fibrosarcoma
NS86 × SIKR(S)		117	Fibrosarcoma
NS86 × SIKR(S)		109	Fibrosarcoma
NS86 × SIKR(S)		81	Fibrosarcoma
NS86 × SIKR(S)		56	Fibrosarcoma
A9 (IMP$^-$ L cell)	57	57	Non-tumourigenic
A9 × SIKR(S)	98		
A9 × SIKR(S)		93	Fibrosarcoma
A9 × SIKR(S)		77	Fibrosarcoma

* A subclone of N18 TG2-13, an HGPRT$^-$ C1300 neuroblastoma cell line.

Hypoxanthine and thymidine have no effect at concentrations up to 1×10^{-3}M and 1.6×10^{-4}M respectively ($\times 10$ the concentration in HAT; see Littlefield, 1964) but aminopterin causes a rapid loss of C cells at and above 2.8×10^{-6}M (i.e. $\times 7$ the normal HAT concentration; see Fig. 1). It is not known to what this effect might be attributable, but it is not reversed by a tenfold increase in hypoxanthine and thymidine concentration, and some other normal cells grow quite satisfactorily in tenfold [illegible]

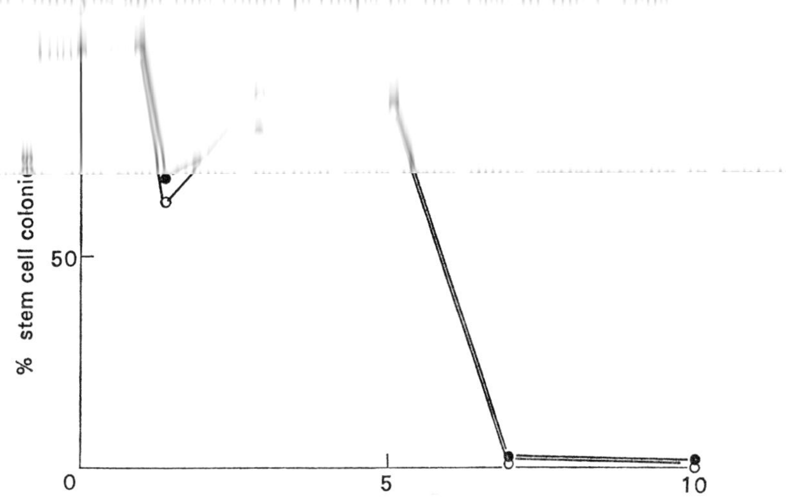

Ratio of experimental HAT or aminopterin concentration to normal level

Fig. 1. Reduction in the percentage of cell colonies containing pluripotent stem cells caused by treatment of the cultures with elevated concentrations of HAT (●——●) or with the equivalent concentrations of aminopterin (○——○). Increased concentrations of hypoxanthine or of thymidine had no effect. After Sit (1973).

The procedure of hybridisation might in itself affect these sensitive cells sufficiently to change their pluripotency. Treatment of SIKR cultures with Sendai virus, in the same way as for virus-stimulated cell fusion, increased the numbers of multinucleate cells. Some of the multinucleate cells grew up into colonies of polyploid cells, which could be recognised by their greater size and isolated from the remaining cells. SIKR–SIKR fused cells with karyotypes of between 71 and 143 were isolated and all of these produced tumours which were fibrosarcomatous and not diversely-differentiating teratomas. Thus there is a strong indication that the other results, in which fibrosarcomatous tumours were produced after fusion of teratoma cells with another cell type, may be the result of the fusion procedure and not necessarily of the interaction between the control mechanisms of the two fused cells. The fusion procedure *per se* may cause restrictions in the cell's developmental ability. This conclusion would be strengthened if one could be sure that the parent cells in the fusion were both pluripotent C cells. SIKR is a heterogeneous cell line containing not only the pluripotent C cells but also non-pluripotent E cells (Martin & Evans, 1974); this will be further discussed below. Sit used a selected subclone of SIKR, SIKR(S), and found that all twenty-five subclones of this line which he isolated in parallel to a cell fusion experiment were pluripotent. Notwithstanding this control, it is possible that in each SIKR–SIKR fusion at least one parent was a determined E cell, and so these experiments should be repeated with a homogeneous pluripotent C cell line.

A variety of unusual karyotypic disturbances were also observed after self-fusion of SIKR cells. For instance, end-to-end fusion of chromosomes, premature chromosome condensation, prophasing of chromosomes in metaphase spreads, and chromatid breaks. Also, remarkably, chromosome pairing and bivalent formation as in a meiotic metaphase was not uncommon (Plate 1*b*). All of these disturbances may be related to mechanisms whereby excess chromosomes are becoming lost from these unbalanced karyotypes, but they also lend strong support to the idea that the nuclear material of the cells is suffering considerable disturbance and it would not be at all surprising if these disturbances disrupt the mechanisms of genetic control in the cell.

Embryonal carcinoma cells, primordial germ cells, early embryo cells and C cells in culture, all have predominantly euchromatic nuclei. On the other hand, all the non-pluripotent derivatives of SIKR (the E cells, drug-resistant mutants which are no longer pluripotent (Evans & Sit, unpublished observations), and the hybrid cells produced by Sendai-stimulated cell-fusion) show multiple heterochromatic condensations. Sit (1973) has suggested that this provides a unifying concept. In normal development strain differentiation may involve (and possibly be the result of) heterochromatinisation of parts of the genetic material thus repressing them and limiting the available genome. Cell hybridisation may lead to heterochromatinisation of some of the genome as a response to the karyotypic disruption of the type seen in SIKR–SIKR fusions, and this heterochromatinisation may in itslf lead to the loss of pluripotency.

Teratoma cell cultures may be subjected to various types of experimental manipulation and the effect upon the cells' pluripotency assayed by re-injection of the cells into a mouse, but in order to be able really to study the processes of cell determination and differentiation it is desirable to find conditions under which these cells differentiate *in vitro*.

Under suitable conditions mouse embryos can develop *in vitro* to a somite stage where most of the processes of cell determination and a number of processes of organogenesis have already taken place (Hsu, Baskor, Stevens & Rash, 1974). Blastocysts explanted into tissue culture will produce a range of cell types (Sherman, this volume). In a similar manner explanted embryoid bodies will give rise to a variety of well- [illegible] of determined cells from an embryo is occurring in monolayer culture.

Gearhart & Mintz (1974) have recently demonstrated that a rise in acetylcholinesterase activity (which may be shown to be present mainly in developing muscle fibres) occurs when embryoid bodies are allowed to spread and differentiate on a plastic surface, but they did not observe such a rise in embryoid bodies cultured by a non-adhesive agar surface. This stimulation of myogenesis by culture *in vitro* might be similar to that observed by Pierce & Verney (1960). Primary cultures of disaggregated cells from solid teratomas also produce a wide variety of cell types *in vitro*. What determination is taking place *in vitro* is difficult to ascertain in these mixed systems. A more defined system is required. It would be preferable to demonstrate cell determination and in-vitro differentiation of a clonal pluripotent teratoma cell line. More than a demonstration of in-vitro differentiation (after all, all these lines very readily differentiate *in vivo*) a means of assay and control of cell determination are needed to investigate the process *in vitro*.

Some differentiation may be observed in aggregates from cloned cultures grown as a floating mass (Finch, 1968; Finch & Ephrussi, 1967), on Millipore filters (Rosenthal, 1968), or on agar, or in methyl cellulose (Evans, unpublished observations). In general, however, few results have been gained with these techniques and a better assay for pluripotency is always re-injection *in vivo*. They are important, however, in demonstrating that these cells are capable of differentiation under culture conditions and without host factors, and it may be that the organ culture type of situation is necessary, not only for the elaboration of histologically recognisable tissues, but also for the initiation of the cellular differentiation. Skreb & Svajger (1973) have recently described organ culture conditions in which good histogenesis may be observed from rat and mouse embryonic shields. These might prove useful for organ culture of teratoma cells.

Whether all determination takes place in *monolayer* culture is uncertain. Jakob *et al.* (1973) reported briefly that their line PCC3 will differentiate well *in vitro* after prolonged re-feeding of a culture. This may represent an organ culture situation as the cells pile up. Lehman, Speers, Swartzendruber & Pierce (1974) have recently reported the isolation of a passaged line of cells from embryoid bodies (not a clonal line) in which C cells are the growing population but in which there are always also three other cell types which they identified as neuro-epithelial cells, smooth muscle and parietal yolk sac cells. These differentiated cells do not, in themselves, grow well in the culture and Lehman *et al.* concluded that they were continuously derived from the C cells. In long-term cultures maintained for 30–50 days without passage, other differentiated cell types were found including striated pulsating muscle, melanocytes, cartilage, glandular structures and keratinising epithelium.

Apart from this it may be noticed that there are two types of clonal culture of teratoma cells which have been described. Kahan & Ephrussi (1970) [Finch & Ephrussi (1967), Finch (1968)] describe their lines as being homogeneous, consisting only of embryonal carcinoma cells (i.e. a pure C culture). The clonal lines of Jakob *et al.* (1973), Bernstine *et al.* (1973) and Jami & Ritz (1974) are similar. Rosenthal *et al.* (1970) and Rosenthal (1968), on the other hand, found that after a short while their clones were always heterogeneous. A colony of C cells became surrounded by spreading granular cells of variable morphology. These are similar to the E cells found by us in CE cultures of SIKR (see below) but differ in that they did not grow independently from the stem cells.

Evans (1972) demonstrated that two types of subclone may be derived from cultures of the clonal line SIKR and these were termed C-type and E-type clones. The C-type clones, which were characterised by colonies of small piling cells, produced teratomas upon re-inoculation into mice, whereas the E-type clones, which were composed of cells growing in a contact-inhibited monolayer, did not, in general, produce tumours at all. Those that did produced not teratomas but a variety of tumours, each with only a single type of tissue which was not identifiable but was not embryonal carcinoma. Martin & Evans (1974) re-examined this system and described the cultures and their growth. The pluripotent C-type clones contain the characteristic small colony-forming cells with large clear nuclei containing a single large nucleolus. These cells, which may be identified with the pluripotent teratoma stem cells – embryonal carcinoma – are called C cells. The C-type clones also contain monolayering cells which are similar to those found in E-type subclones and are called E cells. Thus the whole is identified as a CE culture. Growth studies show that the E cells in E-type clones or in CE cultures are density-inhibited in growth and stop dividing as the culture becomes confluent. The C cells on the other hand continue to grow into piled-up colonies. [illegible]

The E-type subclones are, in the main, contact-inhibited in growth *in vitro* and poorly tumourigenic *in vivo*. This is a parallel with the non-malignancy of the differentiated tissues of a teratoma *in vivo* (Pierce, Dixon & Verney, 1960). E cells may, however, undergo a spontaneous transformation *in vitro* to a non-contact-inhibited type which readily forms tumours. All the E-type tumours observed in this second study (Martin & Evans, 1974) were homogeneous fibrosarcomas. Because CE cultures on subcloning give rise either to more CE cultures, or to pure E cultures, and as the E cells in CE cultures could be seen to be normal contact-inhibited cells (although isolated E cell cultures rapidly underwent a malignant transformation), we suggested that the C cells had given rise to the E cells in the CE cultures. Moreover we considered that this transition represented cell determination.

In order to study cell determination *in vitro*, a means of assay for the C cells and for determined cell types is required. Three types of assay have been studied and applied to following the dynamics of the mixed cultures of SIKR.

(1) *Alkaline phosphatase*

Damjanov, Solter & Skreb (1971) have shown that embryonal carcinoma cells *in vivo* are characterised by a high level of alkaline phosphatase and Bernstine *et al.* (1973) have recently demonstrated a high level of this enzyme in embryonal carcinoma cells *in vitro* but not in cultures of non-malignant somatic cells derived from embryoid bodies. We have also been using this enzyme as a means of identification of C cells *in vitro*. The E cells have a low alkaline phosphate content. A culture can be trypsinised and the cells stained using a diazo method (Burstone, 1962). The proportion of stained cells in the sample is counted after fixing and mounting the cells in glycerol. Using this method the ratio of C to E cells can be followed during the growth of a culture (see Fig. 2*b*). As expected from growth studies, the C cells are seen to overgrow the E cells after the

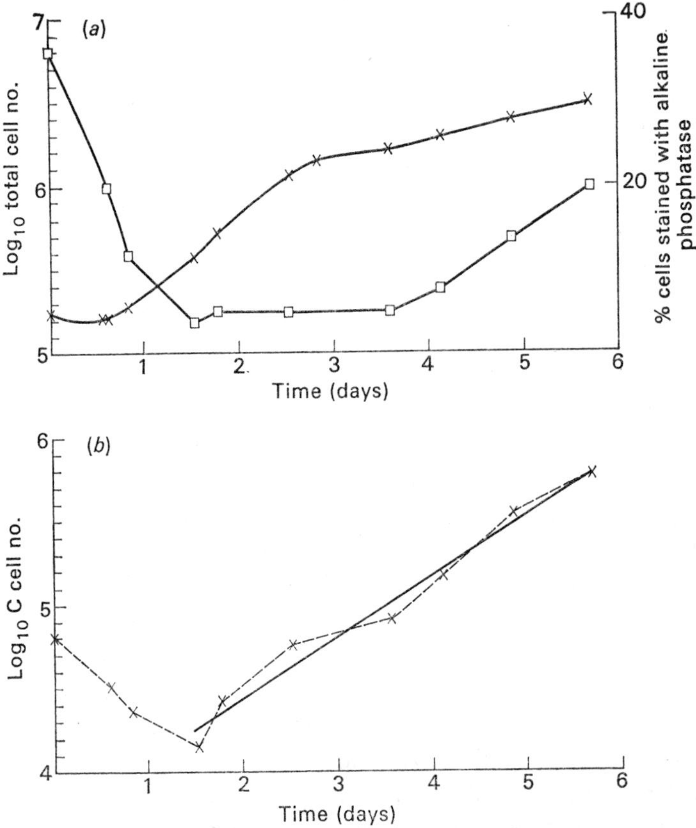

Fig. 2. Growth of a culture of SIKR over 6 days. The percentage of C and E cells in the culture was assayed by alkaline phosphatase staining. (*a*) x——x, $\log_{10}$ total number of cells (note biphasic growth curve); □—□, % of cells which are alkaline phosphatase stained (i.e. % of C cells). (*b*) x – – – x $\log_{10}$ number of C cells in the culture. The solid line is a least square regression line for the points after 24 hours and demonstrates that the C cell number increases exponentially after 36 hours.

culture reaches confluence. On the other hand the proportion of C cells falls progressively over the first two days of culture. Either C cells are giving rise to E cells at this time or there is a very considerable death of C cells. It is difficult to obtain cultures with a sufficiently high initial proportion of C cells to discriminate unequivocally between these two possibilities.

(2) *Cell surface antigenic determinants*

The C and E cells may also be recognised by cell surface immunological markers (P. Stern & M. J. Evans, unpublished observations). We have prepared an antibody preparation, anti-C, by immunisation of syngeneic mice with lethally γ-irradiated SIKR, which is similar to that described [illegible] pluripotent teratoma cell lines (pure C). The E cells present in SIKR cultures were not labelled with the anti-C serum.

In contrast, the E cells of SIKR cultures could be shown to express the cell surface alloantigen θ (thy-1-2), both by immunofluorescence and by cytoxicity. After specific cytotoxic killing of the E cells from a SIKR culture a residual population of cells remained which were considerably enriched for anti-C immunofluorescent staining. There was considerable loss of C cells, which were very sensitive to complement cytotoxicity and to the lengthy manipulative procedure involved. In order to establish whether or not the two cell surface antigens were on non-overlapping populations of cells, anti-θ-C3H conjugated to fluorescein and anti-C conjugated to rhodamine were used in direct immunofluorescence tests. The expression of the antigens recognised by these two reagents was shown to be mutually exclusive. There are, however, cells which consistently do not label with either reagent. We have used anti-C and anti-θ to follow the growth curve of a SIKR culture. The results are similar to those using alkaline phosphatase as a marker.

(3) *Focus assay*

An entirely different assay procedure for viable C cells and E cells has been developed by Gail Martin (unpublished observations) which exploits the ability of the C cells to overgrow a contact-inhibited monolayer. A simple focus assay follows only the C cells present. If a constant number of SIKR cells is plated together with an increasing number of 3T3 cells the number of foci of C cells increases to a plateau (Fig. 3). This closely follows the Poisson distribution expected if each C cell must fall within 75 μm of a 3T3 cell in order to form a colony. This suggests a contact or local feeding effect of the 3T3 cells. The question is what happens to the C cells at low 3T3 density? Do they become contact-inhibited E cells or do they merely die? This is being investigated by assaying the E cells as well as the C cells. At the moment no clear indication of a C to E transition is available.

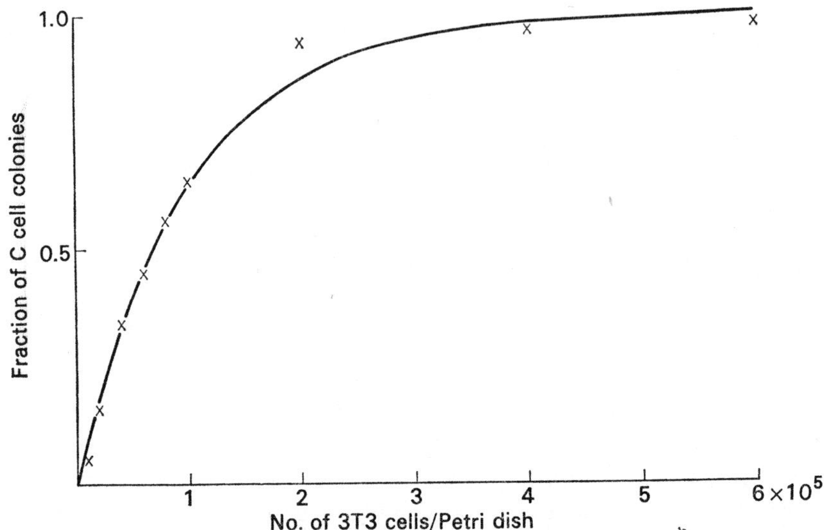

Fig. 3. Focus assay for C cells. Fraction of maximum number of C cell colonies formed when a cell sample is plated with increasing numbers of 3T3 cells. Solid line is the theoretical Poisson distribution on the assumption that each C cell has a probability of 1 of being sufficiently near a 3T3 cell to form a colony at a 3T3 density of 1×10^5 3T3 cells per Petri dish.

Recently it has been observed that if SIKR is subcloned on a feeder layer and C-type colonies free from E cell contamination are re-fed for six weeks instead of being picked and passaged, a variety of differentiated cell types arises directly from the C colony. They are seen as a halo of differentiated cells migrating out of the colony. Cells of fibroblastic and a variety of epitheloid morphologies, neural cells, and pulsating muscle may be observed and the cultures are very similar to those derived from an explanted embryoid body. The E cells in SIKR CE clones may arise in this way.

It has now been possible to isolate specific single cell clones from SIKR by microdrop cloning and maintain them as pure C cultures. (G. Martin, unpublished observations.) Under suitable conditions a variety of types of differentiated cell may be obtained from these clones (Plates 2 and 3), providing absolute proof of the in-vitro cell determination and differentiation of these cells. Whether these clones give rise to E cells closely analogous to those in SIKR CE subclones, and the conditions for this transition are under current investigation.

Teratoma cell lines of restricted developmental potency may be useful in dissection of the system of cell determination and may provide a simplification for analysis of cell determination and differentiation *in vitro*. A cell line has been developed from a tumour subline of OTT6050- [illegible] duce a variety of cell types as well as the stem cell. A very few θ-positive cells with a fibroblastic morphology are found, islands of epithelioid cells which are not C cells are formed and cells with networks of long processes which closely resemble neuroblasts and early neurones in culture are seen. Electron microscopy of these neurone-like cells shows stacked endoplasmic reticulum and neurotubules (Plate 4). Although more work is needed to characterise the neurone-like cells and their mode of production in the culture, this looks a promising system both for the analysis of the cell determination and for the production of cultures of neural cells *in vitro*.

I should like to thank Dr Hung Sit and Dr Gail Martin for allowing me to include reference to unpublished work, Gail Martin for much useful and continuing discussion and Dr Ruth Bellows and Dr Robert Tresman for their electron microscopical studies. Mrs P. Beverage, Mrs M. Reynolds and Mr R. Moss are thanked for their excellent technical assistance. Part of the original work reported here was supported by a grant from the Cancer Research Campaign.

REFERENCES

Artzt, K. & Bennett, D. (1972). A genetically caused embryonal ectodermal tumour in the mouse. *Journal of the National Cancer Institute*, **48**, 141–58.

Artzt, K., Dubois, P., Bennett, D., Condamine, H., Babinet, C. & Jacob, F. (1973). Surface antigens common to cleavage embryo and primitive teratocarcinoma cells in culture. *Proceedings of the National Academy of Sciences, USA*, **70**, 2988–92.

Auerbach, R. (1972). The use of tumours in the analysis of inductive tissue interactions. *Developmental Biology*, **28**, 304–9.

BENNETT, D., GOLDBERG, E., DUNN, L. C. & BOYSE, E. A. (1972). Serological detection of a cell surface antigen specified by the *T* (Brachyury) mutant gene in the house mouse. *Proceedings of the National Academy of Sciences, USA*, **69**, 2076–80.

BERNSTINE, E. G., HOOPER, M. L., GRANDCHAMP, S. & EPHRUSSI, B. (1973). Alkaline phosphatase activity in mouse teratoma. *Proceedings of the National Academy of Sciences, USA*, **70**, 3899–903.

BUONASSIS, V. G., SATO, G. & COHEN, A. I. (1962). Hormone producing cultures of adrenal and pituitary origin. *Proceedings of the National Academy of Sciences, USA*, **48**, 1184–90.

BURSTONE, M. S. (1962). *Enzyme Histochemistry and its Application in the Study of Neoplasms*. New York & London: Academic Press.

COLE, R. J., EDWARDS, R. G. & PAUL, J. (1966). Cytodifferentiation and embryogenesis in cell colonies and tissue cultures derived from ova and blastocysts of the rabbit. *Developmental Biology*, **13**, 385–90.

DAMJANOV, I. & SOLTER, D. (1974*a*). Host-related factors determine the outgrowth of teratocarcinomas from mouse egg-cylinders. *Zeitschrift für Krebsforschung und klinische Onkologie*, **81**, 63–9.

DAMJANOV, I. & SOLTER, D. (1974*b*). Experimental teratoma. *Current Topics in Pathology*, **59**, 69–130.

DAMJANOV, I., SOLTER, D., BELICZA, M. & SKREB, N. (1971). Teratomas observed through extrauterine growth of seven-day mouse embryos. *Journal of the National Cancer Institute*, **46**, 471–80.

DAMJANOV, I., SOLTER, D. & SKREB, N. (1971). Enzyme histochemistry of experimental embryo-derived teratocarcinomas. *Zeitschrift für Krebsforschung und klinische Onkologie*, **76**, 249–50.

DAVIDSON, R. L. & EPHRUSSI, B. (1965). A selective system for the isolation of hybrids between L cells and normal cells. *Nature, London*, **205**, 1170.

EDIDIN, M., PATTHEY, U. L., MCGUIRE, E. J. & SHEFFIELD, W. D. (1971). An antiserum to 'embryoid body' tumour cells that reacts with mouse embryos. In *Proceedings of the First Conference Workshop on Embryonic and Fetal Antigens in Cancer*, ed. N. G. Anderson & J. H. Coggin, p. 23a. Oak Ridge, Tennessee: USAEC.

ENGELHARDT, N. V., POLTORANINA, V. S. & YAZOVA, A. K. (1973). Localisation of alfa-fetoprotein in transplantable teratocarcinomas. *International Journal of Cancer*, **11**, 448–59.

EPSTEIN, C. J. (1970). Phosphoribosyltransferase activity during early mammalian development. *Journal of Biological Chemistry*, **245**, 3289–94.

EVANS, M. J. (1972). The isolation and properties of a clonal tissue culture strain of pluripotent mouse teratoma cells. *Journal of Embryology and Experimental Morphology*, **28**, 163–76.

FINCH, B. W. (1968). Multiple potentialities of clonal cultures of mouse testicular teratoma cells: their retention during *in vitro* culture and extinction upon somatic hybridisation. PhD Thesis, Western Reserve University.

FINCH, B. W. & EPHRUSSI, B. (1967). Retention of multiple developmental potentialities by cells of a mouse testicular teratomacarcinoma during prolonged culture *in vitro* and their extinction upon hybridisation with cells of permanent lines. *Proceedings of the National Academy of Sciences, USA*, **57**, 615–21.

GEARHART, J. D. & MINTZ, B. (1974). Contact-mediated myogenesis and increased acetylcholinesterase activity in primary cultures of mouse teratomacarcinoma cells. *Proceedings of the National Academy of Sciences, USA*, **71**, 1734–8.

Guénet, J. L., Jakob, H., Nicholas, J. F. & Jacob, F. (1974). Teratocarcinome de la souris: étude cytogénétique de cellules à potentialités multiples. *Annales de Microbiologie (des Annales de l'Institut Pasteur)*, **125A**, 135–51.

Hsu, Y.-C., Baskor, J., Stevens, L. C. & Rash, J. E. (1974). Development *in vitro* of mouse embryo from the two-cell egg stage to the early somite stage. *Journal of Embryology and Experimental Morphology*, **31**, 235–45.

Jakob, H., Boon, T., Gaillard, J., Nicholas, J. F. & Jacob, F. (1973). Teratocarcinome de la souris: isolement, culture et propriétés de cellules à potentia- [illegible]

[illegible] *Institute*, **44**, [illegible]

Kahan, B. W. & Levine, L. (1971). The occurrence of serum fetal α-1 protein in developing mice and murine hepatomas and teratomas. *Cancer Research*, **31**, 930–6.

Kleinsmith, L. J. & Pierce, G. B. (1964). Multipotentiality of single embryonal carcinoma cells. *Cancer Research*, **24**, 1544–51.

Lehman, J. M., Speers, W. C., Swartzendruber, D. E. & Pierce, G. B. (1974). Neoplastic differentiation: characteristics of cell lines derived from a murine teratocarcinoma. *Journal of Cell Physiology*, **84**, 13–28.

Littlefield, J. (1964). Selection of hybrids from mating of fibroblasts *in vitro* and their presumed recombinants. *Science*, **145**, 709.

Martin, G. R. & Evans, M. J. (1974). The morphology and growth of a pluripotent teratocarcinoma cell line and its derivatives in tissue culture. *Cell*, **2**, 163–72.

Moog, F. (1965). In *Biochemistry of Animal Development*, ed. T. Weber, vol. 1, pp. 307–65. New York & London: Academic Press.

Pierce, G. B. (1970). Epithelial basement membrane: origin development and role in disease. In *Chemistry and Molecular Biology of the Intercellular Matrix*, ed. E. A. Balasz, pp. 471–506. New York & London: Academic Press.

Pierce, G. B., Dixon, F. J. & Verney, E. L. (1960). Teratocarcinogenic and tissue-forming potentials of the cell types comprising neoplastic embryoid bodies. *Laboratory Investigations*, **9**, 583–602.

Pierce, G. B. & Verney, E. L. (1960). An *in vitro* and *in vivo* study of differentiation in teratocarcinomas. *Cancer*, **14**, 1017–29.

Rosenthal, M. D. (1968). The *in vitro* growth and differentiation of clonal populations of multipotential mouse cells derived from a transplantable testicular teratocarcinoma. MA Thesis, Brandeis University.

Rosenthal, M. D., Wishnow, R. M. & Sato, G. H. (1970). *In vitro* growth and differentiation of clonal populations of multipotential mouse cells derived from a transplantable testicular teratocarcinoma. *Journal of the National Cancer Institute*, **44**, 1001–14.

Rubin, H. (1970). Overgrowth stimulating factor released from Rous sarcoma cells. *Science*, **167**, 1271–2.

Sit, K. H. (1973). A study of the maintenance and loss of the pluripotent state of the mouse teratoma stem cells. PhD Thesis, University of London.

Skreb, N. & Svajger, A. (1973). Histogenic capacity of cat and mouse embryonic shields cultivated *in vitro*. *Wilhelm Roux Archiv für Entwicklungsmechanik der Organismen*, **173**, 228–34.

Smith, J. B. (1970). α-Fetoprotein: occurrence in certain malignant diseases and review of clinical applications. *Medical Clinics of North America*, **54**, 797–803.

Stevens, L. C. (1958). Studies of transplantable testicular teratomas of strain 129 mice. *Journal of the National Cancer Institute*, **20**, 1257.

Stevens, L. C. (1968). The development of teratomas from intratesticular grafts of tubal mouse eggs. *Journal of Embryology and Experimental Morphology*, **20**, 329–41.

Stevens, L. C. (1970). The development of transplantable teratocarcinomas from intratesticular grafts of pre- and post-implantation mouse embryos. *Developmental Biology*, **21**, 364–82.

Weiss, P. A. (1967). In *Cell Differentiation. Ciba Foundation Symposium*, ed. A. V. S. de Reuch & J. Knight, p. 13. London: Churchill.

EXPLANATION OF PLATES

Plate 1

(*a*) A great diversity of differentiated cells migrate from an explanted embryoid body. Multinucleate contractile muscle cells, nerve cells and others, less clearly identifiable, are shown here.

(*b*) Metaphase spread from a SIKR–SIKR hybrid cell in mitosis showing pairing between homologous chromosomes. (After Sit, 1973.)

Plate 2

(*a*) Thick araldite section of cells at the edge of a differentiating colony of cells. The culture was initiated from a pure clonal line of C cells and subclonal colonies grown up after plating the cells on feeder layers. After 3–4 weeks a halo of differentiated cells may be observed around the colonies. This section shows a variety of cell types but in particular long neural cell processes.

(*b*) Electron micrograph of cell processes seen in (*a*).

Plate 3

(*a*) Electron micrograph of a parietal yolk sac cell. (*b*) Electron micrograph of embryonal carcinoma C cell. Both are from the preparation shown in Plate 2(*a*).

Plate 4

Neurone-like cells in a culture from the neural-only teratoma cell line NK1. (*a*) Cells in culture; (*b*) ultrastructure of cell processes.

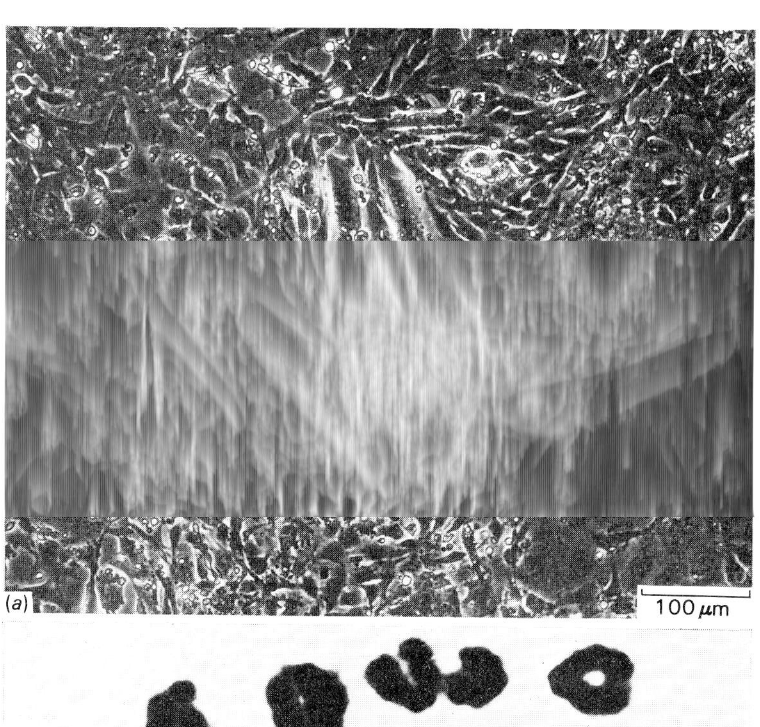

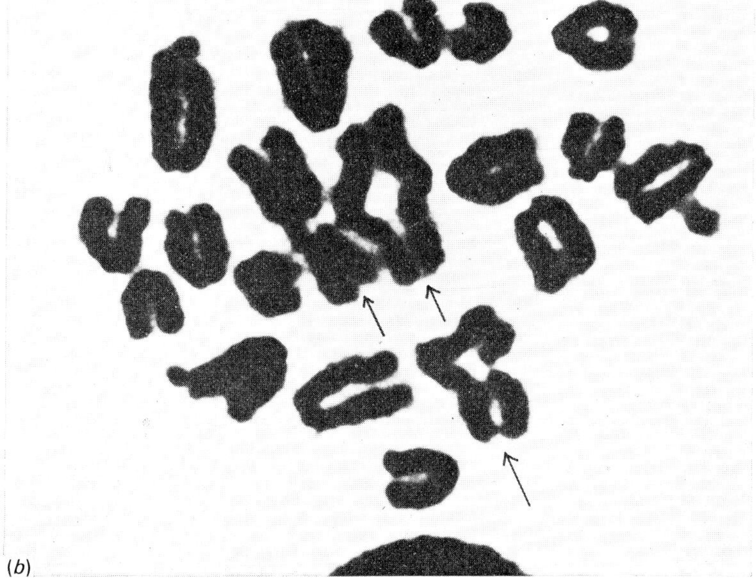

(Facing p. 284)

PLATE 2

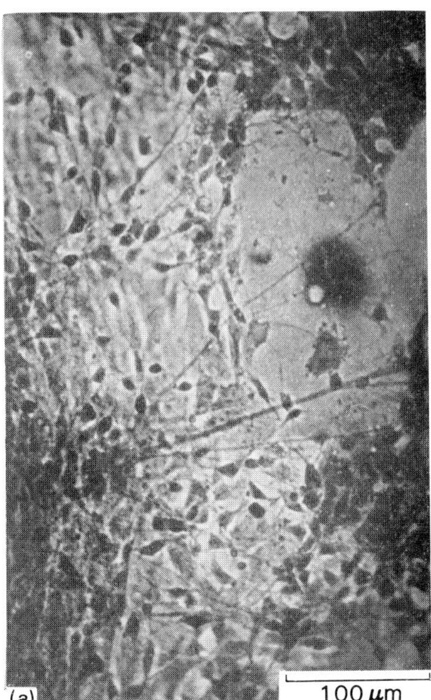

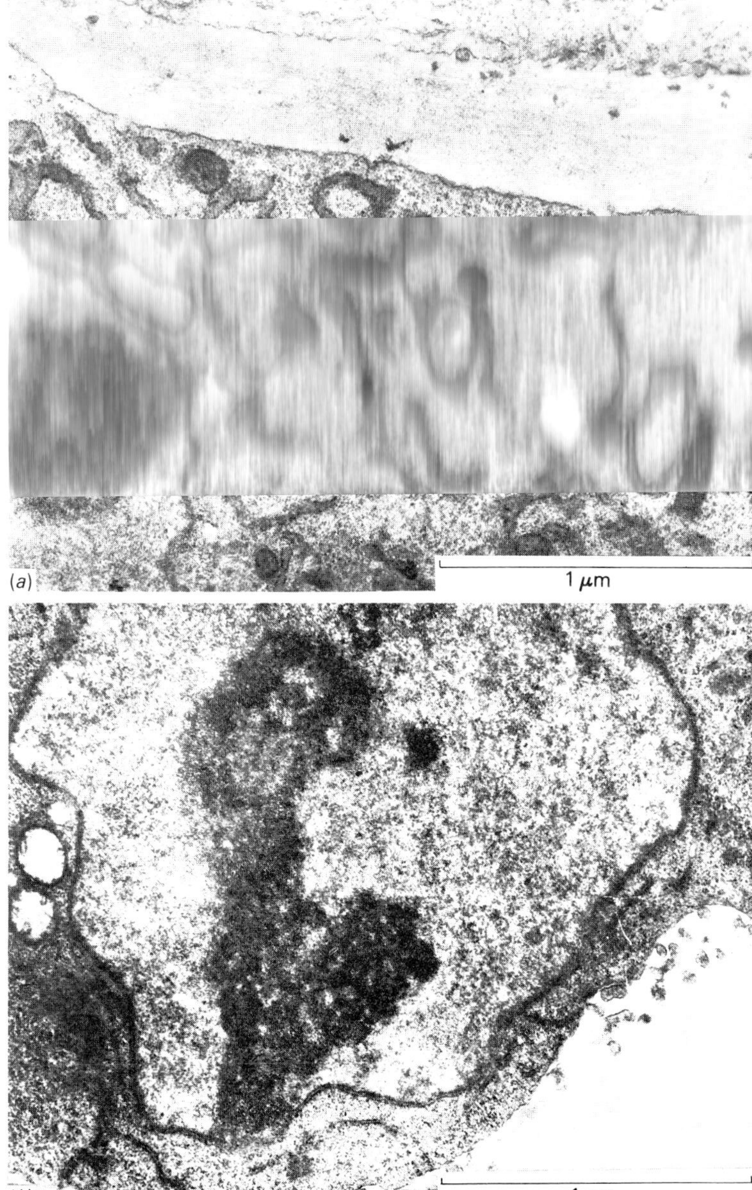
(a)
1 μm
(b)
1 μm

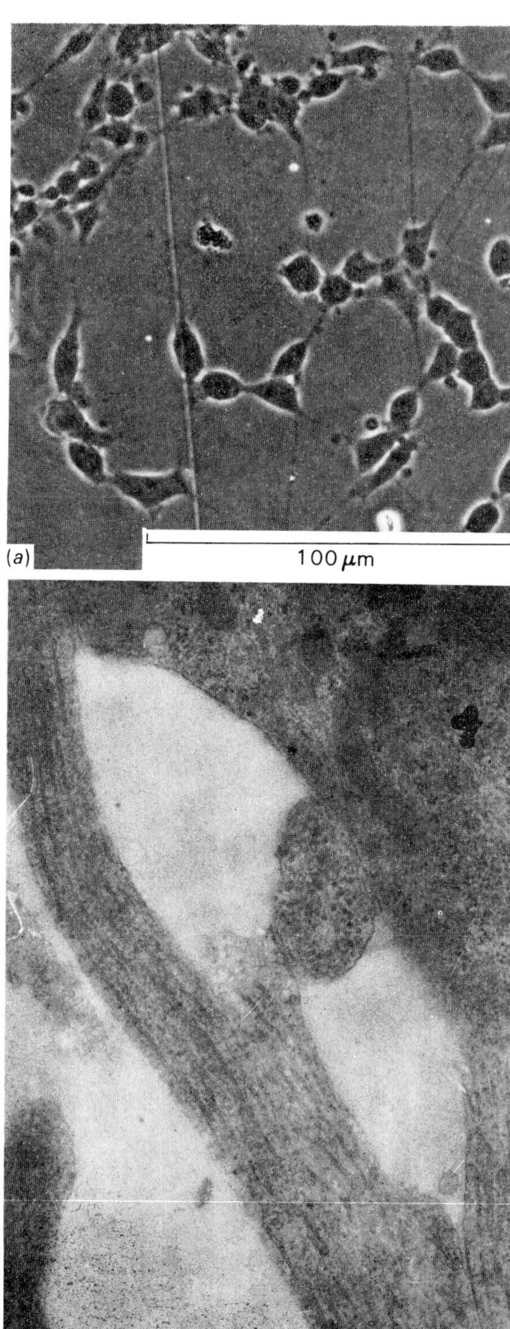
(a)
100 μm
(b)
1 μm

THE TIME IN DEVELOPMENT AT WHICH GROSS GENOME UNBALANCE IS EXPRESSED

BY C. E. FORD

Medical Research Council's External Staff,
[illegible]

[illegible] ally as those identifiable by conventional light microscopy. 'Gross genome unbalance' therefore embraces all deviations from a balanced euploid karyotype detectable by standard cytogenetic methods. No study of which I am aware has correlated biochemical events with genome unbalance in embryos. The criteria of 'expression' will therefore be restricted to morphological properties in the wide sense.

Haploidy, triploidy and tetraploidy are excluded by definition. They will, therefore, not be considered despite the disturbances of development they engender and despite the current interest aroused by such developments (all concerned with the mouse) as the introduction of methods for inducing haploid parthenogenesis (Graham, 1970; Tarkowski, Witkowska & Nowicka, 1970), the description of the specific developmental consequences of triploidy (Wroblewska, 1971) and the remarkable discovery by Snow (1973) that tetraploid embryos could be produced by treatment of cleaving eggs with cytochalasin B and that some could develop, apparently normally, to term.

Interest in the developmental consequences of gross genome unbalance has a distinguished ancestry: it may be traced back to Boveri's classic analysis (1902, 1907) of the consequences of the formation of multipolar spindles in dispermic (presumptively triploid) sea-urchin eggs. Between the wars the topic received much attention from plant cytogeneticists and Drosophilists. Only within the last decade has the improvement of cytogenetic methods made it possible to commence parallel studies in mammals. But now the interest of the medical profession in the causes and treatment of infertility on the one hand, and the development of special mouse stocks on the other, are combining to provide both the incentive and the opportunity for rapid development. My purpose will be to review the relevant information from mammalian species, giving special attention to the point in development at which the effect of genome unbalance can first be detected.

Sex chromosome unbalance presents a special case. It is well known that aneuploidy of the sex chromosomes in general has little effect on the phenotype or viability and may even be compatible with fertility. The mammalian Y chromosome carries the genetic information which normally determines, early in embryogenesis, that the indifferent gonad shall become a testis. Its independent effect on the soma can be studied in the agonadal, endocrinologically neutral human females with 46,XY and 46, XX pure gonadal dysgenesis, and may prove to be negligible (Ford, 1970). The effect of missing or additional X chromosomes, provided a minimum of one is present, is very largely regulated by the mammalian X chromosome inactivation system (see Lyon, 1972 for review), though the severe mental retardation and skeletal malformation of the human 49,XXXY male as well as the minor anatomical stigmata, poor growth and infertility presented by 45,X cases of Turner's syndrome (Hamerton, 1971) are sharp reminders that the regulation is incomplete. I shall confine my further attention to autosomal unbalance.

The small supernumerary autosomes seen in some individuals of some populations of some species doubtless have a regulatory function at the population level but have no obvious effect on the phenotype. They also constitute a special case and will be disregarded.

A final exclusion is the phenomenon of 'dominant lethality' (i.e. increased death of embryos following exposure of a parent to ionising radiation or a chemical mutagen) since effects of genome unbalance are confounded with possible effects of the inducing agent on viability not mediated through genome changes. Dominant lethality in the mouse has received special attention from Bateman (see Bateman & Epstein, 1971 for review).

It could be a matter of debate whether gross genome unbalance confined to the normal autosomes is ever wholly without phenotypic effect. Certainly in the cases to be considered it is invariably associated either with death *in utero* or (particularly in man) birth of a malformed or otherwise handicapped individual with a diminished expectancy of life.

RECIPROCAL TRANSLOCATIONS IN MICE

J. D. Snell commenced a genetic study of the descendants of X-irradiated male mice and reported his results in a series of papers starting in 1933. It seems that he was endeavouring to follow Muller and produce radiation-induced mutations in a mammalian species. He did indeed find good, formal Mendelian mutations, but they were 'dominant semi-steriles', that is, individuals with sharply reduced fertility (relative to sibs) that transmitted this property to half their descendants of both sexes. Belling (1928) had then recently put forward his hypothesis of 'segmental interchange' (reciprocal translocation) to account for dominant semi-sterility in plants [illegible] adopted this hypothesis. By dissecting at mid-term the female partners of matings between heterozygous semi-sterile and normal, Snell & Picken (1935) then showed that the reduction in litter size was attributable to the death of about half the embryos shortly after implantation. They thereby provided the first demonstration of the serious disturbance of development that may result from gross chromosome unbalance in mammals. The case, however, was inferential and direct verification by studying the chromosomes of the embryos was not then possible for technical reasons. Even though suitable methods are now available, direct evidence that the zygotic karyotypes are indeed as predicted does not appear to have been published.

Snell later (1941, 1946) published genetic evidence that his semi-sterile mice were, as predicted, heterozygous for reciprocal translocations and Koller (1944), using other stocks, demonstrated the presence of the expected quadrivalent configurations at diakinesis in primary spermatocytes.

Carter, Lyon & Phillips (1955, 1956) identified a further series of reciprocal translocations in the progeny of irradiated males and studied them extensively. They obtained accurate estimates of the relative loss of zygotes by setting up pairs of matings consisting of heterozygote with normal, and normal sib of the tested heterozygote with normal, and then dissecting the female partners in mid-term to determine the proportion of live and dead implants as Snell & Picken had done. They also counted the numbers of corpora lutea. The relative fertility of the heterozygotes (compared with their normal sibs) was found in no case to be significantly greater than 0.50, the actual range being from 0.36 to 0.59 in males and from 0.32 to 0.44 in females, depending upon the particular translocation concerned. Subsequently J. L. Hamerton and I (unpublished results) studied meiosis in male heterozygotes from the same series of translocation stocks and were able to show that, if all meiotic products formed spermatozoa and all spermatozoa were potentially functional regardless of genome unbalance, 0.50 was the upper limit of frequency of balanced zygotes to be expected, and that reductions below 0.50 could be anticipated depending upon the specific frequency of adjacent-2 disjunction at first anaphase. (In adjacent-2 disjunction, homologous centromeres of the quadrivalent move to the same pole at first anaphase so that *all* the meiotic products are necessarily unbalanced.) This might be expected to vary as between one translocation and another and also as between male and female meiosis. The results obtained by Carter, Lyon & Phillips were therefore consistent with the assumption that gross genome unbalance did not influence the capacity of the gamete to mature and function. Furthermore the fact that effectively all the excess loss of zygotes in the matings of heterozygote by normal occurred subsequent to implantation, gave a first hint that the effect of genome unbalance on zygotic development might not, in general, be expressed until after implantation.

The capacity of unbalanced gametes of both types, ova as well as spermatozoa, to function quantitatively has been demonstrated in mice for three specific pairs of *complementary* unbalanced genomes that combine to give balanced zygotes (Searle, Ford & Beechey, 1971). The method was to intercross males and females both heterozygous for the same reciprocal translocation, the one partner being homozygous for a recessive marker gene located distal to the point of exchange and the other homozygous for the corresponding normal allele. The occurrence among the progeny of individuals exhibiting the recessive phenotype indicated that *both* recessive genes were received from the one parent (implying a gamete that was duplex for the chromosome segment concerned) and none from the other (implying a complementary nulloplex gamete). At present the quantitative case is confined to the distal segments of the six autosomes involved in the three reciprocal translocations, T(2.8)26H, T(1.13)70H, and T(5.13) 264Ca. The functional competence of gametes in which the proximal segment is represented twice or not at all was demonstrated in parallel intercrosses by using recessive marker genes located on these segments, but in this situation qualitative evidence only is obtainable. The method could in principle be extended to cover the whole autosomal genome.

Snell (1946) had reported that the relative fertility of males heterozygous for his translocation T(5.8)a (now written in terms of chromosomes rather than linkage groups as T(2.4)Sn) had relative fertilities in the range 0.60–0.63. The difference from 0.50 was significant and (viewed retrospectively) contrary to expectation. It was possible to infer from meiotic studies that the chromosome segments exchanged in this particular translocation were very short. (Evans & Burtenshaw (unpublished observations) have recently confirmed this in banded preparations). The two major classes of unbalanced gametes produced by heterozygotes would then each be duplex for one of the very short segments and nulloplex for the other. The possibility of a delayed lethal effect of one or both of the two corresponding zygotic classes was therefore [illegible] until birth but died before weaning. The genetic evidence gives no hint that any are capable of surviving to breed and there is therefore no essential conflict with the hypothesis.

I am unaware of any other instance in the mouse in which individuals with unbalanced, but euploid, genomes are capable of survival until after birth. On the other hand some of the (tertiary) trisomic progeny of mice heterozygous for markedly unequal translocations (i.e. in which a very long segment is exchanged for a very short one) may survive for some time after birth and may even be fertile (Lyon & Meredith, 1966; de Boer, 1973). In all these cases the extra translocation chromosome was appreciably shorter than the smallest autosome, so that the amount of extra genetic material present in triplicate was presumptively much less, in a crude quantitative sense, than in any primary autosomal trisomic.

The reciprocal translocations in the mouse just discussed were all radiation-induced. A few have arisen spontaneously (Lüning & Searle, 1971) and spontaneous examples with similar properties have also been reported in the rat (Bouricius, 1948) and pig (Henricson & Bäckström, 1964).

RECIPROCAL TRANSLOCATION AND SPONTANEOUS ABORTION IN MAN

Reliance on the mouse model would have led one to anticipate that heterozygosity for a reciprocal translocation in either partner of a fertile human couple would result in an erratic sequence of normal live births and spontaneous abortions, with the latter somewhat exceeding the former. However, although reciprocal translocations have sometimes been identified in habitual aborters or their husbands, the frequency is low (e.g. Kaosäar & Mikelsäar, 1973; Nuzzo, Giorgi, Zuffardi & Dambrosio, 1973). Some have been discovered fortuitously while examining karyotypes for other purposes (e.g. Jacobs *et al.*, 1970). But the great majority of reciprocal translocations known in man have been found as a result of the investigation of mentally retarded children with multiple congenital abnormalities (see Ford & Clegg, 1969, for review).

Many of these children had grossly unbalanced karyotypes and frequently one parent was a balanced heterozygote. But however the translocation was identified, live-born children with unbalanced karyotypes, and spontaneous abortions in excess of the basic rate (estimated to be 15%) almost invariably amounted to appreciably less than half the clinically recognised pregnancies in the matings of heterozygotes (of either sex) to normal. There are, therefore, two major differences from the mouse model in respect of unbalanced zygotic genomes in the progenies of heterozygotes; (1) they are recorded much less frequently than would have been expected, and (2) they are often associated with live birth. The first is a purely cytogenetic issue and does not concern us here. The second poses a developmental problem: why should foetuses with apparently equivalent (quantitative) genetic deviations from the norm die in mid-gestation in the mouse but commonly survive to term in man and enjoy a modest expectation of post-natal life? This difference is most unlikely to be a spurious one attributable solely to differences in ascertainment. Competition between foetuses in the polytoccous species and its (usual) absence in the monotoccous species is a formally possible explanation, but there may be others and the problem has not been investigated.

All biological systems are prone to error. Even the most exquisitely regulated disjunction of the chromosomes on the spindle at mitosis and meiosis fails sometimes. It was therefore to be expected that some human zygotes would receive more than the normal number of chromosomes and others less. The questions were essentially quantitative ones: how frequent would these events be, what effect would specific types of chromosome unbalance have on the rate of embryonic development, and when would the affected individuals die? Classical cytogenetics would have predicted death of unbalanced zygotes so early in development that there was either no overt evidence of pregnancy, or brief pregnancy followed by spontaneous termination.

An enormous effort has gone into the determination of the karyotypes of human spontaneous obortuses. This work, and related studies, was reviewed at a symposium held in Paris in September, 1973 (Boué &

Thibault, 1973). Nearly all spontaneous abortion occurs before the end of the third month and Boué & Boué (1973), the principle contributors, have confined their attention to abortions occurring within this period. The procedure involves the collection of specimens under difficult conditions and the growth of healthy cultures for karyotype determination from material that is often partly macerated. Nevertheless the technical and [illegible] 60%. The increased incidence of abnormality compared with the results of earlier investigators, notably Carr (1970, 1971), is probably attributable to several factors including improved success in culture (karyotypically normal specimens may be more likely to give successful cultures), effective elimination of induced abortuses from the samples studied, and the inclusion of a more-nearly representative proportion of the earliest abortions (in which the highest frequency of karyotypic abnormality is found).

Boué & Boué (1973) summarised the results obtained by nine different groups. The data were obtained entirely from industrial countries of the North Temperate Zone and almost exclusively from subjects of Caucasian origin. Whether social, climatic and ethnic factors might have an influence is unknown. Though the proportions of total abnormal to normal karyotypes recorded by each group varied considerably, the relative proportions of the different classes of abnormality showed concordance just short of strict statistical homogeneity. These data are combined in Table 1 to give an indication of overall incidence of the different types. A point obscured by the Table is that there were only two autosomal monosomics; all the remaining monosomics were X-monosomics (i.e. they exhibited the 45, X karyotype, which, if the foetus survives to birth, is typically associated with Turner's syndrome). Non-disjunction at either of the meiotic divisions should generate nullosomic $(n-1)$ gametes with the same frequency as disomic $(n+1)$ gametes, and if these are all potentially functional, autosomal monosomic zygotes should be as frequent as autosomal trisomic zygotes. If lagging of a bivalent or chromosome on the spindle should occur sometimes, with eventual exclusion from both daughter nuclei, monosomics could even be more frequent than trisomics at fertilisation. Are the expected monosomic zygotes formed but eliminated so early that

Table 1. *Types of chromosomal abnormality identified in human spontaneous abortuses and their relative frequencies*

Type	Number of observations	%
Monosomics	262	19
Trisomics	703	51
Triploids	239	18
Tetraploids	76	6
Mosaics & rearrangements	79	6
Total	1359	100

After Boué & Boué (1973) and Therkelsen *et al.* (1973).

they do not appear among the spontaneous abortuses? Or are they not formed at all?

It is frequently speculated that many human zygotes could be lost between fertilisation and the clinical recognition of pregnancy with no more sign than a delayed period, and further, that such early losses could include additional chromosomal abnormalities, the 'missing' monosomics among them. The only relevant evidence appears to be the publication of Hertig, Rock & Adams (1956) who reported the recovery of the products of conception from the oviducts and uteri of 34 out of 211 hysterectomy patients. The specimens ranged '. . . from a 2-day, 2-cell tubal egg to a 17-day well implanted ovum . . .'. Twenty-one of the specimens were stated to be normal and thirteen were '. . . abnormal in one or more ways'. This investigation was carried out before the development of modern cytogenetic methods and we do not know whether the morphological abnormalities observed, or even any part of them, were consequential upon genome unbalance. Direct evidence of an increased incidence of chromosomal abnormality in the early human zygote must wait upon a repeat of this study, combined with cytogenetic examination of the recovered embryos.

In Table 2, 679 of the 703 trisomics shown in Table 1 are partitioned between the autosomal groups that provided the extra chromosome. (The remaining twenty-four included twenty-three double trisomics and one in which the group was not identified.) The observed frequencies differ considerably from those expected if frequency were linearly related to the number of pairs in the group. These differences might be attributable to different rates of non-disjunction between one group and another, to different probabilities of survival as zygotes, or to a combination of the two. The new staining procedures which permit the identification of individual chromosomes should make it possible in the future to estimate the specific frequency of occurrence of each of the twenty-two possible types of primary autosomal trisomy.

Table 2. *Partition of primary trisomic types from Table 1 between chromosome [illegible]*

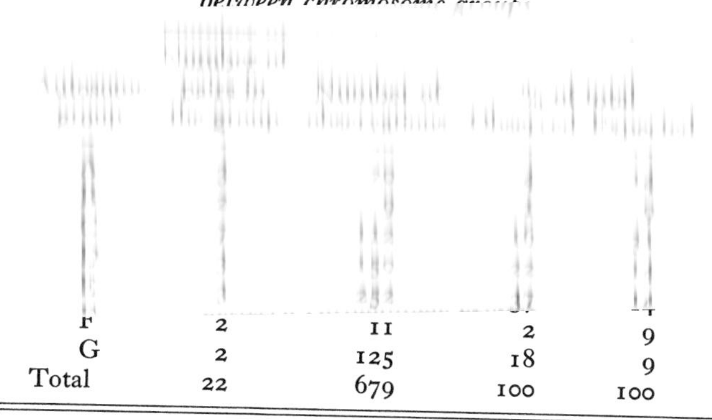

[illegible]	[illegible]	[illegible]	[illegible]	[illegible]
F	2	11	2	9
G	2	125	18	9
Total	22	679	100	100

Phillipe & Boué (1970) have conducted an anatomical study of the products of conception and the placenta in many of the cases examined and subsequently reported by Boué & Boué (1973) and have contended that each type of chromosomal abnormality is associated with a specific mean period of development of the zygote and a specific period of retention *in utero*. The latter figure is the difference between the estimated gestational age at which development ceased and the gestational age at which the abortion occurred. Although the authors apparently favour the concept of an abrupt termination of the life of a normally developing foetus, their data do not appear to exclude the possibility that development may be retarded before it ceases altogether. Were it otherwise, our central question would now be answered for certain types of gross genome unbalance in man. The reported 'durées de developpement' may then not be absolute estimates of gestational age at death but rather estimates, referred to a standard scale, of the degree of development attained at death.

However interpreted, the results obtained by Phillipe & Boué are of great comparative interest. For example, the mean 'period of development' of trisomics of group A is given as 3.2 weeks, of group B 3.7 weeks, of group D 4.9 weeks, of group F 3.1 weeks, and of group G 5.3 weeks (Boué & Boué, 1973). The chromosomes of group F are only marginally longer than those of group G and only one-quarter or one-third of the length of the chromosomes in groups A and B. Yet the F trisomics are shown as having the shortest 'period of development' of all primary trisomics. This suggests that the F chromosomes may carry genetic information of such importance that dosage changes have a disproportionately great effect on the development of the zygote. The rarity of F trisomics compared with expectation (Table 2) might then be due to a high risk of death and resorbtion or expulsion of the embryo before pregnancy is clinically recognisable. However, the chromosomes of group F are small metacentrics which give meiotic bivalents almost invariably with a single terminal chiasma in each arm, and in the absence of the information obtained by Phillipe & Boué, the rarity of F trisomics might plausibly have been attributed to exceptionally regular disjunction at meiosis.

In concluding this section we may estimate the total wastage of zygotes attributable to chromosomal abnormality in man and compare it with what is known in other mammalian species. The overall rate of spontaneous abortion in Caucasian populations is generally considered to lie in the range 15% (Warburton & Fraser, 1964) to 20% (Inhorn, 1967) of clinically recognised pregnancies. Estimates of the frequency of chromosomal abnormality of all kinds in spontaneous abortuses range from 36% (Carr, 1970) to 60% (see above). Chromosomal abnormality is then implicated in between 5.4% and 12% of recognised pregnancies. But some trisomics and effectively all autosomal monosomics are likely to be eliminated without detection. Also, autosomal monosomics should be at least as frequent as trisomics in the zygotic population. Since trisomics constitute rather more than 50% of all chromosomal abnormality identified in the spontaneous abortuses, the proportion of human zygotes that are chromosomally abnormal could be as high as 20%. It would not therefore be surprising if the incidence of chromosomal abnormality in early human embryos produced by in-vitro fertilisation (see Fowler & Edwards, 1973) should be as great as this. In the absence of cytogenetic studies of early natural embryos obtained at hysterectomy (see above) there would be no way of partitioning chromosomal abnormality in excess of that expected in clinically recognised pregnancy between components equivalent to natural early zygotic losses and attributable to the process of in-vitro fertilisation and culture itself.

The limited comparative data from other mammalian species are summarised in Table 3. They suggest that the total incidence of zygotic chromosomal abnormality may be very much less than in man. They also indicate that errors of fertilisation (producing haploids and probably most triploids) and at mitosis during cleavage (producing tetraploids and mosaics) may be the major contributory factors, rather than those of meiotic disjunction (producing monosomics and trisomics) as in man. However, man differs from the five species of Table 3 in three major biological respects: he is a primate, monotoccous, and (comparatively) very long-lived, and any one might have an important influence on the vital processes of meiosis, fertilisation and mitosis in the cleavage divisions of the zygote. Nevertheless, the uneasy suspicion remains that the very high incidence of chromosomal abnormality, particularly non-disjunctional abnormality, in human zygotes may be another 'disease of civilization'. Much further work will be required to exclude this possibility.

[illegible]

[illegible]	[illegible]	[illegible]	[illegible]	[illegible]	[illegible]	[illegible]	[illegible]	[illegible]	[illegible]
[illegible]	[illegible]	[illegible]				[illegible]		[illegible]	[illegible]
[illegible]	[illegible]	[illegible]				4	3	2	10.2
Chinese hamster	4–8-cell eggs	126	1		2	3			4.8
Mouse	8–11-day embryos	607		3		5	1	4	2.1
Rabbit	$5\frac{1}{2}$-day blastocysts	463		4		8		11	5.0

Data from Butcher & Fugo (1967) for rat, McFeely (1967) for pig, Schmidt & Biukert (1973) for Chinese hamster, Ford & Evans (1973) for mouse and Fechheimer & Beatty (1974) for rabbit. Nineteen of the twenty 'others' were mosaics of various kinds; one (in the pig) was a deletion.

PRIMARY AUTOSOMAL MONOSOMY AND TRISOMY IN THE MOUSE

The extensive data from human spontaneous abortuses are valuable and informative in some respects, but limited in others, and for further consideration of our problem we have to turn to experimental material. The development of efficient techniques for determining the karyotypes of mouse embryos before implantation (Tarkowski, 1966) and after implantation (Evans, Burtenshaw & Ford, 1972) makes a new approach possible in this species. The reciprocal translocations mentioned earlier, however, are not ideal material because the unbalanced karyotypes produced by heterozygotes are always simultaneously duplicated and deficient and though abnormal zygotes could be obtained in large numbers, interpretation of the results would not be straightforward. The position has been changed completely by the introduction of the 'tobacco' mouse, *Mus poschiavinus*, as a laboratory animal.

Mus poschiavinus is a relict species confined in nature, so far as is known, to Val Poschiavo in southeastern Switzerland. An account of the circumstances of its rediscovery is given by Gropp, Tettenborn & von Lehman (1970). Its special cytogenetic value stems from the fact that it is partially interfertile with laboratory stocks of *Mus musculus* yet differs in having seven pairs of large metacentric chromosomes in the place of fourteen pairs of acrocentric chromosomes. The diploid number in *M. poschiavinus* is therefore twenty-six, compared with forty (all acrocentrics) in *M. musculus*, the difference being attributable to seven independent Robertsonian translocations that have become homozygous. Meiosis and gametogenesis are normal in F_1 hybrids of both sexes, except that seven trivalent associations are formed by the pairing of two *M. musculus* acrocentrics with each *M. poschiavinus* metacentric chromosome. This meiotic evidence of homology between individual *M. musculus* acrocentrics and individual arms of *M. poschiavinus* metacentrics is supported by identity of banding patterns (Zech, Evans, Ford & Gropp, 1972).

The trivalent associations are liable to undergo irregular segregation (non-disjunction), the metacentric chromosome and one acrocentric going to one pole and the remaining acrocentric to the other. Gametes may then be produced with one extra or one missing chromosome arm and give rise to trisomic and monosomic zygotes respectively. Each of the seven trivalents in the F_1 male has an independent chance of undergoing irregular segregation at AI and the consequences are directly observable at MII as a high frequency of deviant counts (Tettenborn & Gropp, 1970). It is convenient to express all counts in terms of the number of arms, one for an acrocentric chromosome and two for a metacentric chromosome. All karyotypes can then be referred to the same scale, the normal diploid genome having a count of forty regardless of the number of metacentric chromosomes present. All trisomics would then have a count of forty-one, monosomics one of thirty-nine, a balanced haploid genome twenty, and so on.

In a preliminary study (Ford, 1972), two F_1 males were mated to normal females, some of which were allowed to go to term, some killed for mid-term embryos and some killed on the fourth day of gestation to provide pre-implantation embryos. Karyotypes of as many as possible of the live-born and embryonic progeny were determined and counts at MII were subsequently obtained from the sires.

The results were very clear: 79 pre-implantation embryos gave counts in the range 34 to 42, 113 mid-term embryos (eleventh to sixteenth day) gave counts from 40 to 44, whereas 40 (sic) live-born young all gave normal counts of 40. Three of the four low counts in the range 34 to 36 from pre-implantation embryos were probably artefactual and attributable to damage during preparation; there was only a single metaphase suitable for counting in each of them. In the fourth, however, the count of 35 was determined by two independent observers on five technically good mitoses and is secure; also, the cell count was within the range recorded for other embryos of the same clutch and there was no hint of cytological abnormality. Counts in 500 cells at MII were symmetrically distributed about a mode of 20 and ranged from 15 to 25. The two count distributions, pre- [illegible] classes.

This experiment was set up primarily to test the possibility of selection against unbalanced gametic genomes during the processes of maturation, transport and fertilisation. None was anticipated and none was found. The data therefore give further support to the assumption that the male gamete can mature and function regardless of genome unbalance; they also imply that the early development of the zygote may be unaffected by unbalance contributed by the paternal genome.

The results indicated that all zygotes with unbalanced genomes may survive up to the time of implantation, that the hypoploid embryos are rapidly eliminated thereafter, and that the hyperploid embryos survive longer but die before birth. This result has been broadly confirmed in a more extensive investigation which has shown that rare monosomic embryos with apparently normal phenotype (presumed to be XO) may survive until the sixteenth day; that the hyperploid embryos, as might have been expected, are steadily reduced in frequency throughout gestation; and that rare trisomic embryos may be carried to term and survive for a few hours of post-natal life (Ford, Evans, Burtenshaw, Clegg, Gropp & Giers, unpublished results). Comparison of corpora lutea counts with numbers of implantation sites indicated that losses of embryos before implantation may have been no greater than in normal matings.

To examine these questions in more detail, seven separate stocks of mice were established, each carrying a single one of the seven metacentric chromosomes of *Mus poschiavinus*. These are designated Rb1Bnr to Rb7Bnr respectively. Heterozygous males from each of the lines were mated to normal females and the type of investigation carried out with the F_1 males was repeated. The observations are incomplete but two interim accounts have been published (Ford & Evans, 1973*a*, *b*). The data analysed so far showed that there was a small but non-significant excess of trisomic over normal embryos before implantation; that the ratio of implantation sites to corpora lutea was slightly lower than in controls, though the difference is of doubtful significance; and that, as expected, the aneuploid post-implantation embryos were almost entirely trisomics. They also confirmed that the rare trisomics among the newborn only survived birth for a few hours at most. The post-implantation embryos were divided for analysis into two groups corresponding to gestational days 8–11 and 12–15 (counting the day on which the plug was observed as day 0). Nine monosomic embryos were identified, seven of them in the 8–11 day period. Two of these were not obviously abnormal morphologically; they had no Y chromosome and were assumed to be XO. The remaining five all presented the same abnormal phenotype, namely a small ball of cells with a rugose surface about 250 μm in diameter. A Y chromosome was present in four and as they were all sired by Rb(9.14)6 males it is surmised that all were monosomic with respect to the same autosome, either chromosome 9 or chromosome 14.

Direct estimates of the frequency of non-disjunction at AI were obtained from the counts at MII. Aneuploid counts at MII are very infrequent in males with entirely normal bivalents at meiosis. More importantly, non-disjunction directly attributable to the trivalent is indicated by counts of 21 including the metacentric. Counts of 19 without the metacentric could have originated by non-disjunction of the trivalent, by non-disjunction of a bivalent, or as a result of damage to the cell during preparation and the loss of one chromosome. Nevertheless, the near-equality of counts of 21 plus the metacentric, and of 19 without it, combined with the negligible frequency of other aneuploid classes, demonstrates that the contribution of factors other than the segregating trivalent itself was minimal. (It is possible, particularly at MII, for a metacentric chromosome to simulate two acrocentric chromosomes and vice versa. The presence or absence of the metacentric was therefore always ascertained before counting commenced, to exclude possible bias in interpretation.) Parallel indirect estimates of non-disjunction were obtained from the combined frequencies of monosomic and trisomic pre-implantation embryos. The data from the post-implantation embryos were also used to derive minimal estimates of non-disjunctional frequency by supposing the (usually) absent monosomic class to have been originally as frequent as the trisomic class. There was a marked decrease in the proportion of trisomic to normal embryos between 8–11 days and 12–15 days, so data from the former period were used. The results obtained so far are summarised in Table 4. The good agreement between the independent estimates for the same Robertsonian translocation indicate that the nominally minimal estimates derived from the post-implantation embryos are good ones and that few, if any, trisomic embryos can die before day 11.

Table 4. *Non-disjunction in male mice heterozygous for single metacentric chromosomes*

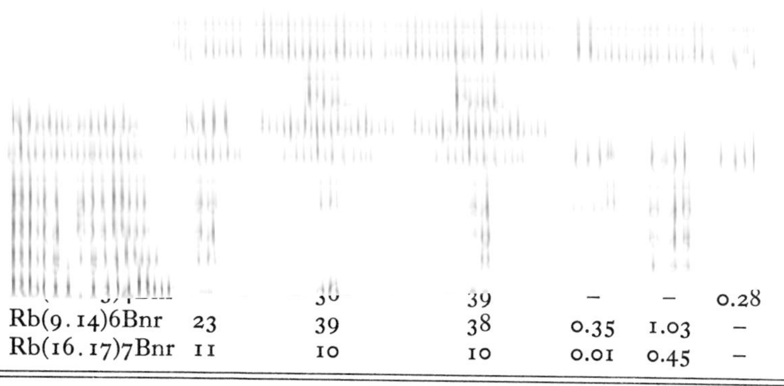

[illegible]	[illegible]	[illegible]	39	–	–	0.28
Rb(9.14)6Bnr	23	39	38	0.35	1.03	–
Rb(16.17)7Bnr	11	10	10	0.01	0.45	–

After Ford & Evans (1973*b*).

* Between MII and pre-implantation embryos.

† Between MII and post-implantation embryos.

‡ Between pre-implantation and post-implantation embryos.

A stringent measure of concordance is given by 2 × 2 homogeneity tests. The results of these are also given in Table 4. The test between the MII data and the pre-implantation data was made by combining the aneuploid classes (n–1, n+1) and (2n–1, 2n+1) and testing against euploid counts. The tests of the MII counts and, separately, the pre-implantation counts against the post-implantation counts were confined to the hyperploid and euploid classes. None of the χ^2_1 values obtained was significant at the 0.05 level and, more importantly, the combined χ^2 for all tests is not significant ($\chi^2_9 = 4.33$, $P > 0.8$).

These results, combined with the data obtained from the F_1 experiments, and supported in part by the investigations of reciprocal translocation heterozygotes and the translocation intercross experiment described earlier, strongly imply that gross genome unbalance does not prejudice the capacity of a spermatid to mature into a functional sperm, with the same chance of effecting fertilisation as a normal sperm, or the capacity of the resultant zygote to develop, at the very least to the late morula stage. It is a striking illustration of the latter point that a zygote with only thirty-five chromosome arms, that is, simultaneously monosomic for five different chromosomes, should have been capable of developing *pari passu* with the other eggs of the clutch for three days and yet have shown no sign of abnormality.

The data also indicate that all or nearly all monosomic embryos normally implant (or at least evoke a decidual reaction), but (if they do indeed implant) that in general they die very shortly afterwards. The maximum survival so far observed (in embryos monosomic for 9 or 14) is to day 11. Since there are two possible monosomics for each metacentric chromosome under test, this conclusion applies (at present) minimally to six, maximally to twelve, of the nineteen possible primary autosomal monosomic types.

The survival of primary autosomal trisomics is much better. Again our data apply minimally to six, maximally to twelve, of the nineteen possible types. They provide no evidence of death before day 11, but rapid elimination thereafter. (Death is defined operationally as a lack of sufficient mitotic cells for the karyotype to be established, although development may have ceased some time previously and the embryo may already be in the process of resorbtion.) On the basis of present evidence, then, the time of first expression of gross genome unbalance must lie between the late-morula stage (say 32 cells) and the immediate post-implantation period; earlier in monosomics, later in trisomics. This conclusion, which is based on morphological criteria, applies directly to the mouse alone. Nevertheless it is concordant with the data from man considered in the previous section.

There are two obstacles in the way of determining this time more exactly for specific monosomics and trisomics. These are that the embryo is at its most inaccessible stage during the period concerned, and that, with the method we have used, two monosomic types and two trisomic types are always confounded. (The preparative method does not lend itself to banding procedures that would permit individual identification of chromosomes.) However the second difficulty at least can be overcome by the use of an ingenious cytogenetic device introduced by White, Tjio, van de Water & Crandall (1972). They found that in males heterozygous for two different Robertsonian translocations with an arm in common, a chain of four chromosomes was regularly formed in meiosis with acrocentrics at both ends and the two metacentrics in the middle, and that irregular disjunction frequently took place at AI such that the two metacentrics went to one pole and the two acrocentrics to the other. The chain quadrivalent could be represented symbolically as A – A.B – B.C – C and the two complementary non-disjunctional products as A.B plus B.C (disomic for B) and A plus C (nullosomic for B). Mating such doubly heterozygous males to normal females could then be expected to result in the formation of specific monosomic and trisomic zygotes with a frequency sufficient for their properties to be studied.

The stocks used by White *et al.* were Rb(5.19)1Wh and Rb(9.19)163H. Monosomics and trisomics of the smallest autosome, No. 19, would then be expected in crosses to normal. The same authors have now defined some of the properties of trisomy 19 (White, Tjio, van de Water & Crandall, 1974*a*, *b*) but have made no mention of monosomy. The same principle has been employed by Gropp & Kolbus (1974), who studied trisomics of [illegible] highly specific morphological property of embryos trisomic for chromosome 12.

The apparent capacity of cleaving embryos with unbalanced karyotypes, monosomic as well as trisomic, to develop as far as the 32-cell stage without evident disturbance deserves comment. From the evidence we have we could say that gene dosage in the range 1–2–3 is equivalent in its effect on the course of development at least up to this stage. Three possible explanations may be considered. First, a proportional genome unbalance that influences cellular events at the transcriptional and translational levels may yet require the passage of time before it is expressed as overtly retarded or abnormal development. A second formal, but unlikely, possibility is the presence of a short-lived regulatory system that ensures that one copy of a given gene is sufficient for immediate requirements and that the effect of additional copies is suppressed. The third possibility is that the ovulated oocyte may contain sufficient long-lived messenger RNA produced by the diploid maternal genome to meet the needs of protein synthesis during the first three days of development.

The second and third possibilities would both be excluded if zygotes with zero dosage of particular chromosomes should fail to cleave. Morris (1968) recovered pre-implantation embryos from XO females after $3\frac{1}{2}$ days gestation and reported that approximately 25% were morphologically abnormal, many having apparently ceased development in the 2-blastomere stage. He thought that these were the anticipated class of YO embryos. However, this supposition was not supported by direct cytogenetic evidence and an alternative explanation for the observation may be possible. If the interpretation is correct it would mean that the zygote wholly devoid of an X chromosome fails in development very much earlier than any of the primary autosomal monosomic types of which we have knowledge. In any case it would be desirable to examine the fate of autosomal nullosomics. Theoretically it should be possible to produce nullosomic zygotes by intercrossing double Robertsonian heterozygotes of the type just considered. The data of White *et al.* (1972) and Gropp & Kolbus (1974) imply that yields of between 3% and 4% nullosomic zygotes could be anticipated. This experiment is in progress.

REFERENCES

BATEMAN, A. J. & EPSTEIN, S. S. (1971). Dominant lethal mutations in mammals. In *Chemical Mutagens: Principles & Methods for their Detection*, ed. A. Hollander, vol. 2, pp. 541–68. New York: Plenum Press.

BELLING, J. (1928). A working hypothesis for the segmental interchange between non-homologous chromosomes in flowering plants. *California University Publications in Botany*, **14**, 335–43.

DE BOER, P. (1973). Fertile tertiary trisomy in the mouse (*Mus musculus*). *Cytogenetics and Cell Genetics*, **12**, 435–42.

BOUÉ, J. & BOUÉ, A. (1973). Anomalies chromosomiques dans les avortements spontanés. In *Les Accidents Chromosomiques de la Reproduction*, ed. A. Boué & C. Thibault, pp. 29–55. Paris: Institut National de la Santé et de la Recherche Médicale.

BOUÉ, A. & THIBAULT, C. (eds.) (1973). *Les Accidents Chromosomiques de la Reproduction.* Paris: Institut National de la Santé et de la Recherche Médicale.

BOURICIUS, J. (1948). Embryological and cytological studies in rats heterozygous for a probable reciprocal translocation. *Genetics*, **33**, 577–87.

BOVERI, T. (1902, 1907). Quoted by E. B. Wilson. In *The Cell in Development and Heredity*, 3rd edition, 1925, pp. 917 *et seq.* New York: Macmillan.

BUTCHER, R. L. & FUGO, N. W. (1967). Overripeness and the mammalian ova. II. Delayed ovulation and chromosome anomalies. *Fertility and Sterility*, **18**, 297–302.

CARR, D. H. (1970). Chromosome abnormalities and spontaneous abortions. In *Human Population Cytogenetics*, ed. P. A. Jacobs, W. H. Price & P. Law, pp. 103–18. Edinburgh: Edinburgh University Press.

CARR, D. H. (1971). Chromosomes and abortion. *Advances in Human Genetics*, **2**, 201–57.

CARTER, T. C., LYON, M. F. & PHILLIPS, R. J. S. (1955). Gene-tagged chromosome translocations in eleven stocks of mice. *Journal of Genetics*, **53**, 154–66.

CARTER, T. C., LYON, M. F. & PHILLIPS, R. J. S. (1956). Further genetic studies of eleven translocation stocks in the mouse. *Journal of Genetics*, **54**, 462–73.

EVANS, E. P., BURTENSHAW, M. D. & FORD, C. E. (1972). Chromosomes of mouse embryos and newborn young: preparations from membranes and tail tips. *Stain Technology*, **47**, 229–34.

FECHHEIMER, N. S. & BEATTY, R. A. (1974). Chromosomal abnormalities and sex ratio in rabbit blastocysts. *Journal of Reproduction and Fertility*, **37**, 331–41.

FORD, C. E. (1970). Cytogenetics and sex determination in man and mammals. *Journal of Biosocial Science*, Supplement **2**, 7–30.

FORD, C. E. (1972). Gross genome unbalance in mouse spermatozoa: does it influence the capacity to fertilize? In *The Genetics of the Spermatozoon*, ed.

R. A. Beatty & S. Gluecksohn-Waelsch, pp. 359–69. Edinburgh: Beatty & Gluecksohn-Waelsch.

FORD, C. E. & CLEGG, H. (1969). Reciprocal translocations. *British Medical Bulletin*, **25,** 110–14.

FORD, C. E. & EVANS, E. P. (1973*a*). Robertsonian translocations in mice: segregational irregularities in male heterozygotes and zygotic unbalance. In *Chromosomes Today*, ed. J. Wahrman & K. R. Lewis, vol. 4, pp. 387–97. New York & Toronto: John Wiley. (Also, Jerusalem: Israel Universities Press.)

[illegible]

[illegible] (1974). Exencephaly in the syndrome of trisomy No. 12 of the foetal mouse. *Nature, London*, **249,** 145–7.

GROPP, A., TETTENBORN, U. & VON LEHMAN, E. (1970). Chromosomenvariation vom Robertson'schen Typus bei der Tabakmaus, *M. poschiavinus*, und ihren Hybriden mit der Laboratoriumsmaus. *Cytogenetics*, **9,** 9–23.

HAMERTON, J. L. (1971). *Human Cytogenetics*, vol. 1. New York & London: Academic Press.

HENRICSON, B. & BÄCKSTRÖM, L. (1964). Translocation heterozygosity in a boar. *Hereditas*, **52,** 166–70.

HERTIG, A. T., ROCK, J. & ADAMS, E. C. (1956). A description of 34 human ova within the first 17 days of development. *American Journal of Anatomy*, **98,** 435–93.

INHORN, S. L. (1967). Chromosomal studies of spontaneous human abortions. In *Advances in Teratology*, ed. D. H. M. Wollham, vol. 2, pp. 37–99. London: Logos Press.

JACOBS, P. A., AITKEN, J., FRACKIEWIECZ, A., LAW, P., NEWTON, M. S. & SMITH, P. G. (1970). The inheritance of families in man: data from families ascertained through a balanced heterozygote. *Annals of Human Genetics*, **34,** 119–36.

KAOSÄAR, M. E. & MIKELSÄAR, A. V. N. (1973). Chromosome investigation in married couples with repeated spontaneous abortions. *Humangenetik*, **17,** 277–283.

KOLLER, P. C. (1944). Segmental interchange in mice. *Genetics*, **29,** 247–63.

LÜNING, K. G. & SEARLE, A. G. (1971). Estimates of the genetic risks from ionizing irradiation. *Mutation Research*, **12,** 291–304.

LYON, M. F. (1972). X-chromosome inactivation and developmental patterns in mammals. *Biological Reviews*, **47,** 1135.

LYON, M. F. & MEREDITH, R. (1966). Autosomal translocations causing male sterility and viable aneuploidy in the mouse. *Cytogenetics*, **5,** 335–54.

MCFEELY, R. A. (1967). Chromosome abnormalities in early embryos of the pig. *Journal of Reproduction and Fertility*, **13,** 579–81.

MORRIS, T. (1968). The XO and OY chromosome constitutions in the mouse. *Genetical Research*, **12,** 125–37.

NUZZO, F., GIORGI, R., ZUFFARDI, O. & DAMBROSIO, F. (1973). Translocation t(1p+;2q-) associated with recurrent abortion. *Annales de Génétique*, **16,** 211–214.

PHILLIPE, E. & BOUÉ, J. F. (1970). Placenta et aberrations chromosomiques au cours des avortements spontanés. *Presse Médicale*, **78**, 641–6.

SCHMIDT, W. & BIUKERT, F. (1973). Cytogenetic studies in preimplantation embryos of Chinese hamsters. *Mutation Research*, **21**, 233–4.

SEARLE, A. G., FORD, C. E. & BEECHEY, C. (1971). Meiotic disjunction in mouse translocations and the determination of centromere position. *Genetic Research*, **18**, 215–35.

SNELL, G. D. (1941). Linkage studies with induced translocation in the mouse. *Genetics*, **26**, 169.

SNELL, G. D. (1946). An analysis of translocations in the mouse. *Genetics*, **31**, 157–80.

SNELL, G. D. & PICKEN, D. I. (1935). Abnormal development in the mouse caused by chromosomal unbalance. *Journal of Genetics*, **31**, 213–35.

SNOW, M. H. L. (1973). Tetraploid mouse embryos produced by cytochalasin B during cleavage. *Nature, London*, **244**, 513–15.

TARKOWSKI, A. K. (1966). An air-drying method for chromosome preparations from mouse eggs. *Cytogenetics*, **5**, 394–400.

TARKOWSKI, A. K., WITKOWSKA, A. & NOWICKA, J. (1970). Experimental parthenogenesis in the mouse. *Nature, London*, **226**, 162–5.

TETTENBORN, U. & GROPP, A. (1970). Meiotic nondisjunction in mice and mouse hybrids. *Cytogenetics*, **9**, 272–83.

THERKELSEN, A. J., GRUNNET, N., HJORT, T., JENSEN, O. M., JONASSON, J., LAURITSEN, J. G., LINDSTEN, J. & PETERSEN, G. B. (1973). Studies on spontaneous abortions. In *Les Accidents Chromosomiques de la Reproduction*, ed. A. Boué & C. Thibault, pp. 81–93. Paris: Institut National de la Santé et de la Recherche Médicale.

WARBURTON, D. & FRASER, F. C. (1964). Spontaneous abortion risks in man: data from reproductive histories collected in a Medical Genetics Unit. *American Journal of Human Genetics*, **16**, 1–27.

WHITE, B. J., TJIO, J.-H., VAN DE WATER, L. C. & CRANDALL, C. (1972). Trisomy for the smallest autosome of the mouse and identification of the T1Wh translocation chromosome. *Cytogenetics*, **11**, 363–78.

WHITE, B. J., TJIO, J.-H., VAN DE WATER, L. C. & CRANDALL, C. (1974*a*). Trisomy 19 in the laboratory mouse. I. Frequency in different crosses at specific developmental stages and relationship of trisomy to cleft palate. *Cytogenetics and Cell Genetics*, **13**, 217–31.

WHITE, B. J., TJIO, J.-H., VAN DE WATER, L. C. & CRANDALL, C. (1974*b*). Trisomy 19 in the laboratory mouse. II. Intra-uterine growth and histological studies of trisomics and their normal litter mates. *Cytogenetics and Cell Genetics*, **13**, 232–45.

WROBLEWSKA, J. (1971). Developmental anomaly in the mouse associated with triploidy. *Cytogenetics*, **10**, 199–207.

ZECH, L., EVANS, E. P., FORD, C. E. & GROPP, A. (1973). Banding patterns in mitotic chromosomes of tobacco mouse. *Experimental Cell Research*, **70**, 263–8.

SEX REVERSAL IN THE MOUSE AND OTHER MAMMALS

BY B. M. CATTANACH

MRC Radiobiology Unit, Harwell, Oxon. OX11 0RD

[illegible] gonadal rudiment to develop as a testis (Miller, 1938). Partial gonadal sex reversal results when male chick embryos are treated with oestrogen (Erickson & Pincus, 1966) and when embryos of opposite sex develop within a single egg and form vascular interconnections (Lutz & Lutz-Ostertag, 1959). It can be induced in the Virginia opossum by oestrogen treatment of the pouch young (Burns, 1961). This flexibility in sexual development is not found in eutherian mammals; functional sex reversal never happens naturally (Short, 1972) and even partial gonadal sex reversal cannot be induced experimentally with steroids (Burns, 1961). However, evidence of a genetically controlled form of sex reversal exists and it is this that is the subject of the present article.

It is generally accepted that the Y chromosome determines male development in the majority of mammals. Thus, the single X, or XO condition has been observed in several species and found to have an essentially female phenotype (Cattanach, 1974). In addition, the phenotypes of other sex chromosome aneuploid conditions, e.g. XXY, have demonstrated that the presence of a Y normally causes testis formation and, hence, male development (Jost, 1955), irrespective of the numbers of X chromosomes (Cattanach, 1974). A disconcerting finding, however, has been the discovery that normal or near-normal male development can occur under certain circumstances in the apparent absence of the Y. In man, there have been quite a large number of cases of XX males reported and the XO male has also been described (de la Chapelle, 1972; Polani, 1973). However, all of these have been isolated cases and could have been mosaics with a hidden Y-bearing cell line, e.g. XX/XXY. More impressive evidence of male-type development in the absence of the Y comes from the other species. In mice, goats and pigs, autosomallyinherited conditions which cause a partial or complete sex reversal of chromosomal females have clearly been demonstrated and, although the possibility that Y–autosome translocations are responsible cannot be ruled out with certainty, these cases raise the possibility that male development may not be entirely dependent upon the Y.

SEX REVERSAL IN MICE

The most complete form of sex reversal so far recognised is that in the mouse (Cattanach, Pollard & Hawkes, 1971). The condition is inherited as an autosomal dominant and causes genotypic females to develop as phenotypic but sterile males. The responsible factor is called *Sex reversed*, symbol *Sxr*, and it is transmitted by XY males. The inheritance can readily be followed with the use of an X-linked marker such as *Tabby* (*Ta*), when it can be seen that about half the XX progeny of carrier XY males develop as phenotypic males, rather than as females, and breeding tests on the XY progeny show that about half carry *Sxr* and half do not (Table 1).

Table 1. *Production of* Sxr *mice utilising* Tabby *as an X chromosome marker*

Mating	Progeny			
XX♀♀ × XY♂♂	XX♀♀	XX♂♂	XY♂♂	
+/+ × *Ta/Y, Sxr/+*	*Ta/+*	*Ta/+*	+/Y	+/Y, *Sxr/+*
Ta/Ta × *+/Y, Sxr/+*	*Ta/+*	*Ta/+*	*Ta*/Y	*Ta*/Y, *Sxr/+*

The important question concerning *Sxr* is whether it is a genic change in an autosome or whether it is a Y–autosome translocation. Thus, in the data to be presented it is important to note just how similar are the effects of *Sxr* and the Y chromosome.

XX males

Several hundred XX progeny of XY males carrying *Sxr* have now been investigated. All have been either normal, fertile females, i.e. non-*Sxr*, or phenotypically normal but sterile males, i.e. XX, *Sxr/+*. No intersex animals have yet been found and dissections have always revealed the presence of normal male accessory reproductive organs, barring a reduced testis size. Therefore, it may reasonably be concluded that *Sxr* causes complete sex reversal. In most crosses the ratio of females to XX males has been close to 1 : 1 and this would indicate there is no loss of viability associated with the sex reversed condition. The animals themselves appear to be quite normal, they behave like males, they readily mate females and, in so doing, cause vaginal plug formation. Thus, the only clear abnormality is that relating to the sterility, i.e. the small size of the testis (only about one-fifth that of normal).

Histological studies have revealed the cause of the sterility. Just as in [illegible] male type; none have yet been observed to enter meiotic prophase at this time as do the presumptive XX, or female, germ cells in the testis of the XX/XY chimaera (Mystkowska & Tarkowski, 1970). This would indicate that *Sxr* is causing functional sex reversal of the germ cells as well as causing testis formation, despite the presence of two X chromosomes. The loss of germ cells from the XX testis appears to occur primarily about the time of birth, i.e. when the genocytes are dividing and differentiating to form spermatogonia. Spermatogonial cell types have been observed in the XX testis 2 to 5 days after birth but generally they do not survive beyond the 10th day. Survival may depend to some extent upon genetic background because in our current *Sxr* stock young XX males are quite frequently found which show a spermatogonial-type cell in a few tubules of their testes (Plate 1*b*). In such tubules these presumptive germ cells are present in numbers well in excess of normal and may even fill the whole lumen. There is no indication that these cells may enter meiosis and their abnormal behaviour may be an indication of events that eventually lead to their absence from the testis. To my knowledge the equivalent situation is not found in the XXY mouse.

A quite different but intriguing situation is also to be found in the XX testis and this has its counterpart in the XXY animal. Germ cells actively undergoing spermatogenesis may be seen in a few tubules of the testes of some animals (Plate 2*a*), and cytological studies have shown that when meiotic cells are present they are chromosomally not XX but XO (Lyon & Glenister, 1973). The equivalent situation in the XXY mouse is the occurrence of actively dividing meiotic cells in limited areas of the testis. These are thought to be chromosomally XY (M. F. Lyon, personal communication). It may be concluded that in both situations an early loss of one X chromosome, probably at or before the gonocyte stage, allows germ cell survival and this leads to the occurrence of clones of viable spermatogonia and meiotic cells in the seminiferous tubules.

XO males

The XO condition in the mouse, as in most mammals, is normally female (Cattanach, 1974). In contrast to the human XO it does not show any phenotypic abnormalities (Welshons & Russell, 1959; Cattanach, 1962; Morris, 1968) and it is fertile, although its reproductive lifespan is shorter than the normal (Lyon & Hawker, 1973). When bred from, it produces XO progeny, but only one-third of the expected number are recovered and the currently available evidence suggests that this results partly from a preferential segregation to the polar body of the haploid chromosome set lacking an X (Kaufman, 1972) and partly from an early pre-natal loss of the XO class (Morris, 1968).

Sxr can readily be combined with the XO chromosome constitution by crossing XO females with XY males carrying *Sxr*. Half the XO progeny then inherit *Sxr* from the father and these develop as phenotypically normal males. They differ from XX males in only one major respect. Their testis size is larger and germ cells are present (Plate 2*b*). Unfortunately, these mice are still sterile. Although spermatozoa are produced, they are generally very few in number and show morphological abnormalities. Histological sections of the testes show that spermatogenesis tends to break down about the time of the meiotic divisions. So far as the nature of *Sxr* is concerned, these germ cell abnormalities could mean that *Sxr* lacks some function normally carried out by the Y chromosome during spermatogenesis. However, the presence of the univalent X might itself be responsible for the spermatogenic abnormalities, since it is known that a variety of different sex chromosomal and autosomal changes can bring about a breakdown of meiosis (Cattanach *et al.*, 1971). Another interpretation derives from the recently discovered effect of *Sxr* upon the XY male, but this will be discussed later.

One other point concerning XO males that should be noted is that their frequency of occurrence is no higher than that of XO females (Table 2). If Morris (1968) is correct that there is an early pre-natal loss among XO females, then it may be concluded that the presence of *Sxr* in the early embryo does not compensate for the absence of the Y; i.e. some function possessed by the Y chromosome is lacking in *Sxr* such that the XO male is as liable to pre-natal loss as the XO female.

Table 2. *Progeny of cross of XO females with XY,* Sxr/+ *males*

XO ♀♀	XO, *Sxr*/+ ♂♂	XX ♀♀	XX, *Sxr*/+ ♂♂	XY ♂♂ and XY, *Sxr* +♂♂
35	34	63	55	115

It is now evident that males carrying *Sxr* generally have smaller testes than their non-*Sxr* sibs (Table 3) and those with the smallest testes are very often sterile. The implication of this for the interpretation of *Sxr* can be considered in at least two ways: (1) The spermatogenic abnormalities could be equated to those found in the XYY male. This would again demonstrate how *Sxr* parallels the action of the Y chromosome although it must be noted that the degree of effect is considerably less. Thus, while most XY males carrying *Sxr* are fertile, all XYY males so far detected in the mouse have shown such an extensive breakdown of spermatogenesis that most probably all would have been sterile (Cattanach & Pollard, 1969; Searle & Phillips, 1971; Rathenburg & Muller, 1973; Searle *et al.*, 1974). (2) The spermatogenic abnormalities could be considered comparable to those seen in the XO male. This might mean that *Sxr* is itself responsible for the spermatogenic failure in XO males and that its less severe effect in XY animals results from the presence of the Y chromosome.

It can be concluded from the data so far presented that *Sxr* fulfils most of the functions of the Y chromosome. Generally, these data are compatible with the view that *Sxr* represents a Y–autosome translocation

Table 3. *Comparison of testis weights of XY ♂♂ and XY,* Sxr/+ *males*

XY ♂♂			XY, *Sxr*/+♂♂		
No. ♂♂	Mean wt (g)	Range	No. ♂♂	Mean wt (g)	Range
12	0.107	0.090–0.120	16	0.075	0.050–0.102

and there is now some new information that would support this interpretation. Bennett and her colleagues have recently found that XX males possess Y-, or male, antigen. It would seem logical to conclude that this means that *Sxr* is a piece of Y chromosome which has been translocated to an autosome and which carries both the male-determining region and the Y histocompatibility locus. However, there are other lines of evidence which are at variance with this concept.

The evidence against *Sxr* being a Y–autosome translocation is primarily cytological and includes the results of investigations upon meiotic and mitotic cells. Thus, meiotic studies on XY males carrying *Sxr* have shown that X–Y pairing is normal and there is no indication of an association with any autosomal bivalent. Similarly, meiotic studies upon XO males have invariably demonstrated the presence of a univalent X. Autoradiographic techniques have shown that the X and Y chromosomes associate by their distal ends (Kofman-Alfara & Chandley, 1970) and, since it is likely that the male-determining region of the Y is closer to the centromere, the apparent absence of any association between an autosomal bivalent and the X or Y only indicates that *Sxr* mice do not carry a translocation involving the greater part of the Y. It is more important to know whether a more proximal region could have been translocated and, at the present time, the cytological data do not support this possibility. Mitotic chromosome studies using both the C-banding and Q-banding techniques have failed to demonstrate the existence of a piece of Y chromosome in an autosome (Cattanach, 1974; M. Nesbitt, personal communication).

One other point that could be considered is that all Y–autosome translocations so far detected in the mouse have been associated with complete sterility (Cachiero, 1971; Léonard & Deknudt, 1969; A. G. Searle, personal communication). This observation might support the view that *Sxr* is not a translocation.

The cytological data perhaps do not entirely eliminate the possibility that *Sxr* represents a very small region of the Y chromosome translocated to an autosome, but the inheritance of *Sxr* raises other difficulties for the Y–autosome translocation hypothesis. If the male-determining region of the Y is now located in an autosome, the Y that remains should be deficient for this region. Apart from the fact that non-*Sxr* XY animals which segregate out from *Sxr* matings appear to have quite a normal Y, they are undoubtedly male. If indeed a translocation were involved it would be necessary to accept that the male-determining region can be split into two parts and that each part, when isolated, retains a male-determining function. This difficulty could be overcome, however, if the original rearrangement had been a chromatid, rather than a chromosomal exchange, such that a complete Y chromosome remained (Fig. 1). In this connection it will be important to see if non-*Sxr* XY males from the *Sxr* stock possess the so-called Y-antigen. If they do, the presence of the antigen in XX males need not support the Y–autosome translocation hypothesis; the antigen might well be considered to be a secondary conse- [illegible]

[illegible] chromosomes except numbers 10, 16, 18, and possibly also number 3 (Cattanach & Moseley, 1973; Searle & Beechey, 1973; Cachiero & Russell, 1974), have proved negative. Tests with the tobacco mouse Robertsonian translocations marking the centromeric regions of chromosomes 1, 3, 4, 5, 8, 12, 14, 15, 16 and 17 have also failed to demonstrate linkage. However, tests with the remaining tobacco mouse Robertsonian translocation, Rb(11 . 13)4Bnr, which marks chromosomes 11 and 13, have yielded an unexpected finding. An analysis of the progeny of the initial cross has clearly demonstrated that XY males carrying both *Sxr* and the Robertsonian translocation are completely sterile. All such animals studied have had very small testes, little larger than those of XX males, and have produced very few spermatozoa. Meiotic studies have shown that spermatogenesis must break down primarily after the first meiotic division for many meiotic I cells can be found. Among these only one anomaly has been seen; approximately 30% of cells show a failure of X–Y association and this was observed in each of three animals. Despite this failure of association, neither chromosome showed any indication of homology with an autosomal bivalent, or with the chain configuration which comprised the Robertsonian translocation and the homologous acrocentric chromosomes 11 and 13. The implication of this finding is not clear but it might be concluded that *Sxr* is located on one or other of these two chromosomes. The appropriate linkage tests do not yet support this contention, however. *Sxr* shows no linkage with *Re*, which marks the distal region of chromosome 11, nor with *Xt*, which marks the proximal region of chromosome 13, and current tests with *wa*-2 and *vt* which mark the proximal regions of chromosome 11 also promise to be negative. Thus, if *Sxr* is located on one of these chromosomes, the distal region of chromosome 13 would

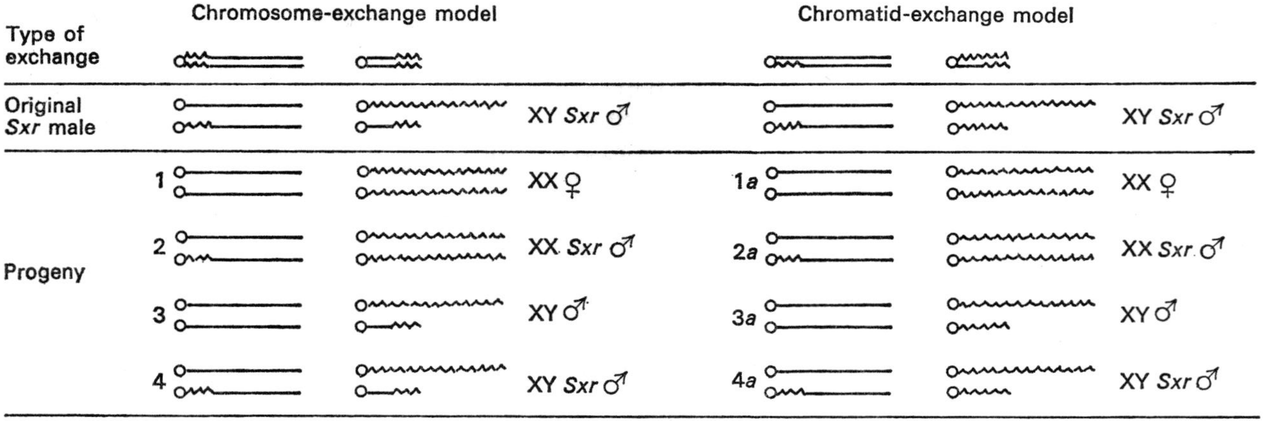

Type of exchange	Chromosome-exchange model		Chromatid-exchange model	
Original *Sxr* male		XY *Sxr* ♂		XY *Sxr* ♂
Progeny	1	XX ♀	1*a*	XX ♀
	2	XX *Sxr* ♂	2*a*	XX *Sxr* ♂
	3	XY ♂	3*a*	XY ♂
	4	XY *Sxr* ♂	4*a*	XY *Sxr* ♂

Fig. 1. Origin and inheritance of *Sxr* on the basis of Y–autosome chromosome and chromatid exchanges. By the chromosome exchange model, maleness is conferred by both parts of the divided Y (cf. 2 and 3). By the chromatid exchange model, maleness is conferred by the translocated segment of Y (2*a*) and by the complete Y (3*a* and 4*a*). Straight lines indicate autosomes, wavy lines indicate X and Y chromosomes.

seem to be the only possible place. Tests with other translocations marking these chromosomes are now under way to see if they too form sterile combinations with *Sxr*.

A final important point that must be taken into consideration when discussing the nature of *Sxr* is that similar genetic systems causing sex reversal have been found in other mammalian species. The view has been [illegible]

SEX REVERSAL IN GOATS

The sex reversed condition recognised in goats (Hamerton *et al.*, 1969; Soller, Padeh, Wysoki & Ayalon, 1969) resembles that of the mouse insofar as it shows an autosomal inheritance and causes testicular development in genetic females. It differs from the mouse in two important respects, however: first, the inheritance is that of an autosomal sex-limited recessive, rather than that of a dominant, and second, masculinisation is generally incomplete, i.e. most animals develop as intersexes. Affected animals show great variability in the degree of sex reversal, ranging from almost-normal females to almost-normal males with scrotal testes. Internally, they all possess testes, or occasionally ovotestes, and show varying degrees of Wolffian duct stimulation and Mullerian duct inhibition. All these intersexes are chromosomally XX (Basrur & Coubrough, 1964; Hamerton *et al.*, 1969; Soller *et al.*, 1969) and all are homozygous for the dominant gene for hornlessness, known as *Polled*. Thus, the factor responsible for the sex reversal is either the *Polled* gene itself or it is very closely linked to *Polled*.

The histological picture of the testis of the XX intersex goat closely resembles that of the sex reversed XX mouse. Germ cells are normally absent in the adult, and the tubules contain only Sertoli cells. In older animals the testes tend to become atrophic, with hyalinisation of the tubules, Leydig cell hyperplasia and often a high tumour incidence. Primordial germ cells have been found in the foetal XX intersex, as in the XX male mouse, and the loss of these cells is thought to occur at about the time they would normally be entering meiosis in the ovary. For this reason it has been concluded that sex reversal does not operate upon the germ cells of the goat (Short, 1972). However, Basrur and her colleagues (Basrur & Coubrough, 1964; Basrur & Kanagawa, 1969) have clearly demonstrated that there is active meiosis in some animals, with spermatids and even spermatozoa being formed. Vascular anastomosis is not uncommon in goats and, hence, it has been proposed that these animals were in fact freemartins with an XY germ line (Short, 1972). The mouse data suggest an alternative interpretation, however. The meiotically active cells could very well be sex reversed XO cells derived from an original XX population by the spontaneous loss of one X chromosome, and the morphological abnormalities of the spermatozoa observed in these intersex goats are consistent with this interpretation. There is thus a case for sex reversal of the germ cells of the goat as in the mouse.

XY male goats which are homozygous for *Polled* are not obviously affected, but some are sterile as a result of occlusion of the epididymis in the caput region. The testes of these animals may show areas of normal spermatogenesis and areas where the tubules have become distended with the seminiferous epithelium undergoing atrophy (Hamerton *et al.*, 1969). This has not yet been observed in the XY mouse carrying *Sxr*, but sterile mice of this genotype have not been extensively investigated.

Basrur (1969) has sought to explain the sex reversal found in goats on the basis of a Y–autosome translocation that had become established during the evolution of the species. However, several other genetic changes would also need to have occurred to account for the incompleteness of the sex reversal and the lack of any effect in the heterozygote. Sex reversal in the goat would seem to be more simply explained on the basis of a genic change.

SEX REVERSAL IN PIGS

The sex reversal found in pigs is even less complete than that in goats (Cantwell, Johnston & Zeller, 1958; Johnston, Zeller & Cantwell, 1958; Makino, Sasaki, Sofuni & Ishikawa, 1962; Gerneke, 1964, 1967; Hard & Eisen, 1965; Hard, 1967; Somlev, Hanson-Melander, Melander & Holm, 1970). The gonads may often be ovotestes and ovaries may even be found in the intersex animals. The inheritance is not well understood and environmental factors such as litter size play an important role. As in the mouse and goat, germ cells are not present in the testes or testicular regions of ovotestes of the sex reversed XX pig but pre-ovulatory follicles are found in the ovaries or ovarian parts of the ovotestes. In view of the phenotypic expression it is not clear whether this really means that sex reversal does not affect the germ cells. The principal conclusion to be drawn from the pig data is that the sex reversal cannot readily be understood in terms of a Y–autosome translocation.

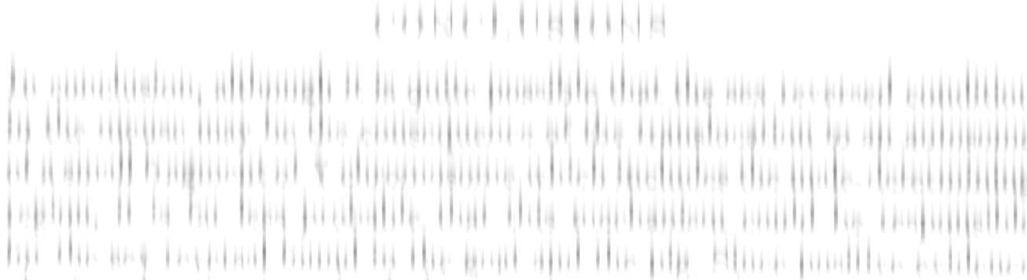

of a translocation is so far lacking in the mouse and other aspects of the *Sxr* data are not easily reconciled with the existence of a translocation, it would at present seem the more probable that *Sxr* represents a genic change at some locus concerned with sex determination. If this could be established, sex reversed mice would provide an ideal tool for helping to unravel some of the problems of sex determination. If, instead, evidence of a Y–autosome translocation is eventually found, sex reversed mice may then provide a useful model for studying the developmental pathology of the XXY syndrome.

REFERENCES

Basrur, P. K. (1969). Some thoughts on the association of the *Polled* trait and intersexuality in goats. *Annales de Génétique et de Sélection Animale*, **1**, 439–46.

Basrur, P. K. & Coubrough, R. I. (1964). Anatomical and cytological sex of a Saanen goat. *Cytogenetics*, **3**, 414–26.

Basrur, P. K. & Kanagawa, H. (1969). Anatomic and cytogenetic studies on 19 hornless goats with sexual disorders. *Annales de Génétique et de Sélection Animale*, **1**, 349–78.

Burns, R. K. (1961). Role of hormones in the differentiation of sex. In *Sex and Internal Secretions*, 3rd edition, ed. W. C. Young, vol. 1, pp. 76–158. Baltimore: Williams & Wilkins.

Cachiero, N. L. A. (1971). Cytological studies of sterility in sons of mice treated with mutagens. *Genetics*, **68**, s8–9.

Cachiero, N. L. A. & Russell, L. B. (1974). Possibility that *Sl* is in chromosome 10. *Mouse News Letter*, **50**, 52.

Cantwell, G., Johnston, E. F. & Zeller, J. H. (1958). The sex chromatin of swine intersexes. *Journal of Heredity*, **49**, 199–202.

Cattanach, B. M. (1962). XO mice. *Genetical Research*, **3**, 487–90.

Cattanach, B. M. (1974). Genetic disorders of sex determination in mice and other mammals. In *Birth Defects*. Proceedings of the Fourth International Congress, 2–8 September 1973, ed. A. G. Motulsky & W. Lenz, pp. 129–41. Amsterdam: Excerpta Medica.

CATTANACH, B. M. & MOSELEY, H. J. (1973). Assignment of LG IV. *Mouse News Letter*, **48**, 31.

CATTANACH, B. M. & POLLARD, C. E. (1969). An XYY sex-chromosome constitution in the mouse. *Cytogenetics*, **8**, 80–6.

CATTANACH, B. M., POLLARD, C. E. & HAWKES, S. G. (1971). Sex reversed mice: XX and XO males. *Cytogenetics*, **10**, 318–37.

CHAN, S. T. H. (1970). Natural sex reversal in vertebrates. *Philosophical Transactions of the Royal Society, London*, Ser. B, **259**, 59–71.

CREW, F. A. E. (1923). Studies in intersexuality. II. Sex-reversal in the fowl. *Proceedings of the Royal Society, London*, Ser. B, **95**, 256–78.

DE LA CHAPELLE, A. (1972). Nature and origin of males with XX sex chromosomes. *American Journal of Human Genetics*, **24**, 71–105.

ERICKSON, A. E. & PINCUS, G. (1966). Modification of embryonic development of reproductive and lymphoid organs in the chick. *Journal of Embryology and Experimental Morphology*, **16**, 211–29.

GERNEKE, W. H. (1964). The karyotype of a gonadal male pig intersex. *South African Journal of Science*, **60**, 347–52.

GERNEKE, W. H. (1967). Cytogenetical investigation on normal and malformed animals, with special reference to intersexes. *Onderstepoort Journal of Veterinary Research*, **54**, 219–300.

HAMERTON, J. L., DICKSON, J. M., POLLARD, C. E., GRIEVES, S. A. & SHORT, R. V. (1969). Genetic intersexuality in goats. *Journal of Reproductive Fertility*, Supplement **7**, 25–51.

HARD, W. L. (1967). The anatomy and cytogenetics of male pseudohermaphroditism in swine. *Anatomical Record*, **157**, 255.

HARD, W. L. & EISEN, J. D. (1965). Phenotypic male swine with female karyotype. *Journal of Heredity*, **56**, 255–8.

JOHNSTON, E. F., ZELLER, J. H. & CANTWELL, G. (1958). Sex anomalies in swine. *Journal of Heredity*, **49**, 255–61.

JOST, A. (1955). Modalities in the action of gonadal and gonad-stimulating hormones in the foetus. *Memoirs of the Society for Endocrinology*, **4**, 237–48.

KAUFMAN, M. H. (1972). Non-random segregation during mammalian oogenesis. *Nature, London*, **238**, 465–6.

KOFMAN-ALFARO, S. & CHANDLEY, A. C. (1970). Meiosis in the male mouse. An autoradiographic investigation. *Chromosoma*, **31**, 404–20.

LÉONARD, A. & DEKNUDT, G. (1969). Etude cytologique d'une translocation chromosome Y–autosome chez la souris. *Experientia*, **25**, 876–7.

LUTZ, H. & LUTZ-OSTERTAG, Y. (1959). Free-martinisme spontane chez les oiseaux. *Developmental Biology*, **1**, 364–76.

LYON, M. F. & GLENISTER, P. H. (1973). Spermatogenesis in *Tfm*/+, *Sxr*/+. *Mouse News Letter*, **49**, 28.

LYON, M. F. & HAWKER, S. G. (1973). Reproductive lifespan in irradiated and unirradiated chromosomally XO mice. *Genetical Research*, **21**, 185–94.

MAKINO, S., SASAKI, M. S., SOFUNI, T. & ISHIKAWA, T. (1962). Chromosome condition of an intersex swine. *Proceedings of the Japan Academy*, **38**, 686–9.

MILLER, R. A. (1938). Spermatogenesis in a sex-reversed female and in normal males of a domestic fowl, *Gallus domesticus*. *Anatomical Records*, **70**, 155–89.

MORRIS, T. (1968). The XO and OY chromosome constitutions in the mouse. *Genetical Research*, **12**, 125–37.

MYSTKOWSKA, E. T. & TARKOWSKI, A. K. (1970). Behaviour of germ cells and

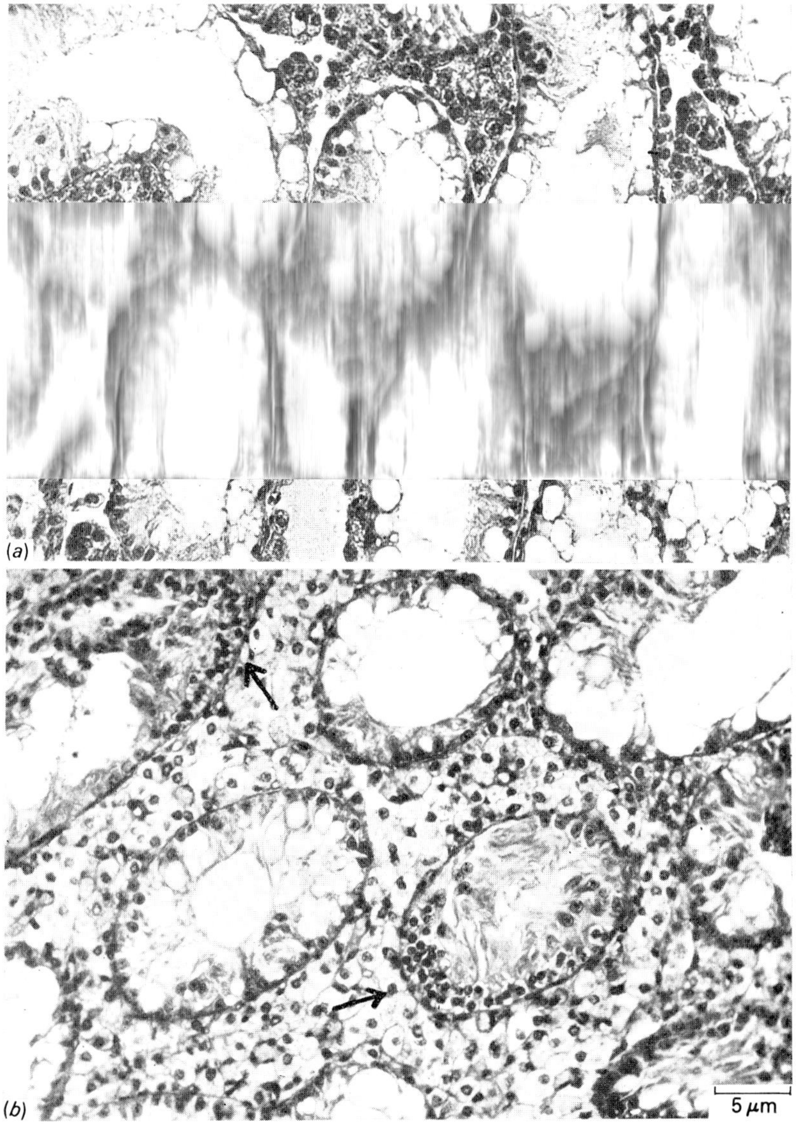

(*Facing p. 316*)

PLATE 2

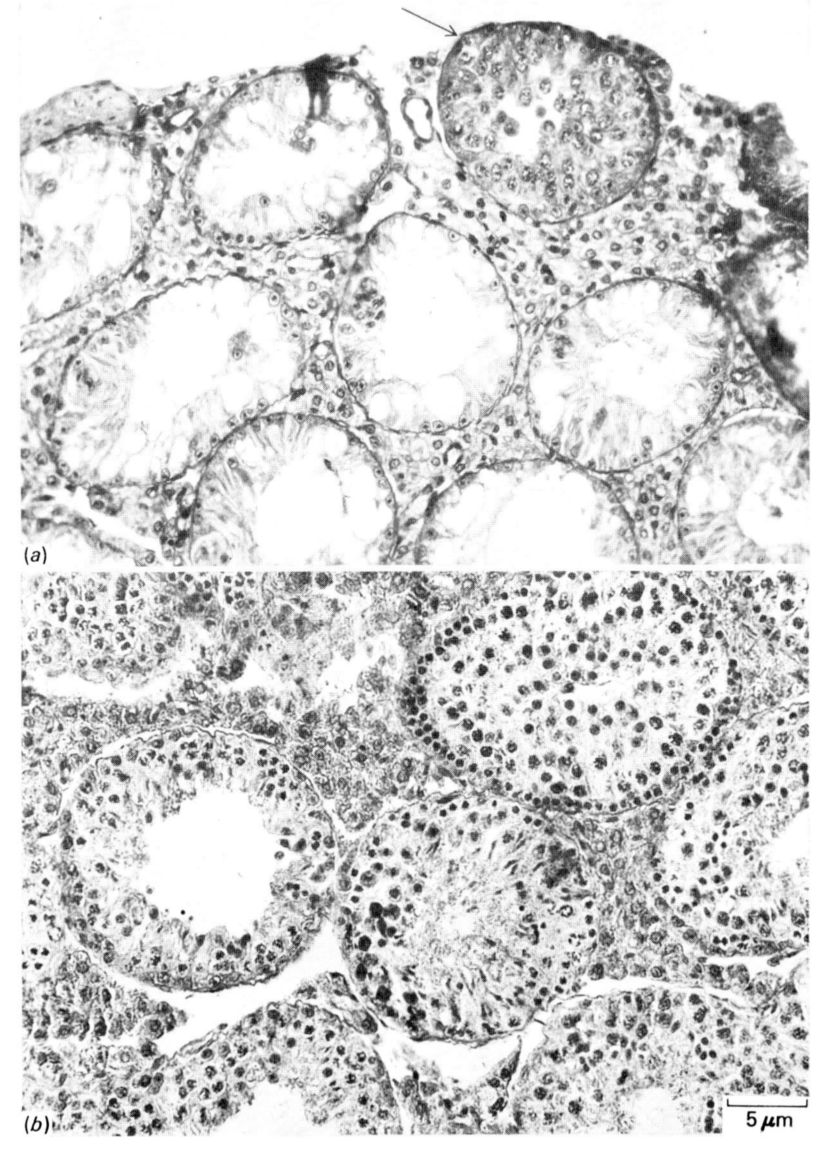

sexual differentiation in late embryonic and early post-natal mouse chimaeras. *Journal of Embryology and Experimental Morphology*, **23**, 395–405.

Ohno, S. (1967). *Sex Chromosomes and Sex-linked Genes*. Berlin: Springer-Verlag.

Polani, P. E. (1973). Errors of sex determination and sex chromosome anomalies. In *Gender Differences: their Ontogeny and Significance*, ed. C. Ounsted & D. C. Taylor, pp. 13–39. Edinburgh & London: Churchill Livingstone.

Rathenberg, R. & Muller, D. (1973). X and Y chromosome pairing and disjunction in a male mouse with an XYY sex-chromosome constitution. *Cytogenetics and Cell Genetics*, [illegible]

[illegible]

& S. Gluecksohn-Waelsch, pp. 325–45. Edinburgh & New York: Beatty & Gluecksohn-Waelsch.

Slizynski, B. M. (1964). Cytology of the XXY mouse. *Genetical Research*, **5**, 328–9.

Soller, M., Padeh, B., Wysoki, M. & Ayalon, N. (1969). Cytogenetics of Saanen goats showing abnormal development of the reproductive trait associated with the dominant gene for polledness. *Cytogenetics*, **8**, 51–67.

Somlev, B., Hanson-Melander, E., Melander, Y. & Holm, L. (1970). XX/XY chimerism in leukocytes of two intersexual pigs. *Hereditas*, **64**, 203–10.

Welshons, W. J. & Russell, L. B. (1959). The Y-chromosome as the bearer of male determining factors in the mouse. *Proceedings of the National Academy of Sciences, USA*, **45**, 560–6.

EXPLANATION OF PLATES

Plate 1

(*a*) Section of testis of an XX male.

(*b*) Section of testis of a young XX male showing presumptive germ cells.

Plate 2

(*a*) Section of testis of an XX male showing some tubules with active meiosis.

(*b*) Section of testis of an XO male.

INDUCTIVE INTERACTIONS IN MORPHOGENESIS

BY L. SAXÉN AND M. KARKINEN-JÄÄSKELÄINEN

Third Department of Pathology, University of Helsinki.
[illegible]

[illegible] inherited by the daughter cells, in which they cause a differential reading of the genome (mosaic development). Extracellular control involves factors asymmetrically and unevenly distributed in the cell population in question, the most obvious being the position of the cells within the colony and their intercellular relations. A good example of the latter type of mechanism has been described in certain colonies of unicellular organisms. A colony of *Pleodorina californica* consists of two distinct cell types, somatic and germinative, which are represented in the colonies in a constant ratio. The somatic cells develop in the central area of the colony, whereas the peripheral cells develop into the germinative type. If this basic organisation within the colony is disrupted at an early stage of its growth, differentiation of the two cell types is severely impaired. This indicates the significance of position for the subsequent fate of the cells (Gerisch, 1959). Recent experiments on early mammalian embryos have yielded comparable results. All cells of a 4-cell mouse embryo are 'totipotent' and can either develop into trophoblastic cells or give rise to the inner cell mass of the embryo. In experimental combinations of two such embryos, the cells of one, if allowed to become completely surrounded by those of the other, all become part of the inner cell mass (Hillman, Sherman & Graham, 1972). These observations indicate that at this stage mammalian embryonic cells are not determined by cytoplasmic factors, their future fate being decided by their position and their relation to other cells in the embryo.

The situation in the 4- to 8-cell stage might be taken as the earliest demonstration of developmentally significant intercellular communication. There is today every reason to suppose that the whole course of embryogenesis following this early determinative stage is guided by similar inductive interactions between cells with different developmental histories brought into close contact by morphogenetic movements.

This view is not, however, based on a thorough understanding of such interactive processes throughout the course of development, but on numerous scraps of knowledge derived from studies on isolated and relatively simple events in embryos at different stages. Thus, our present understanding of inductive interactions is still far from complete, and a variety of apparently different processes are lumped together under this term. A search for a common determinative factor, the 'organiser', proved a failure in the thirties, and today we may fall into the same trap in searching for common denominators for the various types of inductive interactions. In what follows, we shall therefore approach the problem in the light of some relatively well-known interactive processes, and shall stress that they appear to operate through different mechanisms. They should also be considered as single steps in a complicated train of interactive events leading to the synchronised development of an organism. Their only common denominator is their vital significance for normal development, and their exact temporal and spatial localisation in the developing embryo.

BIOLOGY AND BIOCHEMISTRY OF INDUCTION

The basic experimental technique for studying intercellular communication as a prerequisite for cell differentiation and morphogenesis is to dissociate the heterotypic cell populations, and subsequently to culture them either separately or after reassociation. The results of such studies in various model systems may justify a few general comments, to be illustrated with examples.

Comments on biology

Inductive tissue interactions constitute a continuous chain of processes where more specific, determinative events may alternate with less specific, permissive steps: the hepatic endoderm becomes determined at an early stage of development by the hepato-cardiac mesoderm, and no other tissue can replace this inductor. The morphogenesis that results in the formation of typical hepatic cords, in contrast, can be supported by various heterologous mesenchymes, whereas the onset of the specific function of these cells, glycogen synthesis, again depends on the presence of the homologous hepatic mesenchyme (Le Douarin, 1968). Correspondingly, evagination of pulmonary buds from the embryonic trachea can be initiated by various heterologous mesenchymes, but homologous bronchial mesenchyme is required for the subsequent branching and normal morphogenesis of the bronchi (Wessells, 1970; Spooner & Wessells, 1970).

True determinative events, where the prospective significance of a given tissue can be altered by a heterologous inductor, have rarely been demonstrated experimentally. In addition to the 'totipotent' amphibian gastrula ectoderm (see Saxén & Toivonen, 1962), embryonic epidermis [illegible] which cannot be changed but may still require permissive stimuli from neighbouring cells. Thus, metanephric kidney mesenchyme does not develop secretory tubules when separated from the ureter bud (Grobstein, 1953*a*), but has already acquired a definite 'kidney bias', and no other embryonic mesenchyme can respond to the ureter inductor by forming tubules (Saxén, 1970). Yet, *in vitro*, a great number of heterologous embryonic tissues can release the 'kidney bias' and trigger tubule formation (Grobstein, 1955; Unsworth & Grobstein, 1970).

Such experimental results imply that during embryogenesis the developmental alternatives open to a cell become gradually restricted, and that this restriction is due to certain determinative interactions. Early embryonic cells (and perhaps some rapidly renewing older tissues) require such specific stimuli, and can be deflected experimentally from their normal developmental pathways. Later, when the spectrum of developmental alternatives becomes still narrower, the cells require different exogenous stimuli which may be merely supportive in providing ideal mechanical or chemical environments for morphogenesis and growth. Such a stage can be demonstrated, for instance, in the pancreas of a 30-somite mouse embryo. Branching and chemodifferentiation of the pancreatic epithelium normally require its mesenchymal counterpart, but this can be replaced by practically any heterologous mesenchyme or even by cell-free embryo preparations (Wessells & Cohen, 1966; Rutter *et al.*, 1968).

Comments on biochemistry

Not a single signal substance operative in normal tissue interactions has been isolated and characterised, and the mode of action and target site of such factors are wholly speculative. Intercellular exchange of actual informative molecules has repeatedly been suggested (Brachet, 1940; Niu, 1956; Temin, 1971), but although transfer of molecules with molecular weights of up to 1000 has been demonstrated in certain experimental conditions (Pitts, 1971), the morphogenetic significance of such events remains to be shown.

Recently, Slavkin and his group have focused attention on some RNA-containing vesicles in the extracellular matrix between the interacting tissue components in the tooth rudiment (enamel epithelium and odontoblasts). This matrix has been reported to exert a morphogenetic action on cells grown on it after isolation. This effect is abolished by ribonuclease treatment, which also leads to the disappearance of the vesicles (Slavkin *et al.*, 1969, 1970; Slavkin, 1972). The idea that these vesicles transport information between the interacting tissues is tempting, but so far not substantiated by direct experimental evidence.

Another component of the intercellular material associated with morphogenesis is the glycosaminoglycans. The salivary epithelium, like several other glandular organs, is dependent on its mesencyhmal stroma during branching and morphogenesis. Dissociation of these two components inhibits the branching process, but this will be resumed immediately if the epithelium and mesenchyme are reassociated. If collagenase is used to remove the intercellular collagen during the separation process, the epithelium retains its original lobular configuration and continues its branching and morphogenesis uninterrupted after reassociation. If, in addition, hyaluronidase is used to remove the extracellular glycosaminoglycans, the epithelium loses its lobular structure and starts branching in the mesenchyme only after a considerable lag period, apparently allowing new glycosaminoglycans to be synthesised at the morphogenetically significant distal ends of the epithelium (where branching starts). The authors conclude that normal salivary morphogenesis requires the presence of acid mucopolysaccharide–protein within the epithelial basal lamina (Bernfield, Banerjee & Cohn, 1972).

In addition to transmissible inductors and substances present in the extracellular matrix, certain 'template' molecules at the surface of interacting cells were postulated long ago by Weiss (1947). Such complementary molecules could be of the antigen–antibody type or could have an enzyme–substrate relation. But even though they are known to be involved in various interactions between cells, they have recently evoked little interest among developmental biologists. The reason is obviously that all the evidence accumulated during recent years seems to suggest relatively long-range transmission of signals and to exclude the possibility that the signal could be transmitted by intimate cell-to-cell relations allowing interactions between complementary molecules associated with the cell membrane. Hence, it appears useful to review some recent observations by our group, which suggest that in certain interactive systems the situation calls for re-evaluation.

[illegible]

[illegible] tion of the lens from the overlying ectoderm by the optic cup is a good example of this. On the other hand, in many instances the population of induced cells extends a certain distance beyond the inductor and constitutes a 'field' of identically induced cells. Wolpert (1969) has calculated that such a field of cells receiving a morphogenetic signal may be up to 100 cell layers thick. So, while dealing with the problem of transmission of inductive signals, we have to face two questions: how is the message transmitted from the inductor to the responding cell, and how does the 'induction wave' spread over a considerable distance within the target cell population? Three alternative mechanisms seem conceivable and should be discussed in some detail:

(1) Long range transmission (diffusion) of stable signal molecules.
(2) Induction by intimate contact between cells and spread of the wave by homoiogenetic (assimilatory) induction.
(3) Induction by close contact followed by migration of the target cells towards the periphery of the field.

In what follows, we shall examine these three alternatives in one model system for morphogenetic tissue interactions, namely the mammalian metanephric kidney. The method is based on the original work of Grobstein (1953*b*, 1956), and the model system has been used by our group since 1960 (Saxén *et al.*, 1968; Saxén, 1971).

During the early development of the metanephric kidney, the ureter bud grows into the loose mesenchymal blastema, where it branches dichotomously and induces the formation of secretory tubules around the tips of the ureter tree. Here, as in many epithelio-mesenchymal interactions, the first visible response of the mesenchymal target cells is their condensation around the inductor. The primary condensate extends over an area of the order of 100 μm, corresponding to 6–8 cell layers, as measured from time-lapse cinefilm pictures (Saxén, Toivonen, Vainio & Korhonen, 1965; Saxén & Wartiovaara, 1966).

The maximum distance for the spread of the tubule induction effect can also be measured directly in certain experimental conditions. As first shown by Grobstein (1956, 1957), the induction can travel through porous Millipore filters for considerable distances. When filters of relatively large pore size were used, the effect could be observed through multi-layered filters with a total thickness of 60 μm, but thicker filters prevented induction. In our experiments, the percentage of inductions through a double filter with an average thickness of 57 μm was 70%, whereas all experiments in which the tissues were separated by a triple filter (average thickness 86 μm) gave negative results. Another experimental design, however, yielded somewhat different results. If, instead of increasing the thickness of the filter, the mass of mesenchymal cells was increased in either the horizontal or vertical direction from the inductor, tubules were detected at greater distances from the inductor. Fig. 1 shows our experimental design in two series of experiments. When, after three to four days' growth, such cultures were fixed and analysed in serial sections, abundant tubules were detected in the mesenchyme, the most distant ones being located some 150 μm from the inductor. We concluded, therefore, that both during normal development of the whole kidney and in our experimental conditions the 'induction wave' extends over distances of the order of 100 to 150 μm. Bearing this in mind, we examined the three alternative transmission mechanisms in our model system.

Long-range diffusion of signals

The 'diffusion hypothesis' explaining transmission of inductive signals was put forward by Holtfreter (1955), when working on the primary induction of amphibian gastrula ectoderm. We showed that, in accordance with his hypothesis, the effect responsible for the neuralisation of the ectoderm could pass through Millipore filters (Saxén, 1961, 1963). Recent electron microscopic examination of filters allowing the passage of induction has not revealed any cytoplasmic material in the filters. Nuclepore polyacryl filters with an average pore diameter of 0.15 μm form no barrier to neural induction by the dorsal lip, although no indications of membrane-coated cytoplasmic processes have been detected in the pores (Tarin, Toivonen & Saxén, 1973, and unpublished observations). These observations strongly suggest that the 'diffusion hypothesis' should be seriously

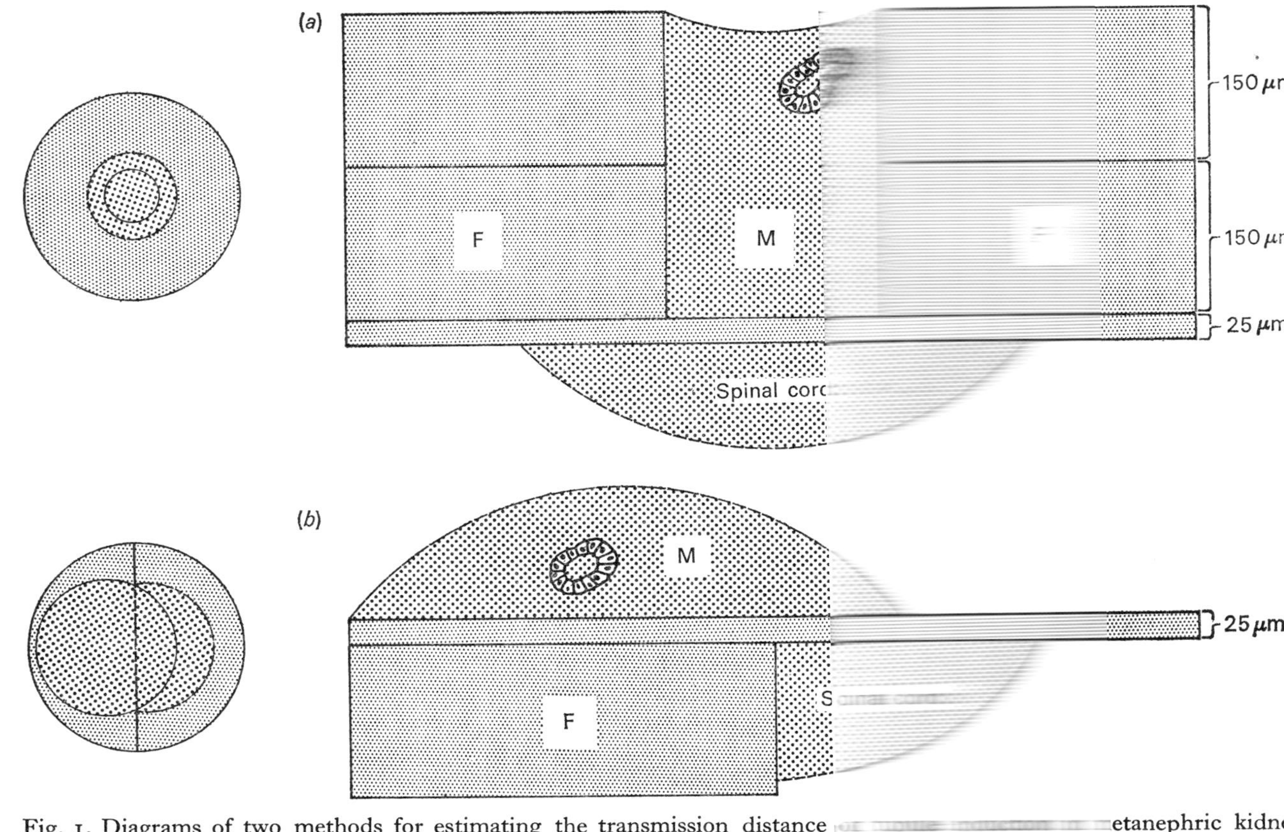

Fig. 1. Diagrams of two methods for estimating the transmission distance [illegible] etanephric kidney mesenchyme. On the left, assembly viewed from above; on the right, s[illegible]. M, metanephric mesenchyme; F, supportive filters.

considered in this induction system, where the effect can pass over 30 μm or more without cytoplasmic contacts. The results and the hypothesis seem to be inapplicable to the tubule induction system (see below) and the two sets of experimental results thus suggest that different interactive processes operate through different mechanisms.

The first observation in the tubule induction system that was difficult to reconcile with the diffusion hypothesis came from experiments designed to measure the transmission velocity of the induction signal in Millipore filters. From comparison of the minimum induction time (i.e. time of transfilter contact required for the completion of tubule induction) in single- and double-filter experiments, we concluded that the transmission velocity of the signal was of the order of 2 μm/h. Both theoretical calculations and actual measurements of components with molecular weight from 250 to 10^8 suggested that transmission of the inductive signal was extremely slow and could hardly be explained by free transport of molecules (Nordling, Miettinen, Wartiovaara & Saxén, 1971).

The Millipore filters used in these experiments were subsequently examined by light microscopy after glutaraldehyde fixation and embedding in Epon. Abundant cytoplasmic material was detected throughout the filter after prolonged culture. The two frontiers of stainable material penetrating into the filter from the interacting tissues met some 18 to 24 hours after being placed on the filters. This coincides with the minimum induction time previously estimated by Nordling *et al.* (1971). Further characterisation of this material in the Millipore filters is in progress, but certain other filters are more useful for such studies. Thin polyacryl filters (Nuclepore) with various pore sizes are commercially available. In contrast to the tortuous channels of varying bore penetrating Millipore filters, Nuclepore filters have straight channels of fairly uniform size. Electron microscopic analysis of such a filter permitting transmission of kidney tubule induction always revealed cytoplasmic processes extending right across the filter and close association of membranes from both sides. In contrast to this, pores with an average diameter of 0.15 μm prevented induction and were devoid of any cytoplasmic processes (Wartiovaara, Nordling, Lehtonen & Saxén, 1974).

From this series of experiments we are inclined to conclude that experiments making use of filter membranes interposed between interacting tissues do not necessarily exclude the existence of intimate cytoplasmic contacts, and do not yield conclusive evidence in favour of the hypothesis of long-range transmission of signal substances. In all our experiments on transfilter induction of kidney tubules, a good correlation has been established between the penetration of cytoplasmic processes through the filter and the actual transmission of inductive signals. Our present working hypothesis is therefore based on induction of kidney tubules through close cell-to-cell contacts, and excludes alternative (1) in the list on p. 323. Consequently, the alternative transmission mechanisms (2) and (3) should be explored.

[illegible]

[illegible] the spread of the induction wave in our kidney model system, where tubules could be detected up to 150 μm from the inductor. When induced metanephric mesenchymal cells with a chromosomal marker were mixed with uninduced mesenchymal cells carrying a normal karyotype, all cells in the tubules which developed two to three days later showed the nuclear marker. The result indicates that only cells originally induced differentiated and did not pass the message on to uninduced, competent cells (Saxén & Saksela, 1971).

Hence, alternative (2) is excluded in the kidney system, and leaves us with the third and last possible mechanism for transmission of signals.

Migration of induced cells

Time-lapse cinefilms clearly show that cells in the metanephric blastema are not fixed but in constant (random?) movement (Saxén *et al.*, 1965). These films do not reveal whether oriented, long-range migration of cells occurs, as single cells cannot be followed in their three-dimensional motility in the thick explants. Various types of experiments were therefore designed to investigate whether mesenchymal cells can migrate from the inductor/mesenchyme interphase deeper into the mesenchyme and thus carry the inductive stimulus over the field.

Transfilter experiments were performed according to the scheme in Fig. 2. After 24 hours of transfilter contact with the inductor (spinal cord), by which time induction is known to be completed, the mouse mesenchyme was stripped off by gentle mechanical pulling. Small groups of cells one or two layers deep are usually left on the filter after this procedure. These cells are probably firmly attached to the filter, with their cytoplasmic

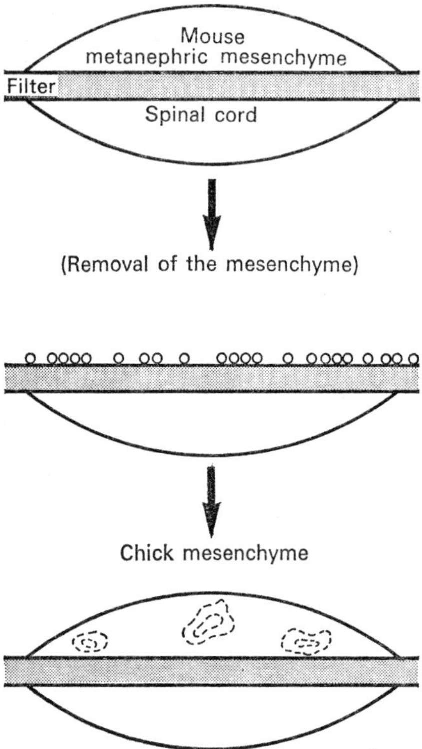

Fig. 2. Scheme of the experimental design for studies on the migration of induced metanephric cells into heterologous mesenchyme. Mouse metanephric mesenchyme was stripped off after 24 hours, leaving behind a thin layer of cells to be coated with heterologous cells. The total culture time varied between 2 and 5 days.

processes deep in the pores (Plate 1). Chick mesenchyme from various embryonic organs was added over the cells remaining on the filter, and the explants were subcultured for another one to four days. As induction was known to have been completed during the first 24 hours, the cells remaining on the filter were expected to develop tubules in the chick mesenchyme. The question was, therefore, where in the mesenchyme the tubules would be found and whether the cells would be able to become detached from the filter and migrate into the heterologous mesenchyme. The main series of experiments made use of various chick mesenchymes because their cells are easy to distinguish from mouse cells by their smaller and paler nuclei (Lash, 1963). Mouse cells could thus be detected in the chick mesenchyme even if they did not form tubules. Fixation and examination of the explants after different intervals of time (Plate 2) clearly showed that the mouse mesenchymal cells, after becoming detached

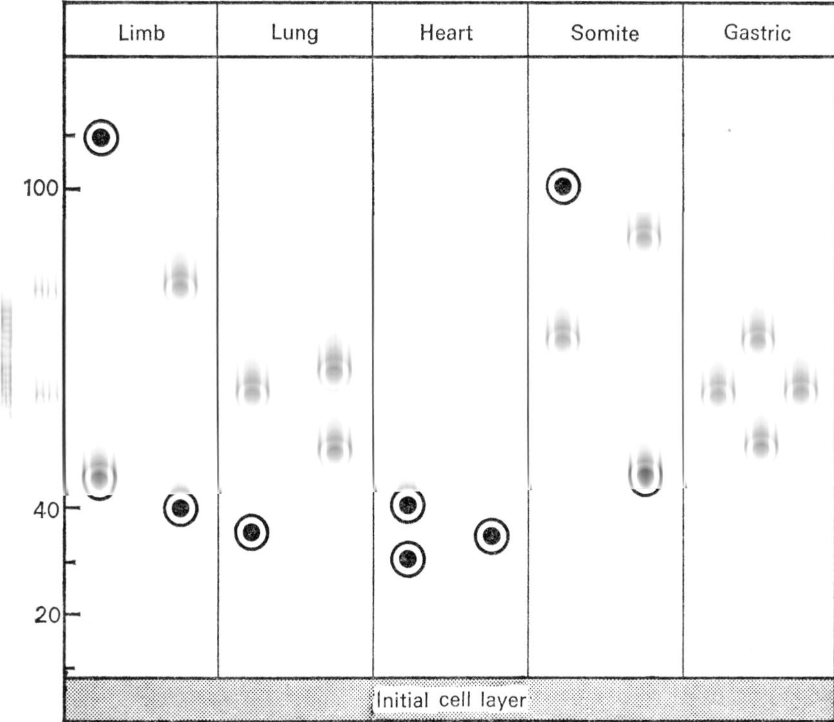

Fig. 3. Measured migration distances of the induced metanephric cells into various chick mesenchymes. Each dot represents the maximum distance of the mouse cells from the filter in one experiment.

from the filter, recognised like cells, and aggregated to form typical secretory tubules (Plate 3*a*). The distance of the tubules from the initial cell layer on the filter (migration distance) was measured with an ocular micrometer and the results are presented in Fig. 3. The inference is that after an initial period of induction the cells had broken their contacts with the inductor, migrated into the heterologous mesenchyme and aggregated there to form secretory tubules. The maximum distance travelled varies somewhat with different types of coating mesenchyme, but may exceed 100 μm.

Once it had been established that the induced cells do migrate in a heterologous mesenchyme, intraspecific experiments were performed. Salivary mesenchyme was dissected from 13-day mouse embryos, freed from its epithelium and used as a coating tissue for induced metanephric cells as in the experiments illustrated in Fig. 2. We knew from previous experiments that under these circumstances salivary mesenchyme never developed tubule structures and that any tubules which developed must have originated from the induced metanephric cells left on the filter after the metanephric mesenchyme had been stripped off. Such an experiment is illustrated in Plate 3(*b*) with an explant fixed after 3 days' subculture, and, again, migration of kidney cells deep within the heterologous mesenchyme could be demonstrated. Eleven such experiments were made and the mean maximal distance travelled was 47 μm, with a range of 30–70 μm.

Both these series of experiments demonstrating the migration and aggregation of induced tubule cells in heterologous mesenchymes can be criticised for being wholly artificial, so that the results obtained might not be applicable to the normal situation *in vivo*. Therefore, a third series of experiments was planned, where normal tissue components, the ureter bud and metanephric mesenchyme, were grown in direct recombinations. Use was made of the natural marker in the cells of the Japanese quail. All the cells of these embryos contain a large nucleolar body staining deeply in Feulgen's reaction and making the cells easy to distinguish from chick cells (Le Douarin & Barq, 1969). The ureter bud from 5½ day quail embryos was dissected from its mesenchymal blastema, so that only a thin layer of mesenchymal cells was left on its surface. Each bud was then placed on a supportive filter and coated with fragments of metanephric mesenchyme dissected from chick embryos at the corresponding developmental stage. These combined explants were fixed and examined in serial section after various culture periods.

As recently shown, chick and quail mesenchymal cells readily mix, indicating their mobile state (Armstrong & Armstrong, 1973). The mesenchymal quail cells originally retained on the isolated ureter bud could be demonstrated to be at various depths in the chick metanephric mesenchyme, where they frequently aggregated and formed tubules (Plate 4). We conclude that here too, in experiments where the normal situation *in vivo* may be better simulated, induced mesenchymal cells migrate over distances corresponding to several cell layers.

Thus, having now, as we believe, excluded the first two transmission mechanisms listed on p. 323, we interpret all observations on the migratory activities of induced metanephric cells as favouring alternative (3), i.e. induction by close cell-to-cell contact and spread of the effect by cell migration.

SUMMARY

Inductive tissue interactions constitute one of the main processes regulating differentiation and morphogenesis. Their specificity varies from determinative events to merely supportive, permissive interactions, and they appear to operate through different mechanisms. No signal substances transmitting inductive stimuli have been identified, although both transmissible molecules and surface-associated compounds have been studied. In some inductive events transmission may take place over considerable distances without cytoplasmic contact, or compounds present in the extracellular matrix may be involved.

In the induction process leading to the formation of kidney tubules the [illegible]

REFERENCES

ARMSTRONG, P. B. & ARMSTRONG, M. T. (1973). Are cells in solid tissues immobile? Mesonephric mesenchyme studied *in vitro*. *Developmental Biology*, **35**, 187–209.

BERNFIELD, M. R., BANERJEE, S. D. & COHN, R. H. (1972). Dependence of salivary epithelial morphology and branching morphogenesis upon acid mucopolysaccharide–protein (proteoglycan) at the epithelial surface. *Journal of Cell Biology*, **52**, 674–89.

BRACHET, J. (1940). Etude histochimique des protéins au cours du développement embryonnaire des Poissons, des Amphibiens et des Oiseaux. *Archive Biologique, Liège*, **51**, 167–202.

COOPER, G. W. (1965). Introduction of somite chondrogenesis by cartilage and notochord: a correlation between inductive activity and specific stages of cytodifferentiation. *Developmental Biology*, **12**, 185–212.

DEUCHAR, E. M. (1970). Neural induction and differentiation with minimal numbers of cells. *Developmental Biology*, **22**, 185–99.

GERISCH, G. (1959). Die Zelldifferenzierung bei *Pleodorina californica* Shaw und die Organisation der Phytomonadienkolonien. *Archiv für Protistenkunde*, **104**, 292–358.

GROBSTEIN, C. (1953*a*). Inductive epithelio-mesenchymal interaction in cultured organ rudiments of the mouse. *Science*, **118**, 52–5.

GROBSTEIN, C. (1953*b*). Morphogenetic interaction between embryonic mouse tissues separated by a membrane filter. *Nature, London*, **172**, 869–72.

GROBSTEIN, C. (1955). Inductive interaction in the development of the mouse metanephros. *Journal of Experimental Zoology*, **130**, 319–40.

GROBSTEIN, C. (1956). Trans-filter induction of tubules in mouse metanephrogenic mesenchyme. *Experimental Cell Research*, **10**, 424–40.

GROBSTEIN, C. (1957). Some transmission characteristics of the tubule-inducing influence on mouse metanephrogenic mesenchyme. *Experimental Cell Research*, **13**, 575–87.

HILLMAN, N., SHERMAN, M. I. & GRAHAM, C. (1972). The effect of spatial arrangement on cell determination during mouse development. *Journal of Embryology and Experimental Morphology*, **28**, 263–78.

HOLTFRETER, J. (1955). Studies on the diffusibility, toxicity and pathogenic properties of 'inductive' agents derived from dead tissues. *Experimental Cell Research*, Supplement **3**, 188–209.

KOLLAR, E. J. & BAIRD, G. R. (1970). Tissue interaction in embryonic mouse tooth germs. II. The inductive role of the dental papilla. *Journal of Embryology and Experimental Morphology*, **24**, 173–86.

KRATOCHWIL, K. (1972). Tissue interaction during embryonic development. General properties. In *Inductive Tissue Interactions and Carcinogenesis*, ed. D. Tarin, pp. 1–47. New York & London: Academic Press.

LASH, J. W. (1963). Studies on the ability of embryonic mesonephros explants to form cartilage. *Developmental Biology*, **6**, 219–32.

LE DOUARIN, N. (1968). Modifications morphologiques et fonctionnelles des hépatocytes d'oiseaux adults associés au mésenchyme métanéphritique d'embryons de poulet. *Comptes rendus Académie des Sciences, Paris*, Ser. D, **266**, 2283–6.

LE DOUARIN, N. & BARQ, G. (1969). Sur l'utilisation des cellules de la Caille japonaise comme 'marqueurs biologiques' en embryologie expérimentale. *Comptes rendus Académie des Sciences, Paris*, Ser. D, **269**, 1543–6.

MCLOUGHLIN, C. B. (1961). The importance of mesenchymal factors in the differentiation in chick epidermis. II. Modification of epidermal differentiation by contact with different types of mesenchyme. *Journal of Embryology and Experimental Morphology*, **9**, 385–409.

NIU, M. C. (1956). New approaches to the problem of embryonic induction. In *Cellular Mechanisms in Differentiation and Growth*, ed. D. Rudnick, pp. 155–71. Princeton: The University Press.

NORDLING, S., MIETTINEN, H., WARTIOVAARA, J. & SAXÉN, L. (1971). Transmission and spread of embryonic induction. I. Temporal relationships in transfilter induction of kidney tubules *in vitro*. *Journal of Embryology and Experimental Morphology*, **26**, 231–52.

PITTS, J. D. (1971). Molecular exchange and growth control in tissue culture. In *Ciba Foundation Symposium on Growth Control in Cell Culture*, ed. G. E. W. Wolstenholme & J. Knight, pp. 89–105. London & Edinburgh: Churchill-Livingstone.

RUTTER, W. J., KEMP, J. D., BRADSHAW, W. S., CLARK, W. R., RONZIO, R. A. & SANDERS, T. G. (1968). Regulation of specific protein synthesis in cytodifferentiation. *Journal of Cellular Physiology*, supplement 1, **72**, 1–18.

SAXÉN, L. (1961). Transfilter neural induction of amphibian ectoderm. *Developmental Biology*, **3**, 140–52.

SAXÉN, L. (1963). The transmission of information during primary embryonic induction. In *Biological Organization at Cellular and Supercellular Level*, ed. R. J. C. Harris, pp. 211–27. New York & London: Academic Press.

SAXÉN, L. (1970). Failure to demonstrate tubule induction in a heterologous mesenchyme. *Developmental Biology*, **23**, 511–23.

SAXÉN, L. (1971). Inductive interactions in kidney development. In *Control Mechanisms of Growth and Differentiation, XXV Symposium of the Society for Experimental Biology*, ed. D. D. Davies & M. Balls, pp. 207–21. London: Cambridge University Press.

SAXÉN, L., KOSKIMIES, O., LAHTI, A., MIETTINEN, H., RAPOLA, J. & WARTIOVAARA, J. (1968). Differentiation of kidney mesenchyme in an experimental model system. *Advances in Morphogenesis*, **7**, 251–93.

Saxén, L. & Saksela, E. (1971). Transmission and spread of embryonic induction. II. Exclusion of an assimilatory transmission mechanism in kidney tubule induction. *Experimental Cell Research*, **66**, 369–77.

Saxén, L. & Toivonen, S. (1962). *Primary embryonic induction*. New York & London: Academic Press.

Saxén, L., Toivonen, S., Vainio, T. & Korhonen, P. (1965). Untersuchungen über die Tubulogenese der Niere. III. Die Analyse der Frühentwicklung mit der Zeitraffermethode. *Zeitschrift für Naturforschung*, **20b**, 340–3.

Saxén, L. & Wartiovaara, J. (1966). Cell contact and cell adhesion during tissue [illegible]

[illegible]

[illegible]

cellular matrix low molecular weight methylated RNAs. *Developmental Biology*, **23**, 276–96.

Spooner, B. S. & Wessells, N. K. (1970). Mammalian lung development: interactions in primordium formation and bronchial morphogenesis. *Journal of Experimental Zoology*, **175**, 445–54.

Tarin, D., Toivonen, S. & Saxén, L. (1973). Studies on ectodermal–mesodermal relationships in neural induction. II. Intercellular contacts. *Journal of Anatomy*, **115**, 147–8.

Temin, H. M. (1971). The protovirus hypothesis: speculations on the significance of RNA-directed DNA synthesis for normal development and for carcinogenesis. *Journal of the National Cancer Institute*, **46**, III–VII.

Unsworth, B. & Grobstein, C. (1970). Induction of kidney tubules in mouse metanephric mesenchyme by various embryonic mesenchymal tissues. *Development Biology*, **21**, 547–56.

Wartiovaara, J., Nordling, S., Lehtonen, E. & Saxén, L. (1974). Transfilter induction of kidney tubules: correlation with cytoplasmic penetration into Nuclepore filters. *Journal of Embryology and Experimental Morphology*, **31**, 667–82.

Weiss, P. (1947). The problem of specificity in growth and development. *Yale Journal of Biology and Medicine*, **19**, 235–78.

Wessells, N. K. (1970). Mammalian lung development: interactions in formation and morphogenesis of tracheal buds. *Journal of Experimental Zoology*, **175**, 455–66.

Wessells, N. K. & Cohen, J. H. (1966). The influence of collagen and embryo extract on the development of pancreatic epithelium. *Experimental Cell Research*, **43**, 680–4.

Wolpert, L. (1969). Positional information and the spatial pattern of cellular differentiation. *Journal of Theoretical Biology*, **25**, 1–47.

EXPLANATION OF PLATES

Plate 1

Electron micrograph of a mesenchymal cell on a 'TA' Millipore filter after 24 hours' culture. Cytoplasmic processes (C) 'fix' the cell on the filter (F), whereas the contact with the other mesenchymal cells has remained loose. Magnification $\times$ 13 600.

Plate 2

Three stages from experiments performed according to the scheme in Fig. 2.

(*a*) Mesenchymal cells on the filter immediately after the bulk of the mesenchyme has been stripped off.

(*b*) After 24 hours the cells have detached themselves from the filter, migrated into the supporting chick gastric mesenchyme (they are distinguishable by their small, pale nuclei after Feulgen staining), and become arranged in pretubular aggregates.

(*c*) By 48 hours distinct tubules have been formed from the mouse cells, now separated from the filter by several layers of chick cells.

Plate 3

(*a*) A well-shaped mouse kidney tubule in chick lung mesenchyme after a total culture time of 5 days. The experiment was performed according to the scheme shown in Fig. 2.

(*b*) Kidney tubules in the mouse salivary mesenchyme used to coat the metanephric cells retained on the filter after stripping off the metanephric mesenchyme.

Plate 4

Section of an explant made by combining a quail ureter bud (U) (with contaminant mesenchymal cells) and chick metanephric mesenchyme (C). The quail cells with their large nucleolar bodies (deeply stained with Feulgen stain) have migrated to the chick mesenchyme and become arranged in pretubular aggregates (A) or have developed small tubules (T).

PLATE I

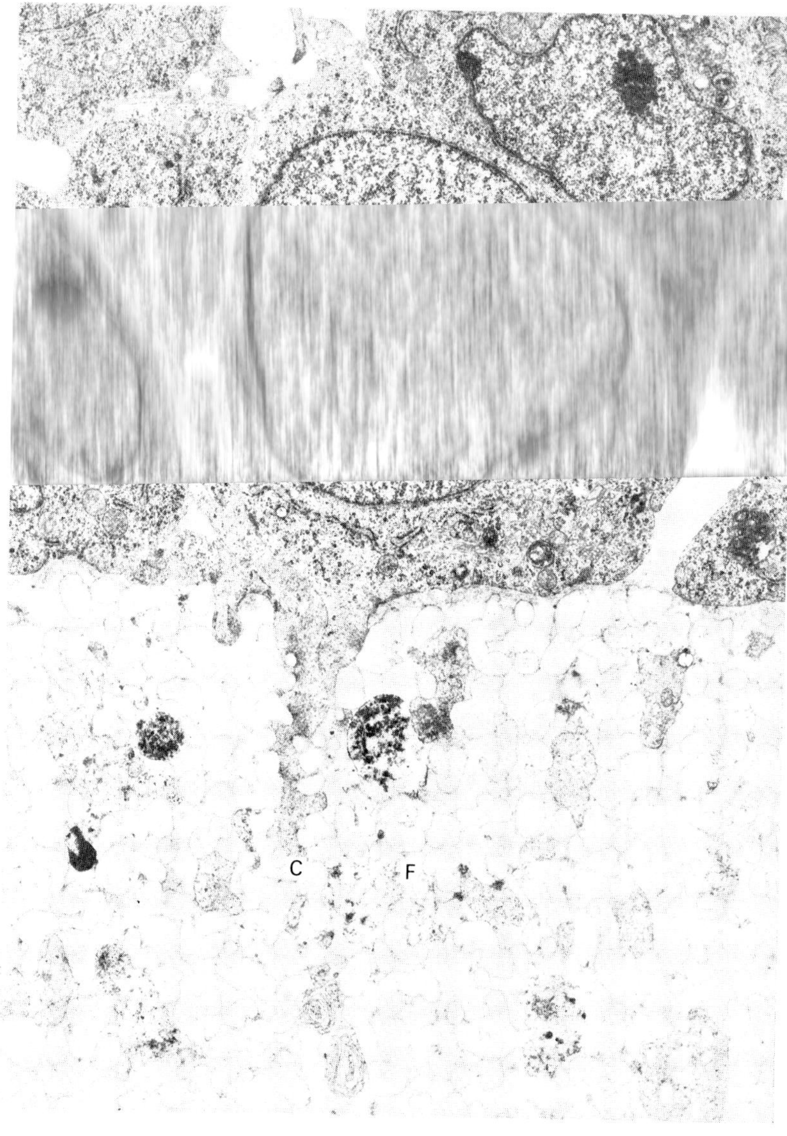

(*Facing p. 334*)

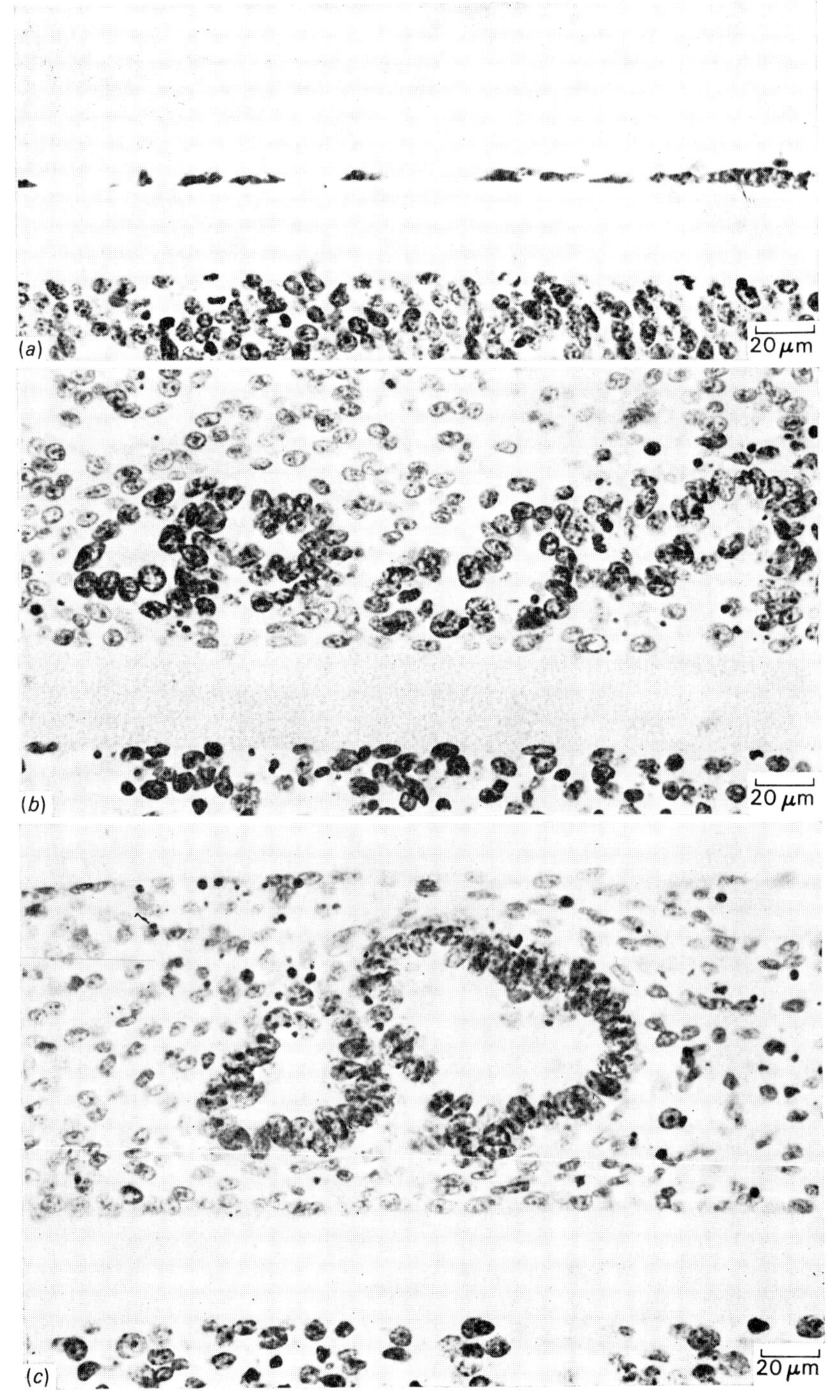
(a)
20 μm
(b)
20 μm
(c)
20 μm

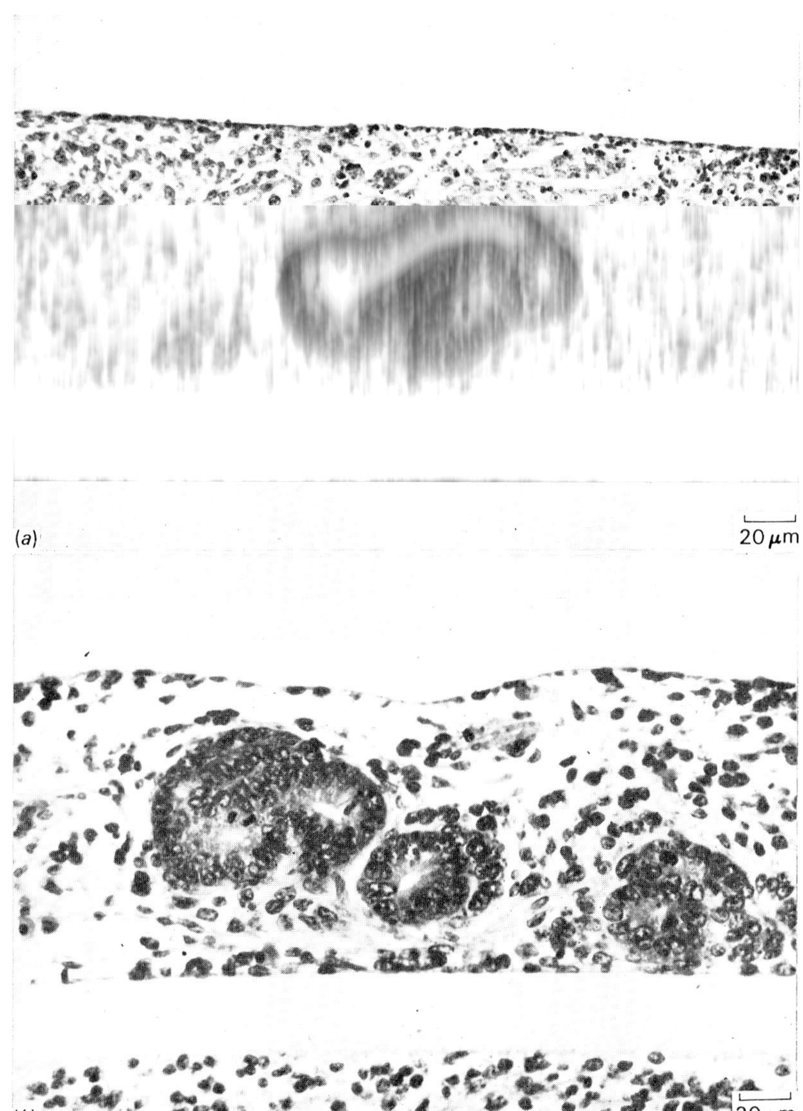
(a)
20 μm
(b)
20 μm

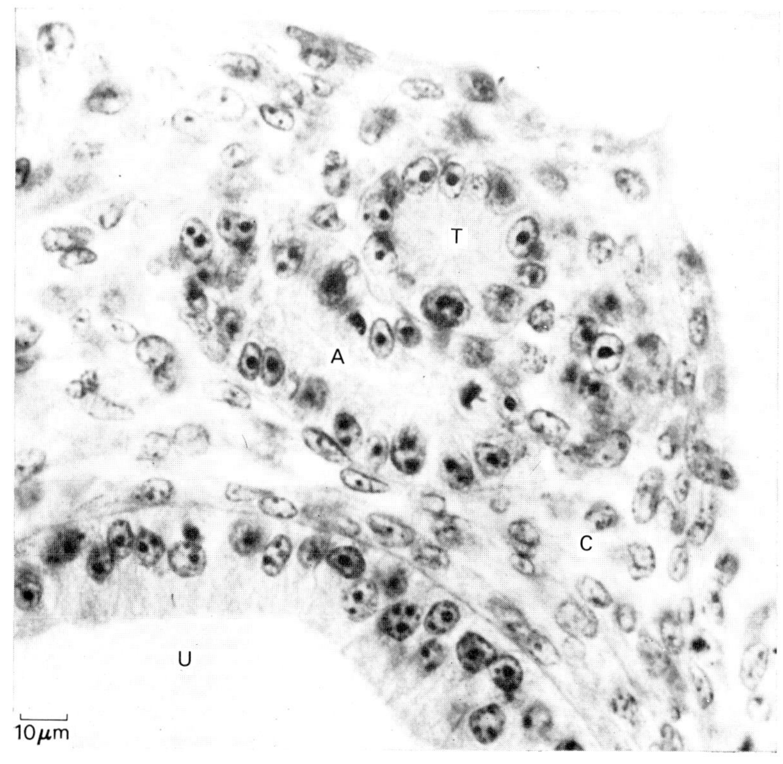
T
A
C
U
10μm

REGULATORY FUNCTIONS OF MICRO-ENVIRONMENTAL AND HORMONAL FACTORS IN PRE-NATAL HAEMOPOIETIC TISSUES

BY R. J. COLE

[illegible]

[illegible] stem cell, are characteristically sensitive to growth stimulatory substances specific for each line of haemopoietic differentiation. These progenitor cells in turn give rise to differentiating cells, the erythroblasts, myeloblasts, megakaryoblasts and lymphoblasts which mature to form the characteristic and functional end cell of each series (Fig. 1).

The ordered progression of cells through each of these compartments, directed towards production of optimum numbers of mature end cells for varying physiological states of the whole organism, depends on interactions between three types of control.

Short range interactions between haemopoietic cells, and with the stromal cells of haemopoietic tissues, define the micro-environment within which haemopoietic differentiation takes place and may determine the overall direction of such differentiation. Within this micro-environment, differentiating cells react with two types of circulating factors,

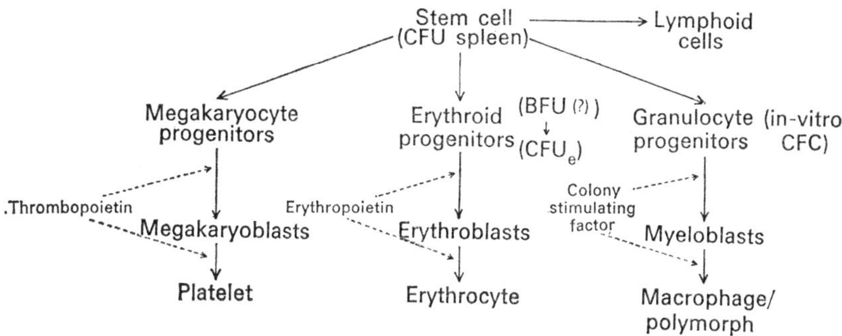

Fig. 1. Cell lineages within the haemopoietic system, and the sites of action of specific growth regulating factors.

those which stimulate proliferation and/or differentiation and those which cause a negative feedback depression of these processes.

FUNCTIONAL ANALYSIS OF THE HAEMOPOIETIC SYSTEM

The development of techniques which permit clonal analysis of the multipotent stem cell, and of progenitor cells of both the granulocytic and erythroid series in the mouse and other mammals, have led to considerable advances in our understanding of haemopoietic regulation. The in-vivo spleen colony assay for mouse haemopoietic stem cells was developed by Till & McCulloch (1961) and has been very widely employed. Suspensions of cells being tested for their haemopoietic potential are injected intravenously (10^3–10^5 cells) into lethally irradiated mice, or congenitally anaemic WW^v mice, and seven to ten days later the recipient spleens removed and fixed. The number of macroscopically visible nodules on the spleen surface is proportional to the number of multipotent stem cells in the original suspension. Since each nodule is a colony produced from a single stem cell the absolute number of stem cells present in any tissue can be estimated.

Attempts to develop in-vivo assays for multipotent stem cells have not so far been successful.

Methods for the growth of cell clones from granulocytic progenitors in semi-solid media are now well established, following original work by Pluznik & Sachs (1965) and Bradley & Metcalf (1966) (see Metcalf & Moore (1971) for full description of methods) and allow reliable estimates of the tissue content of such cells to be made. The multiplication and differentiation of such cells are dependent on the presence of colony stimulating factor (CSF) in the culture medium. This material is produced by a wide variety of cell types, and also secreted in urine, and is closely related to the humoral factor which regulates granulopoiesis in intact mammals.

The precursor to the erythroid colonies which form in semi-solid media is less well characterised than the granulocytic progenitor cell. Discrete colonies of erythroblasts develop after two to three days in plasma clot or methyl cellulose cultures seeded with bone marrow, spleen or foetal liver cell suspensions and provided with erythropoietin (Stephenson, Axelrad, McLeod & Shreeve, 1971; Iscove, 1974). Since the number of such progenitor cells is reduced in the bone marrow of plethoric mice (Gregory, McCulloch & Till, 1973) it appears that this culture method detects a relatively late cell within the progenitor cell compartment More recent studies (Axelrad, Mcleod, Shreeve & Heath, 1974) suggest that continued exposure to high erythropoietin doses elicits colony formation from an earlier precursor so that various stages of erythroid commitment may be investigated in clonal cultures.

[illegible]

[illegible] via the blood stream, can multiply and give rise to differentiating descendants. In some situations such a micro-environment may be physically defined by a boundary of cells, by the creation of a functional region by the diffusion of humoral factors which cease to be effective a short distance (i.e. a few cell diameters) from their source, or by availability of suitable surfaces for cell attachment, in the form of cell coats or other forms of extra-cellular matrices.

The mouse spleen, although the most complex, is probably the most studied haemopoietic organ within which the role of micro-environmental influences can be examined, since it contains multipotent stem cells and differentiating populations of erythroid, granulocytic, megakaryocytic, lymphoid and antibody-forming cells which can be examined by the spleen colony method outlined above (Curry & Trentin, 1967). Erythroid colonies appear after four to five days in the red pulp of the spleen, usually as a mass of differentiating erythroblasts surrounded by relatively immature cells. Granulocytic colonies develop in the spleen trabeculae and sub-capsular regions and are predominantly neutrophilic, and megakaryocytic colonies also appear beneath the splenic capsule. The relative proportions of these colony types at particular stages after grafting are characteristic, and largely independent of the tissue source of the colony forming units. Erythroid colonies predominate, being twice as frequent as granulocytic, megakaryocytic or mixed colonies, which are usually present in approximately equal proportions. Each type of colony can be shown to contain colony forming units which can produce all colony types when retransplanted (Lewis & Trobaugh, 1964). Such behaviour is a strong indication that the direction of differentiation is a property of the cellular environment within which the colony forming cell lodges, rather than an intrinsic property of each colony forming cell (though this may be influenced to a limited extent by the effects of a previous micro-environment). Continued growth of colonies of initially pure cell type leads to the formation of 'mixed' colonies with the formation of the second cell type usually at the edge of the initial cell mass. This is most convincingly explained as a result of the spread of the expanding mass of cells from one 'haemopoietic inductive micro-environment' and their encroachment on another, changing the direction of stem cell differentiation (Wolf & Trentin, 1968). This conclusion is strongly supported by studies in which implants of spleen and marrow served as host tissue, and where the predominantly granulocyte-determining influence of the marrow micro-environment continued to be expressed so that clonal colonies overlapping implant boundaries showed clear demarcation into erythroid and granulocytic areas.

Since isolated erythroid and granulocytic progenitor cells can proliferate and differentiate *in vitro*, provided that specific growth regulatory substances are present, it appears that the most important regulatory role of the haemopoietic micro-environment must be fulfilled by its effects on the multipotent stem cell, or during the transition from multipotent stem cell to progenitor cell. Unfortunately, because of the relative infrequency of stem cells (1–3 /10^5 spleen cells) and lack of adequate ultrastructural characterisation, structural studies have added little to our understanding of the basis of such interactions. However, it is reasonable to suppose that cells of the reticulo-endothelial system provide the physical foundation of the haemopoietic micro-environment. Although of course fibroblasts and collagen are important in the supporting framework of splenic tissue, fixed reticulum cells can be distinguished from them by the presence of many cell surface extensions, and are themselves organised into definable tissue substructures. One of the most visually striking cell associations seen in yolk sac, foetal liver, spleen and bone marrow is the aggregation of erythroblasts around a large, central cell, often described as a 'macrophage' (Bessis & Breton-Gorius, 1962; Orlic, Gordon & Rhodin, 1965). It has been suggested that this cell has trophic functions transferring ferritin to the developing erythroblasts, but it seems probable that still earlier erythroid precursors enjoy a similar relationship with such 'macrophage'-like cells. However, the interrelationship of this cell type with the erythropoietic inductive micro-environment is a matter of conjecture. In fact, the radio-resistance and cellular stability of such micro-environments (Knospe, Blom & Crosby, 1968) argue against the direct involvement of macrophages derived from multipotent stem cells, lathough the well-established role of the macrophages in promoting cellular interactions renders it a suitable candidate for a regulatory role, perhaps at a stage later than commitment, with the fixed cells of the adventitial reticular-cell network concerned with the initial stages of micro-environmental effects. Such effects could be the result of qualitative differences in such cells or based on quantitative differences such as cell density (Plate 1).

[illegible] supposed cellular basis of the bone marrow micro- [illegible] differentiation than the spleen.

Haemopoietic micro-environments therefore appear to be specifically organised for the production of particular classes of cell, and to exert their regulatory role by promoting the specific differentiation of previously uncommitted and potentially multipotent stem cells or by their self-replication. Such micro-environments can be tentatively identified as small and specialised regions within haemopoietic tissues which may also have the ability to selectively retain or localise differentiating cells, and which can respond to varying demands for haemopoiesis, both intrinsically, and in response to circulating regulatory substances.

THE PROLIFERATIVE STATES OF PRECURSOR CELL POPULATIONS AND THE ROLE OF SPECIFIC REGULATOR SUBSTANCES

During steady-state haemopoiesis, stem cells from adult haemopoietic tissues are very resistant to [^{3}H]thymidine, indicating either that these cells are multiplying very slowly (e.g. a cell cycle of 30–40 h) or that a majority of stem cells are in a non-proliferating, G_0 state, from which they must be triggered before stem cell depletion can be repaired (Lajtha, Oliver & Gurney, 1962). Most recent studies indicate that a variety of multiplicative states exist within the stem cell compartment, and that the potential for self-renewal and degree of commitment to differentiative pathways are related to the cell cycle state of these subpopulations, which can in some cases be separated from one another on the basis of cell size or density (Metcalf & Moore, 1971). Some progress has recently been made in identifying agents which can trigger G_0 stem cells into the cell cycle. Stimulation of β adrenergic sites leads to activation of the adenyl cyclase system which in turn initiates DNA synthesis, and similar sensitivity has been found at cholinergic sites (Byron, 1974) although it is not known whether particular receptors can influence the direction of subsequent differentiation.

In contrast to the multipotent stem cell, cells which have reached the progenitor stage of erythroid differentiation, and are responsive to erythropoietin, are extremely sensitive to the effects of [^{3}H]thymidine and other cycle-specific cytotoxic agents and are therefore in fairly rapid cell cycle even in the absence of erythropoietin (Millar, Constable & Blackett, 1974). When the suspensions of haemopoietic cells are injected into irradiated polycythaemic mice, i.e. mice with depleted erythropoietin levels, there is a reduction in macroscopic spleen colony numbers. This is due to a failure of cells to mature both in pure erythroid colonies and in the erythroid portions of mixed colonies. Closer examination reveals that such colonies consist, in plethoric mice, of masses of 'undifferentiated' cells with large pale nuclei, large nucleoli, and abundant cytoplasm (Trentin, 1970), which can be caused to differentiate into erythroblasts by administration of erythropoietin. Since the expected number of non-erythroid colonies is not exceeded in plethoric mice these studies provide additional evidence that the commitment towards erythroid differentiation during the stem cell–progenitor cell transition is not dependent on erythropoietin, and also that erythropoietin has no direct effect on multipotent stem cells.

Erythropoietin

In post-natal mammals, hypoxia is the fundamental physiological factor influencing erythropoiesis, through the availability of erythropoietin, now identified as a glycoprotein hormone of molecular weight approximately 45 800. An inactive precursor to this hormone, 'erythropoietinogen', exists in plasma, and is converted to the effective form as a result of activation by erythrogenin, which is produced in the kidney (Gordon *et al.*, 1973). Current studies indicate several points at which effective erythropoietin levels may be controlled. Plasma erythropoietin levels exert a negative feedback control over the production of erythropoietinogen, and the level of this substrate appears to be a major rate-limiting factor in the production of erythropoietin. The rate of production and release of erythrogenin from the kidney is proportional to degree of hypoxia and in turn may determine the level of serum erythropoietin. Hypertransfusion is associated with a decrease in renal erythrogenin activity, with which a specific inhibitor may be involved, in addition to the direct effects of increased renal oxygen supply (Whitcomb & Moore, 1965).

Detailed discussions of the developmental aspects of the relationship between the kinetic behaviour of erythroid cells and their differentiative potential will be found in the first symposium of this series (Cole & [illegible] 1971); haem synthetic enzymes (Necheles & Rai, 1969; Freshney & Paul, 1971) as well as cell population changes *in vitro*. In the erythroid progenitor compartment the main effects of erythropoietin are to enhance the rate of proliferation and to bring about a new pattern of epigenetic expression, perhaps by initiating transcription of a co-ordinated 'battery' of genes, expression of which leads to transition to the clearly recognisable proerythroblast stage. The maintenance of this transition *in vitro*, and probably *in vivo*, is absolutely dependent on the continued presence of erythropoietin. Once the early erythroblast state is expressed, high levels of erythropoietin shorten the cell cycle, and accelerate the rate at which precursors to haemoglobin are formed and functional haemoglobin molecules accumulate. The transition from progenitor to erythroblast therefore marks absolute changes in only some aspects of cellular properties, and the progression of erythropoietic differentiation is most profitably viewed as a restriction of some cellular activities and potentials, and enhancement of others, with some events being causally related and others occurring in parallel.

Colony stimulating factor

CSF, the tissue fluid factor which promotes the growth of granulocyte/macrophage colonies, resembles erythropoietin in that it is glycoprotein of molecular weight about 45 000. However, in contrast to erythropoietin it appears to be produced by, or released from, a very wide variety of tissues, *in vivo* and *in vitro*, and slight differences in properties are detectable in CSFs from certain organs. Since CSF is found at physiological levels in normal serum, raised level during infection and reduced levels in germ-free mice, it is reasonable to suppose that it has a significant role *in vivo*. Since CSF alone is unable to stimulate the formation of granulocyte colonies from multipotent stem cells, and the in-vitro colony forming cell population in bone marrow is subject to rapid depletion when exposed to [^{3}H]thymidine or hydroxyurea (Richard, Shadduck, Howard & Stohlmann, 1970) the CSF-sensitive precursor to in-vitro colonies is probably at the progenitor stage. The incidence of granulocytic progenitors (in-vitro CFCs) in mouse bone marrow drops rapidly when the animals are bled, while erythroid cells increase, but in polycythaemic mice, with suppressed erythropoiesis, the incidence of in-vitro CFCs is very high (Metcalf & Moore, 1971). These observations support the view that both erythropoietin-sensitive cells and in-vitro CFCs are the immediate products of multipotent stem cells, and that the direction of differentiation depends on physiological demand. In common with erythropoietin, CSF both enhances multiplication of its target cell, and shows concentration-dependent effects on the cell cycle of differentiating granulocytes *in vitro* (Metcalf, 1973). Developing colonies deprived of CSF do not survive.

A general view of the regulation of post-natal haemopoiesis, developed from experimental evidence outlined above, may therefore be stated as follows. The population of multipotent stem cells is small but heterogenous, e.g. with respect to size (affecting nuclear/cytoplasmic ratio, cell surface/volume ratio, etc.) and cell cycle state. Individual stem cells may be directed towards production of differentiated descendants or self-replication by a multiplicity of interacting factors including random intrinsic events resulting from their heterogeneity, and micro-environmental events mediated by specific cell surface receptors. The net result of these interactions is to ensure that depletions of stem cells, or within the populations of their immediate descendants, are recognised and counteracted. The immediate products of multipotent stem cells are committed to fixed pathways of differentiation; this commitment is strengthened as differentiation proceeds and occurs as a result of concentration-dependent interactions with specific hormonal factors, e.g. erythropoietin and CSF, which continue to influence differentiation for several cell cycles and states of transition. The rates of production, activation and destruction of these substances reflect the physiological requirements of the whole organism.

ARE THERE NEGATIVE FEEDBACK REPRESSORS OF HAEMOPOIESIS?

It is conceivable that rapid cessation of synthesis of substances such as erythropoietin and CSF, coupled with rapid breakdown or excretion, would provide an adequate restraint on excessive activity within haemopoietic progenitor and later cell populations. There is, however, a considerable amount of data suggesting that more direct interference with proliferation and differentiation occurs and is mediated by circulating inhibitors which prevent overproduction of differentiated cells. Such dual control has the theoretical advantage of offering greater stability, tending to dampen oscillations within control circuits. Such inhibitors [illegible] terminally differentiated cells could inhibit differentiation of earlier precursors preventing their release from haemopoietic tissue, and therefore limiting cell production through the micro-environment mechanisms reviewed earlier.

Several studies have indicated the presence of erythropoiesis inhibiting factors in the plasma of mammals with suppressed erythropoiesis (Whitcomb & Moore, 1965) in high altitude dwellers brought to sea level (Reynaforje, 1968) and in neo-natal humans (Skjaelaan, Halvorsen & Seip, 1971). An 'erythropoiesis inhibiting factor' can be purified from human urine (Lindemann, 1971) and has a molecular weight similar to that of erythrocyte 'chalone' prepared directly from erythrocytes (Kivilaakso & Rytomaa, 1971). A granulocytic chalone has also been isolated (Rytomaa & Kiviniemi, 1968). The cellular specificity of erythrocytic and granulocytic 'chalones' has been critically re-examined and lack of effect of granulocytic chalone on DNA synthesis in erythroblasts, and vice versa, confirmed in carefully controlled studies, in which marked depression of DNA synthesis in homologous cell precursors was found (Bateman, 1974).

PRE-NATAL HAEMOPOIESIS

Active haemopoiesis begins in the yolk sac of mouse embryos during the 7th day of gestation. At this stage of development the yolk sac consists of a ring of extra-embryonic splanchnopleure, and blood islands appear as thickenings of the inner mesodermal layer where it is in contact with extra-embryonic endoderm. During the 9th and 10th days of development the blood islands become connected by a capillary network, the circulation is initiated, and the islands mature into sinusoids in direct continuity with the intra-embryonic circulation. The erythrocytes produced by the yolk sac are restricted to the large primitive generation, whose cytoplasm matures without nuclear extrusion. These cells also produce a characteristic sequence of haemoglobin types, containing at least three types of globin chain which are not found in the 'adult' ($\alpha_2\beta_2$) haemoglobin, produced by all other erythropoietic cells (Kovach, Marks, Russell & Epler, 1967). Since the yolk sac contains multipotent stem cells, which can give rise to erythroblasts synthesising adult haemoglobin when transplanted to adult recipients (Auerbach, 1968), it is reasonable to suppose that the selection of clones of cells capable of synthesising 'embryonic' haemoglobins and non-proliferation of clones synthesising adult haemoglobin are functions of the yolk sac micro-environment. The mouse yolk sac is somewhat unusual since it supports the differentiation of only primitive generation erythroblasts, and no other differentiated haemopoietic cells are formed. However, the pattern of sensitivity to erythropoietin appears to be similar to that exhibited during the development and regression of erythropoiesis within the foetal liver. Dissociated cell cultures from relatively early yolk sac stages increased their rate of haem synthesis *in vitro* when exposed to erythropoietin (Bateman & Cole, 1971) but erythropoietin had no effect on later yolk sac stages (when the ratio of haemoglobin content to embryo weight reached a higher level) when these were cultured intact (Cole & Paul, 1966). Cultures of avian area vasculosa have also been shown to respond to erythropoietin (Malpoix, 1967).

Multipotent stem cells (supporting erythroid, myeloid and lymphoid development) have been identified in yolk sac from the 8th day of development (Moore & Metcalf, 1970) and although their seeding efficiency is rather low they appear to have considerably greater self-replicative potential than stem cells from subsequent stages of development. The yolk sac also contains granulocyte/macrophage progenitor cells, which form CSF-dependent colonies *in vitro*.

Haemopoiesis becomes apparent in the foetal liver on the 10th day of development, with the initiation of erythropoiesis, followed by low levels of granulocyte and megakaryocyte production. This haemopoietic micro-environment is created as the result of inductive interactions between an endodermal diverticulum from the foregut and mesenchyme surrounding the vitelline veins of the septum transversum. These interactions result in the formation of hepatic cords and hepatocytes from the endodermal elements, and connective tissue stroma and vascular endothelium from the mesenchyme. In contrast to the preceding yolk sac stage hepatic erythropoiesis is mainly extra-vascular, with close contact between haemopoietic and endodermal cells (Rifkind, Chui & Epler, 1969).

There is now available very detailed information on the kinetics of erythropoiesis in the mouse foetal liver (Paul, Conkie & Freshney, 1969; Tarbutt & Cole, 1970) and there is little doubt that erythropoietin is an essential component in the regulation of foetal hepatic erythropoiesis, not only in the mouse but in a variety of other mammals. In addition [illegible]

[illegible] (Tarbutt & Cole, 1970). The majority of erythroid colony forming cells are dependent on the continued presence of erythropoietin, but the liver also contains a proportion of cells which appear to be already 'triggered' *in vivo*, and do not require continued exposure to high levels of erythropoietin to express their potential for growth.

Although there is only limited granulocytic differentiation in foetal liver, *in vitro*, colony forming cells dependent on CSF are relatively frequent. Recent studies (Moore & Williams, 1973) indicate that mouse foetal liver CFCs differ from adult spleen and marrow CFCs in that they are larger, are in more rapid cell cycle and are more biased towards macrophage–monocyte differentiation *in vitro*, consistent with the frequency of macrophages in foetal liver. It is likely that micro-environmental influences have an important role in establishing and maintaining these differences.

Multipotent stem cells reach maximum numbers in foetal liver from the 14th day of gestation, but are first detectable on the 10th day of gestation with a doubling time of about 8 hours between days 10 and 14 (Silini, Pozzi & Pons, 1967; Moore & Metcalf, 1970). The number of stem cells in the liver declines slowly after birth, but small numbers may persist into adult life in some strains of mice. The stem cells of foetal liver are considerably more sensitive to [^{3}H]thymidine than those in adult haemopoietic tissue and functional distinctions between foetal and adult stem cells have been found. Foetal liver stem cells can proliferate for longer in adult spleen than bone marrow derived stem cells, which they eventually outgrow (Micklem *et al.*, 1972), and foetal stem cells also promote more rapid erythroid regeneration (Kubanek *et al.*, 1971).

A number of lines of evidence indicate that the multipotent stem cells (and their descendants) in the foetal liver are direct descendants of those in the yolk sac, and that the liver in the early embryo is seeded by stem cells which migrate in the circulation. The numbers of stem cells and CFCs in embryonic blood is unusually high (Moore & Metcalf, 1970). Seven-day-old mouse embryos cultured without yolk sac show no haemopoietic development, but isolated fragments of yolk sac show normal development of blood islands.

The initiation of haemopoiesis in the mouse spleen on the 15th day of gestation may also be the result of migration of stem cells and perhaps progenitor cells from the foetal liver. Erythropoiesis is established in the early spleen, but by 17 days it is predominantly granulopoietic, with small numbers of lymphocytes appearing just before birth, when haemopoietic activity declines to be largely replaced by lymphoid development. Erythropoiesis is responsive to erythropoietin in both foetal rat spleen (Cole *et al.*, 1968) and foetal mouse spleen in organ culture (Cole, unpublished data).

The marrow cavity of the long bones first appears in the 16–17 day mouse foetus and rapidly becomes populated by granulopoietic cells, and can be shown to contain both multipotent stem cells and granulocyte macrophage precursors. Erythropoiesis first becomes apparent in mouse femurs at the time of birth and appears to be organised in separate foci from granulopoietic areas. The creation of the haemopoietic micro-environment within the marrow results from the penetration of the shaft cartilage by mesenchymal buds, and cannot occur if the mesenchyme is removed. As in the case of liver and spleen, the marrow cavity is colonised by migration of multipotent stem cells and perhaps progenitor cells, which in the mouse reach the periosteal mesenchyme during the 15th day of development suggesting that the foetal liver is the source of these immigrant cells (Petrakis, Pons & Lee, 1969; see Fig. 2).

MUTANT GENES IN THE STUDY OF FOETAL HAEMOPOIETIC MICRO-ENVIRONMENTS

Mice homozygous for mutant alleles at the 'Steel' (*Sl*) locus show varying degrees of impairment of haemopoietic function, and their survival depends on which members of this multiple allelic system are present. A proportion of Sl^d/Sl^d and Sl/Sl^d mice survive into adult life and are characterised by severe macrocytic anaemia, high plasma erythropoietin levels, resistance to exogenous erythropoietin, susceptibility to sudden

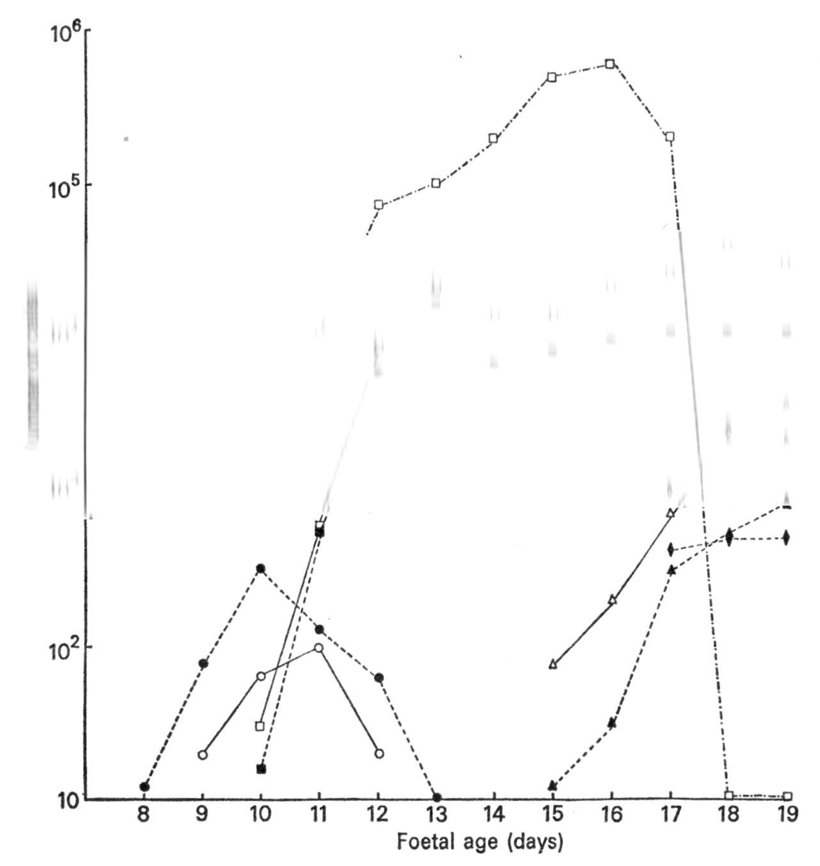

Fig. 2. The numbers of multipotent stem cells (CFC $_{spleen}$), granulocyte/macrophage colony forming cells (CFU_c) and erythroid colony forming cells (CFU_e) in pre-natal yolk sac, liver, spleen and bone marrow. The numbers of stem cells have been corrected for the following seeding efficiencies in spleen colony assays; yolk sac and 10–12 day liver 5%, 13–15 day liver 7.5%, 16–19 day liver and spleen 10%, bone marrow 20%, and on the assumption that one femur contains 10% of total bone marrow. The cloning efficiencies of granulocytic and erythroid colony forming cells have not been determined. (Data from Silini *et al.*, 1967; Moore & Metcalf, 1970; and Cole, unpublished data.)

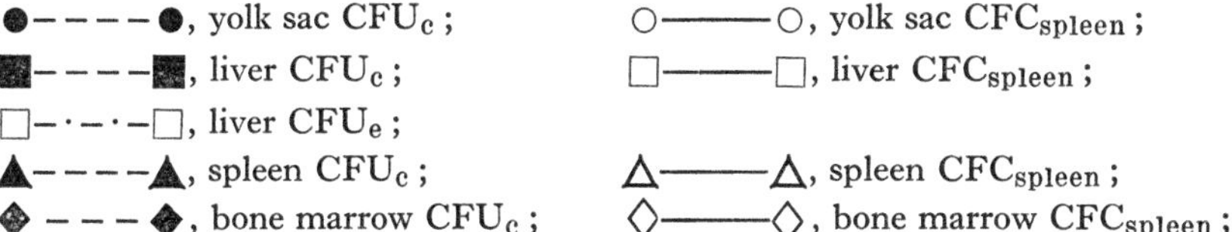

hypoxia, increased radio-sensitivity, lack of hair pigmentation and sterility due to absence of germ cells within the gonads. Since the haemopoietic tissues of such mice are incapable of supporting the normal development of transplanted non-mutant isogenic haemopoietic stem cells, while stem cells from 'Steel' mice can develop normally in 'non-Steel' haemopoietic tissue, the primary effect of the 'Steel' locus appears to be on those cells which constitute the tissue micro-environment necessary for haemopoiesis to proceed. The perturbation of haemopoiesis in pre-natal 'Steel' mice is, therefore, a potentially useful tool in the further analysis of the role and constitution of the haemopoietic micro-environment. The Sl^j gene used in our experiments is intermediate between *Sl* and Sl^d in the severity of its effects; Sl^j/Sl^j foetuses survive until birth and $Sl^j/+$ foetuses compensate to normal haematological values.

Although Sl^j/Sl^j mice show normal increases in foetal weight, hepatic erythropoiesis is severely restricted. The maximum cell number of Sl^j/Sl^j livers is one-third, and the highest proportional content of erythroblasts one-half that of normal mice. This erythropoietic failure is reflected by abnormalities in the circulation, so that during the last quarter of gestation the haemoglobin content per unit weight of embryo in Sl^j/Sl^j mice is one-third to one-quarter the normal value. However, in spite of these quantitative abnormalities, the temporal progression of erythropoiesis is similar in mutant embryos to that found in normals. The highest proportion of erythroblasts is found on the 13th day of gestation, and changes in relative proportions of early and late erythroblasts also follow the normal pattern. The spleens of late Sl^j/Sl^j foetuses are small and pale, with little erythropoietic activity, and development of the bone marrow is also restricted (Fig. 3).

Dissociated cell cultures of livers from normal foetuses respond to erythropoietin by increased haem synthesis up to the 16th day of gestation, early erythroblasts persist *in vitro* in the presence of erythropoietin, and mature erythroblasts are produced. In contrast, in cultures from Sl^j/Sl^j livers, while erythropoietin slows the rate of decline in haem synthesis which characterises cultures deprived of stimulation, there is no enhancement of synthesis, very limited maintenance of early erythroblasts, and no accumulation of late erythroblasts (Cole, Tarbutt, Cheek & White, 1974). However, the rate of haem synthesis per erythroblast is similar in both Sl^j/Sl^j and normal liver cell cultures, indicating that the biochemical basis of haemoglobin synthesis is unimpaired by *Sl* genes. The inability of the haemopoietic micro-environment within the Sl^j/Sl^j foetal liver to support normal erythropoiesis is therefore clearly expressed *in vitro* even in dissociated cell cultures, although it is not yet clear whether this is due to a reduction in the numbers of an early erythroid precursor cell, an effect imprinted on erythroid cells before explantation, a continuing interaction between erythroid and non-erythroid cells remaining separated *in vitro*, or whether re-aggregation between erythroid

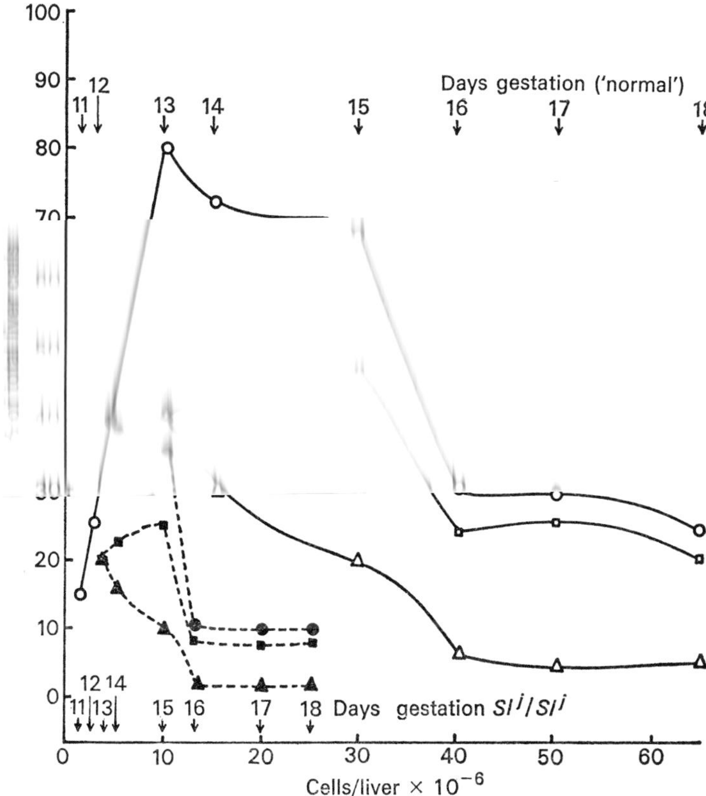

Fig. 3. Erythropoietic development of normal and Sl^j/Sl^j foetal livers. Normal livers: ○——○, total erythroblasts; △——△, early erythroblasts; □——□ late erythroblasts. Sl^j/Sl^j liver: ●– – –●, total erythroblasts; ▲– – –▲, early erythroblasts; ■– – –■, late erythroblasts. The arrows indicate the total number of cells in the liver for each day of development.

and non-erythroid cells occurs during culture. Erythropoietin responsiveness of Sl^j/Sl^j liver cell cultures (in terms of haem synthesis) can be 'rescued' by exposure to late (non-erythropoietin responsive) normal liver cells, providing evidence for faulty cellular interactions within pure Sl^j/Sl^j cultures as well as indicating a method by which such interactions may be analysed in greater detail. Cultured intact, normal and Sl^j/Sl^j foetal livers behave comparably to dissociated cell cultures, and reach similar levels of haem synthesis, on a per liver basis, with the relative difference between mutant and normal being maintained, showing that disaggregation of Sl^j/Sl^j livers cannot release erythroblasts from any inhibitory effects of their abnormal micro-environment (Fig. 4).

The maximum complement of erythroid colony forming cells in normal

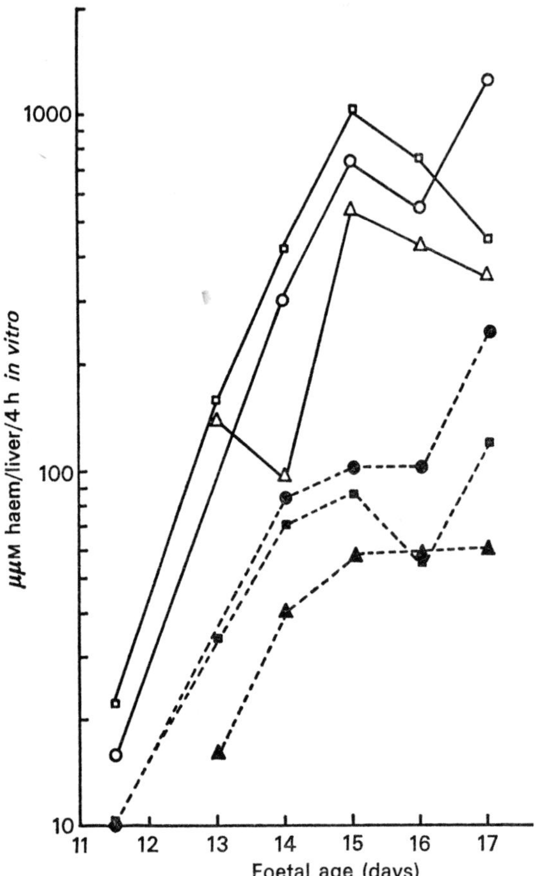

Fig. 4. Erythropoietin sensitivity in terms of haem synthesis of normal and *Sl^j/Sl^j* foetal liver cells *in vitro*. Values are computed per liver complement of erythroblasts. Normal (○———○) and *Sl^j/Sl^j* (●– – –●) liver cells labelled 0–4 h. Normal (□———□) and *Sl^j/Sl^j* (■ – – – ■) liver cells cultured with erythropoietin and labelled for 25–9 h. Normal (△———△) and *Sl^j/S^j* (▲ – – – ▲) liver cells cultured without erythropoietin and labelled for 25–9 h.

liver is found on day 16 of gestation when they represent 1.5–2% of total liver cells. The ratio of colony forming cells to erythroblasts is relatively high in early foetal liver (4–5 per 100), falls to about half this level for the next 3–4 days, and rises again briefly on day 16. The accumulation of erythroid colony forming cells in *Sl^j/Sl^j* foetal livers is severely restricted, and at maximum is less than 20% of the normal value. Except on days 11–12 the ratio of erythroid colony forming cells to recognizable erythroblasts is similar in *Sl^j/Sl^j* and normal livers, suggesting that the transit of cells from progenitor to differentiating cell compartments is not impaired by the Sl^j/Sl^j micro-environment, although the high values in early livers may indicate a temporary lag during the establishment of foetal liver erythropoiesis. *In vivo* the cell cycle time of Sl^j/Sl^j early erythroblasts is normal but that of late erythroblasts is significantly lengthened (Cole *et al.*, 1974).

Normal proportions of granulocyte/macrophage colony forming cells are found within the total liver cell population of Sl^j/Sl^j foetuses, but the [illegible]

[illegible] stroma is normal, but proliferation of stem cells, commitment towards erythroid differentiation, and proliferation of erythroid precursors are markedly reduced. The tissue element which appears to be defective has been tentatively identified as a relatively radiation resistant, non-migratory cell, common to all adult haemopoietic tissues (Fried *et al.*, 1973). Since this cell is non-migratory, dissociated cell suspensions are unable to 'cure' 'Steel' anaemia, but this can be alleviated by implants of intact haemopoietic tissue. Haemopoiesis in the mouse foetal liver is established by migration of multipotent stem cells from the yolk sac on the 9th–10th day of gestation. The Sl^j/Sl^j pre-natal lesion could be the result of failure of proliferation of stem cells in the yolk sac or of reduced migration, or lodgement, or of reduced proliferation of stem cells and/or their immediate products in the foetal liver. Studies of adult 'Steel' mice suggest that the last alternative is most likely. Pre-natal haemopoietic organs offer many advantages for study of the phenomena discussed here, particularly their ability to develop normally in culture, and it is expected that this property, linked with the variety of techniques outlined here, will lead to a much more detailed description of the cellular and molecular bases of haemopoietic micro-environments and their ability to direct differentiation.

NEGATIVE FEEDBACK CONTROL OF PRE-NATAL ERYTHROPOIESIS

The kinetics of pre-natal erythropoiesis suggest that its regulation results from complex interactions between the proliferation of early precursor cells, the availability of micro-environments within which they can differentiate, the presence or absence of erythropoietin, and the presence or absence of humoral factors exerting negative feedback control on multiplication or differentiation. Marked changes in several features of blood formation occur midway through the period of hepatic erythropoiesis (i.e. the 15th–16th day of gestation) which may recapitulate a similar series of events during erythropoiesis in the yolk sac and which are explicable in terms of interactions between erythropoietin and a negative feedback regulator: the length of cell cycle in early and late erythroblasts alters, erythroblasts pass from erythropoietin responsiveness *in vitro* to non-responsiveness, and there is a five- to sixfold increase in the population doubling time of circulating erythrocytes (Paul *et al.*, 1969; Tarbutt & Cole, 1970). The concentration of erythrocytes and haemoglobin concentration rises rapidly until day 16, with a steady value of about 65% of the adult value then being maintained until birth. The foetal content of haemoglobin per unit of body weight reaches a peak on day 17 at 85% of adult value. Both the daily increment in haemoglobin, relative to the number of haemoglobin synthesising cells (erythroblasts and reticulocytes) per foetus, and the rate of haem synthesis in newly explanted erythroblasts decline as the red cell content of the peripheral blood rises, with a marked

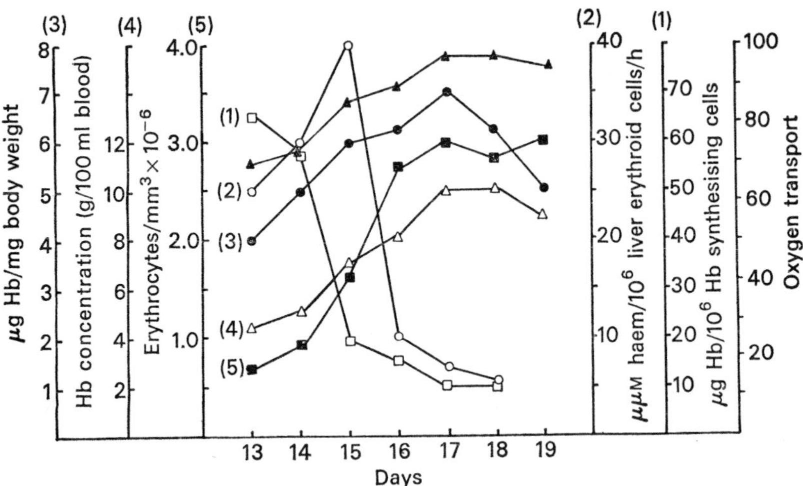

Fig. 5. The relationship between oxygen transport (blood flow × O_2 capacity, Thorling & Erslev, 1968) expressed as a percentage of the normal adult value; (▲——▲) and various pre-natal haematological parameters. □——□, μg haemoglobin (Hb)/10^6 haemoglobin synthesising cells; ○——○, μμM haem synthesised/10^6 liver erythroid cells/h; ●——●, μg haemoglobin/mg of foetal body weight; △——△, Hb concentration (g/100 ml blood); ■——■, erythrocyte concentration (erythrocytes/mm³ × 10^{-6}).

change of rate on day 16. Both the rate of transport and the relative concentration changes achieved during transplacental maternal–foetal exchanges are critically dependent on the rate of blood flow in the foetal placenta, which is inversely proportional to haematocrit. The optimum erythrocyte concentration in the foetal blood may therefore be physically determined by the placental capillary bed and its relationship to maternal [illegible]

We have recently been able to show (Cole, unpublished data) that material eluted *in vitro* from mature erythrocytes exerts a concentration-dependent inhibition on haem sythesis in foetal liver erythroblasts *in vitro*. This material also counteracts the rise in RNA synthesis which precedes haem synthesis in erythropoietin stimulated erythroblasts without

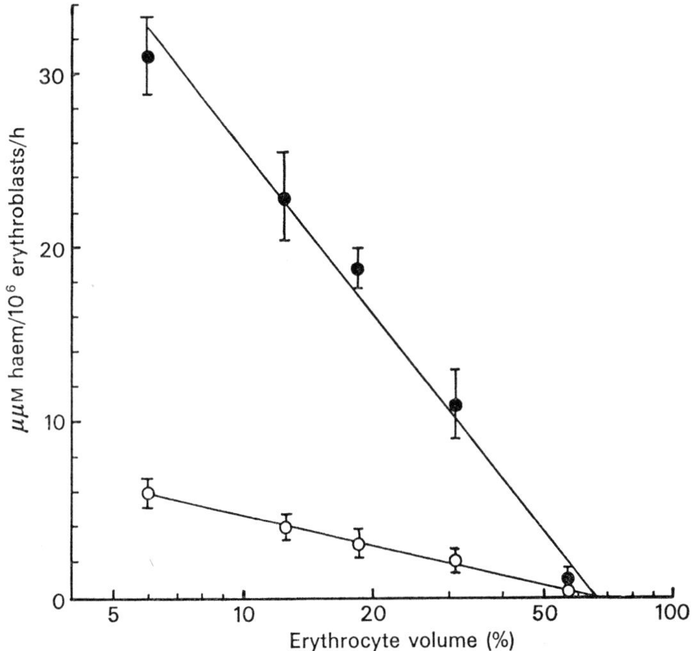

Fig. 6. The effects of the products of varying concentrations of erythrocytes on haem synthesis by foetal liver erythroblasts in the presence (●) and absence (○) of erythropoietin.

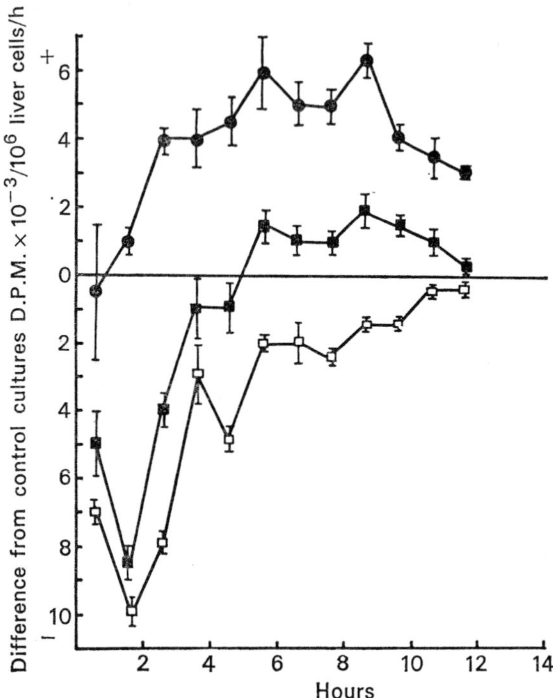

Fig. 7. The effect of erythrocyte products on the stimulation of RNA synthesis by erythropoietin in liver erythroblasts *in vitro*. ●———●, with erythropoietin, no erythrocyte factors; ■———■, with erythropoietin and erythrocyte factors; □———□, with erythrocyte factors, no erythropoietin. RNA synthesis measured by incorporation of [^{3}H]uridine.

inhibiting DNA synthesis or cell multiplication (Figs. 6 and 7). Material with similar properties is also present in serum from polycythaemic mice. This material, therefore, appears to restrict differentiation rather than having a 'chalone'-like effect. Interference with the maturation of erythroblasts within the foetal liver is likely to slow down their release and restrict the production of earlier cells due to competition for space, and will potentiate the effects of progressive restriction in the erythropoietic potential of the liver micro-environment which also occurs. There is, therefore, both circumstantial and experimental evidence for a directly acting negative control mechanism particularly effective late in gestation, which together with reductions in the level of erythropoietin consequent on rising tissue oxygenation, ensures strict regulation of erythropoiesis during a critical stage of development.

Original studies on pre-natal haematopoiesis reported here were supported by the Medical Research Council. It is a pleasure to acknowledge the skilled collaboration of Mrs G. Garlick, Mrs S. White and Mr T. Regan.

REFERENCES

AUERBACH, R. (1968). Some aspects of tissue interaction *in vitro*. In *Epithelial–Mesenchymal Interactions*, ed. R. Fleischmajer & R. E. Billingham, pp. 200–7. Baltimore: Williams & Wilkins.

AXELRAD, A. A., McCLEOD, D. L., SHREEVE, M. M. & HEATH, D. S. (1974). Properties of cells that produce erythrocytic colonies *in vitro*. In *Hemopoiesis* [illegible]

[illegible]

BRADLEY, T. R. & METCALF, D. (1966). The growth of mouse bone marrow cells *in vitro*. *Australian Journal of Experimental Biology and Medical Sciences*, **14**, 287–300.

BULLOUGH, W. S. (1973). Chalone control systems. In *Humoral Control of Growth and Differentiation*, ed. J. LoBue & A. S. Gordon, pp. 1–24. New York & London: Academic Press.

BYRON, J. (1974). Molecular basis for the triggering of haemopoietic stem cells into DNA synthesis. In *Hemopoiesis in Culture*, ed. W. A. Robinson, pp. 91–100. Washington: DHEW Publication NIH 74–205.

CHUI, D. H. K., DJALDETTI, M., MARKS, P. A. & RIFKIND, R. A. (1971). Erythropoietin effects on fetal mouse erythroid cells. I. Cell populations and hemoglobin synthesis. *Journal of Cell Biology*, **51**, 585–95.

COLE, R. J., HUNTER, J. & PAUL, J. (1968). Hormonal regulation of pre-natal haemoglobin synthesis by erythropoietin. *British Journal of Haematology*, **14**, 477–88.

COLE, R. J. & PAUL, J. (1966). The effects of erythropoietin on haem synthesis in mouse yolk-sac and cultured foetal liver cells. *Journal of Embryology and Experimental Morphology*, **15**, 245–60.

COLE, R. J. & TARBUTT, R. G. (1973). Kinetics of cell multiplication and differentiation during adult and prenatal haemopoiesis. In *The Cell Cycle in Development and Differentiation*, ed. M. Balls & F. S. Billett, pp. 365–95. London: Cambridge University Press.

COLE, R. J., TARBUTT, R. G., CHEEK, E. M. & WHITE, S. L. (1974). Expression of congenital defects in the haemopoietic micro-environment: pre-natal erythropoiesis in anaemic 'Steel' (Sl^j/Sl^j) mice. *Cell and Tissue Kinetics*, **7**, 489–503.

CURRY, J. L. & TRENTIN, J. J. (1967). Hemopoietic spleen colony studies. I. Growth and differentiation. *Developmental Biology*, **15**, 395–413.

DJALDETTI, M., PREISLER, H., MARKS, P. A. & RIFKIND, R. A. (1972). Erythropoietin effects on fetal mouse erythroid cells. II. Nucleic acid synthesis and the erythropoietin sensitive cell. *Journal of Biological Chemistry*, **247**, 731–5.

FRESHNEY, R. I. & PAUL, J. (1971). The activities of 3 enzymes of haem synthesis during hepatic erythropoiesis in the mouse. *Journal of Embryology and Experimental Morphology*, **26**, 313–22.

FRIED, W., CHAMBERLIN, W., KNOSPE, W. H., HUSSEINI, S. & TROBAUGH, F. E.

(1973). Studies on the defective haematopoietic micro-environment of Sl/Sl^d mice. *British Journal of Haematology*, **24**, 643–50.

GORDON, A. S., ZANJANI, E. D., GIDARI, A. S. & KUNA, R. A. (1973). Erythropoietin: the humoral regulator of erythropoiesis. In *Humoral Control of Growth and Differentiation*, ed. J. LoBue & A. S. Gordon, pp. 25–50. New York & London: Academic Press.

GREGORY, C. J., MCCULLOCH, E. A. & TILL, J. E. (1973). Erythropoietin progenitors capable of colony formation in culture: state of differentiation. *Journal of Cellular Physiology*, **81**, 411–20.

GROSS, M. & GOLDWASSER, E. (1970). On the mechanism of erythropoietin induced differentiation. 7. The relationship between stimulated DNA synthesis and RNA synthesis. *Journal of Biological Chemistry*, **245**, 1632–6.

HARRISON, P. R., CONKIE, D. & PAUL, J. (1973). Role of cell division and nucleic acid synthesis in erythropoietin induced maturation of foetal liver cells *in vitro*. In *The Cell Cycle in Development and Differentiation*, ed. M. Balls & F. S. Billett, pp. 341–54. London: Cambridge University Press.

ISCOVE, N. (1974). Human urinary erythroid colony stimulating activity. In *Hemopoiesis in Culture*, ed. W. A. Robinson, pp. 44–52. Washington: DHEW Publication NIH 74–205.

KIVILAAKSO, E. & RYTOMAA, T. (1971). Erythrocyte chalone, a tissue specific inhibitor of cell proliferation in the erythron. *Cell and Tissue Kinetics*, **4**, 1–10.

KNOSPE, W. H., BLOM, J. & CROSBY, W. H. (1968). Regeneration of locally irradiated bone marrow. 2. Induction of regeneration in permanently aplastic medullary cavities. *Blood*, **31**, 400–5.

KOVACH, J. S., MARKS, P. A., RUSSELL, E. S. & EPLER, H. (1967). Erythroid cell development in fetal mice: ultrastructural characteristics and hemoglobin synthesis. *Journal of Molecular Biology*, **25**, 131–42.

KUBANEK, B., RENCRICCA, N., PORCELLINI, A. & STOHLMAN, F. (1971). The effect of plethora and erythropoietin on erythropoiesis in heavily irradiated recipients receiving foetal liver cells. In *The Regulation of Erythropoiesis and Haemoglobin Synthesis*, ed. T. Travnicek & T. Neuwirt, pp. 157–64. Prague: Universita Karlova.

LAJTHA, L. G., OLIVER, R. & GURNEY, C. W. (1962). Kinetic model of a bone marrow stem cell population. *British Journal of Haematology*, **8**, 442–60.

LEWIS, J. P. & TROBAUGH, F. E. (1964). Haemopoietic stem cells. *Nature, London*, **204**, 589–90.

LINDEMANN, R. (1971). Erythropoiesis inhibiting factor. I. Fractionation and demonstration of urinary E.I.F. *British Journal of Haematology*, **21**, 623–31.

MALPOIX, P. (1967). Effects of erythropoietin in chick embryos. *Biochimica et Biophysica Acta*, **146**, 181–4.

METCALF, D. (1973). Colony stimulating factor (CSF). In *Humoral Control of Growth and Differentiation*, ed. J. LoBue & A. S. Gordon, pp. 91–118. New York & London: Academic Press.

METCALF, D. & MOORE, M. A. S. (1971). *Haemopoietic Cells*. Amsterdam: North-Holland.

MICKLEM, H. S., FORD, C. E., EVANS, E. P., OGDEN, D. A. & PAPWORTH, D. S. (1972). Competitive *in vivo* proliferation of foetal and adult haematopoietic cells in lethally irradiated mice. *Journal of Cellular Physiology*, **79**, 293–8.

MILLAR, J. L., CONSTABLE, T. B. & BLACKETT, N. M. (1974). Delayed response to erythropoietin in polycythaemic animals treated with cytotoxic agents. *Cell and Tissue Kinetics*, **7**, 363–70.

MOORE, M. A. S. & METCALF, D. (1970). Ontogeny of the haemopoietic system.

Yolk sac origin of *in vivo* and *in vitro* colony forming cells in the developing mouse embryo. *British Journal of Haematology*, **18**, 279–96.

MOORE, M. A. S. & WILLIAMS, N. (1973). Analysis of proliferation and differentiation of foetal granulocyte–macrophage progenitor cells in haemopoietic tissue. *Cell and Tissue Kinetics*, **6**, 461–76.

NECHELES, T. F. & RAI, U. S. (1969). Studies on the control of haemoglobin synthesis. The *in vitro* stimulating effect of a 5 βH steroid metabolite on haem formation in human bone marrow cells. *Blood*, **34**, 380–4.

[illegible]

REYNAFORJE, C. (1968). Humoral regulation of erythropoietic depression of high altitude polycythaemic subjects after return to sea level. *Annals of the New York Academy of Sciences*, **149**, 472–4.

RICKARD, K. A., SHADDUCK, R. R., HOWARD, D. E. & STOHLMANN, F. (1970). A differential effect of hydroxyurea on hemopoietic stem cell colonies *in vitro* and *in vivo*. *Proceedings of the Society for Experimental Biology and Medicine*, **134**, 152–6.

RIFKIND, R. A., CHUI, D. & EPLER, H. (1969). An ultrastructural study of early morphogenetic events during the establishment of fetal hepatic erythropoiesis. *Journal of Cell Biology*, **40**, 343–65.

RYTOMAA, T. & KIVINIEMI, K. (1968). Control of granulocyte production. I. Chalone and anti-chalone; two specific humoral regulators. *Cell and Tissue Kinetics*, **1**, 329–40.

SILINI, G., POZZI, L. V. & PONS, S. (1967). Studies on haemopoietic stem cells of the mouse foetal liver. *Journal of Embryology and Experimental Morphology*, **17**, 303–18.

SKJAELAAN, P., HALVORSEN, S. & SEIP, M. (1971). Inhibition of erythropoiesis by plasma from newborn infants. *Israel Journal of Medical Science*, **7**, 857–60.

STEPHENSON, J. R., AXELRAD, A. A., MCLEOD, D. L. & SHREEVE, M. M. (1971). Induction of colonies of haemoglobin-synthesising cells by erythropoietin *in vitro*. *Proceedings of National Academy of Sciences, USA*, **68**, 1542–6.

TARBUTT, R. G. & COLE, R. J. (1970). Cell population kinetics of erythroid tissue in the liver of foetal mice. *Journal of Embryology and Experimental Morphology*, **24**, 429–46.

THORLING, E. B. & ERSLEV, A. J. (1968). The tissue tension of oxygen and its relation to hematocrit and erythropoiesis. *Blood*, **31**, 332–43.

TILL, J. E. & MCCULLOCH, E. A. (1961). A direct measurement of the radiation sensitivity of normal mouse bone marrow cells. *Radiation Research*, **14**, 213–222.

TRENTIN, J. J. (1970). Influence of hematopoietic organ stroma (hematopoietic inductive microenvironments) on stem cell differentiation. In *Regulation of Hematopoiesis*, ed. A. S. Gordon, pp. 159–86. New York: Appleton Century Crofts.

WHITCOMB, W. H. & MOORE, M. (1965). The inhibitory effect of plasma from hypertransfused animals on erythrocyte iron incorporation in mice. *Journal of Laboratory and Clinical Medicine*, **66**, 641–8.

WOLF, N. S. (1974). Dissecting the hematopoietic microenvironment I. Stem cell lodgement and commitment, and the proliferation and differentiation of erythropoietic descendants in the *Sl/Sl*^{d} mouse. *Cell and Tissue Kinetics*, **7**, 89–98.

WOLF, N. S. & TRENTIN, J. J. (1968). Hemopoietic colony studies. V. Effect of haemopoietic organ 'stroma' on differentiation of pluripotent stem cells. *Journal of Experimental Medicine*, **127**, 205–14.

EXPLANATION OF PLATE

PLATE 1

The erythropoietic micro-environment of 14-day foetal liver. Erythroblasts (Proerythroblasts, PR; basophilic erythroblasts, B; and polychromatic erythroblasts, PO) are maturing in close association with stromal cells (S). These show cytoplasmic extensions which partly envelop the differentiating erythroblasts. An endothelial (E) lined vessel (lower left) contains reticulocytes (R) and an orthochromatic cell (O) on the point of extruding its nucleus.

PLATE I

(*Facing p. 358*)

EARLY DIFFERENTIATION OF THE LYMPHOID SYSTEM

BY M. A. RITTER

Department of Zoology, South Parks Road, Oxford OX1 3PS

[illegible] the blood and lymphatic vessels. The thymus and bursa differ from other lymphoid organs in that: (*a*) they are derived from intestinal epithelium, (*b*) they are the first sites of lymphopoiesis in the embryo, (*c*) they have a high mitotic activity independent of antigenic stimulation and (*d*) thymic and bursal extirpation lead to characteristic anatomical (lymphopaenia in other lymphoid organs) and functional (impairment of immune responses) deficits. Thus the thymus and bursa of Fabricius can be considered as primary, or central, lymphoid organs that act as the source of small lymphocytes for the secondary, peripheral, lymphoid organs such as lymph nodes, Peyer's patches and spleen.

In the peripheral system the small lymphocytes form a recirculating pool in the blood and lymphatic vessels and the secondary lymphoid organs (Gowans & Knight, 1964). The majority of small lymphocytes are long-lived and proliferate in response to antigenic stimulation, giving rise to more lymphocytes in addition to other cell types (immunoblasts, plasmablasts and plasma cells). The function of these cells is to initiate and participate in immune reactions.

Two major types of reaction can be identified: (*a*) those mediated directly by cells (lymphocytes), for example graft rejection and delayed hypersensitivity, and (*b*) those mediated by humoral factors (immunoglobulins produced by plasma cells), the antibody responses. Extirpation experiments show that cell-mediated immunity is thymus dependent (Miller, 1961), whilst that mediated by antibodies is, in the bird, bursa dependent (Warner, 1967). Thus in the bird two primary lymphoid organs are responsible for the production of two classes of lymphocyte (thymus-derived T and bursa-derived B cells) that function in these two types of immunity. In addition, although it is the bursa-derived cells that actually produce the immunoglobulin, the response to many antigens requires the co-operation of T cells which may act either as helpers or suppressors. There is no structural bursa-equivalent in mammals; instead, some peripheral lymphoid tissues may play a bursal type role in the production of B lymphocytes.

The purpose of this article is to define, as far as is possible, the ontogenetic processes leading to the production of immunologically mature T and B lymphocytes. Discussion will, therefore, centre on lymphopoiesis in the primary lymphoid organs. Most studies refer to the situation in the mouse as the availability of many inbred strains makes them a convenient mammalian species to study. However, in certain situations (intravenous injection into the embryo, B cell inductive environment) the avian embryo forms a simpler experimental model. Some additional, non-mammalian, data will therefore be presented.

PRIMARY LYMPHOID ORGANS: GENERAL DEVELOPMENTAL ASPECTS

Early in mouse embryogenesis the thymus develops from the third and fourth pharyngeal pouches, forming a pair of pouch-like structures. Later the pouch form is lost and the paired lobes come to lie just above the heart. The chick thymus has the same origin, but then forms two chains of lobes that lie alongside the jugular veins in the neck. The avian bursa develops as an outpouching of the gut in the cloacal region. The epithelial lining of this sac-like structure becomes increasingly folded, mitotically active thickenings appear and eventually become separated from the lining to form rounded cell aggregates lying in the mesenchyme.

The histological picture is strikingly similar in all primary lymphoid organs. Initially the organ rudiments are composed entirely of large pale-staining epithelial cells. Later a second cell type can be seen, characterised by prominent nucleoli and a large amount of heavily basophilic cytoplasm (time of appearance in the embryo: for mouse thymus, 11 days; chick thymus, 7 days; chick bursa, 14 days). Small lymphocytes first appear two to four days later (mouse thymus, 15 days; chick thymus, 10 days; chick bursa, 16 days).

There are three major questions concerning this development of lymphocytes within primary lymphoid organs: (1) what is the nature and origin of the lymphocyte precursor or stem cell? (2) what are the stages by which the stem cell is converted to an immunologically competent small lymphocyte? and (3) how are such changes mediated? The evidence relating to these topics will be discussed in the following sections of this article.

STEM CELLS: THEIR NATURE AND ORIGIN

Nature of the stem cell

The nature of the lymphocyte precursor has been the subject of much controversy, with proponents of both intrinsic (epithelial) and extrinsic (mesenchymal) theories (Beard, 1900; Hammar, 1909; Auerbach, 1961; [illegible] (CAM) of chick embryos. This system provides a nutritive environment that is isolated from any cell inflow or outflow.

Rudiments at three developmental stages were selected: (*a*) 10 day, with epithelial cells only; (*b*) 11 day, containing epithelial cells and a small number of large basophils, and (*c*) 14 day, with epithelial and many basophilic cells. No lymphocytes were present at any stage. The rudiments were cultured until equivalent in age to 17 days gestation. Plate 1, Plate 2 (*a*) and (*b*) and Table 1 show a comparison of cell types present in the thymus before and after culture. In all rudiments studied (and in a parallel experimental series using chick thymus) there was a positive correlation between the presence and number of basophilic cells and the capacity of the cultured organ to develop lymphocytes. This strongly supports identification of the lymphoid precursor with the large basophilic cell.

Table 1. *A comparison of the cell types present in embryonic mouse thymus rudiments before and after culture in cell-tight diffusion chambers on chick CAM*

Age of rudiment at start of culture	Cell types present at start of culture		Presence of lymphocytes at end of culture
	Epithelial	Basophilic	
10 day	+ + +	–	–
11 day	+ + +	+	+
14 day	+ + +	+ + +	+ + +

The immediate origin of the stem cell was demonstrated in a second experimental situation where 10 day mouse thymus (epithelial cells only) was grafted perirenally in syngeneic hosts. These grafts, in contrast to the cultured 10 day rudiments, underwent good lymphoid development (Plate 2*c*). An inflow of precursor cells from the bloodstream is therefore a prerequisite for thymic lymphopoiesis.

Further information concerning the identity and route of entry of the inflowing cells was obtained using autoradiography (Ritter unpublished results). Chicks were used for this work owing to the feasibility of embryonic intravenous injection. Embryonic chick spleen cells (a known source of stem cells – Moore & Owen, 1967) were labelled *in vitro* with [^{3}H]thymidine and injected intravenously (i.v.) into 9 day chick embryos. The uptake of labelled cells by the thymus was followed. After 6 hours no labelled cells remained in the blood. In the thymus the majority (70%) were in the mesenchyme surrounding the thymic lobes. The proportion of intrathymic labelled cells increased progressively until by 36 hours the majority (69%) were within the thymic lobes (Table 2). Labelled cells were photographed

Table 2. *Migration of labelled cells into the thymus of 9 day chick embryos*

Time after injection (hours)	Labelled cells outside thymus (mesenchyme)	Labelled cells inside thymic lobes	% labelled cells localised within thymic lobes
6	85	36	30
12	34	23	40
18	66	56	46
24	40	44	52
36	14	31	69

Embryos were injected i.v. with [^{3}H]thymidine-labelled spleen cells, and their thymic rudiments sampled at various time intervals for autoradiography. The numbers of labelled cells within the thymic lobes and in the surrounding mesenchyme were scored at each stage. Mean number of background grains per cell was 3: cells with > 10 grains scored as 'labelled'.

before and after grain removal. Two morphological types were identified: basophilic and granulocytic. Labelled large basophils were found in the thymus itself and in the surrounding mesenchyme, often in contact with the basement membrane around the thymic lobes (Plate 3). Labelled granulocytes were mainly confined to the mesenchyme. Similar results were obtained with the bursa. Thus large basophils do migrate from the bloodstream to the thymic rudiment and bursal follicles via the surrounding mesenchyme. This further supports the idea that the basophilic cell is the blood-borne stem cell.

Stem cell origins

What is the source of stem cells prior to their entry into the bloodstream?

In experiments where early mouse embryos and their yolk sacs were cultured either together or singly, Moore & Metcalf (1970) showed the yolk sac to be essential for haemopoiesis, suggesting that this tissue con- [illegible] Later in development the mouse [illegible] of stem cells responsible for seeding all other, secondary, sites has recently been questioned. Experiments with chick/quail chimaeras (F. Dieterlen, personal communication) suggest that this scheme may not be complete. In combinations of chick primitive streak grafted to quail yolk sac, lymphocytes were found to be chick in type. In the reverse combination lymphocytes were quail in type. The original stem cell source may therefore be an intra-embryonic one, followed secondarily by the yolk sac.

Stem cell potentialities

The repopulation of all haemopoietic tissue by a common stem cell supply (cells that repopulate thymus and bursa also repopulate spleen and bone marrow; see Moore & Owen, 1967) has been interpreted as evidence in favour of a multipotential precursor. However, selective migration of different subclasses of stem cell (pre-erythroid, pre-granuloid, pre-lymphoid) could equally well explain the results. More definite evidence for multipotentiality has been obtained in the rat (Nowell, Hirsch, Fox & Wilson, 1970). Rat bone marrow cells, bearing radiation-induced chromosome abnormalities, were injected into lethally irradiated mice. Peripheral blood lymphocytes and erythropoietic spleen colonies were found in several cases to carry the same chromosome abnormality, indicating that they were the progeny of the same stem cell from the donor rat.

Whatever their potentiality, lymphoid precursors are not self-sufficient, but are dependent upon an inductive environment for their maturation into T or B lymphocytes.

THYMUS

T cell maturation

The differentiation of thymus lymphocytes can be assessed in terms of histological appearance, cell surface antigens (H-2, Thy-1, TL), and functional status, i.e. ability to mount a graft-versus-host reaction (G-v-H) and to proliferate in response to the mitogens phytohaemagglutinin (PHA) and concanavalin A (Con A).

Small lymphocytes first appear in the foetal mouse thymus at day 15. By birth the organ has enlarged and become highly lymphoid. Two areas can be distinguished: cortex and medulla. Both contain lymphocytes and epithelial cells, but the former are more densely packed in the cortex.

It has been shown that 95% of thymocytes (lymphocytes within the thymus) have less H-2 antigen on their surface than other cell types. They also carry additional 'differentiation' antigens: Thy-1 (previously termed Θ), in relatively large amounts (Thy-1.1 or 1.2, according to strain), and TL (+ or − according to strain). This major proportion of thymocytes is short-lived and is not immunocompetent. The remaining 5% of thymocytes, which can be separated by their cortisone resistance, are characterised by normal levels of H-2, reduced levels of Thy-1 and loss of TL when present. These lymphocytes are situated in the medulla, are long-lived and are G-v-H, PHA and Con A reactive (Matsuyama, Wiadrowski & Metcalf, 1966; Blomgren & Anderson, 1969; Leckband & Boyse, 1971; Raff, 1971; Stobo & Paul, 1972). Since such properties are all typical of peripheral T lymphocytes it is tempting to conclude that this minor population represents cells that have matured within the thymus and are destined for the periphery. An intrathymic origin is supported by observations of lymphocyte migration *from* but not *to* the thymus (Ernstrom, Gyllensten & Larsson, 1965). Additional support comes from Weissman's (1973) experiments where following subcapsular administration of [^{3}H] thymidine he showed that cells in the cortisone resistant medullary population are the descendants of cortical lymphocytes. However, it is likely that most T lymphocytes never enter this pool but leave the thymus either as newly mature cells or as partially mature cells whose maturation is completed in the periphery. Elliot (1973) has demonstrated that in CBA–T_6T_6 thymus grafts after 217 days in a syngeneic CBA host, of cells responding to PHA stimulation as many as 74% remained of donor type. Hence it is probably only a minority of responsive cells that leave for the periphery. Elliot suggests that the resident subpopulation may be important in maintaining self non-reactivity.

Studies on structural development

In order to study the sequence of morphological changes transforming stem cell to small lymphocyte (Ritter, unpublished observations), inbred embryonic mouse thymic rudiments were grafted perirenally to F1 host mice and inflowing host stem cells in the grafts studied by chromosome marker and surface antigen cytotoxicity techniques (Schlesinger, 1965; Ford, 1966). All cells carry H-2 antigens so anti-H-2 antisera pick up [illegible] until 17 days after grafting. At this time there was a sudden increase in host-derived dividing and Thy-1 positive cells (33% and 43% respectively). The results are summarised in Fig. 1. Thus, following entry into the thymus, stem cells rest for several days before undergoing mitotic division to give rise to Thy-1-bearing small lymphocytes. Other work has shown that the acquisition of Thy-1 is accompanied by a decrease in H-2 and the appearance of TL on the cell surface (Schlesinger, 1970). The development

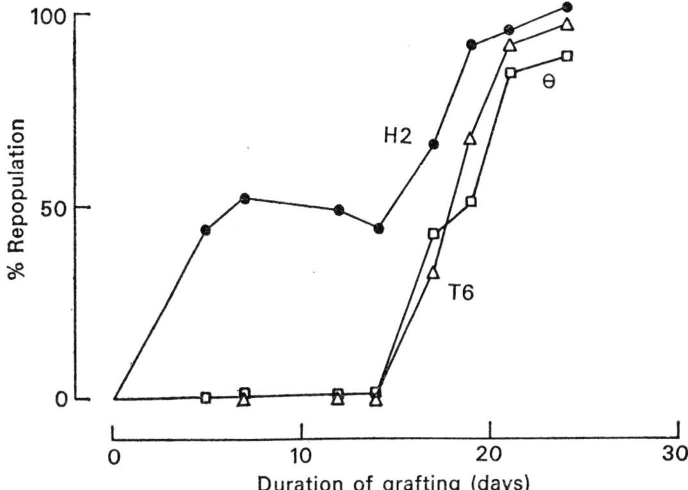

Fig. 1. Thymus of 14 day embryonic mouse grafted perirenally in syngeneic or F1 adult hosts: a comparison of chromosome marker (T_6), H-2 and θ repopulation. Cells bearing H-2 and θ surface antigens were assessed using the Trypan Blue dye exclusion cytotoxicity test. Stem cells bearing H-2 antigen enter the thymus rudiment soon after grafting. Here they rest for several days before undergoing mitotic division to give rise to Θ-positive small lymphocytes.

from basophilic stem cell to small lymphocyte is characterised, therefore, not only by gross morphological changes, but also by alterations in cell surface structures.

Studies on functional development

At what stage during structural differentiation do small lymphocytes acquire immunocompetence?

Certainly peripheral T cells can mount immune reactions. In contrast, only the small cortisone-resistant thymus population is similarly competent. Are these thymus lymphocytes cells that have developed functional maturity within the thymic environment prior to migration? Alternatively, are they lymphocytes initially of thymic origin that have subsequently acquired immunocompetence in the periphery, and have then returned to the thymus?

The origin of the immunocompetent thymus lymphocytes was studied using the G-v-H response (parental type lymphoid cells injected into newborn F1 recipients) as a functional assay for T lymphocytes, the presence of immunocompetent cells in the inoculum being indicated by spleen enlargement in the host (Ritter, 1971). The degree of development of isolated pre-lymphoid (14 day) and early lymphoid (15 day) embryonic mouse thymus was assessed after culture in cell-tight diffusion chambers on the chick CAM. Since by birth some immunocompetent cells are present in the normal thymus, rudiments were cultured until temporally equivalent to newborn (5–7 days culture). The effectiveness of the cultured cells was compared with that of uncultured 14 and 15 day embryonic thymus cells. The finding that the cultured cells, although unreactive before culture, produced a significant splenomegaly demonstrates the acquisition of immunocompetence during the culture period. Spleen enlargement for cultured cells was also increased in comparison with that for normal newborn thymocytes, suggesting a build-up of immunocompetent lymphocytes that in the absence of a cell barrier would have migrated to the periphery (Fig. 2). Thus, solely within the thymus environment stem cells can undergo functional maturation.

In toto, therefore, the evidence indicates an important role for the thymus in providing a suitable inductive environment for the development of both structural and functional maturity. Some lymphocytes leaving the thymus for the first time are thus already equipped to participate in peripheral immune reactions.

Studies in the adult suggest that T lymphocyte development consists of two main parts, occurring either in the thymus or in the periphery. In the thymus and periphery two populations of T lymphocytes, T_1 and T_2,

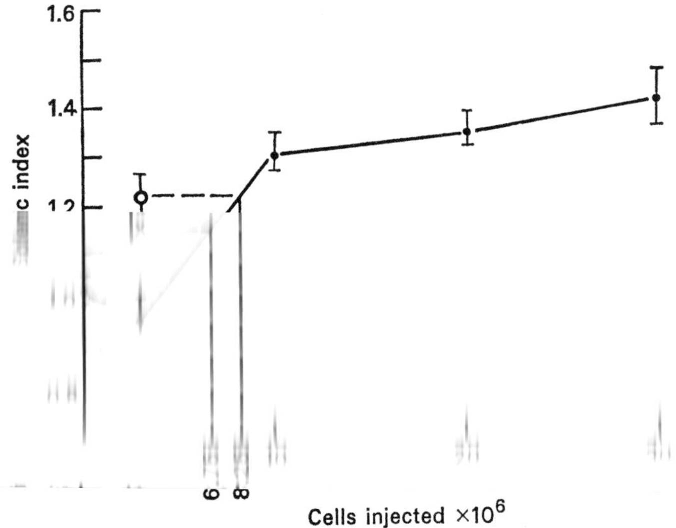

Fig. 2. Graft-versus-host activity of mouse thymocytes. The dose–response curve for normal newborn thymocytes is given by (●——●). In comparison the greater splenic indices produced by 3×10^6 14/15 day embryonic thymus cells cultured for 5–7 days (newborn equivalent) in diffusion chambers on the CAM (O) or in chambers inserted intraperitoneally (x) in syngeneic hosts, are shown. The values plotted are antilog$_e$ (weighted log$_e$ mean splenic index ± 95% confidence limits). Splenic index is defined as experimental relative spleen weight ÷ control relative spleen weight.

have been identified (Jacobsson & Blomgren, 1972; Olsson & Claesson, 1973). T_1 cells are short-lived, migrate to the spleen and do not recirculate. T_2 cells are long-lived, lymph node seeking, do recirculate and have a further reduction of Thy-1 on their surface. T_1 and T_2 lymphocytes are thought to co-operate with each other in certain cell-mediated responses such as G-v-H and mixed lymphocyte reactions (Cantor & Asofsky, 1972).

Inductive environment

How does the thymus mediate its inductive influence on T cell maturation? Two rival theories have been presented. Firstly, that direct contact between stem cell and thymus is necessary ('stromal' effect). Secondly, that the thymus exerts its influence via a humoral factor ('humoral' effect). This factor is found in an extract of the thymus and seems to exert its effect through cyclic AMP (Trainin & Small, 1970; White & Goldstein, 1970; Kook & Trainin, 1974). However, these two influences need not be mutually exclusive; it may be only after a cell has encountered the 'stromal' effect that it becomes sensitive to the action of a humoral factor (Stutman, Yunis & Good, 1969). It is not yet clear which part of development is controlled by which facet of thymus influence.

Epithelial cells in the thymus are thought to be the producers of T lymphocyte inducers, possibly both long-acting 'humoral' and short-acting 'stromal' factors (Mandel, 1970; Mandi & Glant, 1973).

BURSA/BURSA-EQUIVALENT

Little is known of B lymphocyte development in the mammalian embryo, owing to the absence of an obvious equivalent of the avian bursa of Fabricius.

In birds, B lymphocytes are produced in the bursa. They are slightly larger than T lymphocytes and are characterised by the presence of surface and intracytoplasmic immunoglobulin. These cells are able to participate in humoral antibody immune reactions, either within the bursa or after migration to other parts of the body.

Bursal development has been studied *in vitro* (Lebacq & Ritter, unpublished data). Bursae of 12 day embryos, which contain large basophilic cells but no lymphocytes and no cells with intracytoplasmic immunoglobulin, were cultured for 4–7 days in an in-vitro filter well system. After culture the rudiments contained both lymphocytes and cells with intracytoplasmic immunoglobulin (detected by histological and immunofluorescent techniques). This points, as in the thymus, to the importance of a stem cell – inductive environment interaction in the production of immunocompetent cells.

There is no strict mammalian anatomical equivalent to the bursa. In the past many organs have been suggested as functional equivalents, for example, the gut-associated tissues, although none has proved satisfactory (Perey, Cooper & Good, 1968; Fichtelius, 1968). Recent studies on embryonic mouse tissues have been more successful (Owen, Raff & Cooper, 1974, & unpublished results). Fourteen day foetal mouse liver and spleen which possess no lymphocytes and no immunoglobulin-bearing cells were found to be positive for both these characteristics after a period of in-vitro culture. These two tissues, therefore, seem to be providing an inductive environment for B cell maturation. Thus there probably is no single mammalian primary lymphoid organ equivalent to the bursa of Fabricius, but rather, a series of peripheral sites that provide the appropriate microenvironment. The avian bursa may represent an evolutionary specialisation.

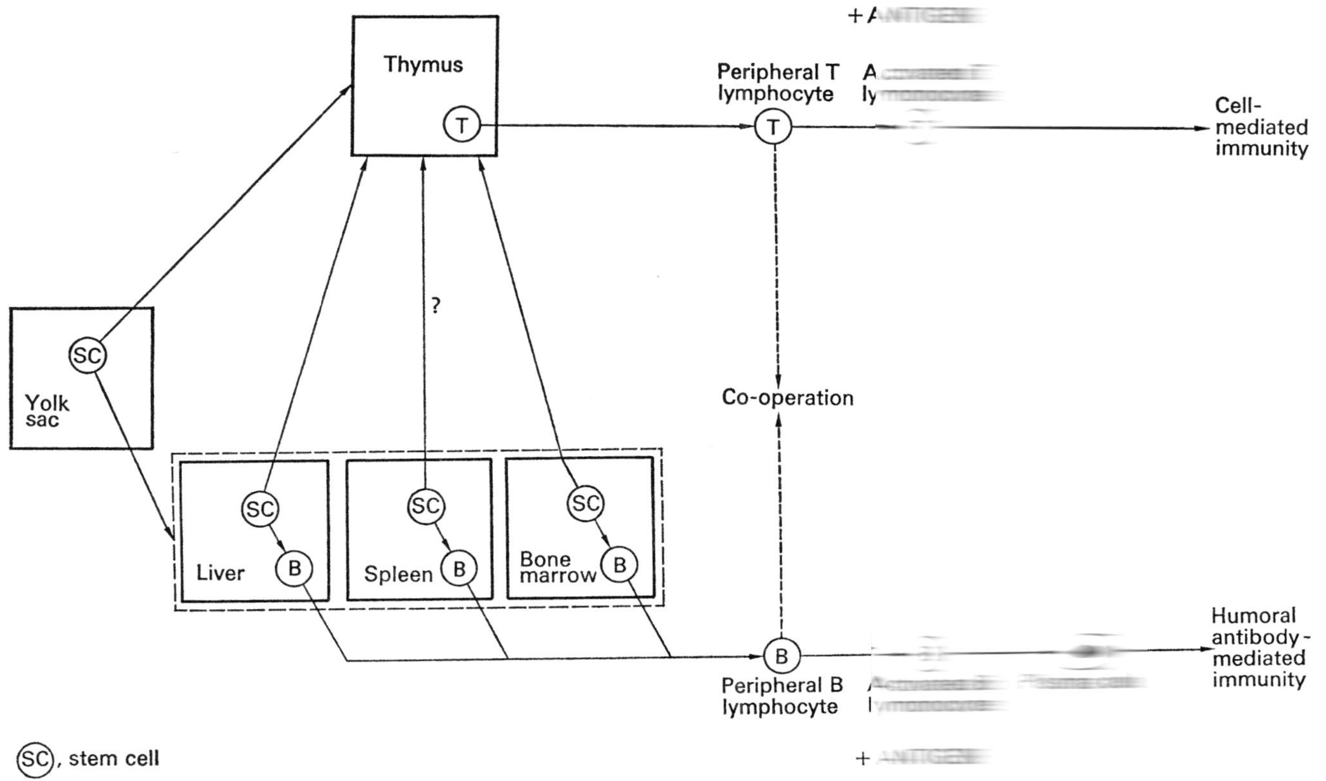

Fig. 3. T and B lymphocyte development in the mam

SUMMARY

The formation of lymphocytes in the mammalian embryo depends upon an interaction between stem cell and inductive environment (see Fig. 3).

The stem cell is probably a large basophil that migrates from sites such as yolk sac, liver and bone marrow to the appropriate inductive organs via the bloodstream. Stem cells within the thymus mature into T lymphocytes which are exported to the periphery. These cells bear differentiation antigens and function in cell-mediated immune responses. Several peripheral sites (e.g. liver and spleen) rather than a single discrete primary organ (as is the avian bursa) are thought to be responsible for B lymphocyte maturation. B cells and their descendants carry surface and intracytoplasmic immunoglobulin and participate in humoral antibody immunity.

REFERENCES

ACKERMAN, G. A. & KNOUFF, R. A. (1964). Lymphocyte formation in the thymus of the embryonic chick. *Anatomical Record*, **149**, 191–216.

AUERBACH, R. (1961). Experimental analysis of origin of cell types in the development of the mouse thymus. *Developmental Biology*, **3**, 336–54.

BEARD, J. (1900). The source of leucocytes and the true function of the thymus. *Anatomischer Anzeiger*, **18**, 550–73.

BLOMGREN, H. & ANDERSSON, B. (1969). Evidence for a small pool of immunocompetent cells in the mouse thymus. *Experimental Cell Research*, **57**, 185–92.

CANTOR, H. & ASOFSKY, R. (1972). Synergy among lymphoid cells mediating the graft-versus-host response. III. Evidence for interaction between two types of thymus-derived cells. *Journal of Experimental Medicine*, **136**, 764–79.

ELLIOT, E. V. (1973). A persistent lymphoid population in the thymus. *Nature New Biology*, **242**, 150–2.

ERNSTROM, U., GYLLENSTEN, L. & LARSSON, B. (1965). *Nature, London*, **207**, 540–1.

FICHTELIUS, K. E. (1968). The gut epithelium – a first level lymphoid organ? *Experimental Cell Research*, **49**, 87–104.

FORD, C. E. (1966). The use of chromosome markers. In *Tissue Grafting and Radiation*, ed. H. S. Micklem & J. F. Loutit, pp. 197–206. New York & London: Academic Press.

GOWANS, J. L. & KNIGHT, E. J. (1964). The route of recirculation of lymphocytes in the rat. *Proceedings of the Royal Society*, Ser. B, **159**, 257–82.

HAMMAR, J. A. (1909). Zur Kenntnis des Teleostierthymus. *Archiv für mikroskopische Anatomie und Entwicklungsmechanik*, **73**, 1–68.

JACOBSSON, H. & BLOMGREN, H. (1972). Studies on the recirculating cells in the mouse thymus. *Cellular Immunology*, **5**, 107–21.

KOOK, A. I. & TRAININ, N. (1974). Hormone-like activity of a thymus humoral factor on the induction of immune competence in lymphoid cells. *Journal of Experimental Medicine*, **139**, 193–207.

LECKBAND, E. & BOYSE, E. A. (1971). Immunocompetent cells among mouse thymocytes: a minor population. *Science*, **172**, 1258–60.

MANDEL, T. (1970). Differentiation of epithelial cells in the mouse thymus. *Zeitschrift für Zellforschung und mikroskopische Anatomie*, **106**, 494–515.

Mandi, B. & Glant, T. (1973). Thymosin-producing cells of the thymus. *Nature New Biology*, **246**, 25.

Matsuyama, M., Wiadrowski, M. & Metcalf, D. (1966). Autoradiographic analysis of lymphopoiesis and lymphocyte migration in mice bearing multiple thymus grafts. *Journal of Experimental Medicine*, **123**, 559–76.

Micklem, H. S., Ford, C. E., Evans, E. P. & Gray, J. (1966). Interrelationships of myeloid and lymphoid cells: studies with chromosome-marked cells transfused into lethally irradiated mice. *Proceedings of the Royal Society*, [illegible]

[illegible]

rat. *Journal of Cellular Physiology*, [illegible]

Olsson, L. & Claesson, M. H. (1973). Studies on subpopulations of theta-bearing lymphoid cells. *Nature New Biology*, **244**, 50–1.

Owen, J. J. T., Raff, M. C. & Cooper, M. D. (1974). *In vitro* generation of B lymphocytes in mouse foetal liver, a mammalian 'bursa equivalent'. *Nature, London*, **249**, 361–3.

Owen, J. J. T. & Ritter, E. A. (1969). Tissue interaction in the development of thymus lymphocytes. *Journal of Experimental Medicine*, **129**, 431–7.

Perey, D. Y., Cooper, M. D. & Good, R. A. (1968). Lymphoepithelial tissue of the intestine and differentiation of antibody production. *Science*, **161**, 265–6.

Raff, M. C. (1971). Evidence for subpopulation of mature lymphocytes within mouse thymus. *Nature, London*, **229**, 182–4.

Ritter, M. A. (1971). Functional maturation of lymphocytes within embryonic mouse thymus. *Transplantation*, **12**, 279–82.

Schlesinger, M. (1965). Immune lysis of thymus and spleen cells of embryonic and neonatal mice. *Journal of Immunology*, **94**, 359–64.

Schlesinger, M. (1970). How cells acquire antigens. *Progress in Experimental Tumour Research*, **13**, 28–83.

Smith, C. (1965). Studies on the thymus of the mammal. XIV. Histology and histochemistry of embryonic and early postnatal thymuses of C57Bl/6 and AKR strain mice. *American Journal of Anatomy*, **116**, 611–30.

Stobo, J. D. & Paul, W. E. (1972). Functional heterogeneity of murine lymphoid cells. II. Acquisition of mitogen responsiveness and of Θ antigen during the ontogeny of thymocytes and T lymphocytes. *Cellular Immunology*, **4**, 367–80.

Stutman, O., Yunis, E. J. & Good, R. A. (1969). Thymus: an essential factor in lymphoid repopulation. *Transplantation Proceedings*, **1**, 614–15.

Trainin, N. & Small, M. (1970). Conferment of immuno-competence on lymphoid cells by a thymic humoral factor. In *Hormones and the Immune Response*, ed. G. E. W. Wolstenholme & J. Knight. *CIBA Foundation Study Group*, **36**, 24–37.

Tyan, M. L. & Cole, L. J. (1963). Mouse foetal liver and thymus: potential source of immunologically active cells. *Transplantation*, **1**, 347–50.

Tyan, M. L. & Cole, L. J. (1965). Bone marrow as the major source of potential immunologically competent cells in the adult mouse. *Nature, London*, **208**, 1223–4.

WARNER, N. L. (1967). The immunological role of the avian thymus and bursa of Fabricius. *Folia Biologica*, **13**, 1–17.

WEISSMAN, I. L. (1973). Thymus cell maturation. *Journal of Experimental Medicine*, **137**, 504–10.

WHITE, A. & GOLDSTEIN, A. L. (1970). Thymosin, a thymic hormone influencing lymphoid cell immunological competence. In *Hormones and the Immune Response*, ed. G. E. W. Wolstenholme & J. Knight. *CIBA Foundation Study Group*, **36**, 3–19.

EXPLANATION OF PLATES

PLATE 1

(*a*) Thymic rudiment of a 10 day mouse embryo. The thymus is composed of pale-staining epithelial cells. There are no basophilic cells at this stage.

(*b*) Thymic rudiment of 11 day mouse embryo. This consists mainly of epithelial cells with a few large basophils (dark-staining cytoplasm). A basophilic cell (arrowed) can be seen immediately adjacent to the basement membrane.

(*c*) Thymus of 11 day mouse cultured for 6 days in a diffusion chamber on the CAM. Only a small number of lymphoid cells (one is arrowed) have developed during the culture period.

All parts of plate at the same magnification.

PLATE 2

(*a*) Thymic rudiment of 14 day mouse embryo. There are many basophilic cells within the rudiment at this stage.

(*b*) Thymic rudiment of 14 day mouse embryo cultured in a diffusion chamber on the CAM for 3 days. The rudiment contains a large number of lymphocytes that developed during the culture period.

(*c*) Thymus of 10 day mouse embryo grafted under the kidney capsule of a syngeneic host for 7 days. There is good lymphoid development.

All parts of plate at the same magnification.

PLATE 3

(*a*) Low power view of labelled cell within 9 day embryonic chick thymus, following i.v. injection of [^{3}H]thymidine-labelled spleen cells.

(*b*) High power view of the same cell.

(*c*) High power view of the same cell after grain removal (using Farmer's solutio n showing it to be a large basophilic cell.

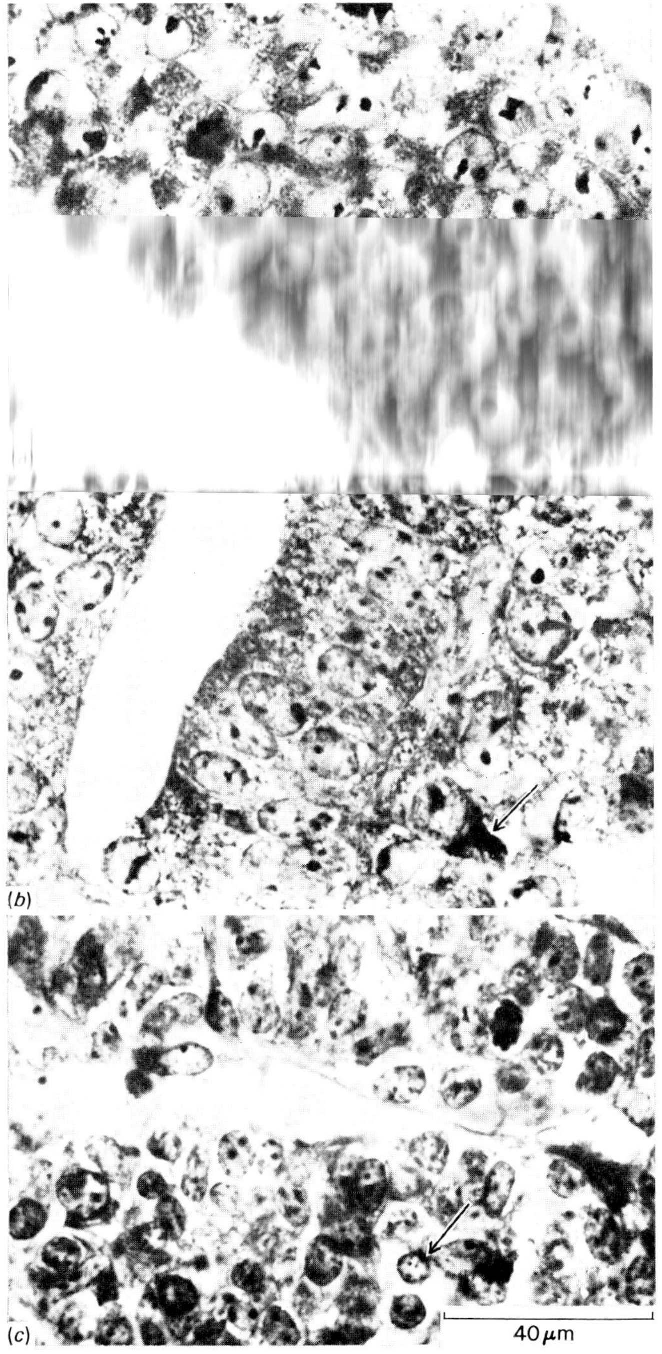

(Facing p. 372)

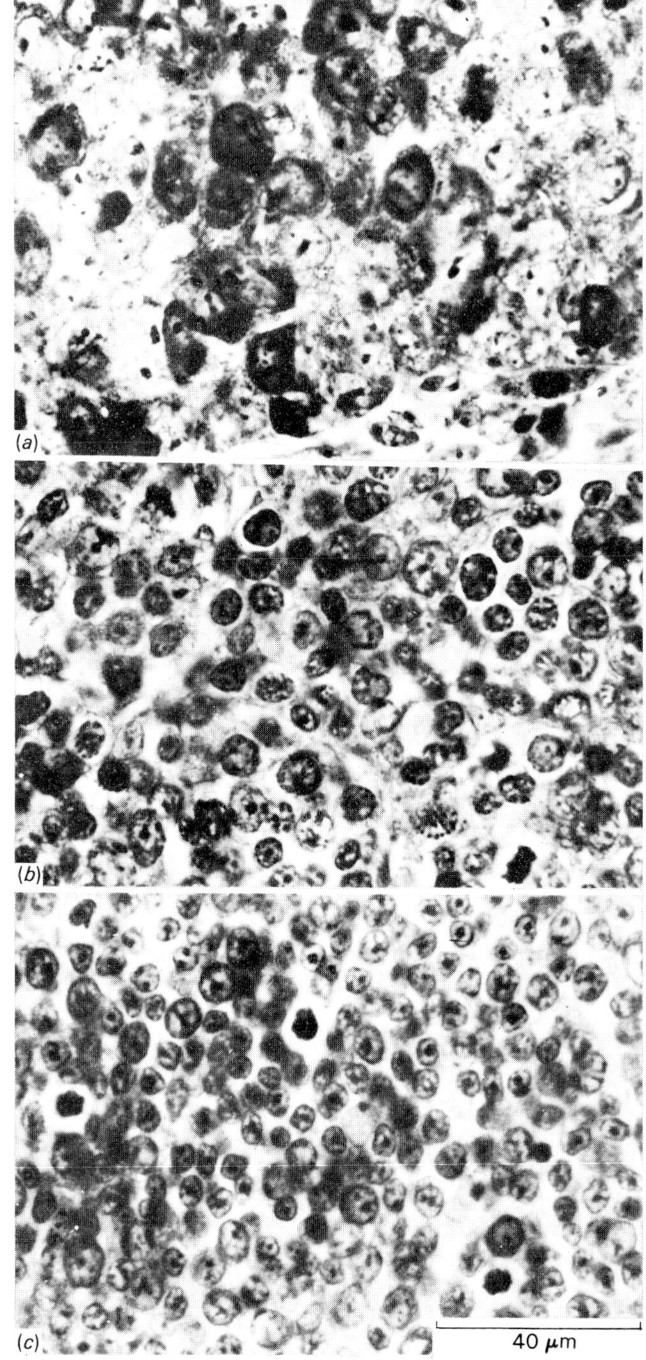
(a)
(b)
(c)
40 μm

PLATE 3

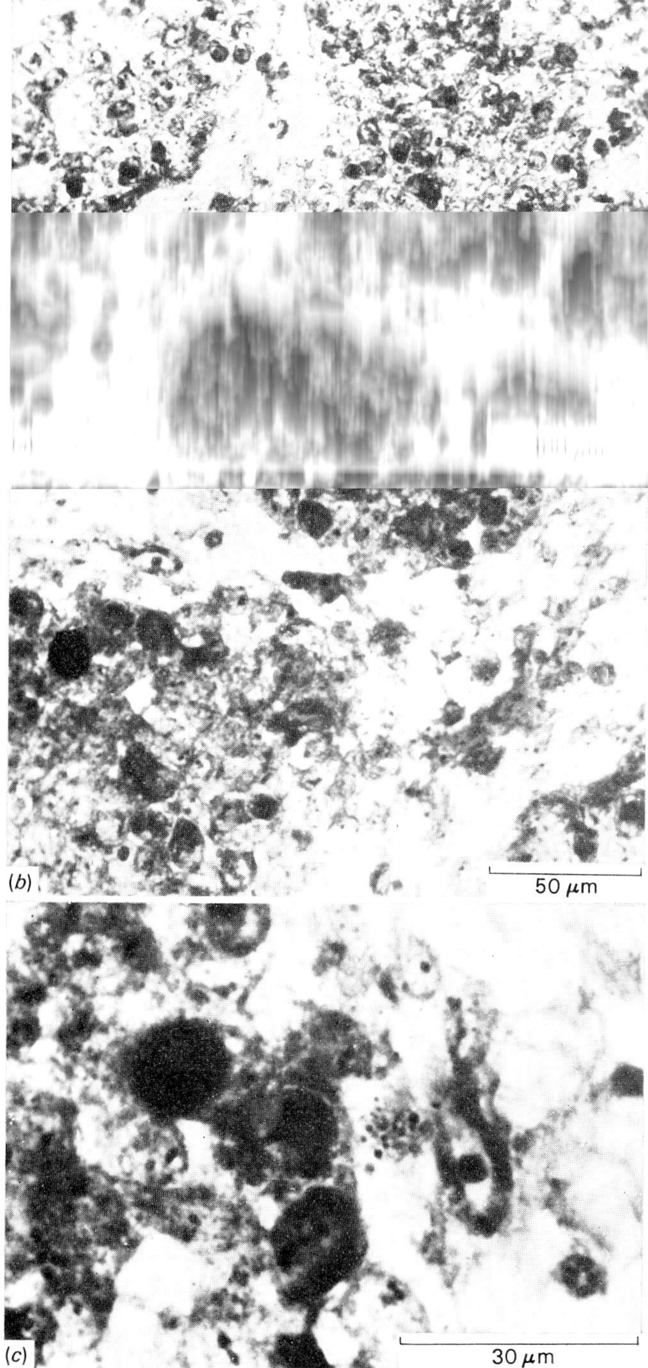

EFFECTS OF GENETIC ATHYMIA ON DEVELOPMENT

BY E. M. PANTELOURIS

Biology Department, University of Strathclyde, Glasgow G1 1XW

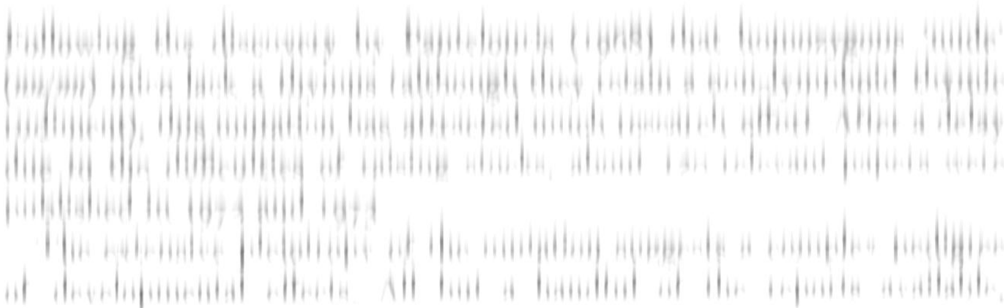

however, deal with the pathology of the lymphoid system and the resulting immunological dysfunctions. These aspects have been reviewed recently (Pantelouris, 1973) and will be left out of the present account, although they too are relevant to early mammalian development. I shall restrict myself to a much less investigated, and hence more speculative aspect, that of hormonal and hormone-related developmental phenomena that arise in the first instance from the absence of a thymus.

THE THYMUS AS AN ENDOCRINE GLAND

Roughly speaking, work appearing during most of the sixties suggested, and circumstantially demonstrated, that the thymus possessed hitherto unsuspected endocrine functions; research in the last few years has been concerned with the identification and properties of the hormone(s) involved.

Metcalf (1962) experimented with thymus grafts from two strains of mice, one (AKR) characterised by high and the other (C3H) by low mitotic rate in the lymphoid thymus cells. The starting point of such experiments at the time was the repair by thymus grafts of the wasting syndrome (runting) produced by neonatal thymectomy. The grafts were placed in hybrids (AKR × C3H) of the same two lines and, as expected, they were eventually repopulated by the recipient's lymphocytes. The mitotic rate, however, exhibited within the thymus transplants by these cells was that characteristic of the donor and not the host. It was thus concluded that the epithelium of the thymus exerts a stimulatory and presumably hormonal action on the lymphoid cells populating it. This hypothesis turned out to be true, although one would perhaps avoid today the use of AKR for this experiment, as this strain is known to suffer from lymphoid cell neoplasia of viral origin. Similar results were obtained by thymus transfers, within an inbred line, from young donors into old hosts.

If it is true that thymus transplants correct the wasting disease of neonatally thymectomised animals by the provision of humoral factors and not by the supply of cells, then such grafts should be effective even through cell-proof chambers. Experiments of this type were carried out by Levey, Trainin & Law (1963), Osoba & Miller (1964) and Defendi & Metcalf (1964). Even within Millipore filter diffusion chambers with pore size as low as 0.01 μm the grafts could prevent the wasting disease (Barclay, 1964).

Another line of evidence is being provided by light and electron microscopists who describe in thymus cells features characteristic of secretory functions (Clark, 1963; Kohnen & Weiss, 1964; Kostowiecki, 1966; Sanel, 1967; Vetters & Macadam, 1973).

The direct approach of preparing active extracts from thymic tissue, mainly calf, has been adopted by several groups of researchers. Their findings were eventually collated at a conference in 1971, from which a book edited by T. D. Luckey (1973) emerged. This contains detailed first-hand accounts and full references and, therefore, only a brief summary will be given in the following paragraphs.

A Rumanian group separated lipid and non-lipid fractions of calf thymus. Hydrolysis and filtration of the non-lipids led to a peptide preparation, TP, which appeared to forestall the effects of thymectomy by stimulating lymphopoiesis, erythropoiesis, glycogenesis and calcium and phosphorus deposition in bone. It is also claimed that it inhibits neoplastic growth.

The lipid, acetone-extracted material yielded three active fractions and a purified 'factor S'. When this was characterised as a steroid it was given the name 'thymosterin', and had effects mainly on the growth of tumour cells.

A protein fraction, homeostatic thymic hormone (HTH), with restitutive activity on thymectomised guinea pigs and chemotactic attraction for lymphocytes, was produced by German workers. Independently, two protein fractions (lymphocyte stimulating hormones, LSH_h and LSH_r) were also being produced in the USA.

The closest approximations to a well-defined thymic hormone, however, have been achieved by two other teams, who named their product 'thymic hormone' (TH) (Trainin & Linker-Israeli, 1967; Trainin & Small, 1970) or 'thymosin' and subsequently 'thymin' (Goldstein, 1968; Goldstein & Hofmann, 1969; White & Goldstein, 1970). Either preparation stimulates the proliferation and immunocompetence of lymphopoietic tissues. The latest, improved method of preparation of thymin from calf thymus has been described by Goldstein (1974).

Initial reports and the recent reviews of the work of the above groups are given below. For TP, Milcu & Potop (1973); Potop & Milcu (1973): for hypocalcemic factor, Ogata & Ito (1944); Mizutani (1973): for HTH, [illegible]

[illegible]

gestions, pursued since 1903 (see Greene, 1969), that there is a connection between thymic dysfunction and myasthenia. In fact, thymectomy has been used as a palliative measure in numerous cases. Experimentally, thymic extracts or grafts were shown to normalise the increased amplitude, characteristic of the disease, of miniature end plate potentials (MEPP). All the same, in Goldstein's view the above property is incidental to the main activity of the hormone: that is, stimulation of mitotic activity of lymphocytes.

It remains to be decided which type of lymphoid cell is the target for thymosin. When bone marrow derived lymphocytes are incubated with the hormone for a short period they acquire certain characteristics of 'thymus-derived T cells', becoming sensitive to antilymphocytic serum, azathioprine and anti-theta serum. The latter two treatments make T cells from animals immunised with sheep red cells unable to form in-vitro 'rosettes' with sheep erythrocytes. (Bach, Dardenne & Davies, 1971*a*; Bach *et al.*, 1971*b*.)

Normal mouse serum may also affect the rosette-forming ability of B cells in a fashion similar to azathioprine. Serum from thymectomised or ageing animals is much less effective – indications that the inhibition is due to a product of the thymus. Furthermore, in accordance with this postulate, sera of four week old nude mice were shown to have no effect (Bach & Dardenne, 1973).

In the view of Rotter, Globerson, Nakamura & Trainin (1973), the bone marrow cells on which thymosin acts are not B cells being transformed to T cells, but in fact T cells or T cell precursors, which do occur in very small numbers in the mouse bone marrow (see Cohen & Claman, 1971).

To summarise, there is no doubt that the epithelial part of the thymus produces a humoral factor which enhances the mitotic rate of lymphocytes and may even advance the differentiation of bone marrow B cells, or at least of T cell precursors, towards full T cell status.

There is evidence that the nude mouse which lacks differentiated T cells (see review by Pantelouris, 1973) is, as expected, also deficient in thymin (Bach & Dardenne, 1973).

The numerous activities claimed for various thymic extracts point to the need for further work in this field.

INTERACTIONS OF THYMUS AND OTHER ENDOCRINES IN DEVELOPMENT

The pituitary–thymus–adrenal axis

Cortisone and cortisol exert a dramatic effect on the weight of the thymus and lymph nodes (Selye, 1950; Weaver, 1955). The loss of thymic weight is widely attributed to lymphocytolysis, and antilymphocytic serum acts synergistically with cortisol (Gunn, Lance, Medawar & Nehlsen, 1970). Munck and coworkers have established by short-term incubation of thymus cells with cortisol *in vitro* that the hormone blocks, in the first instance, the formation from glucose of glucose - 6 - phosphate and, eventually, protein and RNA synthesis. The hormone is claimed to bind to cell nuclei (see Munck, Young, Mosher & Wira, 1971), but lymphoid and epithelial cells have not been separated in these experiments.

There is also some evidence that, as a result of glucorticoid action, thymic lymphocytes may leave the thymus and become sequestered in the bone marrow. This is deduced from the finding that under these conditions the marrow contains antibody-producing cells (Cohen & Claman, 1971). This might explain why the level of serum haemolysin is not necessarily reduced by these hormone treatments (Ferreira, Moreno & Hoecker, 1973).

In fact, this level is dependent on the timing of the hormonal injections. Should cortisone acetate be injected three to four days prior to the immunisation of mice with sheep red cells, the numbers of plaque-forming (i.e. haemolysin-producing) cells of both immunoglobulin classes (IgG and IgM) are depressed. Hormonal injection four days after immunisation affects mainly the IgG class (Elliott & Sinclair, 1968). The converse experiment, i.e. immunisation following adrenalectomy, was performed by Streng & Nathan (1973). The IgG plaque-forming cells were not changed but the IgM producing cells actually increased in numbers.

In contrast to adrenal hormones, hypophyseal factors exert an enhancing activity on the thymus. Hypophysectomy is generally followed by involution of the thymus, as was first shown by Benedict, Putnam & Teel (1930) working on dogs.

A mouse mutation, *dwarf* (*dw*) has provided a natural demonstration of the impact of the pituitary on the thymus during early post-natal development. Homozygous dwarf (*dw*/*dw*) mice suffer after puberty an involution of the thymus, leading eventually to wasting. This can be prevented by [illegible] somatotropin, thyrotropin and thyroxin, used singly or in combination (Pierpaoli, Baroni, Fabris & Sorkin, 1969).

Points still under discussion include whether the pituitary effect is attributable to prolactin rather than somatotropin (Shire, 1973), and whether it is exerted on the thymus directly or via the thyroid (Pierpaoli & Sorkin, 1967; 1968; Pierpaoli *et al.*, 1969).

Hormonal status of athymic animals

There is no doubt that the interrelations of thymus, adrenals and pituitary are complex and important in post-natal and even pre-natal development. The nude mouse, developing as it does without a thymus, would be expected to provide opportunities for unravelling these relations. Such possibilities remain a promise so far, and it becomes necessary to discover and describe, as a first step, the hormonal or hormone-dependent abnormalities that might be related to the athymic condition.

The first clue in this direction has been the report that the 'X zone' of the adrenal cortex (erroneously described at the time as the reticular zone) in nude mice is overdeveloped in comparison with normal (Pantelouris, 1968; Shire & Pantelouris, 1974). It is known that the adrenal cortex is very active in mammals at the perinatal stage and that this activity is dependent on the pituitary. In the rat, for example, the adrenal has as yet no concentric zones on the 16th foetal day, but there are well-formed 'reticular' and fascicular zones by the time of birth. These zones fail to differentiate if the foetuses are decapitated (as a means of hypophysectomising them) on days 18–20 (Kitchell & Wells, 1952; Cohen, 1963). Also strain differences in adrenal activity have been studied in the mouse (Shire & Spickett, 1968).

Despite the failure of the thymic rudiment to develop in the nude foetus, the thyroids and parathyroids – which share with the thymus a common origin from the third and fourth pharyngeal arches – appear to be differentiating normally (Pantelouris & Hair, 1970). Thus there is no sign of tetany in the newborn nude mouse, in contrast to babies with the related Di George syndrome in man; and in two month old nude mice I have found the serum calcium level to be within normal levels (9.97 ± 0.11 mg/100 ml in normal males and 9.95 ± 0.49 in nude males). It is true that in the runting nude mouse, deterioration of the thyroids can be detected histologically. Pierpaoli & Sorkin (1972), after describing this and other symptoms, propose that thyroid deficiency is characteristic of the nude mouse and speculate that this might be due to failure of some hypothetical perinatal stimulus emanating normally from the thymus.

Alternatively, these symptoms can plausibly be seen as reflections of the stress and the elevation of ACTH production associated with wasting disease (Shire & Pantelouris, 1974). Under specific pathogen-free conditions, nude mice have normal levels of plasma corticosterone and of testosterone (Ohsawa *et al.*, 1973).

The gonads of the nude mouse also provide some contributory evidence. Normally, nude mice are infertile and spermatid maturation is disturbed (Flanagan, 1966). The ovaries are deficient in corpora lutea (Shire & Pantelouris, 1974). The development of these ovaries is now being studied quantitatively by Dr Sue Lintern-Moore and myself. The differences between nude animals and their normal littermates in ovary size, stroma development and oocyte numbers become significant only in the second month of life, i.e. at about the time that periodic functioning should be beginning (Table 1). Fertility of both sexes is achieved by the implantation soon after birth of a normal thymus (Pantelouris, 1972).

Before any speculation is made about these facts it is necessary to take into account the results of raising specific pathogen-free nude mice. These grow more or less normally and are often fertile (King, 1973; Poiley, Ovejera, Reeder & Otis, 1973). So long as they are kept under such protected conditions, these animals also avoid the wasting disease, although it is not clear whether this is permanent or not.

In the present context, the information provides no support for the hypothesis that the nude syndrome can largely be attributed to a pituitary-dependent thyroid abnormality. The situation in the dwarf mouse is quite different, for thymic and bone marrow cells cannot prevent wasting unless pituitary hormones are also supplied (Fabris, Pierpaoli & Sorkin, 1971). The nude animal however can be 'restored' by thymus grafts without additional hormone injections.

Table 1. *Numbers of small and growing oocytes in the ovaries of nude and control mice**

Age	Nude			Controls		
	Small	Growing	No. mice	Small	Growing	No. mice
1 month[a]	7143 ± 1013	543 ± 46	5	5452 ± 1032	606 ± 117	4
[illegible]	[illegible]	[illegible]	[illegible]	[illegible]	[illegible]	[illegible]
[illegible]	[illegible]	[illegible]	[illegible]	[illegible]	[illegible]	[illegible]
[illegible]	[illegible]	[illegible]	[illegible]			

[illegible]

Other hormone-dependent characteristics examined in nude mice concern the submaxillary glands. In mice these exhibit normally a striking sexual dimorphism (Lacassagne, 1940; Lacassagne & Chamorro, 1940). The males have larger glands and the cells of acinar tubules are rich in granules and enlarged, hence almost filling the lumen. This histological difference between the sexes has been extended to the levels of nerve growth factor, epidermal growth factor and a variety of enzymes. In the nude mice the histological dimorphism is very subdued but becomes established in animals given thymus grafts (Shire & Pantelouris, 1974).

The submaxillary gland enzymes exhibiting sex differences include amylase. Activity is much higher in males, is reduced by castration and is testosterone propionate dependent (Swigart, Hilton, Dickie & Foster, 1965).

Nude mice also show this sex difference (Table 2) and this might be considered in conjunction with the finding by Ohsawa *et al.* (1973) of normal testosterone levels. As between genotypes, nude males appear to possess higher activity per gram of tissue. If, however, the total content of amylase activity is calculated for the whole gland, the difference disappears.

It was recently reported that nude mice have high glutamic and tyrosine decarboxylase activities in the liver compared with normal. Ontogenetically this arises from a common high level post-natally which becomes reduced later in the normal, but not in the nude, genotype. Neonatal thymus grafts 'normalise' the nude animals in this respect (Pantelouris & MacMenamin, 1973). A study by Davies (1963) may be relevant here. He finds that cortisone raises the level of rat liver tyrosine decarboxylase activity (by facilitating its saturation with coenzyme), whilst thyroxine reduces it.

Table 2. *Amylase activity in serum and in homogenate of the submaxillary gland of 4-month-old normal and nude (ex SPF) mice. (UI* units)*

	Normal ♂	Normal ♀	Nude ♂	Nude ♀
Serum UI ml^{-1}	1.76 ± 0.26	1.71 ± 0.22	1.96 ± 0.60	1.19 ± 0.25
Submaxillary UI g^{-1}	4.28 ± 1.97	2.82 ± 0.02	8.208 ± 3.23	3.87 ± 1.96
Submaxillary total	9.38 ± 4.64	5.36 ± 2.52	8.19 ± 4.98	2.14 ± 1.30

* One UI unit corresponds to a substrate conversion of 1 μmol/min/l.

I have now made comparisons of decarboxylase activity in the submaxillaries. This tissue has even higher activities than the liver, and again the nude animals exceed the normal (Table 3). Removal of the submaxillaries twenty days before the determination does not lower enzymatic activity in the liver of normal mice.

Another enzyme examined is liver histidase. In the normal animal, activity is higher in the mature female than in the young. It is well established that this is greatly enhanced by oestrogen and may be restrained by ovariectomy or by testicular, adrenal and even some hypophyseal hormones (Feigelson, 1968, 1971).

At the age of six weeks, and later at ten weeks, I find that nude females, conventionally raised, do not show the normal rise in histidase activity and in fact even score below the males (Pantelouris, unpublished data; (Table 4)).

This type of enzymatic information requires amplification. Since what is measured may be the net effect of complex hormonal inducing and reducing factors, data of this type cannot form the basis of exact conclusions regarding hormonal status. As straws in the wind, however, they suggest that steroid levels, particularly adrenocortical but also (in the female) gonadal, deviate from normal in early post-natal stages of the athymic animal.

Table 3. *Some measurements of amino acid decarboxylase activity in normal and nude mice aged 11.5 weeks.* (μl CO_2 *evolved/g of tissue/5 min. Substrate: equimolar solution of* L-*glutamic acid and* L-*tyrosine; see Pantelouris & MacMenamin, 1973*)

	Liver	Liver 21 days after removal of submaxillaries	Submaxillary
Males, normal (+/?)	374.2 ± 72.85	337.2 ± 77.54	619.5 ± 20.27
Females, normal (+/?)			662.5 ± 153.4
Males, nude (*nu/nu*)			881 ± 166.8
Females, nude (*nu/nu*)			929.4 ± 139.1

Table 4. *Measurements of liver histidase activity in normal and nude mice*

	5 weeks	7 weeks	10 weeks
Normal (+/?) males	13.08 ± 1.40	10.30 ± 1.22	12.21 ± 2.35
Normal (+/?) females	22.10 ± 2.99	12.41 ± 2.73	23.92 ± 6.90
Nude (*nu/nu*) males	13.12 ± 2.73	14.22 ± 2.42	16.83 ± 8.90
[illegible]	[illegible]	[illegible]	[illegible]

[illegible]

[illegible]

(1). Immunological dysfunctions of the nude mouse are consequential on the failure of the thymic rudiment to develop and produce thymin, the latter being an indispensable differentiation factor for T cells.

As has been demonstrated, the failure of thymic development is primary and not simply a case of premature involution (Pantelouris & Hair, 1970). We now suspect that the failure can be traced even further back into foetal life than was shown in our previous report. Whole 'litters' of heterozygous matings were dissected out of the uterus as early as 10–11 days from conception and serially sectioned. At this stage a confluence is already established between branchial cleft ectoderm and pharyngeal pouch endoderm and mitotic activity in this area of confluence was higher than in the surrounding tissue (Plate 1*a*). Whilst this was the picture presented by fifteen out of nineteen foetuses of this stage, the situation was different in the remaining four. In these, mitotic activity in the same area was very low and there was pycnosis of nuclei and destruction of cells (Plate 1*b*). In two out of these four foetuses, cell degeneration was indicated by an acellular region surrounded by ectodermal cells.

Unfortunately it is not possible at this stage to identify nude and normal foetuses. We suggest that the minority of foetuses with pycnotic thymic anlage might well be the *nu/nu* individuals. If this speculation proves correct, then the pycnosis and cytolysis described must be the earliest observable abnormality leading to the failure of normal thymus differentiation. Perhaps as a result there is a failure of induction by the ectoderm, and the endoderm develops abnormally and in the end produces the non-functional cysts of the thymic rudiment.

Contributory evidence in support of the theory of epithelial cytolysis in the thymic anlage is provided by the electron microscopical study of the nude thymic rudiment by Cordier (1974*a*, *b*). The cysts of this rudiment comprise endodermal elements that have differentiated, or rather dedifferentiated abnormally. They retain, for example, the tendency to form cilia whilst ciliogenesis is no longer occurring in normal animals at similar post-weaning stages. At the same time, epithelial cells are missing.

(2). The nude syndrome can largely be avoided by neonatal thymus grafts or even by specific pathogen-free conditions. There are, however, some qualifications to this generalisation.

In particular, the fertility of the male is restored much more fully than that of the female. Furthermore, even with neonatal thymus grafts, any fertile nude females fail to lactate and to raise their litters (Pantelouris, 1972).

This suggests that there are certain developmental abnormalities that have become irreversible by the time the thymus grafts become effective. Even where grafts are made on the day of birth, it is doubtful whether they are functional before they become vascularised, i.e. after about a week or so. Thus abnormalities that are irreversibly determined prenatally or in the first few days after birth may escape correction.

(3). It is known that a transient androgenic influence at the perinatal stage will frustrate the establishment of the normal cyclic female pattern of sexual function. The first observation was made by Pfeiffer (1936) using testicular grafts in the rat. Subsequent experiments in this field are numerous (see van der Werff ten Bosch, Tuinebreijer & Vreeburg, 1971). The current theory holds that perinatal androgenic episodes act at the level of the hypothalamus to prevent its programming for cyclic (female) gonadotropin function.

Based on the above three lines of argument, I suggest that there is in the nude mouse such a perinatal androgenic excess. This might be of adrenocortical origin. It is in fact well known that perinatally there is substantial corticosterone activity in the normal animal (Davies, 1963); and the greater development and erratic involution of the X zone in nude mice suggest overactivity at the critical period, perhaps, in view of the antagonism between thymus and adrenals, correlated to athymia. Alternatively or concomitantly, the androgenic influence originates under athymic conditions, from the ovary itself. Nishizuka & Sakakura (1971) present reasons for suggesting some androgen production by the ovary interstitial cells, which become hyperplastic following neonatal thymectomy.

An effect such as that postulated here would explain the failure of ovulation in the nude female and of periodic gonadotropic function generally. Lactation may further be made impossible by another irreversible developmental anomaly, namely that of the skin. The pathology of the nude skin attracted the attention of Flanagan (1966) and lately of others (Pantelouris, 1974). It would appear reasonable to suggest that the differentiation of mammary glands is also inhibited, and in a way that cannot be reversed by later treatments.

This working hypothesis may or may not be disproved by the work [illegible]

[illegible]

[illegible]

BACH, J. F., DARDENNE, M. & DAVIES, A. J. S. (1971*a*). Early effect of adult thymectomy. *Nature New Biology*, **231**, 110–11.

BACH, J. F., DARDENNE, M., GOLDSTEIN, A. L., GUHA, A. & WHITE, A. (1971*b*). Appearance of T-cell markers in bone marrow rosette-forming cells after incubation with thymosin, a thymic hormone. *Proceedings of the National Academy of Sciences, USA*, **68**, 2734–7.

BARCLAY, T. J. (1964). Discussion of thymic function. In *The Thymus*, ed. V. Defendi & D. Metcalf, pp. 117–19. Philadelphia: Wistar Institute.

BENEDICT, E. B., PUTNAM, T. J. & TEEL, H. M. (1930). Early changes produced in dogs by injections of sterile active extract from anterior lobe of hypophysis. *American Journal of Medical Science*, **179**, 489–97.

BERNARDI, G. & COMSA, J. (1965). Purification chromatographique d'une préparation de thymus douée d'activité hormonale. *Experientia*, **21**, 416–17.

BERZSONOFF, N. A. & COMSA, J. (1958). Préparation d'un extrait purifié de thymus: Application à l'urine humaine. *Annales d'Endocrinologie*, **19**, 222–7.

CLARK, S. L. (1963). The thymus in mice of strain 129/J studied with the electron microscope. *American Journal of Anatomy*, **112**, 1–33.

COHEN, A. (1963). Correlations entre l'hypophyse et le cortex surrénal chez le foetus de rat. *Archive d'Anatomie microscopique et de Morphologie expérimentale*, **52**, 279–407.

COHEN, J. J. & CLAMAN, H. N. (1971). Thymus–marrow immunocompetence. I. Hydrocortisone-resistant cells and processes in the haemolytic antibody response of mice. *Journal of Experimental Medicine*, **133**, 1026–34.

COMSA, J. (1973). Thymus replacement and HTH, the homeostatic thymic hormone. In *Thymic Hormones*, ed. T. D. Luckey, pp. 39–58. Baltimore: University Park Press.

CORDIER, A. C. (1974*a*). Ultrastructure of the thymus in 'nude' mice. *Journal of Ultrastructural Research*, **47**, 26–40.

CORDIER, A. C. (1974*b*). Ciliogenesis and ciliary anomalies in thymic cysts of nude mice. *Cell Tissue Research*, **148**, 397–406.

DAVIES, V. E. (1963). Effect of cortisone and thyroxine on aromatic aminoacid decarboxylation. *Endocrinology*, **72**, 33–8.

DEFENDI, V. & METCALF, D. (Eds.) (1964). *The Thymus*. Philadelphia: Wistar Institute.

DUQUESNOY, R. J. & GOOD, R. A. (1970). Prevention of immunological deficiency in pituitary dwarf mice by prolonged nursing. *Journal of Immunology*, **104**, 1553–5.

ELLIOTT, E. V. & SINCLAIR, N. R. St C. (1968). Effect of cortisone acetate on 19S and 7S haemolysin antibody. A time course study. *Immunology*, **15**, 643–52.

FABRIS, N., PIERPAOLI, W. & SORKIN, E. (1971). Hormones and the immunological capacity. III. The immunodeficiency disease of the hypopituitarism Snell–Bagg dwarf mouse. *Clinical Experimental Immunology*, **9**, 209–25.

FEIGELSON, M. (1968). Estrogenic regulation of hepatic histidase during postnatal development and adulthood. *Journal of Biological Chemistry*, **243**, 5088–93.

FEIGELSON, M. (1971). Hypophyseal regulation of hepatic histidase during postnatal development and adulthood. I. Pituitary suppression of histidase activity. *Biochimica et Biophysica Acta*, **230**, 296–308.

FERREIRA, A., MORENO, C. & HOECKER, G. (1973). Lack of correlation between the effect of cortisone on mouse spleen plaque-forming cells and circulating antisheep red blood cell haemolysins. *Immunology*, **24**, 607–16.

FLANAGAN, S. P. (1966). Nude, a new hairless gene with pleiotropic effects in the mouse. *Genetical Research*, **8**, 295–309.

GOLDSTEIN, A. L., SLATER, F. D. & WHITE, A. (1966). Preparation assay and partial purification of a thymus lymphocytopoietic factor (thymosin). *Proceedings of the National Academy of Sciences, USA*, **56**, 1010–17.

GOLDSTEIN, G. (1968). The thymus and neuromuscular function. A substance in thymus which causes myositis and myasthenic neuromuscular block in guinea pigs. *Lancet*, **ii**, 119–32.

GOLDSTEIN, G. (1974). Isolation of bovine thymin: a polypeptide hormone of the thymus. *Nature, London*, **247**, 11–14.

GOLDSTEIN, G. & HOFFMANN, W. W. (1969). Endocrine function of the thymus affecting neuromuscular transmission. *Clinical Experimental Immunology*, **4**, 181–9.

GREENE, R. (Ed.) (1969). *Myasthenia gravis*. London: Heinemann.

GUNN, A., LANCE, E. M., MEDAWAR, P. B. & NEHLSEN, S. L. (1970). Synergism between cortisol and antilymphocyte serum. In *Hormones and the Immune Response*, ed. G. E. W. Wolstenholme & J. Knight, pp. 66–99. London: Churchill.

HAND, T., CASTER, P. & LUCKEY, T. D. (1967). Isolation of a thymus hormone, LSH. *Biochemical and Biophysical Research Communications*, **26**, 18–23.

KING, D. (1973). Large scale production of nude mice. *Proceedings of the 1st International Workshop on Nude Mice* (Aarhus, 1973), ed. J. Ryqaard & C. O. Poulsen, pp. 221–6. Stuttgart: Gustav Fischer Verlag.

KITCHELL, R. L. & WELLS, L. J. (1952). Functioning of the hypophysis and adrenals in fetal rats. *Anatomical Record*, **112**, 561–91.

KOHNEN, P. & WEISS, L. (1964). An electron microscopic study of thymic corpuscles in the guinea pig and the mouse. *Anatomical Record*, **148**, 29–57.

KOSTOWIECKI, M. (1966). Secretory component in the thymus of the pregnant white rat. *Zeitschrift für mikroskopische anatomische Forschung*, **76**, 141–83.

LACASSAGNE, A. (1940). Dimorphisme sexuel de la glande sous-maxillaire chez la souris. *Comptes rendus des Séances de la Société de Biologie*, **133**, 180–1.

LACASSAGNE, A. & CHAMORRO, A. (1940). Réaction à la testostérone de la glande sous-maxillaire, atrophiée consécutivement à l'hypophysectomie chez la souris. *Comptes rendus des Séances de la Société de Biologie*, **134**, 223–4.

LEVEY, R. H., TRAININ, N. & LAW, L. W. (1963). Evidence for function of thymic tissue in diffusion chambers implanted in neonatally thymectomized mice. *Journal of the National Cancer Institute*, **31**, 199–217.

LUCKEY, T. D. (Ed.) (1973). *Thymic Hormones*. Baltimore: University Park Press.

LUCKEY, T. D., ROBEY, W. G. & CAMPBELL, B. J. (1973). LSH, a lymphocyte stimulating hormone. In *Thymic Hormones*, ed. T. D. Luckey, pp. 167–84. Baltimore: University Park Press.

METCALF, D. J. (1962). Leukemogenesis in AKR mice. In *CIBA Symposium: Tumour Viruses of Murine Origin*, ed. G. E. W. Wolstenholme & M. O'Connor, pp. 233–61. London: Churchill.

MILCU, S. M. & POTOP, J. (1973). Biologic activity of thymic protein extracts. In *Thymic Hormones*, ed. T. D. Luckey, pp. 97–134. Baltimore: University Park Press.

[illegible]

ceutical Society of Japan, **64,** 332–7.

OHSAWA, N., MATSUZAKI, F., ESAKI, K., TAMAOKI, N. & NOMURA, T. (1973). Endocrine functions of the nude mouse. *Proceedings of the 1st International Workshop on Nude Mice* (Aarhus, 1973), ed. J. Rygaard & C. O. Poulsen, pp. 221–6. Stuttgart: Gustav Fischer Verlag.

OSOBA, D. & MILLER, J. F. A. P. (1964). The lymphoid tissues and immune responses of neonatally thymectomised mice bearing thymus tissues in millipore diffusion chambers. *Journal of Experimental Medicine*, **119,** 177–94.

PANTELOURIS, E. M. (1968). Absence of thymus in a mouse mutant. *Nature, London*, **217,** 370–1.

PANTELOURIS, E. M. (1971). Some hormonal effects on the ontogenesis of enzyme and protein patterns. In *Hormones in Development* (Nottingham Conference, 1968), ed. M. Hamburgh & E. J. W. Barrington, pp. 41–8. New York: Meredith Corporation.

PANTELOURIS, E. M. (1972). Thymic induction and ageing: a hypothesis. *Experimental Gerontology*, **17,** 73–81.

PANTELOURIS, E. M. (1973). Athymic development in the mouse. *Differentiation*, **1,** 437–450.

PANTELOURIS, E. M. (1974). Common parameters of genetic athymia and senescence. *Experimental Gerontology*, **9,** 161–71.

PANTELOURIS, E. M. & HAIR, J. (1970). Thymus dysgenesis in nude (nu nu) mice. *Journal of Embryology and Experimental Morphology*, **24,** 615–23.

PANTELOURIS, E. M. & MACMENAMIN, P. N. (1973). Amino acid decarboxylase activity in nude mouse liver. *Comparative Biochemistry and Physiology*, **45,** 967–70.

PFEIFFER, C. A. (1936). Sexual differences of the hypophyses and their determination by the gonads. *American Journal of Anatomy*, **58,** 195–224.

PIERPAOLI, W. & SORKIN, E. (1967). Relationship between thymus and hypophysis. *Nature, London*, **215,** 834–7.

PIERPAOLI, W. & SORKIN, E. (1968). Hormones and the immunological capacity. I. Effect of heterologous anti-growth hormone antiserum on thymus and peripheral lymphatic tissue in mice. Induction of a wasting syndrome. *Journal of Immunology*, **101,** 1036–43.

PIERPAOLI, W. & SORKIN, E. (1972). Alterations of adrenal cortex and thyroid in mice with congenital absence of the thymus. *Nature New Biology*, **238**, 282–5.

Pierpaoli, W., Baroni, C., Fabris, N. & Sorkin, E. (1969). Hormones and the immunological capacity. II. Reconstitution of antibody production in hormonally deficient mice by somatotropic hormone, thyrotropic hormone and thyroxin. *Immunology*, **16**, 217–30.

Pitot, H. C. & Yatvin, M. B. (1973). Interrelationships of mammalian hormones and enzyme levels *in vivo*. *Physiological Reviews*, **53**, 228–325.

Poiley, S. M., Ovejera, A., Reeder, C. R. & Otis, A. P. (1973). Reproductive behaviour of athymic nude mice in a variety of environments. *In Proceedings of the 1st International Workshop on Nude Mice* (Aarhus 1973), ed. J. Rygaard & C. O. Poulsen, pp. 189–202. Stuttgart: Gustav Fischer Verlag.

Potop, I. & Milcu, S. M. (1973). Isolation, biologic activity and structure of thymic lipids and thymosterin. In *Thymic Hormones*, ed. T. D. Luckey, pp. 205–74. Baltimore: University Park Press.

Rotter, V., Globerson, A., Nakamura, I. & Trainin, N. (1973). Studies on characterisation of the lymphoid target cell for activity of a thymus humoral factor. *Journal of Experimental Medicine*, **138**, 130–42.

Sanel, F. T. (1967). Ultrastructure of differentiating cells during thymus histogenesis. *Zeitschrift für Zellforschung*, **83**, 8–29.

Selye, H. (1950). *Stress*. Montreal: Acta Inc. Medical Publishers.

Shire, J. G. M. (1973). Growth hormone and premature ageing. *Nature, London*, **245**, 215–16.

Shire, J. G. M. & Pantelouris, E. M. (1974). Comparison of endocrine function in normal and genetically athymic mice. *Journal of Comparative Biochemistry and Physiology*, **47A**, 93–100.

Shire, J. G. M. & Spickett, S. G. (1968). Genetic variation in adrenal structure: strain differences in quantitative characters. *Journal of Endocrinology*, **40**, 215–219.

Streng, C. B. & Nathan, P. (1973). The immune response in steroid deficient mice. *Immunology*, **24**, 559–65.

Swigart, R. H., Hilton, F. K., Dickie, M. M. & Foster, B. J. (1965). Effect of gonadal hormones on submandibular gland amylase activity in male and female (S7B1/6) mice. *Endocrinology*, **76**, 776–9.

Trainin, N. (1974). Thymic hormones and the immune response. *Physiological Reviews*, **54**, 272–315.

Trainin, N., Bejerano, A., Strahilevitch, M., Goldring, D. & Small, M. (1966). A thymic factor preventing wasting and influencing lymphopoiesis in mice. *Israel Journal of Medical Science*, **2**, 549–59.

Trainin, N. & Linker-Israeli, M. (1967). Restoration of immunologic reactivity of thymectomised mice by calf thymus extract. *Cancer Research*, **27**, 308–13.

Trainin, N. & Small, M. (1970). Studies on some physicochemical properties of a thymus hormonal factor conferring immunocompetence on lymphoid cells. *Journal of Experimental Medicine*, **132**, 885–97.

Trainin, N., Small, M. & Kimhi, Y. (1973). Characteristics of a thymic humoral factor involved in the development of cell-mediated immune competence. In *Thymic Hormones*, ed. T. D. Luckey, pp. 135–58. Baltimore: University Park Press.

Weaver, J. A. (1955). Changes induced in the thymus and lymph nodes of the rat by the administration of cortisone and sex hormones and by other procedures. *Journal of Pathology and Bacteriology*, **69**, 133–9.

van der Werff ten Bosch, J. J., Tuinebreijer, W. E. & Vreeburg, J. Th. (1971).

PLATE I

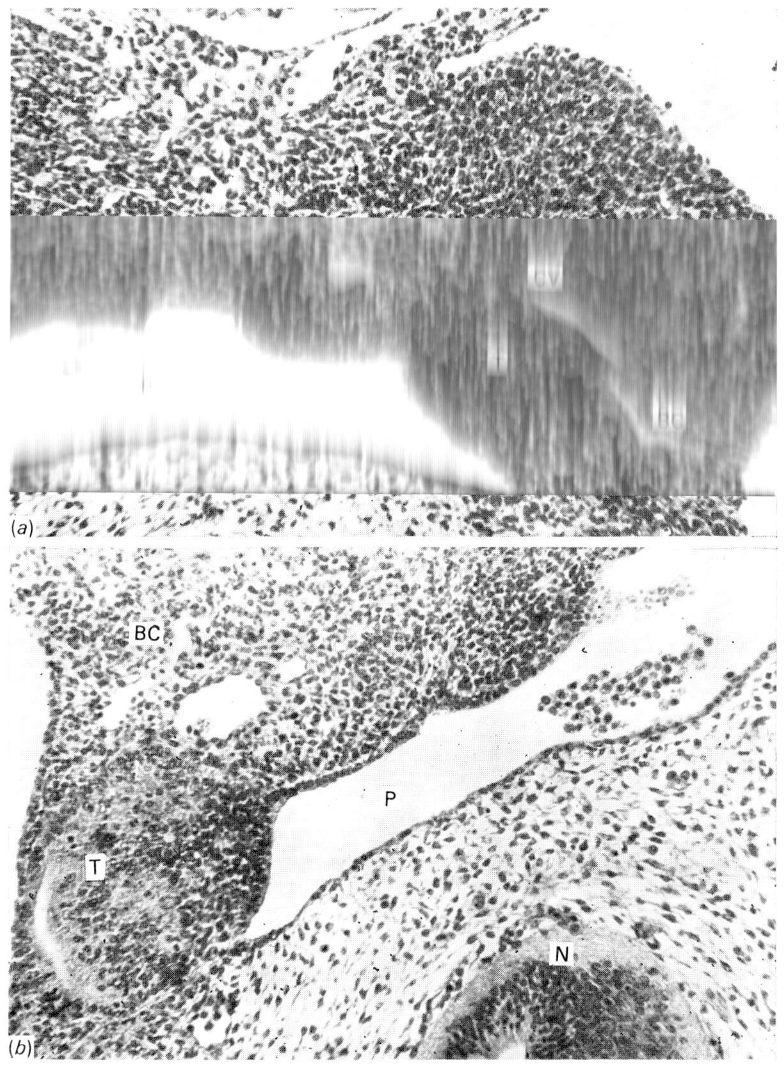

(Facing p. 386)

The incomplete or delayed early-androgen syndrome. In *Hormones in Development*, ed. M. Hamburgh & E. J. W. Barrington, pp. 669–76. New York: Meredith Corporation.

White, A. & Goldstein, A. L. (1970). The role of the thymus gland in host resistance. In *Control Processes in Multicellular Organisms*, ed. G. E. W. Wolstenholme & J. Knight, pp. 210–37. London: Churchill.

Vetters, J. M. & Macadam, R. F. (1973). Fine structural evidence for hormone secretion by the human thymus. *Journal of Clinical Pathology*, **26**, 194–7.

[illegible]

[illegible]

(b) Approximately transverse section through the pharyngeal region of a 10–11 day presumed nude foetus. Note the pycnotic nuclei and the disintegrated cells in the thymic anlage, T. BC, branchial cleft; N, neural tube; P, pharynx.

AMNIOTIC BANDS?

BY J. McKENZIE

Departmental of Developmental Biology, Marischal College, University of Aberdeen, Aberdeen AB9 1AS

[illegible] aspects of developmental biology, although apparently irrelevant, may nevertheless help to explain the problems of congenital abnormality.

One group in particular, congenital or intra-uterine amputations or ring constrictions of the fingers, toes, arms or legs may be an illustration of this point (Plate 1*a*, *b* and *c*). These anomalies are often, but by no means invariably, associated with amniotic bands, the name given to strands of tissue running between the constrictions or amputation stumps and the inner surface of the amnion. However, the presence of these tissue strands has fascinated and, I believe, misled the clinician with regard to the etiology of the constrictions and amputations. I admit to having paid little or no attention to amniotic bands and their associated abnormalities, being quite satisfied to acknowledge their existence and accept the current explanations. It was a specimen of a 12–14-week-old foetus (Plate 2*a* and *b*) which stimulated me to examine the subject more closely. The amniotic band in this case was not attached to any constriction or amputation but to the head of the foetus above the left eye and to the region of the nose. The cranium consisted of a large flabby fluid-filled sac containing degenerate brain tissue. In the left frontoparietal region the bones were missing; elsewhere they had collapsed inwards. The true median hare lip had arisen from defective growth of the median nasal process which, with the remainder of the frontonasal process, was represented by a solid slab of bone. This specimen is by no means unique; Ballantyne (1904), for instance, illustrated an almost identical case – even to the extent of having a median hare lip. Others have been reported with amniotic bands attached to different parts of the face and with abnormal clefts around the mouth.

If there had been no amniotic band in this specimen, I would have been content to suggest some vascular accident or severe necrosis affecting the left side of the skull and the frontonasal process and to let the matter rest there. But the amniotic band intrigued and encouraged me to examine the literature describing such bands and dealing with their etiology. There has certainly been no lack of speculation on this topic. Lennon (1947), for instance, listed ten theories to explain the origin of these attachments of the foetus to the amnion and nearly all of them focussed attention on the amnion as the cause. They included incomplete cavitation of the amniotic cavity leaving strands to bridge the space between amnion and embryo; adhesions of the amnion to the early embryo occurring, persisting and stretching to form anchoring bands; rupture of the amnion with adhesion of the torn edges to the foetus and if the extremities became trapped or caught by the edges of the tear in the amnion, constriction or even amputation of the foetal tissues. The most elaborate theory (Torpin, 1968) visualised rupture of the amniotic membrane with escape of the amniotic fluid and the foetus into the amniochorionic space where strands of fibrous tissue developed, trapping and entangling the foetus, constricting and even amputating digits or limbs. In those unusual cases in which amniotic bands were found attached to the region of the mouth and associated with abnormal clefts, Torpin suggested that the foetus had swallowed the end of a strand which, when drawn tight, cut into the tissues around the mouth to cause the clefts. These are the popular views today on the origin and role of amniotic bands, at least in clinico-pathological circles. Only in embryological texts do Streeter's (1930) views receive more than passing mention:

> Structures that have been mistaken for amniotic bands apparently fall into two categories: (*a*) macerated sheets of epidermis; (*b*) strands of hyalinised fibrous tissue which are the residue of localised areas of defective tissue . . .
>
> In these cases sharply circumscribed areas of limb-bud tissue are of such inferior quality that only imperfect histogenesis occurs. Whether injured in some way or defective from the outset, and the latter is probably true, these areas maintain themselves only in the earlier weeks of pregnancy. Already at the fourteenth week they are found as fibrous moribund masses, surviving after the manner of tissue cultures and sloughing away from the adjoining normal tissues. In older fetuses the signs of disintegrating areas diminish and the normal tissues adjust themselves to the defects, the ectoderm closing in over the raw surfaces. By the time of birth . . . one finds traces of the damage in the form of depressions, grooves or healed stumps, occasionally with slender

residual strands of defective hyalinised material still adhering to the affected regions. It is those that have been mistaken for amniotic bands . . .

The vague nature of the tissue defect or devitalisation which he proposed as the initial lesion but failed to specify, may account for the lack of enthusiasm for Streeter's views. Perhaps it was his misfortune that cell death [illegible] shape the form of organs, by the removal of superfluous cells or by the preparation of a dedifferentiated blastema in histio- and phylogenesis'. Saunders & Fallon (1966) investigated cell degeneration in the limbs of the chick embryo and compared the pattern of its distribution in the chick with that in the duck and mouse. They found evidence for the genetically programmed nature of the cell death. Zones of programmed cell necrosis which they identified in the wings and legs of the chick were thought responsible for modelling these limbs; beginning at the base of the early limb bud, the necrotic zones 'sweep the length of the limb, appearing last in the soft tissues between the digits as their outlines are sculptured'. Attention was drawn to the greater number of dead and dying cells in the interdigital regions of the developing chick foot compared with that of the duck (Fig. 1*a*) and the authors suggested that soft tissue syndactyly arises from failure or derangement of the programme which normally leads to cell necrosis in these positions.

In the mouse, Saunders & Fallon (1966) could find no zones of intensive cell necrosis in conjunction with the formation of proximal contours in the limbs, although abundant cell death could be identified prior to and during the emergence of the digits. After investigating the distribution of areas of cell death in human limbs, Menkes, Deleanu & Ilies (1965) illustrated a similar pattern to that found in the mouse, viz. clearly defined areas in the interdigital regions and a hint that cell death may also be present along the ulnar side of the little finger and the radial side of the thumb (Fig. 1*b*). Both Saunders & Fallon and Menkes *et al.* used Nile Blue vital staining techniques to demonstrate the activity of the macrophages ingesting the necrotic material rather than the necrotic cells themselves.

A series of fixed human embryos of different ages, sectioned at 10 μm

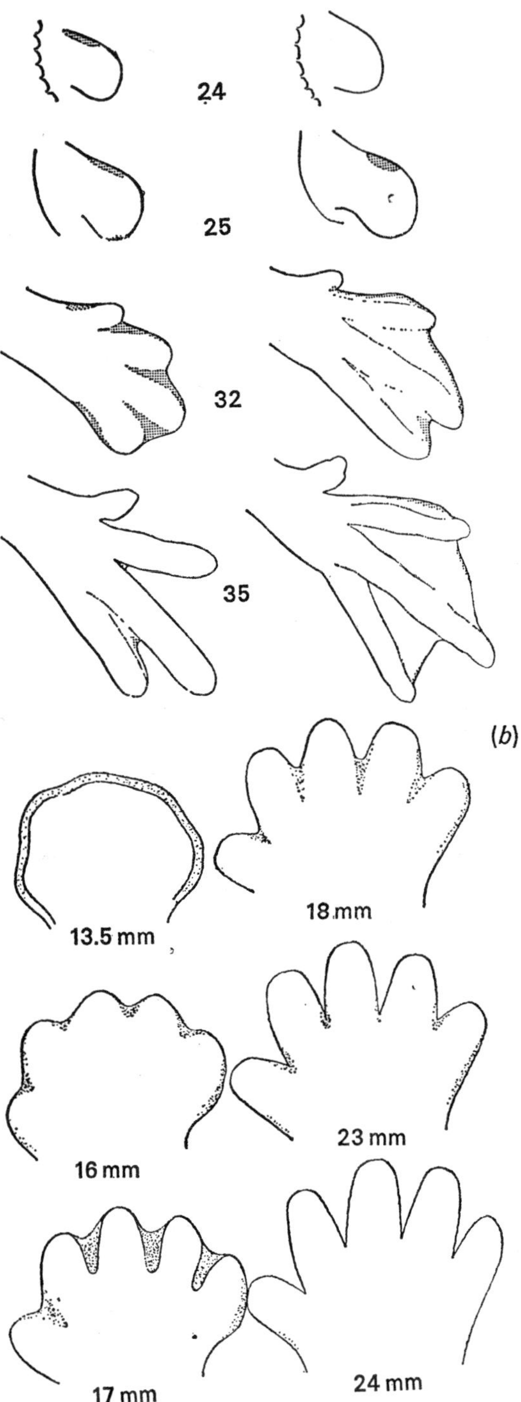

Fig. 1. (*a*) Necrotic zones at different developmental stages of the hind limbs of chick (left) and duck (right). (From Saunders & Fallon, 1966.) (*b*) Areas of cell necrosis (identified by Nile Blue staining) at different stages in the development of the human hand. (From Menkes, Deleanu & Ilies, 1965.)

and stained with haematoxylin and eosin, have now been examined, confirming and extending the findings of Menkes and his co-workers. Embryos of 7.5 mm, 9 mm and 12 mm showed no superficial areas of necrosis comparable with those in the chick limb, but a distinct collection of pycnotic material was seen alongside the main blood vessel in the root of the limb. This feature cannot be related to sculpturing of the limb but the close association of necrotic material with blood vessels was noticeable [illegible] for the flexor muscles of the fingers, related perhaps to the later subdivision of these muscles into a deep and superficial group. The early cartilaginous models of the forearm bones were well-defined at this stage.

A 17 mm and an 18 mm embryo were examined next but the smaller embryo was more advanced in its development than the larger one; for instance the sculpturing of the fingers and the removal of the interdigital tissues were only beginning in the 18 mm embryo, an excellent stage for studying and illustrating zones of cell death in this region (Plate 3*a* and *b*). Pycnotic nuclei are profuse in the 10 μm section illustrated and since this is only one of a series of sections all showing this feature to the same degree, then the number of cells undergoing necrosis must be very large indeed. Whereas at the 14 mm stage of development the necrotic cells were found alongside the marginal blood vessel, between it and the surface ectoderm, at the 18 mm stage they encircled the vessel and conceivably could damage it with resulting haemorrhage. One feature which was not emphasised in previous reports on the distribution of necrotic cells in the human hand is the extension of the necrotic zone along the radial side of the thumb to overlie the wrist and lower end of the radius (Plate 4*a* and *b*). In the 17 mm embryo, the interdigital clefts were well-defined, the concentration of necrotic material was diminished compared with the earlier stage, but the extension of the necrotic zone onto the wrist and forearm was still clear, even encroaching onto the front of the wrist.

In the 20 mm embryo, the amount of pycnotic material was much less; obviously the peak of the cell death programme was past.

Genetic control over the programme of cell death in normal development is obvious from a comparison of the distribution of the necrotic zones in the wing of the chick and the forelimb of the mammal, i.e. they are present in early development at the root of the chick limb but absent at the corresponding stage and from the corresponding position in the limb of mouse and man; even more convincing are the different degrees of necrosis in the feet of the chick and duck.

However, Menkes & Deleanu (1964), Deleanu (1965) and Saunders & Fallon (1966) have shown that the injection of Janus Green into the amniotic cavity of the chick at 6½ days incubation inhibits interdigital cell death in the foot and results in soft tissue syndactyly. If we are to accept that syndactyly of this type may arise from an inhibition or failure (due to genetic or environmental factors) of the programme of cell death in these regions, it is but a short step to a consideration of the effects of exaggerated cell death in these same regions.

In fact, it is easy to visualise the necrotic zones in the interdigital spaces extending until they encircle the digits at one or more levels. The necrosis, of course, affects only the superficial tissues but the final result will depend on the severity of the damage. A minimal degree will give a superficial scar with little or even no constricting effect on the circulation to and from the rest of the digit. Moderate tissue damage and its subsequent scarring will interrupt the venous flow from the finger or toe giving oedematous swelling of the tissues, while maximal damage may constrict or destroy the arterial supply with death of the digit. The stump is healed by the time of birth. Exudate, haemorrhagic or otherwise, from the necrotic tissues will escape into the amniotic fluid; if not completely absorbed the exudate will be organised (and vascularised) into fibrous 'amniotic' strands attached to the constriction or amputation stump and sometimes to the amnion. Alternatively the necrotic tissue of the finger or toe may itself attach to the amnion and eventually draw out an 'amniotic' band during the healing process. Ring constrictions and amputations at the level of the wrist or forearm may also be explained in this way because, as we have seen, the necrotic zone normally extends on to the forearm and could also show exaggeration of the cell death programme.

Two important points arise from the association of programmed cell death in the interdigital zones of the limbs with the abnormalities found there. In the first place, this is unlikely to be the only site in the body where abnormalities may arise from exaggeration or inhibition of normal cell death. Glucksmann (1951) and others have recorded cell death occurring normally in many regions and we must consider the possibility of variations in the intensity of this phenomenon being the basic mechanism for the development of those congenital abnormalities which are usually regarded as hereditary. Lending support to this view are the comparative studies by Saunders & Fallon (1966), Menkes *et al.* (1965) and Menkes, Sandor &

Ilies (1969) on the distribution of necrotic zones in the limbs; their findings suggest that phylogenetic variations are achieved by altering the extent and intensity of the cell death programme. On this basis, and from the observations of Menkes *et al.* (1965) and others, that cell death occurs normally in early development of the first and second visceral arches in the human, we should consider whether hereditary defects in those regions, e.g. cleft lip and palate, micrognathia, may arise from genetic errors mediated [illegible] day of incubation. The same condition can be induced in the chicken by the injection of insulin into the egg during the second day of incubation (Landauer, 1945). These observations, one in the mutant and the other in the phenocopy, suggest that teratogenic agents may act in the same way as gene defects, i.e. via the normal cell death programme, to produce their effects. Perhaps this is the answer to the problem posed by Menkes *et al.* (1969) in discussing the selective localisation of effect in human virus embryopathies. Instead of accepting 'as a provisional explanation . . . the existence of a specific sensibility of some primordia and tissues towards various agents', the authors might have suggested that normally occurring necrotic zones are specifically sensitive in their response to different teratogens.

I am indebted to Professor A. G. M. Campbell, Department of Child Health, Aberdeen, for permission to use the illustrations of congenital amputations (Plate 1), to Dr Jane Gibson, Department of Pathology, Aberdeen for permission to examine and illustrate the foetal specimen (Plate 2), to Professor A. L. Stalker, Department of Pathology, Aberdeen for the stimulating discussions with him on the topic of cell death and to Professor R. J. Scothorne, Department of Anatomy, Glasgow, for so kindly allowing me to examine the serial sections of ten human embryos from his departmental collection.

REFERENCES

BALLANTYNE, J. W. (1904). *Manual of Antenatal Pathology and Hygiene. The Embryo.* Edinburgh: Wm. Green & Sons.

DELEANU, M. (1965). Toxic action upon physiological necrosis and macrophage reaction in the chick embryo leg. *Revue Roumaine d'Embryologie et de Cytologie, Série d'Embryologie*, **2**, 45–56.

GLUCKSMANN, A. (1951). Cell death in normal vertebrate ontogeny. *Biological Reviews*, **26**, 59–86.

LANDAUER, W. (1945). Rumplessness of chicken embryos produced by injection of insulin and other chemicals. *Journal of Experimental Zoology*, **98**, 65–77.

LENNON, G. G. (1947). Some aspects of foetal pathology. *Journal of Obstetrics and Gynaecology of the British Empire*, **54**, 830–7.

MENKES, B. & DELEANU, M. (1964). Leg differentiation and experimental syndactyly in chick embryo. *Revue Roumaine d'Embryologie et de Cytologie, Série d'Embryologie*, **1**, 69–79.

MENKES, B., DELEANU, M. & ILIES, A. (1965). Comparative study of some areas of physiological necrosis at the embryo of man, some laboratory-mammalians and fowl. *Revue Roumaine d'Embryologie et de Cytologie, Série d'Embryologie*, **2**, 161–71.

MENKES, B., SANDOR, S. & ILIES, A. (1969). Cell death in teratogenesis. In *Advances in Teratology*, ed. D. H. M. Woollam, vol. 4, pp. 170–215. London: Logos Press.

SAUNDERS, J. W. Jr & FALLON, J. F. (1966). Cell death in morphogenesis. In *Major Problems in Developmental Biology*, ed. M. Locke, pp. 289–314. New York & London: Academic Press.

STREETER, G. L. (1930). Focal deficiencies in foetal tissues and their relation to intra-uterine amputation. *Contributions to Embryology*, **22**, 1–44.

TORPIN, R. (1968). *Foetal Malformations*. Springfield, Illinois: C. C. Thomas.

ZWILLING, E. (1942). The development of the dominant rumplessness in chick embryos. *Genetics*, **27**, 641–56.

ZWILLING, E. (1945). The embryology of recessive rumpless condition of chickens. *Journal of Experimental Zoology*, **99**, 79–91.

EXPLANATION OF PLATES

PLATE 1

(*a*) Intra-uterine amputations and ring constrictions of the fingers of both hands.

(*b*) Oedema of the ring finger resulting from an intra-uterine constriction.

(*c*) Gross oedema of the left foot arising from intra-uterine constriction at the level of the ankle.

PLATE 2

(*a*) Lateral view of 14-week human foetus showing amniotic band attached to the left frontal region.

(*b*) Face of the same foetus showing median hare lip and the deformation of the frontonasal region.

PLATE 3

(*a*) Transverse section through the tips of the little and ring fingers and intervening web in an 18 mm human embryo.

(*b*) The area outlined in (*a*) shown at a higher magnification. Large numbers of pycnotic nuclei (P) can be seen.

PLATE I

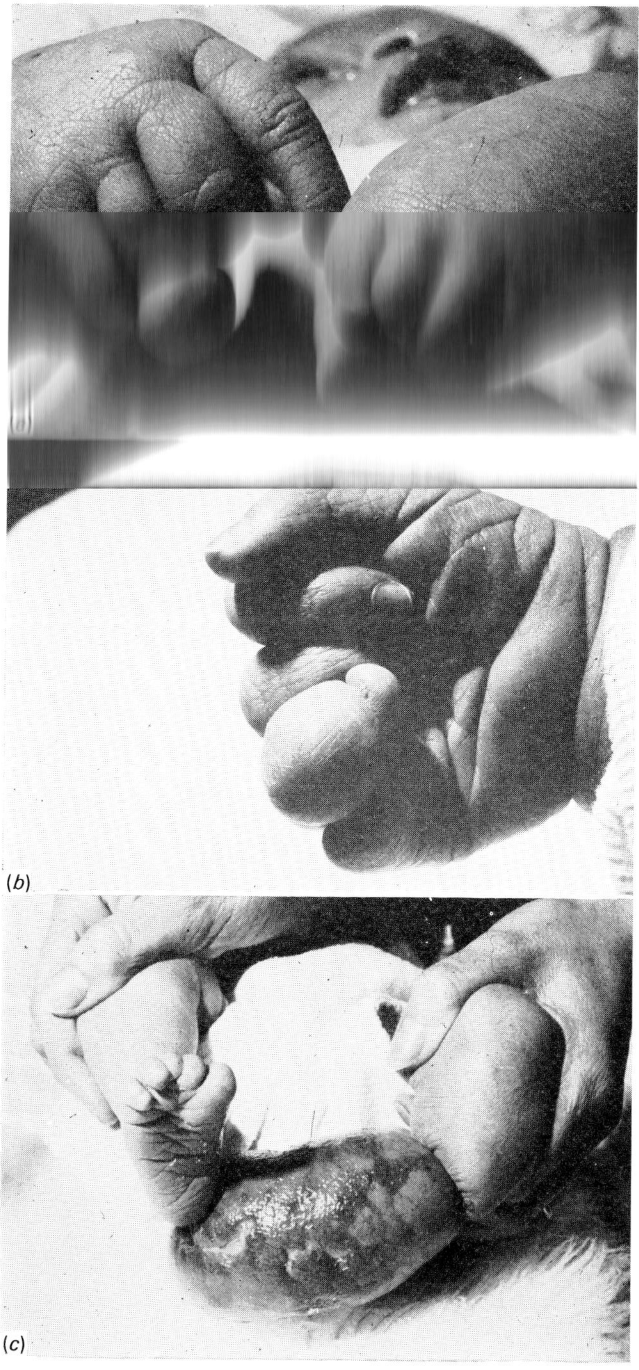

(b)

(c)

(Facing p. 396)

PLATE 2

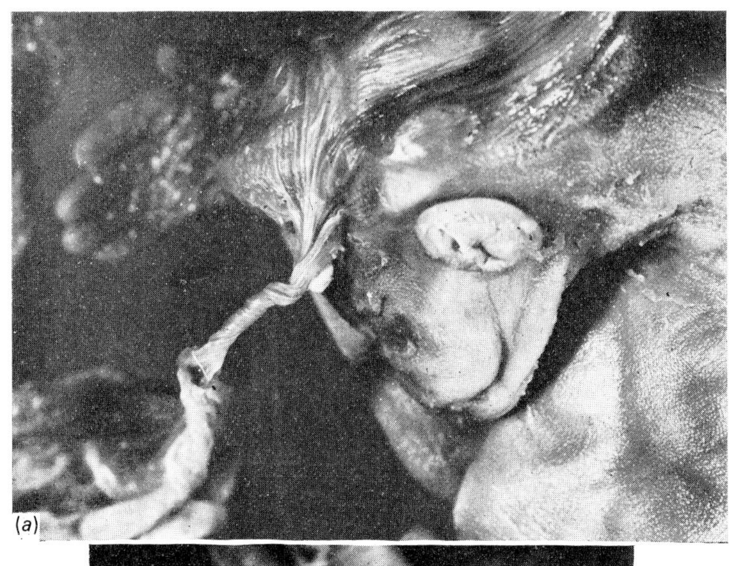
(a)

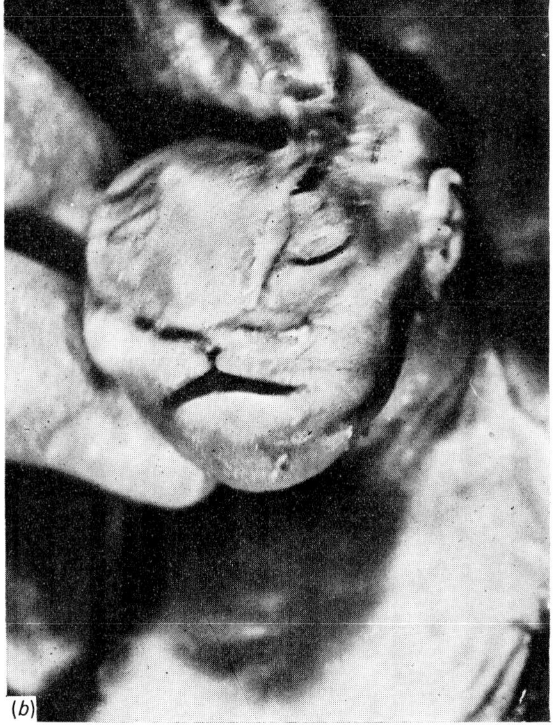
(b)

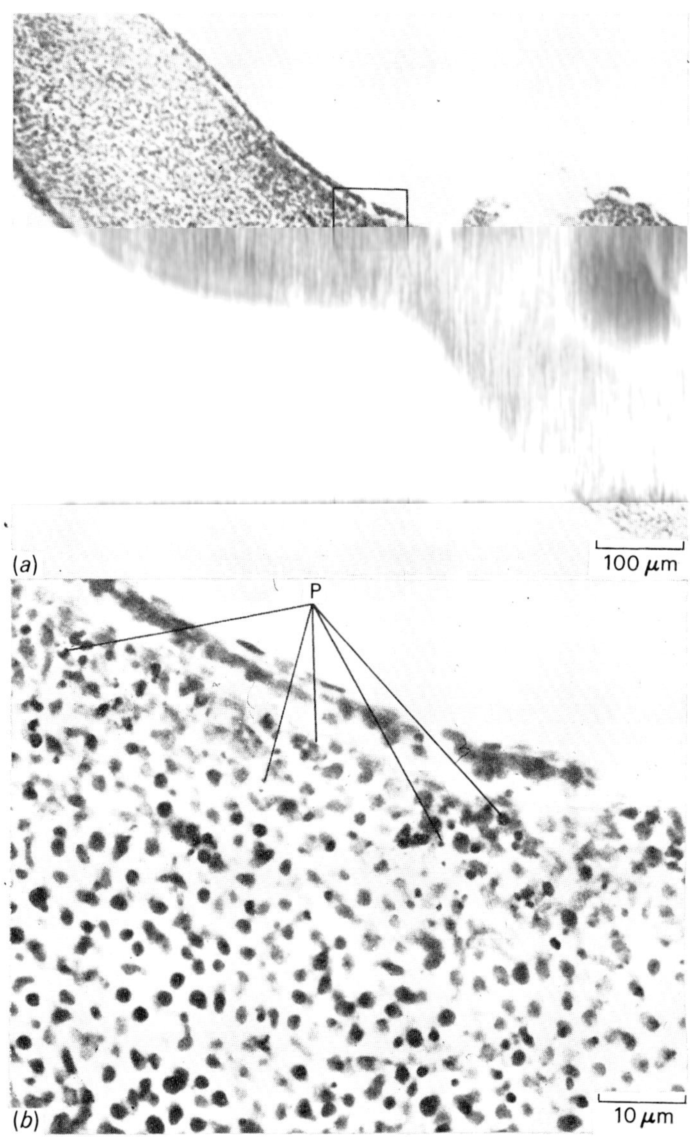
(a)
100 μm
P
(b)
10 μm

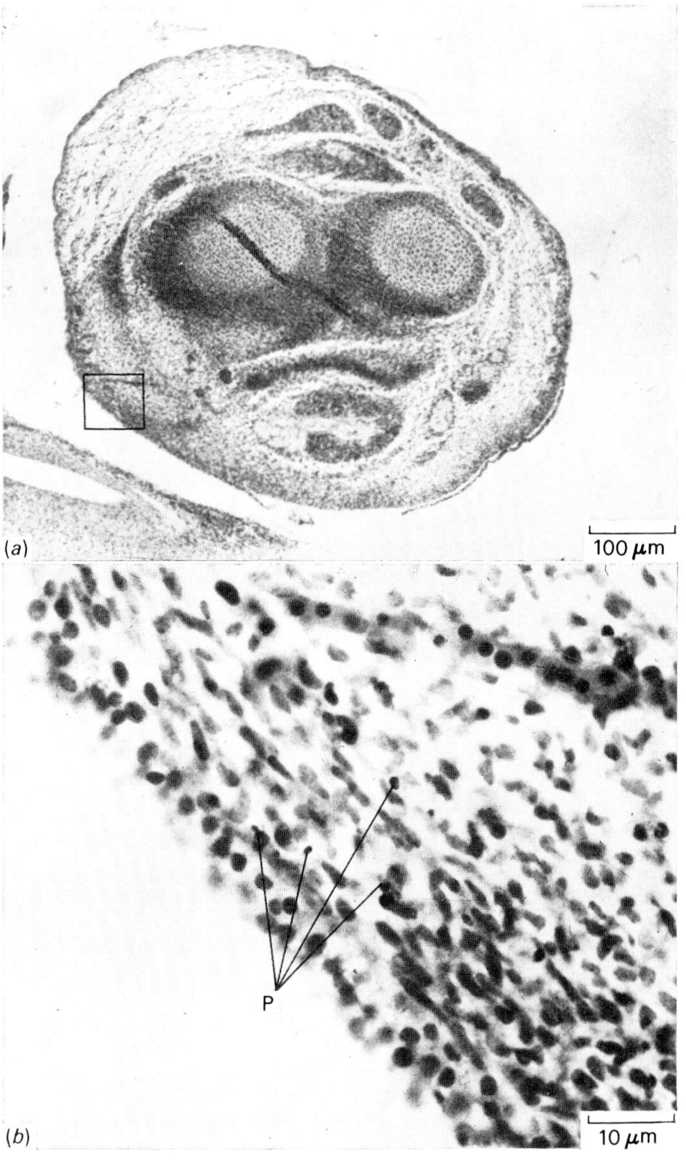
(a)
100 μm
P
(b)
10 μm

PLATE 4

(*a*) Transverse section through lower ends of radius and ulna in a 17 mm human embryo.

(*b*) The area outlined in (*a*) shown at higher magnification. Pycnotic nuclei (P) are easily identified in the region overlying the lower end of the radius.

INDEX

Note: references to animals are to mice unless otherwise stated.